Hartwig Lohse
Matthias Popp

Formulating Adhesives and Sealants

Chemistry, Physics and Applications

2nd revised Edition

VINCENTZ
Providing Professional Information

Cover: noprati – stock.adobe.com

Bibliographische Information der Deutschen Bibliothek
Die Deutsche Bibliothek verzeichnet diese Publikation in der Deutschen Nationalbibliografie; detaillierte bibliografische Daten sind im Internet über http://dnb.d-nb.de abrufbar.

Lohse, Hartwig and Popp, Matthias
Formulating Adhesives and Sealants, 2nd revised Edition
Hanover: Vincentz Network, 2025
European Coatings Library
ISBN 3-7486-0610-9
ISBN 978-3-7486-0610-9

Please ask for our book catalogue
Vincentz Network, Plathnerstr. 4c, 30175 Hanover, Germany
T +49 511 9910-033, F +49 511 9910-029
books@european-coatings.com
www.european-coatings.com

Layout: abavo, Buchloe, Germany
Printed by Gutenberg Beuys, Hanover, Germany

European Coatings Library

Hartwig Lohse
Matthias Popp

Formulating Adhesives and Sealants

Chemistry, Physics and Applications

2nd revised Edition

VINCENTZ
Providing Professional Information

Preface

When we, the new team of authors were asked by Vincentz Network to continue the work of Bodo Möller and Walther Rath and to revise the book, we spontaneously agreed, having come to appreciate the book already years ago as a useful tool for getting started in the formulation of adhesives and sealants. In revising the 1st edition, we have come to appreciate the excellent groundwork laid by Bodo Müller and Walter Rath in preparing the 1st Edition, and we would like to express our sincere thanks to them.

We were aware that a lot had changed since the 1st Edition was published in 2010. Not only have product names and raw material suppliers changed, in many cases as a result of mergers and acquisitions. Due to stricter health and safety and environmental regulations, raw materials have been withdrawn from the market or no longer accepted by adhesive users as ingredients in adhesives. Substitution with other less hazardous products was rarely possible without changing the overall formulation. On the other hand, new raw materials, such as those based on renewable resources, have entered the market.

Like the authors of the 1st Edition, we aim with this 2nd Edition to provide not only a reference work for those interested in adhesives and sealants, but in particular to familiarise chemists, engineers, laboratory technicians and skilled workers involved in the development of adhesive and sealant formulations with the practice of adhesive and sealant formulation.

While adhesives focus on their load-bearing function, sealants are mainly used to seal joints or gaps between components. For both functions, good long-term adhesion to the substrate is essential. Not only are adhesives and sealants similar in composition, but they are often used for both bonding and sealing, or even to fulfil both functions simultaneously. Therefore, a clear distinction between adhesives and sealants is generally neither useful nor possible. We have therefore decided to depart from previous editions and treat adhesives and sealants together in the relevant chapters, without neglecting any differences in formulation.

Part I briefly summarises the basic principles of bonding technology. Part II discusses the chemical basis of the various types of adhesives and sealants, together with information on formulation, manufacturing processes and important areas of application. Wherever possible, this part is supplemented by guide formulations and their explanation and evaluation. As formulations are generally highly proprietary and not publicly available, they are based on formulations published by some raw material manufacturers, in usually older technical literature, in patents, or developed as part of our own work. It should be noted that some of the formulations provided may no longer reflect the current state of the art and may not necessarily meet the requirements of a specific application. Nevertheless, we consider them to be a valuable basis for the targeted development of application-specific adhesives or sealants, taking into account the structure-activity relationships. Part III represents an attempt to look into the future and to identify future challenges in new product development in line with the expected changes in market requirements. Part IV provides general information on the development process for adhesives and sealants. In addition to information on establishing the requirements profile at the beginning of any adhesive or sealant development, the most common test methods for determining processing properties, for quality assurance in the manufacture of adhesives and sealants, and for testing adhesive bonds are listed, each with a brief description.

We are aware that the present edition does not cover all adhesive and sealant systems in their full diversity and would therefore like to encourage experts to send us further formulation examples for publication in a future 3rd Edition of this book, stating the source and a brief description of the application where appropriate.

Patent restrictions or registered trademarks (TM or ®) are not explicitly mentioned. It should also be noted that product and brand names may change in the future as a result of mergers and acquisitions.

Breitenberg and Bremen, Germany
March 2025

Hartwig Lohse & Matthias Popp
hlohse@hdyg.de
matthias.popp@ifam.fraunhofer.de

Exclusion of liability

It should be noted that the contents of this book reflect the personal views of the authors, based on their own knowledge and experience. This does not absolve readers of the responsibility of performing their own tests with respect to the uses and applications of the various processes or products described herein, and/or obtaining additional advice regarding the same.

The statements made in this book do not claim to cover all safety aspects associated with the formulation of adhesives and sealants. It is the responsibility of the respective formulator to observe all relevant legal health, safety and environmental legislation and to determine all applicable regulatory requirements.

All liability of the authors is excluded to the fullest extent permitted by law, subject to all legal interpretations.

Contents

Part I General basics

1 Introduction

Adhesives and sealants have been used since prehistoric times. For example, Stone Age people used birch resin to attach spear and axe points made of stone and in the broadest sense mud, grass, and reeds to seal dwellings from the weather. Adhesive bonding also has role models in nature, an example being the barnacle, which as a free-swimming larva can attach itself to practically all maritime hard substances by means of its own secretion. For a long time, adhesives and sealants were made solely from natural raw materials, mostly in small craft workshops. With the development of macromolecular chemistry and the new synthetic raw materials, bonding as well as sealing experienced a real boom.

Joining of virtually any desired combination of materials

- most plastics,
- both thermosets and thermoplastics,
- various metals and alloys,
- glass, ceramic
- wood,
- paper,
- cardboard,
- …

Construction – design

- light weight design: uniform stress distribution over a large surface allows for higher loads or reduced material thickness
- joining of thin adherends
- cost savings due to reduced material usage (e.g., bonding of local reinforcements)

Retention of the construction material's properties

- little or no heat input allows low tolerance manufacturing minimal surface distortion,
- no mechanical damage e.g., due to required holes

Customized integration of additional functionalities:

- leak proof joints,
- thermal and/or electrical conductivity or isolation,
- contact corrosion avoided
- sound dampening,
- compensation of tolerances or uneven surfaces,
- corrosion protection, etc.
- compensation of different thermal elongation
- compensation of part movements, vibrations

Maintaining the integrity of the adherents

- low-heat joining process, no thermal damage to the adherends like distortion or change in microstructure
- no thermal distortion of adherends
- drilling of holes not required

Figure I-1.1: Advantages of adhesive bonding

Nowadays adhesives and sealants play an essential role in a huge range of consumer, professional and industrial products and adhesive bonding is considered a key technology of the 21st century. There is hardly a branch of industry or trade in which it is not used innovatively. One of the many reasons for the wide range of adhesive applications is the potential of bonding both similar and dissimilar materials while avoiding stress peaks and maintaining their properties, thus providing long-term stability and safety. An overview about the advantages is summarized in Figure I-1.1; like any technology, adhesive bonding has some disadvantages, which are summarized in Figure I-1.2.

Without modern adhesive bonding technology, for example, today's intelligent lightweight construction and the utilization of renewable energy such as solar and wind energy as a contribution to reaching the ambitious climate and environmental goals would be inconceivable.

The transition to a circular economy called for by the European Green Deal, with the aim of improving the repair and recycling capabilities of products without having to compromise on long-term durability and reliability, i.e., durability, is not necessarily in conflict with the use of adhesive bonding technology. Achieving these two seemingly contradictory requirements is certainly not easy and requires the development of specific solutions adapted to the particular need. Here, the formulator of adhesives and, to the same extent, the designer of the bonded component are equally challenged. There are already a considerable number of examples, some of which have been proven for a long time. Examples include the removal of labels before reusing returnable bottles or the replacement of defective display screens on cell phones. DIN/TS 54405:2022-12 [1] provides a systematic description of the processes for debonding.

The European adhesives and sealants market amounted to 4.7 million tons in 2020, with a value of € 15.5 billion. Although sales in 2020 declined by 5 % compared to the previous year (probably caused by global COVID-19 pandemic) in terms of volume and value, it is expected to increase in 2021–2026 at an annual growth rate (CAGR) of 3.6 % to reach 22.2 billion euros. [2]

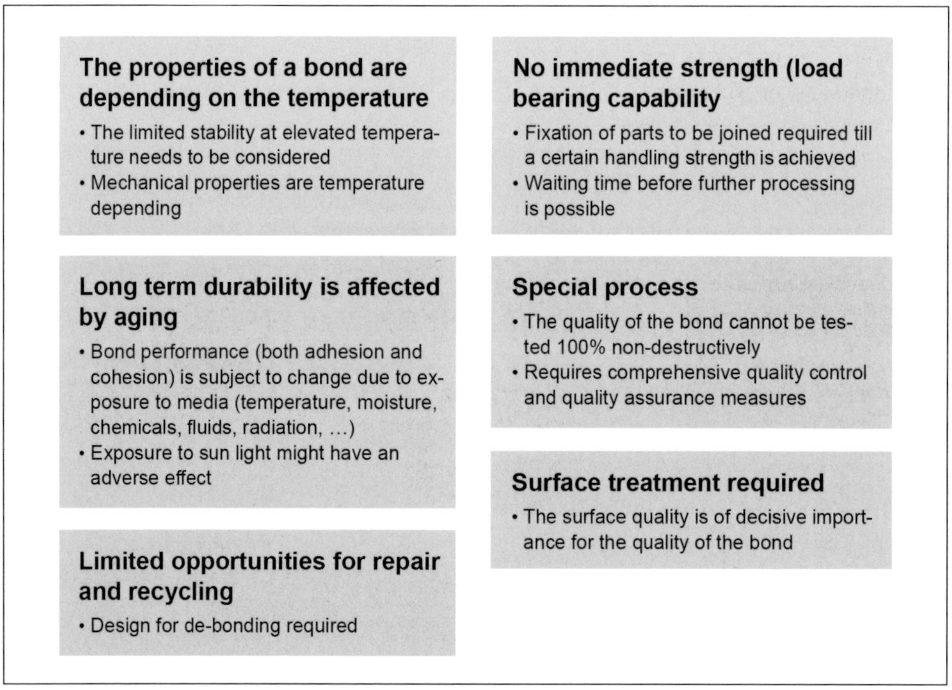

Figure I-1.2: Disadvantages of adhesive bonding

In terms of chemical and applications technology, the field of adhesives and sealants is extremely diversified. Like the previous editions, this edition covers the most important adhesives and sealants based on organic or organosilicon raw materials; inorganic adhesives are not covered; needless to say, the selection presented here is subjective.

From the title "Formulation of Adhesives and Sealants" it is obvious that the focus of this book lies on the materials (chemistry) and not on the basics of joining technology (bonding and sealing), which, by the way, is already well covered by a large number of textbooks. The underlying technological aspects are both multidisciplinary and interdisciplinary by nature. Figure I-1.3 illustrates this interdisciplinarity from the point of view of the disciplines involved, while Figure I-1.4 shows the most important parameters influencing the quality of a bond.

1.1 Definitions

1.1.1 Adhesive and adhesive bonding

An adhesive is defined by EN 923 [3] as a non-metallic substance capable of joining materials by surface bonding (adhesion) and the bond possessing adequate internal strength (cohesion). Adhesive bonding is the process of joining two surfaces together by means of an adhesive.

Typically, adhesives are applied in a liquid-like state to one or both of the parts to be bonded (often also referred to as „adherends" or „substrates"), wet the adherends as a prerequisite for

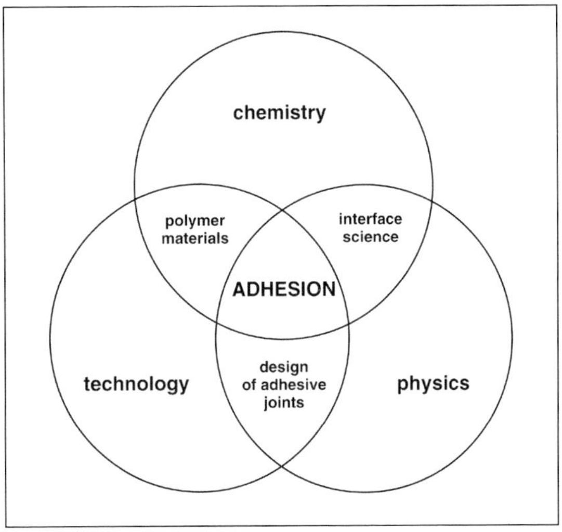

Figure I-1.3: Interdisciplinarity

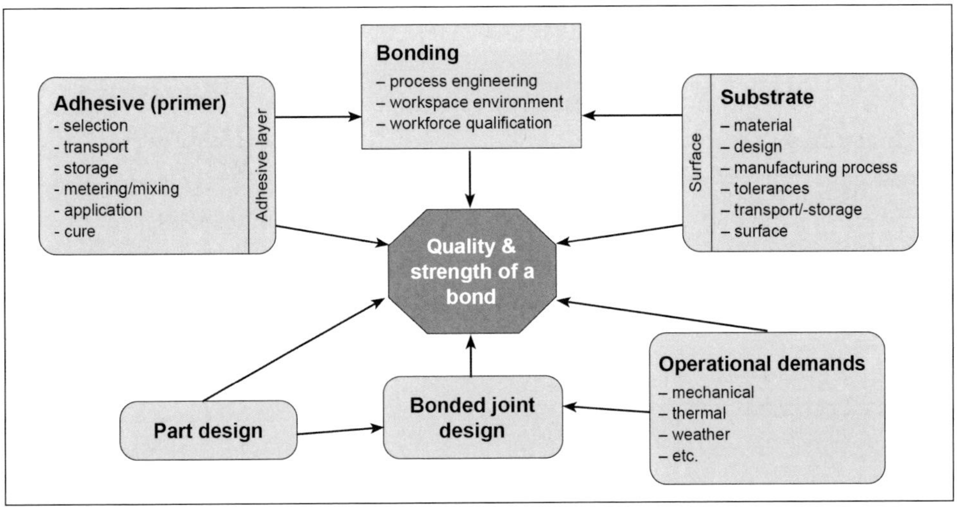

Figure I-1.4: Factors influencing the quality and strength of an adhesive bond

forming adhesion, and then, with the exception of pressure-sensitive adhesives (see Part III), harden (i.e., building cohesion) by a physical or chemical process (curing); in special cases, hardening occurs by a combination of both.

1.1.2 Sealants

A sealant is likewise a non-metallic substance with the ability to wet the surface of the substrate and to build adhesion as well as cohesion. Compared to adhesives the main function is to seal cracks or joints, whereby the adhesive and cohesive properties allow a permanent seal to compensate joint movements without loss of adhesion. Like adhesives, sealants are applied in a liquid state, wet the substrates and then harden (i.e. build cohesion) by a physical or chemical process.

Non-hardening elastic sealants and gaskets produced by cutting from sheet material which transmit only small loads (< 1 MPa) between the sealed structural elements are not subject of this book.

1.1.3 Adhesive sealants

The transition between sealing and bonding is smooth. A distinction based on the achievable strength level alone is not meaningful. Besides strength the ability to compensate joint movements and to fill wider gaps must be considered. It must be decided on a case-by-case basis which function, that of load bearing or that of sealing, is of primary importance. In many cases, it is precisely the ability to fulfil two functions with one material that has led to the use of an adhesive or sealant. Since not only the classification of an application as bonding or sealing, but also the classification of a material as adhesive or sealant is often not clearly recognizable, such materials are often referred to as "adhesive-sealant".

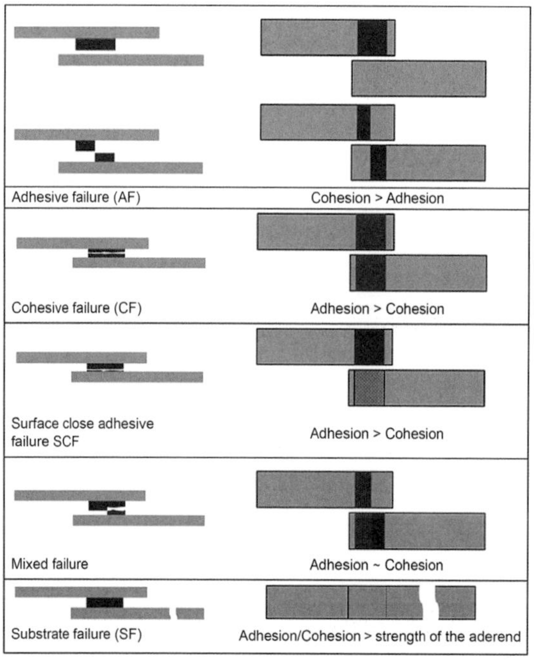

Figure I-1.5: Most important failure modes [4]

1.1.4 Wettability

Wettability as prerequisite to form adhesion is the ability of a liquid (e.g. adhesive or sealant) to spread on a specific solid surface [3] (see Chapter II-2.2).

1.1.5 Adhesion/cohesion

Adhesion is the tendency of dissimilar particles or surfaces to cling to one another while cohesion refers to the tendency of similar or identical particles or surfaces to cling to one another. In terms of adhesives or sealants, adhesion describes the ability of an adhesive or sealant to adhere to a surface. See Chapter I-2 for detailed information about adhesion. Cohesion, on the other hand, describes its internal strength.

The ability of an adhesive or sealant to transfer loads in a bond is determined by two parameters: adhesion to the substrate(s) and cohesion. As in a chain, the weakest link will fail first when a mechanical load is applied. Thus, a distinction is made between different failure modes defined by EN ISO 10365:2022 [4]. The most important ones are explained schematically in Figure I-1.5.

1.2 Classification of adhesives and sealants

1.2.1 Classification according to the mechanism of solidification

When using adhesives and sealants, they must be flowable so that they can wet the substrates as a prerequisite for the formation of adhesion (Chapter I-2). The subsequent solidification denotes and represents the transition from the liquid to the solid state, i.e., the formation of cohesion. This transition can occur by a physical process or by a chemical reaction. An exception here are pressure-sensitive adhesives, which are permanently tacky (typically used for self-adhering labels or adhesive tapes) and which, strictly speaking, do not solidify (see Chapter II-3). Even though there are various other criteria, some of them explained below according to which adhesives can be classified, the classification according to their hardening mechanism has proven itself and became established.

Figure I-1.6 schematically shows the classification of adhesives according to the hardening (setting) mechanism.

In the cured state, the adhesives considered here represent plastics and, like plastics, adhesives can also be classified into one of the plastic classes described by Figure I-1.7.

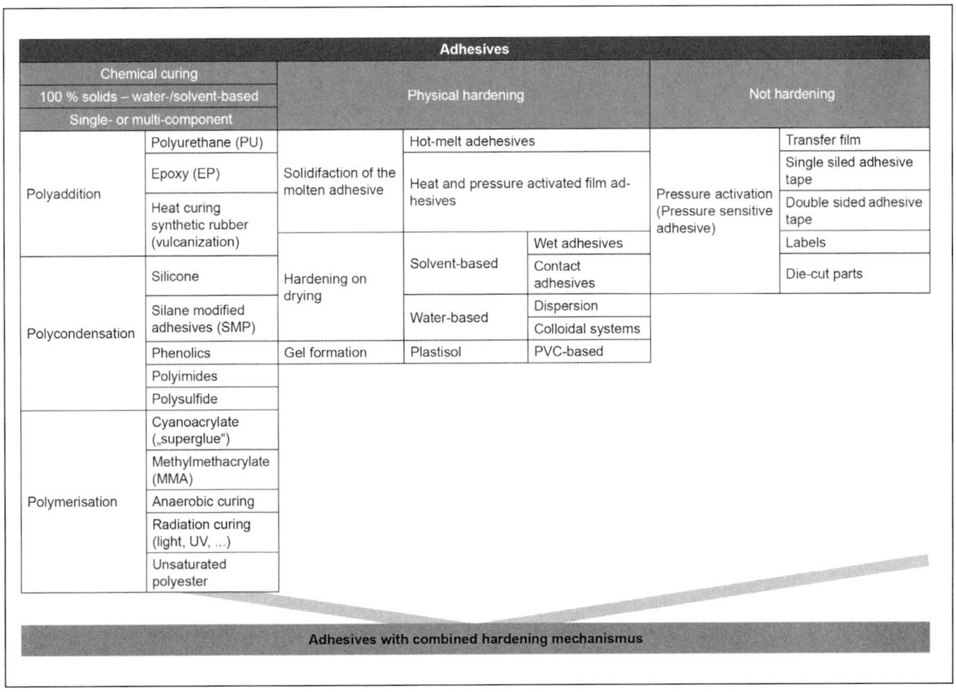

Figure I-1.6: Classification of adhesives by (the most important) hardening mechanisms

Table I-1.1: Classification of adhesives (in the solidified state) into polymer classes

Polymer class	Examples of adhesive and sealant types
Thermoplastic	– Hot-melt adhesives – Solvent based adhesives – Water-based dispersion adhesives (depending on polymer) – Polyimide adhesives – Cyanacrylate adhesives (super glue) – Radiation curing acrylate adhesives – Methyl methacrylate adhesives (MMA) – Pressure-sensitive adhesives
Elastomer	– Water-based dispersion adhesives (depending on polymer) – Plastisol – 1-part moisture curing polyurethane – 1-part heart curing polyurethane – 2-part polyurethane – 1-part rubber-based adhesives – Silicone adhesives – Silane modified adhesives – Polysulfide adhesives – Pressure-sensitive adhesives
Thermoset	– Epoxy adhesives – 2-part polyurethane – Phenolic adhesives – Methyl methacrylate adhesives (MMA) – Anaerobically curing adhesives – Unsaturated polyester adhesives – Radiation curing acrylate adhesives

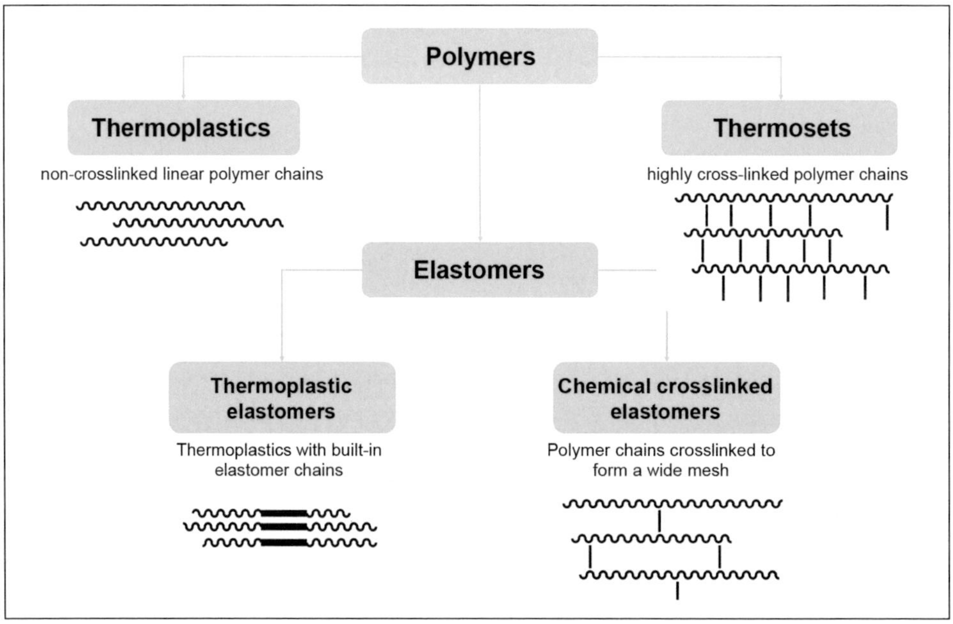

Figure I-1.7: Classification of polymers and simplified illustration of the polymer chains of the different plastic classes

In addition to the type of monomer composition of the polymer, its classification in one of the polymer types mentioned in Figure I-1.6 and Figure I-1.7 determines its properties. Table I-1.1 characterizes the most important adhesives with regard to their classification into the polymer classes thermoplastic, elastomer and thermoset.

1.2.1.1 Physical hardening

The group of physically hardening adhesives and sealants is characterized by the fact that the substance determining the cohesion and adhesion properties, usually a polymer, is already present in its final form when the adhesive is delivered, i.e., the chemical structure does not change during its solidification.

Hardening occurs solely as a result of a physical process. This can be the solidification of the adhesive melted by heating before application, the evaporation of the liquid (organic solvent, water) in which the polymer is dissolved or dispersed, or the diffusion of a plasticizer into the polymer under heating.

The polymers used for the formulation of physically hardening adhesives and sealants all belong to the group of thermoplastics (see Figure I-1.7 and Tabel I-1.1). Due to their structure of entangled linear polymer chains, thermoplastics show the ability to deform. They can be described as tough to brittle at room temperature, meltable, swellable, soluble and showing a tendency to creep. These properties, which typically limits the resistance of joints made with thermoplastic adhesives and sealants, must be considered in the formulation of adhesives and sealants. The polymers used must meet the resistance requirements of the intended adhesive application. On the other hand, thermoplastic adhesives offer the possibility of controlled debonding, e.g., by heating to a temperature above the softening point.

1.2.1.2 Chemical curing

In contrast to physically hardening adhesives and sealants, the polymer chains that determine the properties of the cured adhesive are formed after the adhesive or sealant has been applied by a chemical crosslinking of the monomers and prepolymers present in the as-supplied state.

Prepolymers can be used in the form of commercially available raw materials together with other formulation components in a pure mixing process to form adhesives or sealants or can be formed directly during the manufacture of the adhesive by a chemical reaction of the raw materials involved. From a regulatory point of view (REACH), it should be noted that the newly formed prepolymers may have to be registered evaluated and authorized before the adhesive can be marketed in Europe, i.e., if they do not meet the polymer definition under REACH. It should be noted today that the EU is currently working on extending the REACH regulation to polymers. Similar regulations may also exist outside Europe and must be considered accordingly.

In the case of crosslinking reactions, polyaddition, polycondensation and polymerization are three mechanisms that have to be taken into account, among others, regarding the processing of adhesives and sealants and the selection of an adhesive or sealant for a specific application.

To allow wetting of the substrate surface, the monomers or prepolymers must be present in a liquid state.

The chemical cure leads to an increase in molecular mass and the formation of a more or less pronounced crosslinking. The resulting crosslink density determines whether the adhesive or sealant cures to a thermoplastic, an elastomer, or a thermoset polymer and thus has an influence on the properties of the cured adhesive or sealant.

The specifics of the different variants of physically bonding adhesives are described in Chapter II-1 of the second part of this book.

1.2.1.3 Not hardening adhesives – pressure-sensitive-adhesive tapes and die-cut parts

In the case of pressure-sensitive or self-adhesive products, a distinction is made between those that join two parts, whether with the main function of load transmission or sealing, and labels including self-adhesive films, e.g., for decorative purposes. The decisive component for the adhesive properties of these products is the so-called pressure-sensitive adhesive. These are polymers which, due to their low glass transition temperature, have a permanently tacky surface and are extremely high-viscosity liquids. The main polymer classes used are polyacrylates or synthetic rubbers made permanently tacky by resins and, to a lesser extent, silicones, and polyurethanes.

They are marketed as single-sided or double-sided adhesive tapes, where the adhesive is coated to a carrier or without carrier as a so-called free film (Figure I-1.8) in, as die-cut parts fitting the geometry of the adherends or as self-adhesive labels or films, e.g., for decorative purposes.

The specifics of the different variants of non-hardening adhesives are described in Chapter II-3.

1.2.2 Other classification systems

In addition to the established classification according to the type of solidification, adhesives and sealants are also classified according to other criteria briefly presented below. The aim is to consider the different properties, types of polymers, types of applications, areas of application, etc. of the adhesive.

1.2.2.1 By strength

The internal strength (cohesion) of adhesives and sealants can vary extensively with the formulation. Within a series of adhesives or sealants with the same polymer base, the elasticity, e.g., determined by measuring the elongation at break, decreases with increasing Young's modulus or increasing cohesive strength. Products of high cohesion are usually classified as adhesives, and if

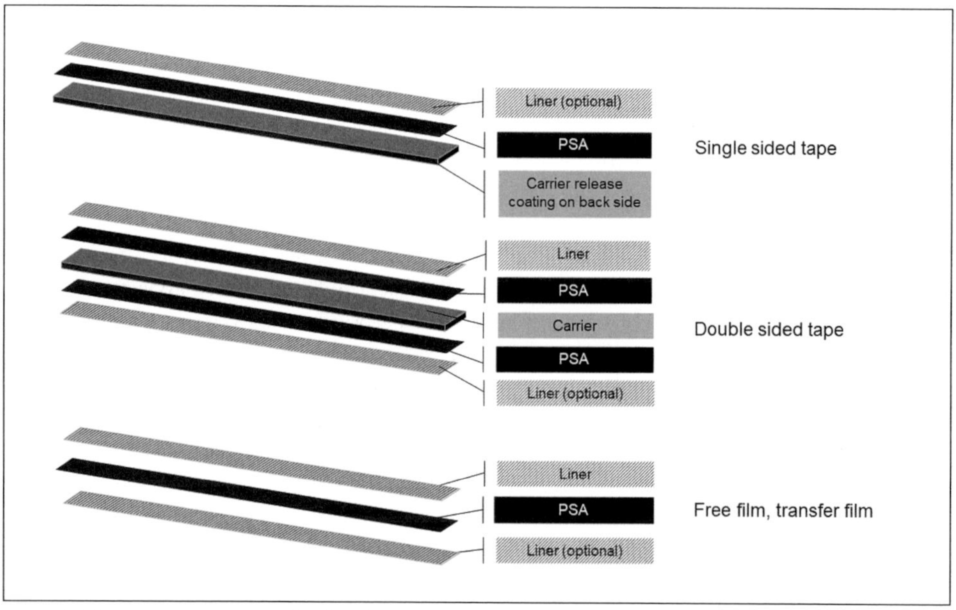

Figure I-1.8: Construction of self-adhesive articles

this property is permanent in an adhesively joined assembly, they are called structural adhesives. Products of low cohesion are classified as sealants, especially if they are able to fill and seal large gaps. As already explained in Part I-1.1 the distinction based on the achievable strength level alone is not sufficient to differentiate between adhesives and sealants. Besides strength the ability to compensate joint movements as well to fill wider gaps must also be considered. Today, a distinction is therefore usually made between low-modulus and high-modulus adhesives or sealants, whereby the boundary between high and low is not generally defined but depends to a certain extent on the adhesive or sealant and also the application.

1.2.2.2 By the substrate they are used to bond/seal

Some adhesives, for example, are particularly suitable for certain kinds of substrates. Some adhesives or sealants provide good adhesion to metals, others to concrete, plastics, glass, wood, etc. Accordingly, they are classified as metal, plastics, glass, wood adhesives etc. In most cases, this classification is not useful since bonding is often used to join different materials together. In addition, each of the above-mentioned material groups contains a large number of different materials, which differ significantly not only from the point of view of bonding technology.

Plastics are a good example; an adhesive that is very suitable for bonding components made of polycarbonate or ABS, for example, is only suitable for polyethylene to a very limited extent. This also applies analogously to the bonding of other material groups such as metals and even the various types of wood.

Needless to mention, a general-purpose adhesive or sealant does not exist even though it might be desirable and some products are marketed as such.

1.2.2.3 By application

Classifying adhesives and sealants by their field of application is a market-oriented approach and is based on the end use of the products and to some extend on the processes typical for this field of application. Automotive adhesives, unsurprisingly, will therefore be used in vehicles. More specifically, the product may be called an automotive body-in-white adhesive so as to clarify the manufacturing step in which it is used. When one talk about a hem-flange adhesive, everyone knows exactly what the product is used for.

Other terms, such as textile-laminating adhesive, shoe-sole adhesive, insulating glass sealant and construction sealant are all self-explanatory. Many of these applications are discussed in the corresponding chapters of this book.

1.2.2.4 By chemistry

Classification of adhesives and sealants by their chemistry essentially describes the chemical composition of the base polymer system used for the final product. Almost all polymers that can be reversibly converted into a liquid state that allows wetting of surfaces either by dissolving in a suitable solvent, including water, by dispersing in water or by melting, or whose monomers or prepolymers are present in a liquid state and are capable of solidifying by means of a controlled chemical reaction, are in principle suitable to produce adhesives and sealants. Accordingly, adhesives and sealants may be based on polyurethane (PU), epoxy (EP), polyamide (PA), ethylene-vinyl acetate-copolymer (EVA), polyvinyl acetate (PVAc), polychloroprene (CR), phenolic resin, acrylic resin or cyanoacrylate etc.

For users with knowledge of adhesives and chemistry, this classification can be quite useful, since the adhesives of a polymer base show to a certain extend properties typical for this group. Even if knowledge of the chemistry of a polymer system provides insight into the properties of the adhesives and sealants in question, it should not be forgotten that the wide range of variations possible in the formulation of such products also leaves room for those with quite atypical properties. As an example, the group of 2-component epoxy adhesives may be mentioned here. They are typically characterized by a high Young's modulus in the low four-digit megapascal range, a high tensile strength in the two-digit megapascal range and a low elongation at break in the single-digit range. In contrast, products with a significantly lower modulus in the low three-digit megapascal range but an elongation at break of 60 % are also available.

Nevertheless, the underlying chemistry is from a R&D point of view naturally the most important criterion for developing an adhesive for a certain application. Hence, R&D specialists know from experience which problem might be solved with an epoxy-based product or where it might be better to use an acrylic system. Besides the setting mechanism, the chemical aspects of adhesives and sealants will play an important role in this book.

1.2.2.5 By the mode of application and hardening

Classification of adhesives according to the type of application and hardening process, is often the simplest and clearest because it addresses the concerns of the user. Besides others, a distinction is made between adhesives that are suitable for bead application and those that can be applied over a wide area, e.g., by spraying, roll coating or using a slot nozzle. Likewise, a distinction is made between adhesives that cure at room temperature, require moisture from the environment and those that require, for example, heat or irradiation with UV light.

1.3 Economic significance of adhesives and sealants

Nowadays, for a variety of reasons, practically every industry or business sector uses adhesive bonding technology as an innovative and reliable joining technology, e.g., to join different materials with long-term stability while preserving their properties. In addition to the actual function of load transmission, further functionalities can be provided by an adhesive bond – for example, the compensation of different thermal expansion of the adherends, electrical and/or thermal conductivity or insulation, corrosion protection, vibration dampening or sealing against liquids and/or gases. Like no other joining technology, adhesive bonding allows the implementation of advanced design through an optimal combination of technological, economic, and ecological aspects. In this respect, adhesive bonding technology is now undisputedly considered a key technology of the 21st century.

Since it is not always possible to make a clear distinction between adhesive and sealant applications, the commercial aspects of the two groups are mainly addressed together in this overview. The information compiled below has been obtained from various publicly available sources [5–12] and is intended to give the reader an impression of the commercial importance of adhesives and sealants.

In contrast to the previously observed steady growth in the global adhesives and sealants market, the year 2020 was characterized overall by a pandemic-related decline of about 7 % in sales. Due to lockdown and other restrictions, this affected products, especially adhesives and sealants for the building and construction sector. On the other hand, the DIY sector has proven to be stable during the Corona pandemic. It is expected that sales in 2021 will recover to the level of 2019 at USD about 63 billion. For subsequent years, in particular due to the increasing use of adhesives and sealants in building and construction, but also in other areas such as transportation, medical and electronics, a further growth is forecasted to just over USD 92 billion in 2029, corresponding to a CAGR of 5 %.

From a regional perspective, Asia-Pacific represents the largest market, followed by North America and Europe at almost equal shares (Figure I-1.9). Asia-Pacific is expected to continue to hold the largest share of the adhesives and sealants market in the future and it is no surprise that China, Japan, and India are the most important countries in this region.

Due to the advantages associated with their use, adhesives are successfully utilized in almost all industrial sectors. Figure I-1.10 provides an overview of the respective shares of the market segments. Applications in paper and packaging and in building and construction account at almost equal percentage for the largest shares. In the building and construction sector, the group of sealants is overrepresented compared to the other market segments.

For environmental, health and safety (HSE) reasons, the market share of solvent-based adhesives and sealants is expected to decline further in favor of solvent-free products. Due to the principle weakness of limited water and moisture resistance of water-based systems, hotmelts are expected to grow disproportionately, also because of their short curing times and the resulting high productivity that can be achieved (see Figure I-1.11).

The reactive adhesives and sealants segment is also expected to experience significant growth. The ability to produce durable and, depending on the requirements, both high-strength as well as elastic bonds, even between different materials, has led to an increasing number of applications in the various industrial sectors.

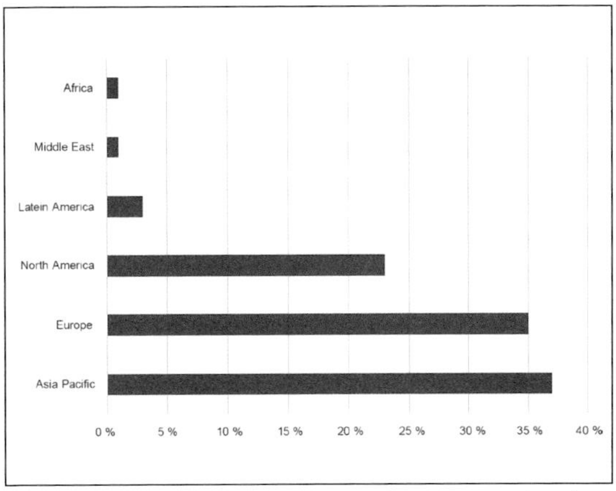

Figure I-1.9: Regional breakdown of adhesive and sealant consumption [5, 9, 10]

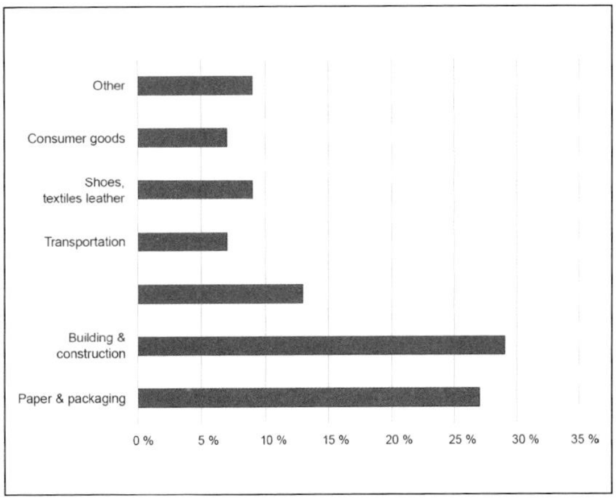

Figure I-1.10: Adhesive and sealant consumption by market [11]

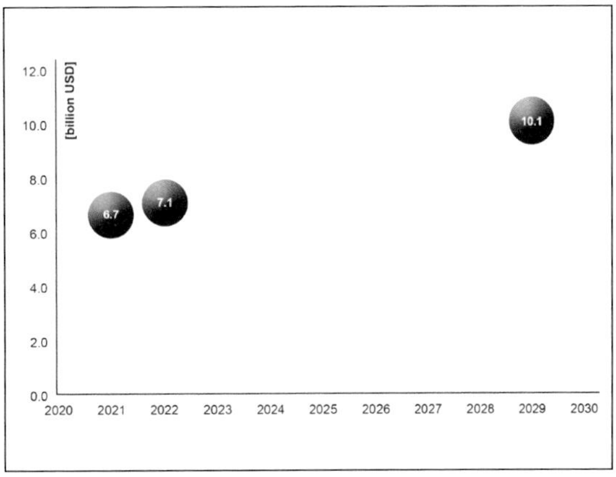

Figure I-1.11: Projected increase in the consumption of hot-melt adhesives [12]

For example, intelligent lightweight construction, which is a prerequisite for complying with the increasingly stringent emission limits of the automotive industry and for achieving acceptable ranges for electric vehicles, would be inconceivable without modern adhesive bonding technology and, in particular, without reactive adhesives.

1.4 References

[1] DIN/TS 54405:2022-12, Construction adhesives – Guideline for separation and recycling of adhesives and substrates from bonded joints, Beuth Verlag, Berlin 2022
[2] FEICA https://www.feica.eu/our-industry 2022-12-08
[3] EN 923:2015-12 Adhesives -Terms and definitions; German version EN 923:2015
[4] EN ISO 10365:2022 – Adhesives – Designation of main failure patterns (ISO 10365:2022), Beuth Verlag, Berlin
[5] Fortune Business Insight, Adhesive and Sealants Market Size – Summary – www.fortunebusinessinsights.com/industry-reports/adhesives-and-sealants-market-101715 – accessed 2023-01-06
[6] H. Pelc, Aktueller Report zum Weltmarkt für Klebstoffe, Springer Professional, 26.05.2020 – www.springerprofessional.de/verbindungstechnik/aktueller-report-zum-weltmarkt-fuer-klebstoffe/18012788 – accessed 2023-01-06
[7] B. Flacke, Klebstoffe: 2022 weltweiter Verbrauchsrückgang, Technischer Handel, 30.05.2022 – www.technischerhandel.com/markt-branche/unternehmen-maerkte/2022/
[8] Grand View Research, Adhesive and Sealants Market Size – Summary – www.grandviewresearch.com/industry-analysis/adhesives-and-sealants-market# – accessed 2023-01-06
[9] Fortune Business Insight, North America Adhesive and Sealants Market Size – Summary – www.fortunebusinessinsights.com/north-america-adhesives-and-sealants-market-105576 – accessed 2023-01-06
[10] FEICA European Adhesive and Sealants Market 2021–2026 – https://www.feica.eu/information-center/market-reports/download/ – accessed 2023-01-06
[11] Ceresana Market Study Adhesives World Report 5th edition, https://ceresana.com/en/produkt/adhesives-market-report-world – accessed 2023-01-06
[12] Fortune Business Insight, Global Hot Melt Adhesive Market – https://www.fortunebusinessinsights.com/industry-reports/hot-melt-adhesives-market-101724 – accessed 2023-01-06

2 Adhesion

The term adhesion has already been defined in Chapter I-1.1 of this book as the tendency of dissimilar particles or surfaces to cling to one another. It is so to say the attraction which exists at the interface of two different solid phases.

The force required to remove an adhesive or sealant from a substrate can be considered as a measure of adhesion. If the force acts in the form of a tensile or shear load, it is given as force/area with the unit MPa or N/mm^2; if, on the other hand, it acts as a peel load, it is given as force/length with the unit N/mm. A prerequisite for this, however, is that in the destructive test the defect occurred exclusively due to a failure of the adhesion forces, i.e. the adhesive detaches completely from the surface of the substrate in the form of a straight adhesive failure (see Figure I-1.5). In case a cohesive, substrate failure or mixed failure occurred it can be assumed that the adhesive strength is higher than determined by the test. Experience shows, however, that adhesive systems practically never separate or fail exactly where the adhesion has occurred. If, during the formation of the bond, the partners have come close enough to each other to form interactions leading to adhesion (see Chapter I-2.1), there is, according to [1], no such phenomenon as pure adhesion failure. Nevertheless, the practical significance of the failure modes described in EN ISO 10365:2022 [2], including adhesive failure, should not be underestimated, as they provide additional information valuable for the evaluation of an adhesive bond in addition to the determined bond strength. Macroscopic detected adhesion failure is usually a sign of suboptimal adhesion, which most often lead to further loss of bond strength and even early failure under aging conditions.

The counterpart to adhesion is cohesion. Cohesion is the attraction which exists within a single phase (in the solidified adhesive or sealant phase). Cohesion is a state in which particles (molecules) of a single substance are held together; it is a special instance of adhesion in which only molecules of the same kind adhere to each other. Figure I-2.1 explains adhesion and cohesion in an adhesive bond.

Note: Throughout this section, it is referred to adhesives, however, similar considerations apply to sealants and to coatings.

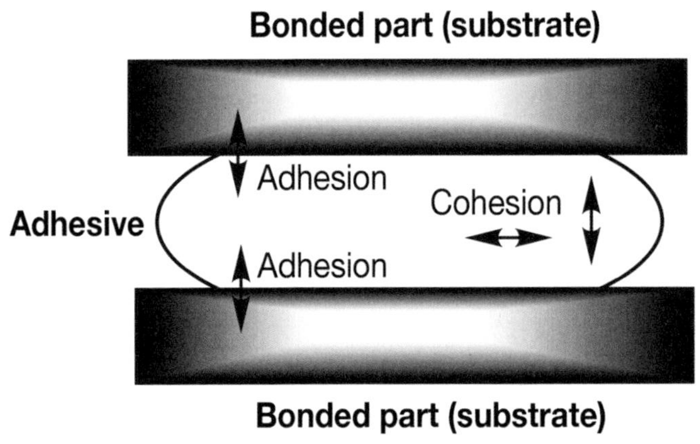

Figure I-2.1: Interaction of adhesion and cohesion in a bond

2.1 Adhesion forces and mechanisms

In view of the fact that the intention of this book is to provide assistance in the formulation of adhesives and sealants, that the phenomenon of adhesion is not only extremely complex but also not yet fully understood, and that the statements of the various adhesion theories are therefore discussed quite controversially, this section is limited to a description summarizing the most important phenomena contributing to adhesion, described in more detail in the respective literature e.g. in [1, 3–6].

Different types of interactions taking place in the atomic and molecular range seem to contribute to the resulting adhesion. It is assumed that these interactions do not only take place at the outermost molecular layer, but can extend over several atomic or molecular layers, depending on the structure and composition of the substrates.

The mechanisms leading to adhesion can be divided into the groups shown in Figure I-2.2 and Table I-2.1. In addition to the range of the respective interactions, Table I-1.1 also indicates the magnitudes of the binding energies measured in each case. If the bond energies are used to calculate the adhesion forces theoretically possible under ideal conditions in relation to a unit area, the values obtained are usually well above the strength values which can be achieved with adhesives. Even if the ideal conditions on which the calculation is based are practically never given in reality, the conclusion is obvious that any of the potential adhesive interactions is capable of forming a bond in an adhesively bonded part. The question seems rather to be how many of these interactions are formed under real conditions.

2.1.1 Primary valence bonds

In the case of the group of primary valence bonds, a distinction is made between the formation of salts (ionic bonds), the formation of chelates and the formation of covalent bonds. All three variants offer by far the highest bond energies and therefore produce strong adhesive bonds. Their contribution is almost undisputed, at least when bonding with chemically reactive adhesive systems, i.e., adhesives curing in the joint by a chemical reaction. The extent to which they play a role in non-reactive adhesives, on the other hand, has not been fully clarified.

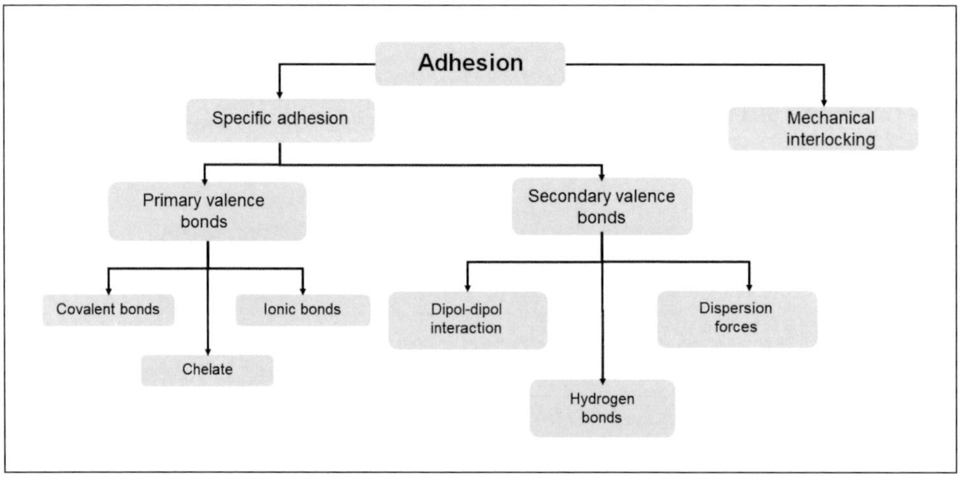

Figure I-2.2: Mechanisms of adhesion

Table I-2.1: Interaction forces in adhesive boundary layers according to [1]

	Primary valence bonds				Secondary valence bonds		
	Ionic bonds	**Chelate**	**Covalent bonds**	**Hydrogen bonds**	**Permanent dipoles**	**Induced dipoles**	**Dispersion forces**
Range	←— 0.1 – 0.2 nm —→			0.3 – 0.5 nm	←— 0.3 – 0.5 nm —→		
Bond energy	600 – 1000 kJ/mol		60 – 700 kJ/mol	≤ 50 kJ/mol	≤ 20 kJ/mol	≤ 2 kJ/mol	0.1 – 40 kJ/mol
Theoretical adhesion forces	30 MPa	5000 MPa	17500 MPa	500 MPa	200 – 1750 MPa	35 – 300 MPa	60 – 360 MPa
Experimentally determined bond strength	←— 15 – 25 MPa —→						

2.1.1.1 Ionic bonds

Ionic bonds (e.g., salt formation) are formed especially on mineral substrates. Examples are metal oxide layers and phosphated metal surfaces. Figure I-2.3 shows the formation of an anionic bond by an acid-base reaction between hydroxyl-groups of the oxide layer of a metal, like aluminium and a carboxyl group of an adhesive. The disadvantage of most ionic bonds is that they are sensitive to moisture as they can be broken by infiltrating water which finally leads to loss of adhesion under wet conditions.

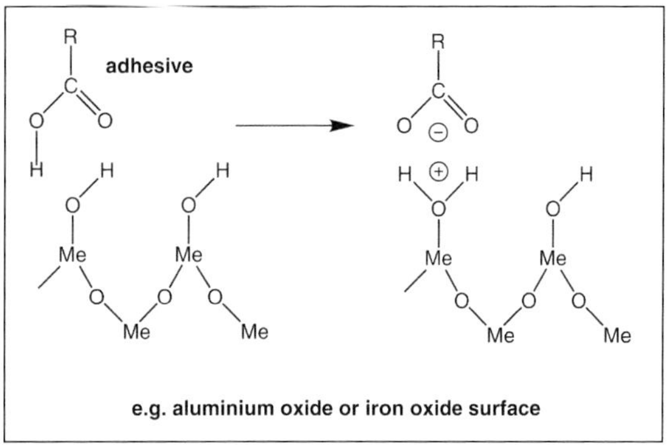

Figure I-2.3: Salt formation on aluminium or iron oxide surfaces (acid-base reaction)

2.1.1.2 Chelate complexes on surfaces

In contrast, chelate complexes formed between metal(oxide) surfaces and adhesive are often largely hydrolytically stable and thus generally lead to bonds with good moisture resistance.

Thus, the good wet adhesion of many amine-cured epoxy resins and thermally cured phenolic resins (resoles) on metal oxide surfaces is attributed to the formation of chelate complexes as shown in Figure I-2.4. [7–8]

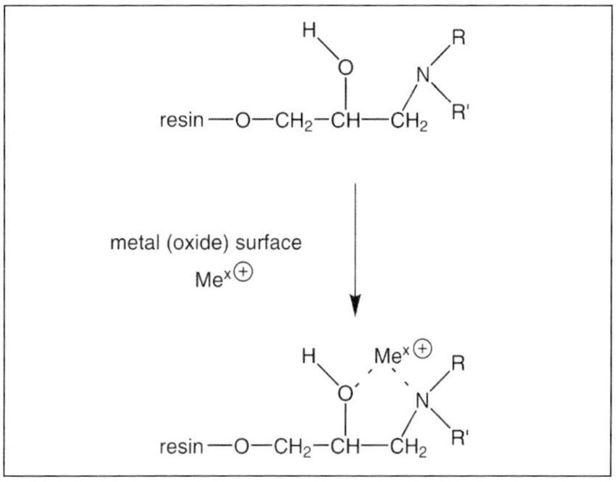

Figure I-2.4: Adhesion of an amine-curing 2-component epoxy adhesive by chelation on aluminium or iron oxide surfaces

According to the principles of complex chemistry, the complexes formed are stable if the bonding angles of the chelate bonds correspond to the bonding angles favourable for the metal atom and the ligands. In other words, if the chelate structure can be built up with as little stress as possible. Thus, the 6-ring chelate complex formed on aluminium surfaces (Figure I-2.5) is very stable, which is considered to be the main reason for the good stability of aluminium bonded with phenolic resins [1]. On the other hand, the 8-ring chelate complex (Figure I-2.5), which also appears plausible on the basis of recent molecular calculations [9], is less stable due to Figure I-2.5 the less ideal bond angles.

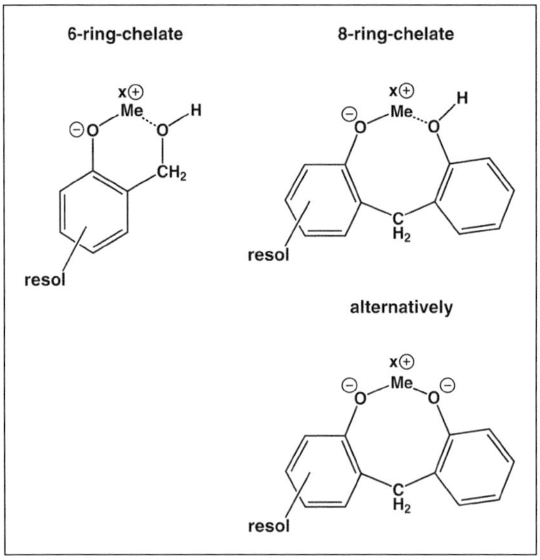

Figure I-2.5: Stable 6-ring chelate complex between hydroxymethylphenol and an aluminium oxide surface compared to the strained, less stable 8-ring chelate complex

For a urethane group, e.g. in a methylene diphenyl isocyanate (MDI)-diol-prepolymer the formation of an ionic complex-like adhesion mechanism (Figure I-2.6) has been proposed [5].

2.1.1.3 Covalent bonds

A further possibility is the formation of covalent bonds on substrates by chemical reaction between reactive groups of the adhesive and suitable functional groups in the surface of the substrate. This may e.g., be hydroxyl- or carboxyl-groups of glass surfaces, in the oxide layer of metals or formed by surface pre-treatments measures of plastic surfaces (e.g., oxidation by flame treatment).

Figure I-2.7 is showing as an example the formation of a covalent bond between the hydroxyl-group of an alu-

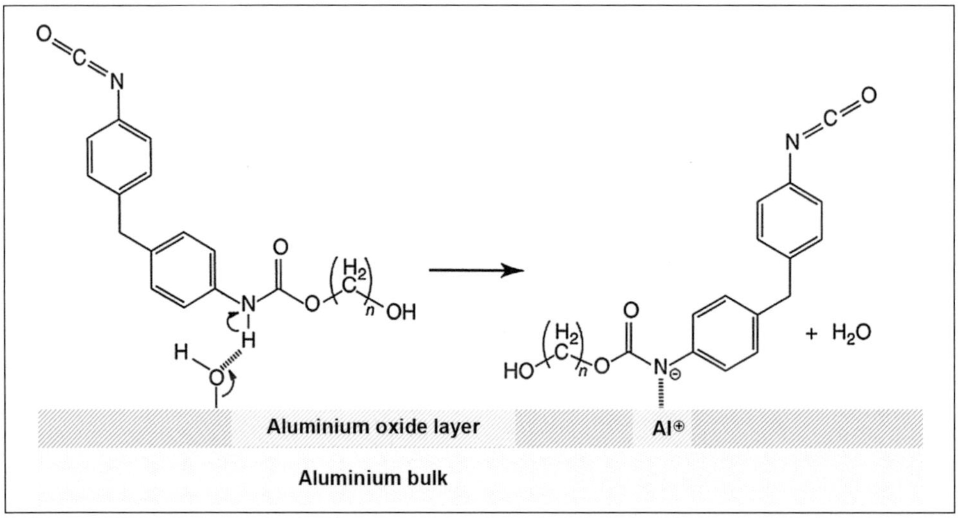

Figure I-2.6: Formation of an ionic complex-like interaction of a methylene diphenyl isocyanate (MDI)-diol-prepolymer on aluminium [5]

minium oxide surface layer and the isocyanate group of diphenyl methane diisocyanate often used as curing agent in a reactive 2-part polyurethane adhesive or sealant [5].

The adhesion-promoting effect of silanes (see also Chapter I-2.3) is an example of the formation of covalent bonds, in this case between the silanol groups formed by traces of water from the alkoxy groups of the silane coupling agent by hydrolysis and the hydroxy- or carboxyl-groups, e.g. of inorganic substrate surfaces. The group R^2- of the silane coupling agent, which is tailored to the adhesive used, allows it to be built into the polymer network of the adhesive that is formed during the curing of the adhesive, so that a strong bond is ultimately formed (Figure I-2.8).

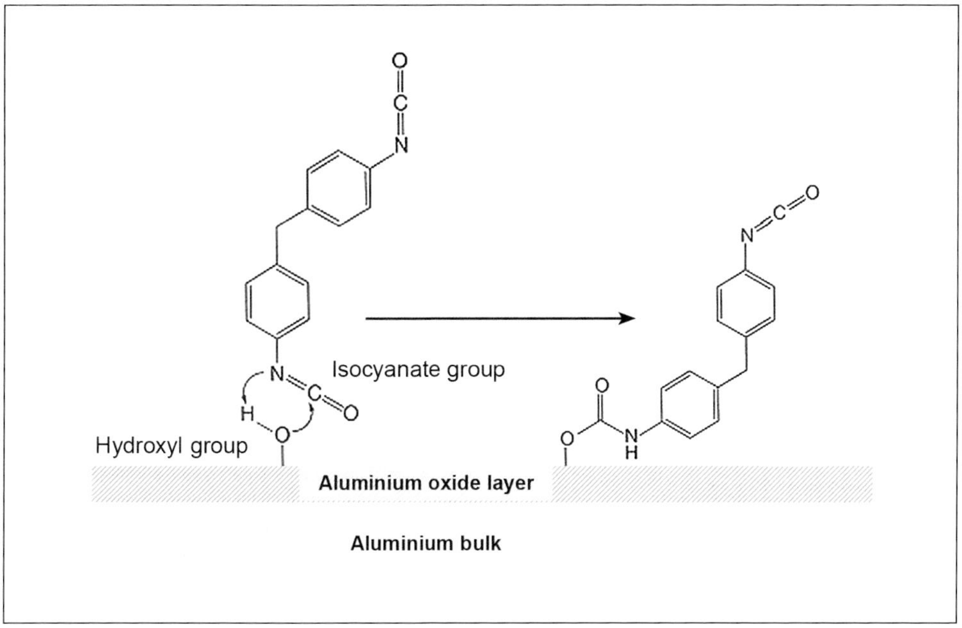

Figure I-2.7: Formation of a covalent bond between hydroxyl group of the aluminium oxide surface layer and the isocyanate group of the diphenyl methane diisocyanate curing agent of a reactive polyurethane adhesive/sealant [5]

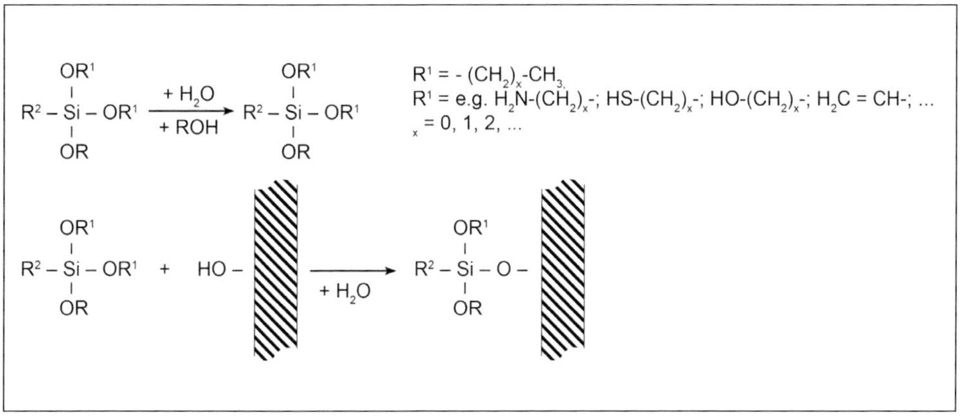

Figure I-2.8: Formation of covalent bonds between silane coupling agent and a surface containing hydroxyl groups

2.1.2 Secondary valence bonds

The weaker, secondary valence bonds (Table I-2.1) are the more common type of specific adhesion. Similar to the formation of primary valence bonds a close contact between adhesive and substrate is required. The following types of secondary valence bonds exist.

2.1.2.1 Dipole forces (dipole-dipole interaction)

In chemistry, dipoles are regions in a molecule or polymer, in which the electron cloud of a covalent bond is not evenly distributed due to a different electronegativity (ability to attract electrons) of the atoms involved but is therefore concentrated at the more electronegative atom. Thus, a negative partial charge is present at the more electronegative atom and a positive partial charge at the less electronegative atom. The greater the difference in the electronegativity, the stronger the dipole. The best-known dipole in chemistry is the water molecule (H_2O). Hydrogen has a low electronegativity of 2.1, while oxygen has a much higher electronegativity of 3.5, so that the water molecule bears a negative partial charge at the oxygen atom and a positive partial charge at the hydrogen atoms (Figure I-2.9).

If two dipoles come close enough to each other, attractive forces occur between the areas with positive and those with negative partial charge. In the case of water, this leads to the fact that, in contrast to hydrogen sulfide H_2S (the electronegativity of sulfur, at 2.4, is significantly lower than that of oxygen with 3.5), water is a liquid at room temperature due to the attractive forces between the dipole molecules and not a gas like hydrogen sulfide.

In terms of the phenomenon of adhesion of an adhesive or sealant to a substrate surface, dipoles of the adhesive or sealant can interact with those on the substrate forming adhesion. An example of such polar groups is the polypropylene glycol frequently used in 2-part polyurethane adhesives or sealants. Here, caused by the, compared to carbon higher electronegativity the oxygen atoms of the polymer chain carry a negative partial charge and the positive partial charge is located in the region of the carbon and hydrogen atoms (Figure I-2.10).

Similar to the adhesive molecules or polymers, shifts in the electron cloud and thus positive and negative partial charges also occur in adherends made of plastics. In the case of polyethylene (PE), which is consists solely of carbon and hydrogen atoms, dipoles are not formed due to the minimal differences in electronegativity. The situation is different with polyvinyl chloride (PVC). Here, every fourth hydrogen atom (electronegativity of 2.2) is replaced by a much more electronegative chlorine atom (electronegativity of 2.8), so that there are negative partial charges on the chlorine atoms and the respective positive partial charge in the area of the carbon and hydrogen atoms (Figure I-2.11). This explains, at least in part, that PVC is much easier to bond than PE.

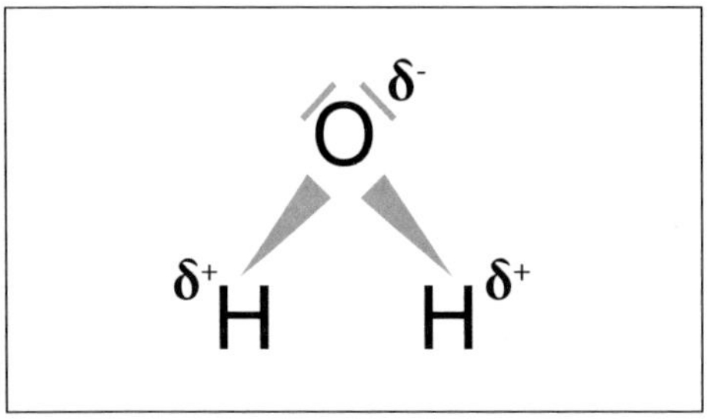

Figure I-2.9: Dipole character of the water molecule

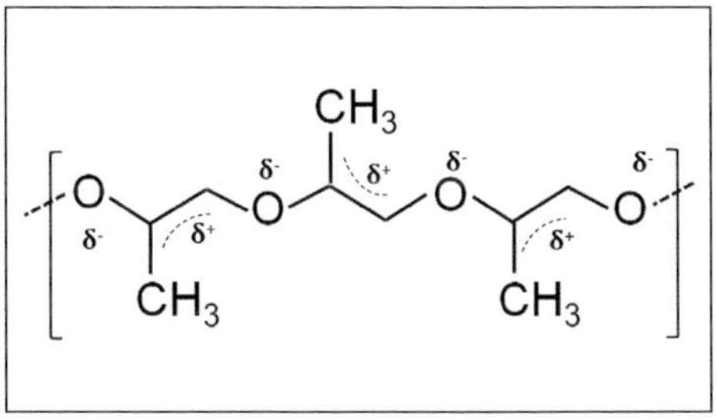

Figure I-2.10: Charge distribution in polypropylene glycol

Dipoles can also be formed by approaching a permanent dipole in intrinsically nonpolar molecules if these, such as the aromatic ring, have delocalized electrons. (Figure I-2.12).

2.1.2.2 Hydrogen bonds

Hydrogen bonds are formed between molecules in which hydrogen atoms (electronegativity of 2.2) are bonded to particularly strongly electronegative atoms, such as oxygen (electronegativity of 3.5) or nitrogen (electronegativity of 3.1) The hydrogen atoms carry a positive partial charge and can interact with the free electron pairs of other electronegative atoms. Figure I-2.13 describes the

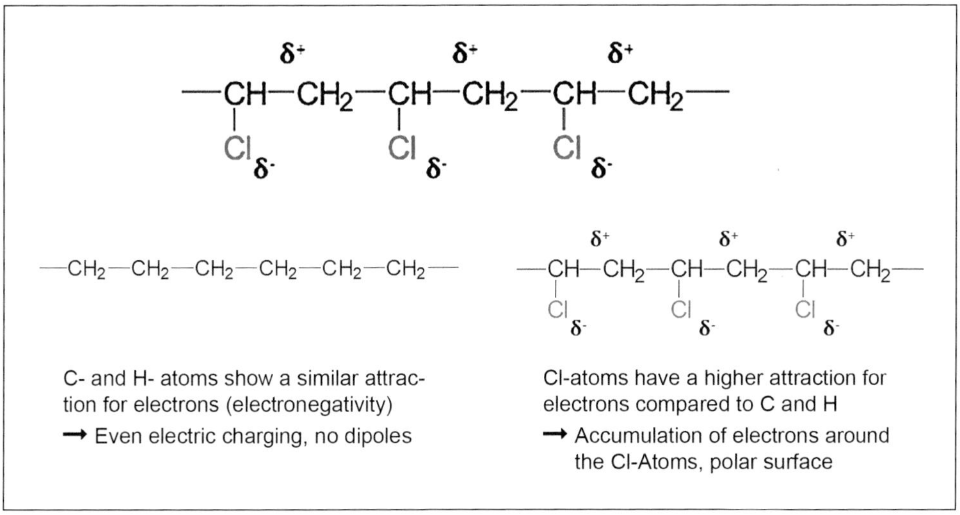

Figure I-2.11: Different charge distribution in PE and PVC due to the different electronegativities of the atoms involved

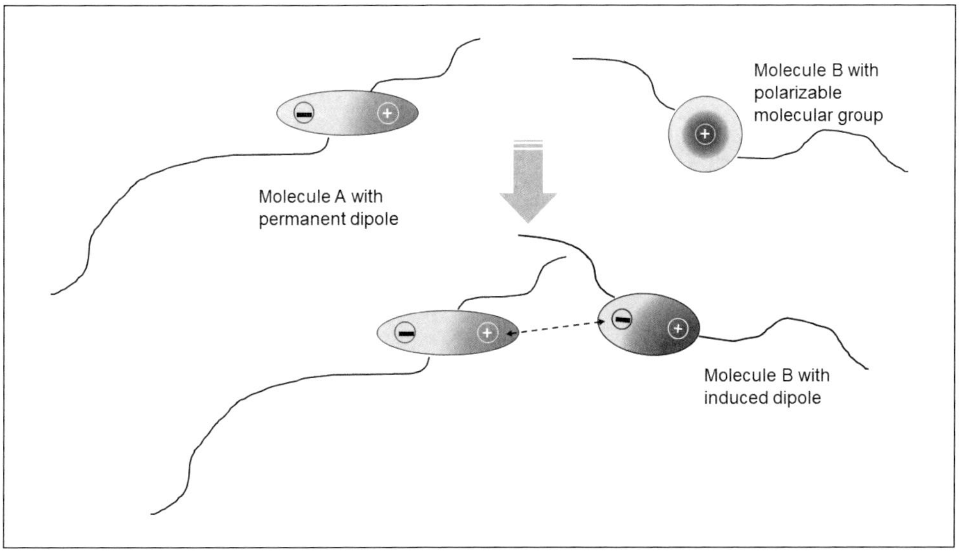

Figure I-2.12: Formation of an induced dipole when a permanent dipole approaches a polarizable molecular group

formation of hydrogen bonds between the carboxy group of an adhesive/sealant and hydroxyl groups of the oxide layer on the surface of aluminium.

A disadvantage of hydrogen bonds is that they have limited resistance to water diffusing into the interface, which might lead to a reduction of adhesion under wet conditions.

2.1.2.3 Diffusion theory

A special case is the often-observed high adhesion between similar and dissimilar surfaces of polymeric materials that are in close contact. The diffusion theory explaining this phenomenon is based on the assumption that polymeric materials, especially in the slightly crosslinked or non-crosslinked state, are not static systems in the sense of the molecular orientation of their basic building blocks. Especially above glass transition temperature, the molecular chains can move between their interlocking points in the form of atomic rotations. In these movements, it is possible that molecular regions may also emerge beyond their original surface and diffuse into an adjacent polymer. It is assumed that adhesion occurs during bonding with contact adhesives (see Chapter II-1.1.3.2, II-1.1.4.1 and II-1.1.4.2) when the parts to be bonded, which are coated on both sides with the well-dried adhesive, are pressed at least briefly under high pressure. As a result of the diffusion of the adhesive polymers into each other, an initial strength is achieved which increases further in the course of time.

In a first step, the adherend surface is dissolved by THF, which greatly increases the mobility of the PVC chains and thus their ability to diffuse. The PVC chains penetrate and interlock with each other, which, after evaporation of the THF, ultimately leads to a strong bond between the parts to be joined. In many cases, PVC or post-chlorinated PVC is additionally dissolved in the THF used as the adhesive, which improves processing and gap-filling capacity due to the increase in viscosity without impairing the bonding mechanism. One application for this is the bonding of rigid PVC drainage pipes.

2.1.2.4 Electrostatic adhesion

This effect is not very important in adhesives technology. Electrostatic adhesion occurs when two different polymers or a polymer (e.g. adhesive) and a metal are in close contact with each other. Differences in electron emission energies leads to electron transfer between the two materials. The electron transfer creates an electric double layer that induces electrostatic attraction.

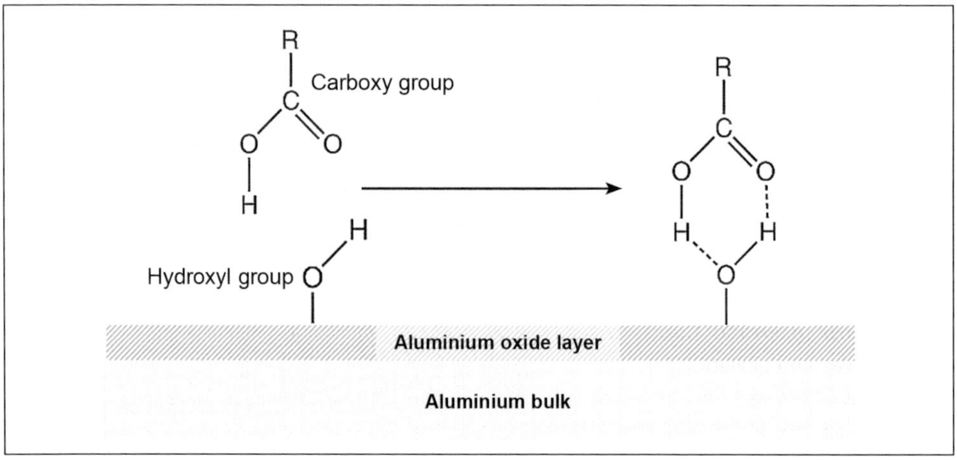

Figure I-2.13: Formation of hydrogen bond on a metal oxide surface

2.1.2.5 Mechanical interlocking

As already shown in the overview diagram of adhesion mechanisms (Figure I-2.2), mechanical entanglement is a special case. In general, a distinction is made between the already described mechanisms of specific adhesion, where the formation of interaction is independent of the geometric shape of the surface, and mechanical adhesion, also called mechanical interlocking, described in the following. Here, a liquid adhesive or sealant first penetrates cavities (voids, roughness including undercuts) of the substrate's surface. In case the adhesive or sealant was able to flow into undercuts, after hardening a form-fit joint, so to speak, by mechanical anchoring is formed.

This can be achieved, for example, by the flow of a sufficiently liquid adhesive or sealant into undercuts deliberately designed during the construction of the components or into undercuts of a very rough surface. Other possibilities are the flow of the adhesive or sealant into a porous material, e.g., a foam or a fibre structure, see Figure I-2.14.

It should be noted at this point that the effect of increasing the surface area by roughening the adherend is often greatly overestimated. The increase in potential bonding sites for interactions between adhesive/sealant and adherend due to the larger real surface area appears to contribute only to a small extent to the sometimes observed higher bond strength. The often observe effect of an increase in bond strength after grinding the surface is in most cases caused more by a removal of an adhesion-unfriendly surface layer than the increase in surface

If the adhesive or sealant is too high in viscosity to completely fill the grooves in the adherend's surface adhesion can therefore only occur at the tips which drastically reduces the "available" surface and as a result the number of potential bonding sites (Figure I-2.15).

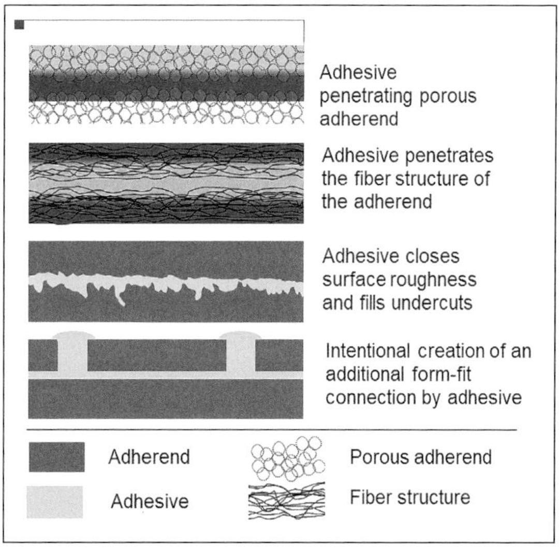

Figure I-2.14: Forms of mechanical interlocking

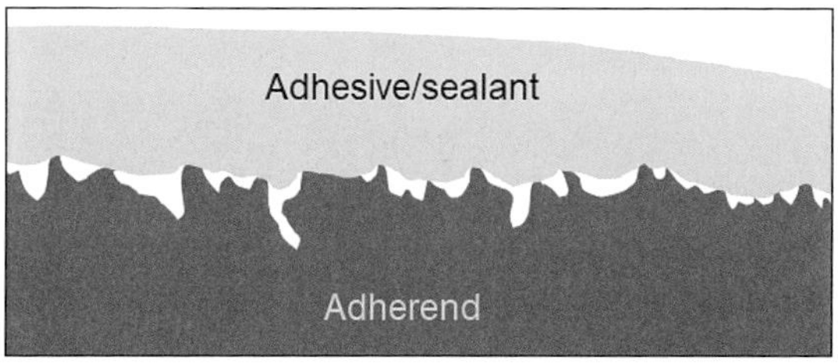

Figure I-2.15: High viscose adhesive/sealant unable to fill the cavities of a rough surface

2.2 Wetting of substrates

Although the phenomenon of adhesion has not yet been fully understood scientifically, it is undisputed that adhesion takes place at the interface between the materials involved. Due to the very small range of the different adhesion forces of less than 1 nm (see Table I-2.1), it is absolutely necessary for the formation of interactions that the adhesion partners come very close to each other. In terms of the utilization of adhesives and sealants, this means that the ability to wet the adherends is a prerequisite for the formation of adhesion, albeit not a sufficient one.

2.2.1 Surface tension and surface free energy

The terms surface free energy (SFE) and surface tension are physically equivalent. SFE is typically used for solid surfaces and surface tension for liquid surfaces. Occasionally, however, surface tension is also used for solids.

The term surface tension or surface free energy refers to the work required to increase the surface area, i.e., it has the unit of work per area (mJ/m^2) or force per wetted length (mN/m) and has the formula symbol γ (small gamma) or, more rarely, σ (small sigma).

Equation I-2.1 $$\frac{Energy}{Area} = \frac{J}{m^2} = \frac{N * m}{m^2} = \frac{N}{m} = \frac{Force}{Length}$$

In a liquid, such as water, all molecules in the bulk phase are uniformly surrounded by neighbouring molecules. The attractive forces acting on these molecules extend equally in all directions of space and cancel each other out (Figure I-2.16). At the water/air interface, things change dramatically, the water molecules are surrounded by other water molecules only at the interface itself and into the direction of the bulk phase. As a result, the forces of attraction do not cancel each other out but force acts on the water molecule towards the inner phase of the liquid (Figure I-2.16).

Being on the surface is therefore energetically less favourable. Without external forces acting, a liquid strives to take the most energetically favourable shape, i.e., the one with the smallest possible surface area. In zero gravity environment liquids would form spherical drops. The drop shape formed on earth under the influence of gravity results from a superposition of the surface tension and the gravitational pull. Surface tension is usually measured using the hanging drop method. For this, the shape of a drop hanging on a needle in a liquid or gaseous phase is measured. The shape of the drop results from the relationship between the surface tension or interfacial tension and gravity. Surface tension values for some exemplarily liquids are provided withTable I-2.2. The greater the surface tension, the greater the cohesion in the respective liquid phase and the stronger are the forces interacting between atoms or molecules. Water molecules associate strongly with each other by means of hydrogen bonds, which generate a high level of cohesion and a high surface tension. The water surface acts like a skin. The less polar a liquid is, the lower is its surface tension.

Unlike liquids, solids are unable to minimize their surface area by deformation. In order to reduce the free energy, they are capable of forming an interface with a liquid, i.e., they can be wetted.

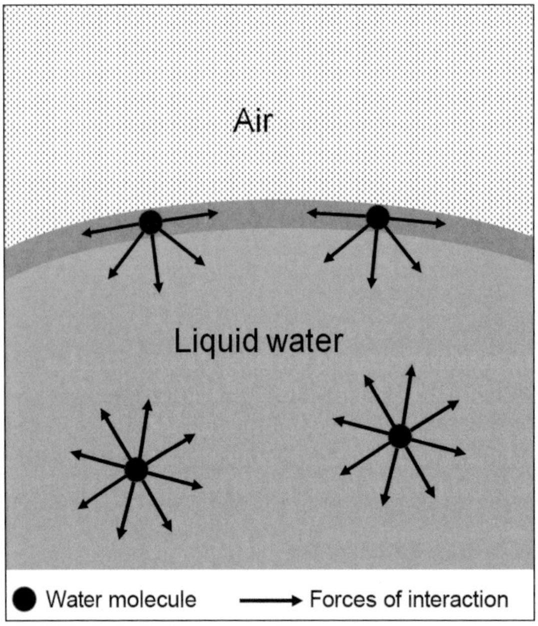

Figure I-2.16: Molecular environment in the volume phase and at the interface to air

2.2.2 Wetting

When a liquid is applied to a substrate surface, a new interface is created between the liquid and the substrate.

The measure of wettability is the contact angle θ, which is usually determined optically as the angle at the intersection of the contour of the liquid droplet with the plane of the surface (see Figure I-2.17). The contact angle results from a force equilibrium of the following three stress or energy components:

γ_S: surface free energy of the solid, γ_S
γ_L: surface tension of the liquid, γ_L
γ_{LS}: interfacial tension between the solid and the liquid

each of which strives to minimize the surface or boundary layer. The relationship between these three components is described by the Young's equation.

Table I-2.2: Surface tension [mN/m] of liquids

Liquid	Surface tension [mN/m]	Source
Mercury (15 °C)	487	[18]
Water	72.75	[18]
Butyl acetate	61.3	[19]
Glycerine	63.0	[18]
Epoxy resins, diglycidyl ether of bisphenol-A (DGEBA) diluted with diglycidyl ether of 1,4 butanediol (80:20)	41.7	[21]
Butyl glycol	65	[20]
Castor oil	39	[23]
Toluene (25 °C)	27.3	[18]
Ethanol	22.3	[18]
Polydimethylsiloxane, surface tension is increasing with molecular weight	15 – 24	[22]
n-Hexane	18.4	[18]
Diethyl ether	17.00	[18]

If not otherwise specified Measurement at 20 °C, more data are available from [24, 25]

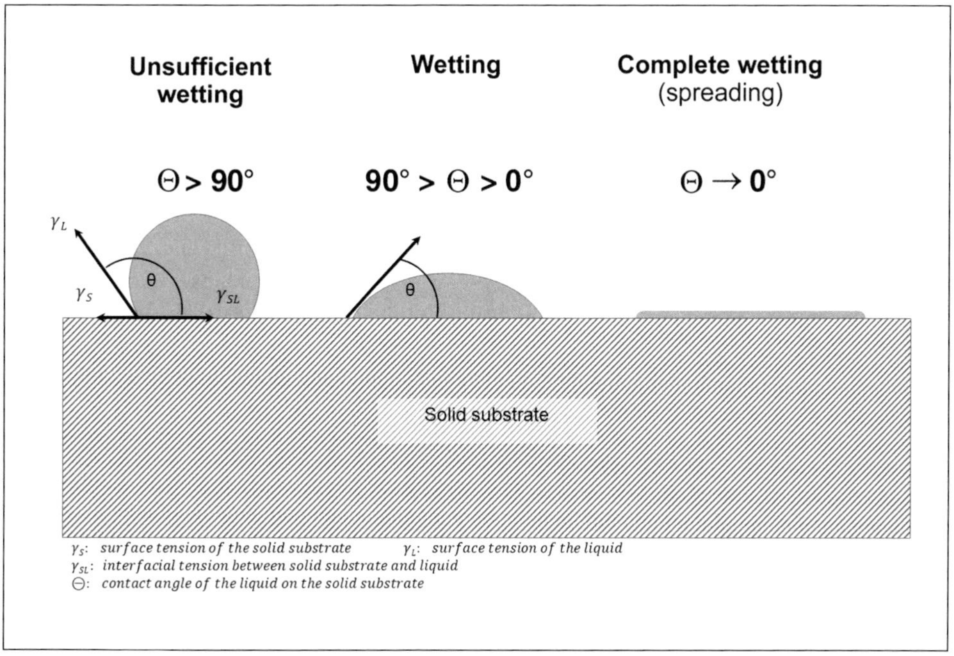

Figure I-2.17: Wetting of a solid by a liquid

Equation 2.2 $\gamma_S = \gamma_{SL} + \gamma_L \cdot \cos\Theta$

In the case of complete spreading ($\Theta = 0°$ and $\cos\Theta = 1$), Young's equation becomes:

Equation I-2.3 $\gamma_S = \gamma_{SL} + \gamma_L$

resp.

Equation I-2.4 $\gamma_L = \gamma_S - \gamma_{SL}$

A liquid, such as an adhesive, cannot wet the surface of a solid substrate unless its surface tension is lower than the surface free energy of the solid ($\gamma_L < \gamma_S$). Thus, if a liquid has a higher surface tension than the substrate, there will be no wetting.

In purely qualitative terms, the basic statement applies that good wetting is evident at a small contact angle, while poor wetting is evident at a large contact angle.

The Young-Dupré's contact angle equation

Equation I-2.5 $W_a = \gamma_L + \gamma_S - \gamma_{SL} = \gamma_L\,(1 + \cos\Theta)$

shows the relationship between the variables of Young's equation (Equation I-2.2) and the work of adhesion. The work of adhesion is the sum of the surface tension of the liquid and the surface free energy of the solid minus the interfacial energy between the liquid and the solid. It is the energy required to separate the interactions between the solid and the liquid phase. With a boundary angle of 180°, i.e., "complete" non-wettability of the solid surface by the liquid, the value of the work of adhesion is zero. [25]

In Chapter I-2.1, the different types of forces causing the cohesion of atoms and molecules were discussed. In particular, a distinction is made between disperse interactions i.e., those based on temporal fluctuations of the electron distribution, and polar interactions i.e. those based on permanent dipoles, and those occurring between permanent and induced dipoles. Accordingly, the surface free energy or surface tension of a component is also composed additively of a dispersive and a polar component [26, 27].

Equation I-2.6 $\gamma = \gamma^D + \gamma^P$

To determine the disperse and the polar part the surface energy of a solid, the contact angles formed on the surface of at least two test liquids are first determined optically. With their known surface tensions including the respective disperse and polar fractions, the polar and disperse fractions of the surface free energy of the solid can be calculated e.g., based on the model of Owens, Wendt, Rabel and Kaelble (OWRK model). In most cases, this is done by the software used by the respective measuring instruments. For more details of the WORK-model see [28].

Table I-2.3 provides the surface free energy data for a number of common solid materials obtained from different sources. The values given in this table for metals differ significantly from

Table I-2.3: Surface free energy [mN/m] of common solid materials

Material	Surface tension [mN/m] γ	γ^D	γ^P	Source
Glass not further specified	43–112			[47]
Microscope glass slide	57.8	18.9	38.9	[41]
Lime-glass	47	–	–	[42]
Steel not further specified	29–56			[47]
Steel S235	39.3	38.90	0.4	[30]
Steel S235 blasted	47.6	45.3	2.2	[30]
Steel S235. electro galvanized	37.6	34.2	3.4	[30]
Iron oxide	28–33			[47]
Stainless steel 1.4401	36.8	35.8	1.0	[30]
Aluminium	33–35			[47]
Aluminium	37.1	35.4	1.7	[30]
PE – Polyethylene not further specified	27–36			[47]
PE – Polyethylene – linear	35.7	35.7	0	[24]
PP – Polypropylene	28.0–34.1			[47]
PP – Polypropylene-isotactic	30.1	30.1	0	[24]
PS – Polystyrene	29.2–42.8			[47]
PS – Polystyrene	40.7	34.5	6.1	[24]
PTFE – Polytetrafluorethylene	15.6–24.8			[47]
PTFE v Polytetrafluorethylene	20.0	18.4	1.6	[24]
PVC v Polyvinylchloride	27.0–44.1			[47]
PVC – Polyvinylchloride	41.5	29.5	2.0	[24]
PMMA – Polymethylmetacrylate	38.7–45.8			[47]
PMMA – Polymethylmetacrylate	41.1	29.6	11.5	[24]
PET – Polyethylene terephthalate	37.4–46.5			[47]
PET – Polyethylene terephthalate	44.6	35.6	9.0	[24]
PBT – Polybutylene terephthalate	45			[47]
PA 6 – Polyamide 6 (polycaprolactam)	38			[47]
PA 6.6 – Polyamide-6.6	46.5	32.5	14.0	[24]
PC – Polycarbonate	29–38.8			[47]
PC – Polycarbonate	34.2	27.7	6.5	[24]
PEEK – Polyetheretherketone	42.1	36.2	5.9	[24]
PDMS – Polydimethylsilicone	19.8	19	0.8	[24]

those of other sources, like [13], which give values in the range of several hundred to a few thousand. The reason for this is that these high values were determined for metals in the molten state i.e., they are not relevant for metal surfaces when it comes to adhesive bonding.

The values also show that the surface free energy is far from being a material constant but is influenced by a number of various parameters.

It is assumed that the polar fractions of a liquid can interact at least preferentially with polar fractions of a solid surface and, conversely, disperse with disperse fractions. It is therefore possible to make a prediction about the quality of the wetting and, to a certain extent, also about the intensity of the resulting adhesive forces by comparing the ratio of the polar and disperse fractions of the surface tension of a liquid with that of the solid surface to be wetted.

The more the disperse and polar fractions match, the more interaction potential there is between the phases, and the better the wetting will be, which is usually reflected in an increased adhesion. This explains, at least in part, the statement that good wetting is a necessary but not necessarily a sufficient condition for the formation of adhesion. [28]

The theory of the so-called wetting envelope is based on this assumption that similar areas can best interact with each other. Provided that the polar and disperse parts of the surface tension and the surface free energy are known, the calculation of the wetting envelope allows to determine whether the liquid phase matches the surface of the solid phase and wetting can take place. This theory has proven useful in questions of wetting of coatings, among others. Unfortunately, adhesives and sealants are often very high viscous liquids or even sag resistant pasty materials whose surface tension is not easy to determine. For example, the geometry of a drop is not determined solely by gravity and surface tension, but also to a large extent by rheological properties. In addition, the surface tension is not constant but changes due to the chemical and physical processes occurring during solidification. [31, 32]

In 2021, a research project was therefore launched with the aim of developing various alternative methods for determining the surface tension of adhesives and systematically investigating the factors that influence the surface tension of adhesives. [33]

2.2.3 Surfaces

If one takes a closer look at solids, one realizes that their surface differs from the bulk material; this is applicable to all materials. In addition, the surfaces of solids are influenced to a not inconsiderable extent by the way in which they were formed and may well change because of external effects.

Consequently, the values provided in Table I-2.3 for the surface free energy of solids can by no means be regarded as material constants for the respective materials, but merely as reference values.

2.2.3.1 Metal surfaces

According to the schematic structure of a metal surface shown in Figure I-2.18, metals, like all other solids, are not homogeneous throughout, but have various very thin layers at the surface. Figure I-2.19 uses the example of aluminium to illustrate this with chemical formulas. It is a simplification of the real situation in that the adsorption layer following the oxide layer is deliberately not included.

If, as in the case of sheet metal production, a cold rolling process has taken place, the undisturbed structure of the bulk metal is initially followed by the so-called cold deformed layer.

This is followed by the oxide layer, which is formed under undefined conditions during storage by a chemical reaction of the metal with oxygen and moisture from the environment. It has therefore an undefined structure, adheres quite firmly to the underlying layers but can have a detrimental effect on adhesion during bonding and should therefore be removed mechanically or chemically in beforehand. If it is not feasible to bond immediately after its removal, it is advisable to build up a defined oxide layer suitable for bonding e.g., by means of an appropriate surface treatment.

ESCA/XPS measurements show that metal surfaces contain in addition to the respective metal and oxygen a significant amount of carbon. Presumably, this layer consists of substances adsorbed from the ambient air, such as water, carbon dioxide, but also traces of organic compounds like hydrocarbons contained in the air [36, 37]. Furthermore, impurities such as dust particles are naturally deposited on the surface and there may also be deliberately applied anti-corrosion oils and residues of processing aids like drawing lubricants on the surface [34].

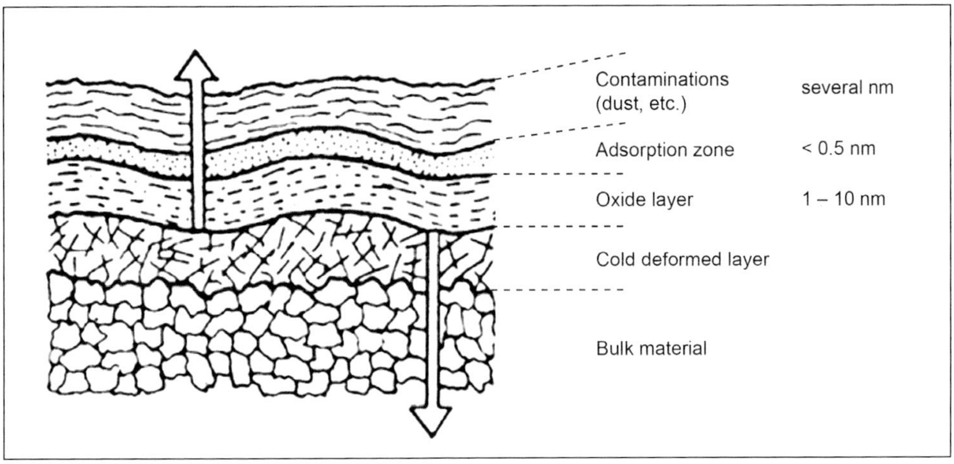

Figure I-2.18: Schematic structure of a metal surface [4]

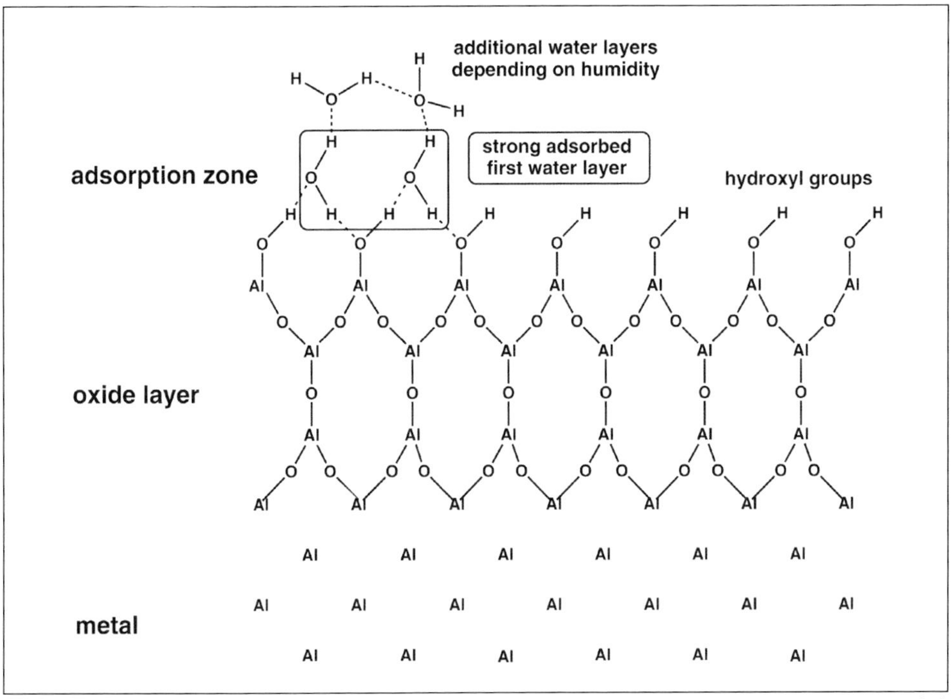

Figure I-2.19: Simplified structure of the oxide layer and the adsorption zone of an aluminium surface [35]

2.2.3.2 Surfaces of plastics

Similar but to some extend even more complicated are the surface structures of engineering plastics [35]. Like metals, the material on the surface differs significantly from the bulk. These differences are the result of the manufacturing or processing conditions and vary depending on the processing conditions and of course also on the type of polymer.

During the moulding process in the manufacturing of plastic parts, the polymers are either in the form of a melt or the polymer is formed by a chemical reaction between the liquid monomers or prepolymers. In both cases, the polymer chains, respectively the resulting polymers, have the ability to align themselves in such a way that an energetically favourable state is achieved. This means that according to the alchemists' maxim "similia similibus solvuntur" (Latin: "Similarity dissolves in similarity") the polar groups preferably align themselves in the direction of a polar environment and, conversely, non-polar groups preferably aligned in the direction of a non-polar environment.

At least in part, this can explain the phenomenon often observed in injection moulded parts of an unusual low surface free energy for the respective polymer and thus poor wettability. Due to the poor compatibility of the polar domains of the polymer chains with the non-polar surface of the mould, which is covered by release agent, the polar polymer domains orientate themselves towards the bulk during moulding. As a result, an accumulation of nonpolar chain segments, the so-called mould skin is formed on the surface (Figure I-2.20).

Furthermore, it must be considered that engineered plastics contain, in addition to the respective polymer, a number of functional ingredients which may influence the bonding properties (see Table I-2.4).

Some of these ingredients show the tendency to migrate. For example, internal release agents accumulate on the surface during the moulding process under the influence of heat and must be removed before bonding or, under certain conditions, can be absorbed into the adhesive during heat-supported curing and neutralized, so to speak. Plasticizers often show even at room temperature a more or less pronounced migration. Depending on the type, they tend to migrate from the bulk of the plastic material to the surface. Even if the plasticizer has been removed from the plastic surface before bonding, migration, now into the boundary layer between adherend surface and adhesive is to be expected, resulting in a significant weakening of the bond. This effect is known, for example, for plasticized PVC and EPDM.

Again, similar to metals an adsorption layer as well as a layer of contaminants like dust are present on polymer surfaces.

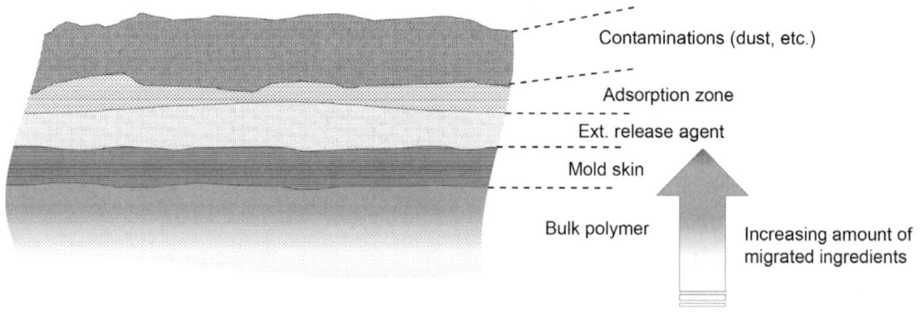

Figure I-2.20: Schematic structure of the surface of a plastic part

Table I-2.4: Additives in engineering plastics

Additive	**Purpose**
Filler	Reduction of raw material cost, improving of specific properties
reinforcing fibres	Improving of mechanical properties
Pigments	Colouration
Optical brighteners	Improving colour and brilliance
Antioxidants	Stabilization against the influence of oxygen, during processing at elevated and during use
Light (UV)-stabilizer	Improving stability against light, especially UV-radiation
Biostabilizers	Prevention biological material deterioration
Antistatic agents	Prevention of static charging
Flame retardants	Prevention or retardation of the spread of fires
Plasticizer (external)	Increasing flexibility/softness
Internal mold release agents	Improving mold release due to accumulation on the surface during the molding process
Anti-blocking agent	Minimize contact between adjacent polymer surfaces

2.2.3.3 Glass surfaces

Glass is a collective term for a group of inorganic amorphous solids consisting mainly of silicon dioxide (SiO_2). Although glass is one of the oldest materials known to mankind, many questions about its atomic composition and structure are still unresolved. In the case of silicate glasses, SiO_4 tetrahedra form the basic structure; in contrast to crystalline quartz, these form a random network. During production of glass Metal ions are intentionally introduced into the network by additives such as sodium-, calcium- or potassium carbonate. The cations break the bonds of the bridging oxygen in the silicon- tetrahedra and the oxygen forms a much weaker ionic bond with a cation. The chemical structure of soda-lime glass (Figure I-2.21), the most common type of glass resembles a silicate network whose bonding conditions, but no long-range order, are comparable to the crystalline state. [38]

Figure I-2.21 clearly shows that the glass surface is favoured for the accumulation of water molecules (formation of hydrogen bonds). Therefore, glass surfaces carry a thin film of moisture depending on the relative humidity. Cations can be dissolved from the surface and replaced by H_3O^+-ions. The resulting OH^--ions lead to an increase in the pH-value, which in turn causes further dissolution of metal ions [40].

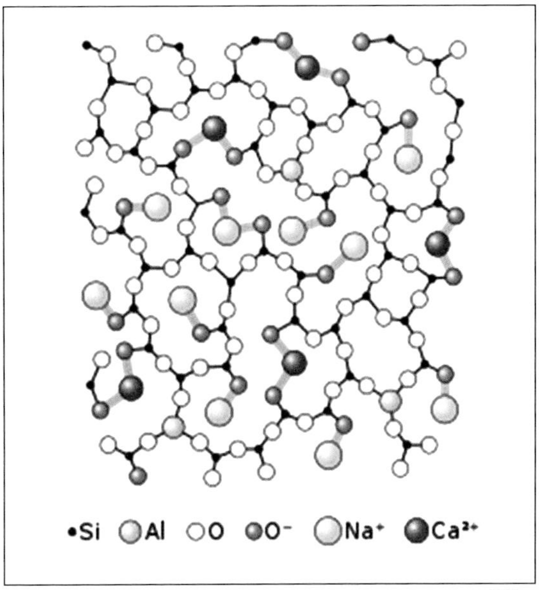

Figure I-2.21: Chemical structure of Soda-lime glass [39]

It can therefore be assumed that the surface of glasses is covered with a hydroxide layer. Due to the dependence of the chemical processes occurring with water on glass surfaces from the respective ambient conditions, this layer is not very well defined and is also subject to changes. This explains the quite different values given in the literature for the free surface energy of glass (see Table I-2.3).

In the case of bonded joints, these irreversible changes in the glass surface, also known as corrosion, which occur under the influence of moisture and may be accelerated by basic fillers contained in the adhesive, lead to adhesive failure in the interface between the adhesive and the glass which explains the observation that bonded joints on glass are without the use of a primer generally not resistant to the effects of moisture. [1]

2.3 Improvement of adhesion through pretreatment

As already explained, the surface of adherends in the untreated state is usually not defined and is therefore of very limited use to produce high-quality bonded joints or seals. It is therefore a good idea to treat the surface to achieve a defined state that allows the adhesive or sealant to build up sufficient adhesion for the intended application. In addition to cleaning, which is unavoidable in most cases, targeted modification of the surface of the adherend should be considered. Figure I-2.22 provides an overview of common technologies.

It is beyond the scope of this book to further explain the different procedures. Therefore, reference is made to the wide-ranging technical literature on the various method like [35, 43, 44]. Since suppliers of adhesives and sealants often offer special primers to improve the adhesion of their products to certain materials, their basic structure and mode of action is briefly explained in the following section.

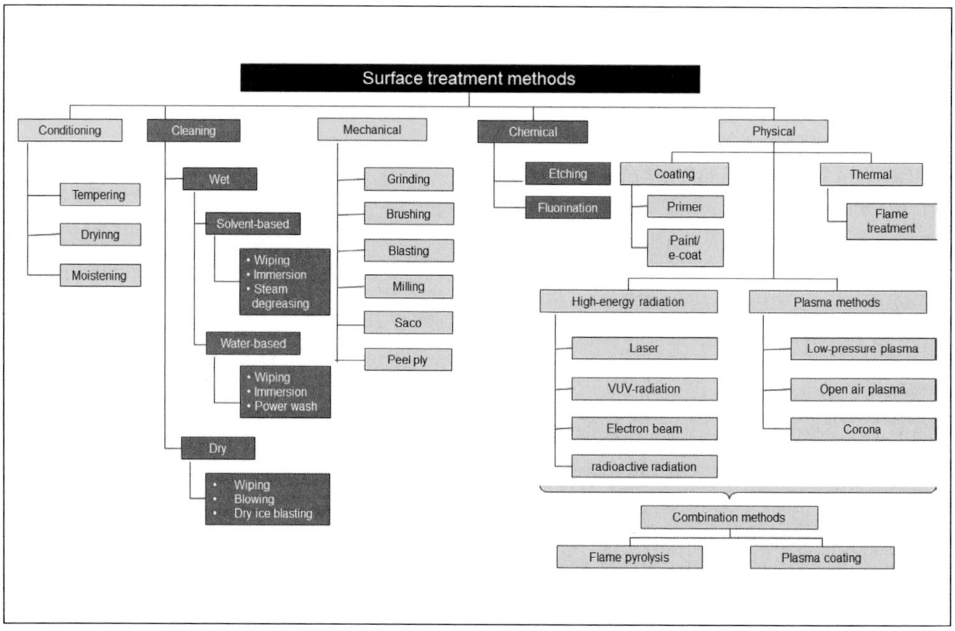

Figure I-2.22: Overview of common surface treatment technologies

2.3.1 Primer and thin adhesive polymer layers

Primers are chemical substances dissolved in suitable solvents, which are applied in a thin layer to the bond area before dispensing the adhesive. The adhesive is applied within a time window which is limited on the one hand by the minimum flash-off time, which is essentially based on the evaporation rate of the solvent, and for most primers also by the maximum flash-off time, i.e. the time after which the primer loses its activity.

The thin film of the primer's active substance remaining on the adherend acts as a bonding bridge between the surface of the adherend and the adhesive or sealant, thereby improving the adhesion of the adhesive or sealant to the adherend and ultimately the strength and long-term durability of the adhesive/sealant.

The mode of action is based on the assumption that the active substance of the primer can interact with the surface of the adherend by means of suitable groups and with the adhesive by means of a different type of groups. This is shown schematically in the upper part of Figure I-2.23 and in the lower part for a primer based on an amino silane as active ingredient as an example for enhancing the adhesion of an epoxy adhesive to aluminium.

Upon application of the primer to the adherent the active molecules are expected to react via one of the two different reactive groups in this example the silanol-group with the adherend forming a strong connection to the adherend's surface. After evaporation of the solvent and application of the adhesive the second reactive group of the primer molecules, in this example an amine group becomes part of the adhesive network with a corresponding group of the adhesive, in this example an epoxy group.

To achieve the best performance in regard to bond strength and durability, it is crucial to create a thin, ideally monomolecular layer of the primer molecules on the surface of the joined part. With increasing layer thickness, the reaction of the primer with the surface of the adherend and with the adhesive no longer necessarily takes place at the same primer molecule. In simplified terms, the strength of the bond is thus increasingly determined by the interactions between the individual primer layers, i.e. the cohesion of the hardened primer.

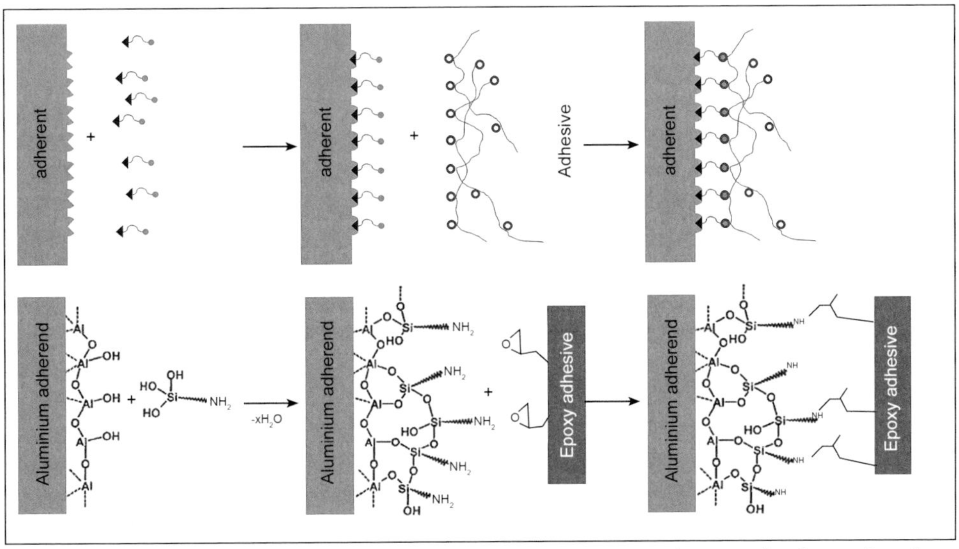

Figure I-2.23: Mode of action of a primer; top schematic, bottom using the example of an amino silane containing primer in an aluminium bond with an epoxy adhesive

It is self-evident and needs no further explanation that when selecting a primer for a bonding task, the active substance must match both the adherend's surface and the adhesive used. However, the solvent also plays a certain role, as it must not only have sufficient dissolving power for the active substance but should also not damage the adherends. In particular, the risk of stress corrosion cracking in plastics should be mentioned here. Of course, the requirements of occupational safety must also be considered, with the solvent being of particular importance, as it is essential that it evaporates during application.

The term primer is used to describe not only materials which, like silanes, are capable of forming covalent bonds between the surface of the adherend and the adhesive, but also products which form an adhesive bridge by forming complex bonds or secondary valence bonds.

Thus, as described above (Figure I-2.5), thermally cured phenolic resols form stable chelate complexes on metal oxide surfaces [7, 8]. The application of thin phenolic resin layers can therefore improve the bondability of metals and, in particular, their resistance to moisture. Furthermore, thermosetting phenolic resins (resols) are well known for their very good adhesion to aluminium, e.g. adhesive bonding in aircraft construction [45, 46].

The use of primers is not limited to metallic materials. Glass bonding, for example, generally requires the use of a primer if good moisture resistance is required [1]. Here, too, silanes are frequently used as active substances. Adhesion to the glass surface is achieved via the silane group, while adhesion to the adhesive requires a group adapted to the adhesive chemistry. Glass primers are often pigmented black, which, compared to a non-pigmented i.e., colourless product, not only facilitates the application of the desired thin layer, but also provides the adhesive (PU adhesives are typically used) with a certain degree of UV protection. Table I-2.5 provides a typical formulation for a glass primer

The bondability of plastic surfaces can also be improved by using suitable primers. For example, primers based on a polyester resin crosslinked by isocyanate groups through atmospheric moisture are used to improve the adhesion of 2-part PU adhesives to SMC (sheet moulded compound, short fibre reinforce polyester-based composite).

Table I-2.6 provides a guide formulation for a primer to enhance adhesion of polyurethane adhesives and sealants on SMC surfaces. When manufacturing the primer, care must be taken to ensure that the process is carried out in a water-free environment, as even traces of water can lead to undesired side reactions with the diisocyanate and the reaction product formed from polymeric diphenylmethan disocyanate (p-MDI) and polypropylene glycol might exhibit deviating properties. The dry primer film which remains after evaporation of the solvent has a maximum open time of up to 8 hours.

The adhesion of various plastics including polypropylene, TPO, and ABS can be improved by a solution of chlorinated polyolefin. Table I-2.7 shows a corresponding guide formulation. Especially in the pretreatment of polypropylene (PP) it has been proven beneficial to support the flash-off of the solvent by a heat treatment. It seems that this supports the effect of incorporation of the chlorinated polyolefin into the PP surface. Using a 2-part PU adhesive, the lap shear strength could be increased from 1.2 MPa for PP cleaned only with isopropanol to 3.5 MPa for the specimens pretreated by a chlorinated polyolefin-based primer.

Table I-2.5: Guide formulation for a glass primer for windscreen bonding

Item	Raw material	Parts by weight [%]
1	Physical drying saturated polyester	10 (solid)
2	Aromatic/aliphatic polyisocyanate based on toluene diisocyanate (TDI)/hexamethylene diisocyanate (HDI)	10 (solid)
3	Carbon black	10
4	Organic solvent mixture	66.5
5	Desiccant, p-toluolsulfonyl isocyanate	0.5
6	Adhesion promoter, glycidyl propyl trimethoxy silane	3
Total		100

Item 1: Polyester, molecular mass 20,000 – 25,000
Item 2: Desmodur HL BA, Covestro AG
Item 4: e.g. ethyl acetate, butyl acetate and 1-methoxypropylacetate-2
Item 5: Additive TI, Borchers
Item 6: Dynasilan GLYMO, Evonik Operations GmbH

Table I-2.6: Guide formulation for a primer to enhance adhesion of polyurethane adhesives and sealant on SMC

Item	Raw material	Parts by weight [%]
1	p-MDI (mixture of diphenylmethan-4,4'-disocyanate with isomers and higher functional homologues	10
2	Polypropylene glycol (MW 2000, OH-content 1.7 %)	10
3	Solvent	80
Total		100

Item 1: Desmodur VKS 20, Covestro AG
Item 2: Arcol Polyol PPG 2000, Covestro
Item 3: e.g butyl acetate

Table I-2.7: Guide formulation for a primer for plastics including PP

Item	Raw material	Parts by weight [%]
1	Chlorinated polyolefine	32.5
2	Acrylate-silicone copolymer	13.0
3	Solvent	54.5
Total		100.0

Item 1: Eastman CP730-1, Eastman Chemical Company
Item 2: Wacker G783, 38 % in Toluene, Wacker Chemie AG
Item 3: Xylene

2.4 References

[1] Brockmann, W., Geiß, P. L. Klingen, K., Schröder, B., Klebtechnik, Verlag Wiley-VCH Weinheim 2005, section 3, p. 11 ff
[2] EN ISO 10365:2022 – Adhesives – Designation of main failure patterns (ISO 10365:2022), Beuth Verlag, Berlin
[3] Habenicht, Kleben-Grundlagen, Technologien, Anwendungen, 4. Auflage, Springer-Verlag, Berlin, Heidelberg, New York, 2002
[4] Rasche, M., Handbuch Klebtechnik, Carl Hanser Verlag, München, Wien, 2012
[5] Possart, W., Adhesion and Interphases: the basics in brief, in W. Possart, M. Brede, Adhesive Joints, Wiley-VCH, Weinheim 2019
[6] Bischoff, C., Possart, W., Adhäsion – theoretische und experimentelle Grundlagen, Akademie Verlag, Berlin, 1983
[7] Kollek, H., C. Matz, C. Adhäsion, No. 12 (1989) p. 27 ff
[8] Müller, B., Surf. Coat. Int., Part B: Coatings Transact. 85 (2002) B2, p. 111–114
[9] Strehmel, V., B. Strehmel, B., Modification of Phenol-Methanal Network Polymers, in Polymer Modification – Principles, Techniques and Applications, Marcel Dekker New York – Basel, p. 575–705 (2000) especially p. 599
[10] commons.wikimedia.org/wiki/File:Benzene_Orbitals.svg, accessed 2025-12-21
[11] Kim, J., Cho, J. and Lim, Y.-S. Bonding urethane reactantas to aluminum surface, Journal of Material Science 40 (2005) 2789-2794
[12] Klingen, J., Fügetechnologie Kleben, Wiley-VCH, Weinheim 2019, p. 45
[13] Habenicht, G., Kleben – Grundlagen, Technologie, Anwendung, 4th ed., Springer-Verlag (2002)
[14] Meichsner, G., Mezger, T., Schröder, J., Lackeigenschaften messen und steuern, Vincentz Network (2003) Chapter 5
[15] Brock, Th., Groteklaes, M., Mischke, P., Lehrbuch der Lacktechnologie, 1st ed., Vincentz Network (1998) p. 173
[16] Wake, W. C., Adhesion and the Formulation of Adhesives, 2nd ed., Applied Science Publishers (1982) p. 55 ff
[17] Krüss GmbH www.kruss-scientific.com accessed 2024-12-21
[18] Wikipedia, en.wikipedia.org/wiki/Surface_tension#Values, accessed 2024-12-21
[19] Safety Data Sheet, Carl Roth GmbH & Co KG Butyl acetate, https://www.carlroth.com/medias/ accessed 2024-12-21
[20] Safety Data Sheet, FRIEDRICH SCHARR KG, Butylglykol, 12.07.2022
[21] Higgins, B. G., Surface tension of a curing epoxy, Journal of Polymer Science Polymer Chemistry Edition, July 1988
[22] www.accudynetest.com/polymer_surface_data/polydimethylsiloxane.pdf, accessed 2023-01-22
[23] www.researchgate.net/figure/Values-of-surface-tension-ST-reported-in-the-literature-at-20C_tbl1_269094854 accessed 2023-01-22
[24] cdn.dataphysics-instruments.com/Downloads/Surface-Tensions-Energies.pdf?v=1.0 accessed 2024/10
[25] Gleich, H., Hartwig, A., Lohse, H.: Das Phänomen der Haftung und warum Vorbehandeln so wichtig ist, in M. Doobe, Kunstoffe erfolgreich kleben, Springer Vieweg, Wiesbaden 2018, p. 9
[26] www.kruss-scientific.com/de-DE/know-how/glossar/freie-oberflaechenenergie, accessed 2023-01-24
[27] www.dataphysics-instruments.com/us/knowledge/understanding-interfaces/dispersive-polar-parts/, accessed 2023-01-24
[28] www.ossila.com/en-eu/pages/a-guide-to-surface-energy, accessed 2023-01-24
[29] Zigura, I., et al. Surface free energy of smooth and dehydroxylated fused quartz from contact angle measurements using some particular organics as probe liquids, JOURNAL OF OPTOELECTRONICS AND ADVANCED MATERIALS, Vol. 15, No. 7-8, July – August 2013, p. 627 - 634 www.researchgate.net/publication/286505242
[30] Schneider, P., Hofmann, J., Wetting and adhesion of liquids on surfaces, www.mpa.uni-stuttgart.de/institut/publikationen/ accessed 2024/10
[31] nanoscience instruments, www.nanoscience.com/techniques/tensiometry/surface-free-energy-sfe /, accessed 2024/12
[32] Krüss Applikationsbericht: Beschichtbarkeit vorhersagen, www.kruss-scientific.com/files/kruss-ar296-de.pdf accessed 2024/10
[33] Herbig, D., et al. Entwicklung eines Verfahrens zur einfachen und schnellen Abschätzung der Adhäsion in Klebverbindungen mittels Wetting Envelope, Fachvortrag Gemeinsame Forschung in der Klebtechnik 28.02./01.03.2023 Frankfurt/Main
[34] Rasche, M., Handbuch Klebtechnik, Carl Hanser Verlag, München, Wien, 2012, p. 136

[35] Petrie, E. M. Handbook of Adhesives and Sealants, McGraw-Hill, New York (2000)
[36] Brockmann, W., Emrich, S., Adhäsion, 44 (2000) No. 9, p. 40–44
[37] Müller, B., Surface Coating International, Part B: Coating Transact. 85 (2002) B2, p. 111-114
[38] de.wikipedia.org/wiki/Glas , accessed 2023-01-27
[39] Von Mrmw – Eigenes Werk, basierend auf: Kalk-Natron-Glas 2D.png:, CC0, commons.wikimedia.org/w/index.php?curid=98920200
[40] Hennemann, Brockmann, Kollek, Handbuch Fertigungstechnologie Kleben, Carl Hanser Verlag, München, Wien, 1992, p. 441
[41] Hejda, F., Solar, P., Kousal, J., Surface Free Energy Determination by Contact Angle Measurements – A Comparison of Various Approaches, WDS'10 Proceedings of Contributed Papers, Part III, 25–30, 2010. ISBN 978-80-7378-141-5 © MATFYZPRESS, www.mff.cuni.cz/veda/konference/wds/proc/pdf10/WDS10_304_f4_Hejda.pdf, accessed 203-01-27
[42] www.tu-chemnitz.de/physik/OSMP/Soft/V_07.pdf, accessed 2023-01-27
[43] Fraunhofer IFAM Adhesive Bonding – the right way, section 5 leitfaden.klebstoffe.com/en/home/
[44] Kollek, H., Reinigen und Vorbehandeln, Vincentz Verlag (1996)
[45] Brockmann, W., Adhäsion 14 (1970), No. 2, p. 52–68
[46] W. Brockmann, W., et al., Adhäsion 30 (1986), No. 9, p. 24–35
[47] www.kruss-scientific.com/de-DE accessed 2012-05-22 (page no longer available)

Part II Adhesives

1 Physically setting adhesives

1.1 Solvent-based adhesives and sealants

1.1.1 Basic principles

Polymers in solution

Solvents are the traditional auxiliaries for applying adhesives. Solvent-based adhesives have a fairly simple composition: Take a polymer and dissolve it in a suitable solvent to produce a polymer solution which is a more or less serviceable adhesive, Figure II-1.1.

All thermoplastic polymers – polymers which are not crosslinked – are suitable if an appropriate solvent can be found for them. However, various additional components, such as resins and additives, still need to be added before a serviceable adhesive is obtained. Various considerations are involved in the choice of solvent (in the adhesives industry, this means an inert liquid which has a boiling point below approximately 200 °C and which can dissolve the polymer and the other ingredients): The first is the nature of the polymer, i.e. the polarity relative to that of the polymer. Others are evaporation rates, safety, threshold limits, flammability, toxicology as well as ecological and economics issues. In many cases, a blend of solvents has to be used [1, 2].

The solubility of polymers in solvents can be predicted. Polar polymers will dissolve in polar solvents and non-polar polymers will dissolve in non-polar solvents. This is true as long as the polymer exhibits no strong tendency to crystallise.

Figure II-1.1: Random polymer model in solution

Better results are obtained by applying the concept of **solubility parameters** δ according to Hildebrand; the idea here is that the parameters of the solvent and the polymer should match as closely as possible. The Hildebrand concept describes the properties of a solvent in terms of three intermolecular forces:

- **dispersive forces** as the weak forces of attraction between all kind of materials
- **dipole or electrostatic forces** arising from permanent dipole moments, and
- **hydrogen bonds** as strong electrostatic forces between electronegative elements, such as fluorine, oxygen and nitrogen on one hand and the hydrogen bonded to one of them on the other.

Table II-1.1: Solubility parameters of solvents [1, 11]

Solvent	δ	**δ dispersive**	**δ polar**	**δ H-bridges**
Heptane	7.4	7.4	0	0
Cyclohexane	8.18	8.18	0	0
Toluene	8.9	8.8	0.7	1.0
Dichloromethane	9.73	8.72	3.1	3.0
Acetone	9.75	7.58	5.1	3.4
Ethyl methyl ketone	9.3	7.77	4.45	2.5
Ethyl acetate	9.08	7.44	2.6	4.5
Diethyl ether	7.61	7.05	1.4	2.5
Ethanol	12.90	7.73	4.3	9.4

Where $\delta^2 = \delta^2_{dispersive} + \delta^2_{polar} + \delta^2_{H\text{-}bridges}$
Unit of δ 1 [Hildebrand] = 1 $[cal^{0.5}cm^{-1.5}]$ = 2.046 $[J^{0.5}\ cm^{-1.5}]$

Solubility parameters are well established for most solvents (see Table II-1.1), polymers (Table II-1.2), and polymer building blocks [1–3]. Solubility is ideal when there is close agreement between the dispersive, polar and hydrogen bonding elements.

Knowledge of the solubility parameters can be a very useful tool in the search for alternative solvents or suitable solvent blends. As shown in Table II-1.3, solvent blends can dissolve polymers, even if the individual solvents are incapable of doing so.

What configuration does a polymer assume in solution?

The most stable and therefore most probable configuration is a disordered one which can be described by the random polymer model. This configuration is the most probable, because in it the polymer has maximum entropy.

When it comes to handling and applying polymer solutions as adhesives, viscosities are critical. Solution viscosities depend on different factors:

Table II-1.2: Solubility parameters of polymers [1, 2, 11]

Polymer	$\delta_{polymer}$
Polytetrafluoroethylene	6.20
Polyisobutylene	7.95
Polyethylene	8.0
Natural rubber	7.9 to 8.3
Polychloroprene	8.2
Polybutadiene	8.34
Polystyrene	9.3
Polymethyl methacrylate	9.27
Polyvinyl acetate	9.37
Butadiene nitrile rubber	9.4 to 9.5
Polyvinyl chloride	9.53
Polyethylene terephthalate	10.7
Phenolic resin	11.5

Table II-1.3: Solubility of polymers in solvent blends [1, 2]

Polymer	$\delta_{polymer}$	**Non-solvent 1**	δ_1	**Non-solvent 2**	δ_2
Polystyrene	9.3	acetone	9.75	cyclohexane	8.18
PVC	9.53	acetone	9.75	CS_2	10
Polychloroprene	8.2	diethyl ether	7.6	ethyl acetate	9.1

The polymers are soluble in combinations with the non-solvents

Table II-1.4: Evaporation rates of solvents [4, 5]

Solvent	Evaporation rate	Boiling point [°C]
Diethyl ether	1	35
Methylene chloride	1.8	40
Acetone	2.1	56
Methyl acetate	2.2	57
Hexane	2.5	69
Ethyl acetate	2.9	77
Benzene	3	80
Heptane	3	98
Butanone	6	80
Methanol	6.3	65
Ethanol	8.3	78
2-Propanol	11	82
Butyl acetate	12	127
Xylene	13.5	136 to 140
Acetic acid	24	118
Water	80	100

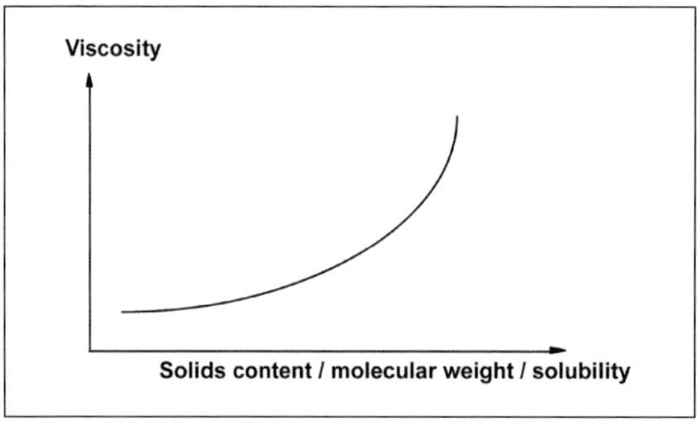

Figure II-1.2: Factors influencing solution viscosities

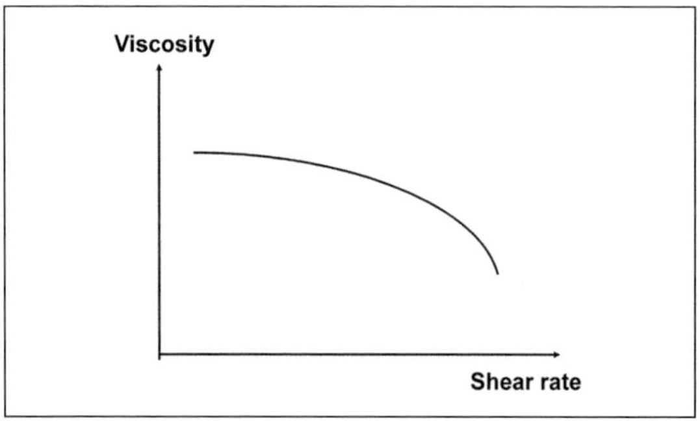

Figure II-1.3: Structural viscosity of polymer solutions

An increase in solids content will lead to an increase in solution viscosity. The random polymers will impede each other as the concentration rises, because they approach each other more and more.

- The solids content of most solvent-based adhesives is limited to 20 to 30 %, expressed in terms of the polymer.
- Besides the solids content, the viscosity is also determined by the solubility of the polymer in the selected solvent. Contrary to what a layman might expect, the solution viscosity increases with increase in solubility.
- Where solubility is good, the random polymer particles will swell more extensively, come into contact sooner and so produce an increase in viscosity. A poor solvent will cause less swelling, and yield solutions that are less viscous. Viscosity also increases with increase in molecular weight of the polymer because longer molecule chains lead to bigger particles.

Figure II-1.2 schematically shows the factors that influence the solution viscosity.

In addition, most polymer solutions exhibit significant pseudoplasticity (Figure II-1.3).

- The viscosity decreases with increase in shear rate because, as a consequence of shearing, the chain-like polymer molecules adapt to the flow conditions by changing from a random shape to a more linear shape.

The evaporation characteristics of solvents is described by their evaporation rate. The evaporation rate is defined as the ratio of the of evaporation time for a given solvent compared to that for diethyl ether at 293 K. As shown in Table II-1.4, the rate is not directly proportional to the boiling point in all cases.

Usually fast solvent evaporation is desirable, because the adhesive develops its bond strength only if a sufficient amount of the solvent has evaporated. On the other hand, fast evaporation can lead to skinning and a shorter open time, which is the time after application of the adhesive within good wetting of the mating substrate is still possible. Fast evaporating solvents often have a low flashpoint that can compromise safety. Nowadays, ecological and mainly toxicological aspects are important aspects of solvent selection. Aromatic solvents, such as benzene and toluene are good solvents for many polymers but, because they are classified as toxic, carcinogenic or teratogenic, they may no longer be used for most industrial applications. Halogenated solvents have been banned as well. Even though they are good solvents for many polymers, they have low boiling points, they have low flammability, which was why they were so widely used. Today, there is reason to believe that many of these solvents are carcinogenic and responsible for the hole in the ozone layer and attendant climate change. Nearly all of these solvents have therefore been replaced either by water-based dispersion technology or by solvent blends based on gasoline. There are also economics aspects to be considered, because the solvent content of solvent-based adhesives ranges from 70 to 90 %. Solvent prices are closely linked to oil prices, and they can change daily.

1.1.2 Manufacturing and formulation

In most cases, an adhesives manufacturer will use **commercially available polymers**. Is it just a question of taking a thermoplastic polymer and dissolving it to obtain the desired adhesive? Sometimes the answer is yes, more often it is no.

In practice, it can take a long time to dissolve polymers, even when a suitable solvent or solvent blend is used. Solubility is a thermodynamic term which is defined by the position of solution equilibria. The rate of dissolution is a kinetic term that is mainly independent of solution equilibria. The rate of dissolution can be increased by raising the temperature and by proper stirring. Efficient stirring is particularly necessary for dissolving crystalline polymers temperatures above the crystalline melting point.

Polymers, such as polychloroprene and nitrile rubber, which have been produced by a free-radical polymerization process, can have a certain content of very high-molecular or even crosslinked polymers. These will dissolve very slowly, if at all. They therefore need to be **masticated** for both laboratory and industrial use. The polymer (shaped in a bale the size of a loaf of bread) is sheared extensively in the presence of mastication aids. As shown in Figure II-1.4, shearing can be performed in large, counter-rotating two-roll mills of variable nip or in powerful kneaders. Mastication causes degradation of very high-molecular and crosslinked polymers, and leads to better, faster dissolution. However, the effect of mastication lasts for a short time, ranging from several hours to a few days. It is believed that free-radicals which form as a result of polymer degradation gradually recombine. The progress of mastication is usually controlled empirically.

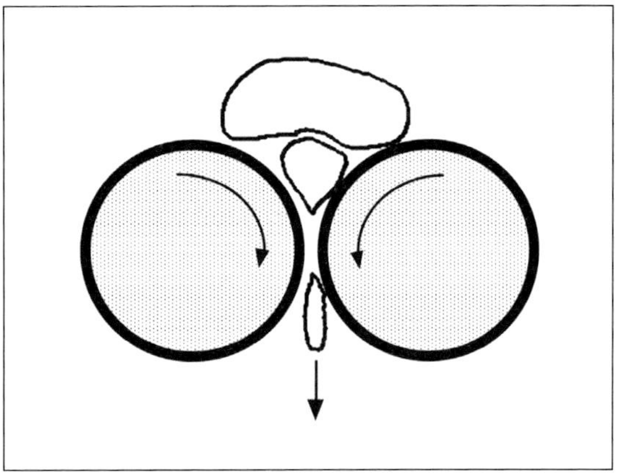

Figure II-1.4: Two-roll mill for masticating polymers

An experienced worker can "feel" if the polymer is "ready" for the subsequent dissolution step. Scientific methods of determining this "readiness" are based on measurements of solution viscosities or comparative measurements of molecular weights by means of size-exclusion chromatography (SEC).

On an industrial scale, dissolution is performed in voluminous churns (up to 20 tons), whereas simple metal or glass vessels equipped with stirrer and reflux condenser are sufficient in the lab. Dissolution can be very time consuming (lasting many hours). High-speed agitators can be used to speed up dissolution and to effect polymer degradation by shearing.

For most small adhesives producers, in-house synthesis of polymers is only possible if the polymerization process can be performed at low pressure. Polyurethanes, polyvinyl acetate and polyacrylates can be made directly in solution by this method. The advantage is that special monomers and monomer combinations can be used to customise specific properties for individual customers.

The polymer solution is formulated with **resins** (up to 50 %, expressed in terms of the quantity of polymer). These can improve tack and adhesion. They also help to adjust the open time and the heat resistance of the adhesive. **Additives** (less than 5 %, expressed in terms of the quantity of polymer) can serve as stabilisers, rheological additives or colorants. Rheological additives are useful for avoiding stringing when polymer solutions are applied through thin nozzles. The use of **fillers** is limited to special cases and requires steps to prevent sedimentation.

The reporting of concentrations follows convention in the rubber industry, where quantities of all substances (without solvents) are quoted relative to the quantity of polymer. The unit is [phr] or parts per hundred rubber with 100 parts of polymer being the reference value.

1.1.3 Application

Components

If the adhesive solution is applied as made to the substrate, it is called a one-component adhesive (1-component). If a second, reactive component (crosslinker, hardener) must be added immediately before application (see Chapter II-2.1), it is called a two-component adhesive (2-component). The two-component (or two-part) process is normally reserved for demanding applications, where the adhesive joint is exposed to high temperatures or a humid environment. For two-component applications, it is important to observe the correct mixing ratio, the pot life (period of time after addition of the second component within which a good adhesive joint can still be obtained) and the full-cure time to allow full development of the end properties.

The adhesive may be applied to either just one of the substrates (one way application) or to both (two way application). The adhesive can be applied by spray equipment (airless or air-assisted), by rolls, brush or coating knife.

Procedure and working technique

The following steps are involved in creating an adhesive bond:

- preparation of the substrate (cleaning, surface treatment) and the adhesive (1-component or 2-component)
- application of the adhesive and wetting of the substrate by the solution
- evaporation of the solvent
- film-formation by the polymer and development of cohesion
- wetting of the substrate by the polymer and development of adhesion

The use of solvent blends can greatly affect application. Small amounts of high-boiling solvents in the formulation can extend the open time and help to improve wetting of the substrate by the polymers. The film-forming process and the development of cohesion start when the solubility of the polymer system is exceeded due to evaporation of solvent. Free solvent and solvent from the random polymer particles evaporate, the polymer particles approach each other and interdiffusion sets in from the edges. This is the first stage in solidification [4]. The polymer forms a film with a disordered structure (at least, in the case of amorphous polymers) resembling that of a non-woven. In the presence of residual solvent and as part of the wetting process, adhesion between the polymer and the substrates starts to develop. Quite often, the surface of plastic substrates is etched by the solvents contained in the adhesive, a process which further boosts adhesion.
Different working techniques can be employed
- wet bonding
- contact bonding
- solvent reactivation
- heat reactivation.

1.1.3.1 Wet bonding

The adhesive is applied to one of the substrates as a solution. While the adhesive is still wet (open time), the second substrate is mated and fixtured until enough solvent has evaporated and the initial bond strength (green strength) developed.

Advantage

+ simple process

Disadvantages

- not suitable for non-porous substrates
- substrate needs to be fixtured during evaporation
- slow development of bond strength
- exposure to solvents
- on certain substrates, solvents can lead to stress corrosion

Field of application

Wet bonding is preferred for porous substrates (at least one must be porous) that can be fixtured and when sufficient drying time is available.

1.1.3.2 Contact bonding

The adhesive is applied to both substrates and the solvent evaporates to a large extent. Subsequently, the two substrates have to be mated within a defined time window under high pressure. At this point, the polymer has already mostly solidified due to crystallisation (green strength), but the polymer chains of both coated substrates still provide a certain degree of mobility, with the result that, in intimate contact and under pressure, they are able to intermingle. Where crystallisation is already too far advanced because of excessive evaporation, diffusion bonding is no longer possible: the adhesive's open time has been exceeded. The final bond strength thus critically depends on the ability of the polymer chains to diffuse within the time window and is the product of the contact time and high pressure. The bonding which occurs at the interface of two identical surface materials is called autohesion. Polychloroprenes are typical solvent-based contact adhesives.

Advantages

+ short fixturing times
+ high green strength immediately
+ also suitable for non-porous substrates
+ application of the adhesive and joining of the substrates can be performed at different times

Disadvantages

- only certain types of polymers can be used
- procedures have to be followed correctly (open time, time window)
- adhesive has to be applied to both substrates (two way)

Field of application

Contact adhesives are used when immediate bond strength is required, e.g. in shoe-making, bonding of flooring materials, lamination processes and DIY applications.

Solvent reactivation

In a solvent-reactivation process, the adhesive is applied to one or both substrates and allowed to dry. Immediately before the substrates are mated, the surface of the dry adhesive is moistened with a small amount of solvent. If the two substrates are then mated under pressure, the mobilised polymer chains can intermingle at the interface and bonding can occur. The small quantity of solvent does not significantly affect the green strength.

Advantages

+ high green strength, comparable to that of contact bonding
+ application of the adhesive and mating of the substrates can be performed at different times

Disadvantages

- additional process step
- solvent is used twice
- where a crosslinker is used (2-component adhesive) the ease of reactivation decreases with time

1.1.3.3 Heat reactivation (thermal reactivation, heat sealing)

In heat reactivation, the adhesive is applied to one or to both of the substrates and the solvent allowed to evaporate. At any chosen time, the surface of the dry adhesive film can be melted by application of heat (e.g. infrared heaters or circulating hot air) because the polymer is thermoplastic. In the hot state, the two substrates are mated quickly under pressure and the molten polymer wets the mate. The thin adhesive joint cools down immediately and the two adherends exhibit high green strength without the need for further fixturing.

Advantages

+ no solvents during bonding
+ brief fixturing times
+ high green strength
+ ease of automation

Table II-1.5: Comparison of the various application methods

	Application	Evaporation	Storage	Activation	Joining
Wet bonding	1/2 mates	non-applicable	open time	non-applicable	fixation slow evaporation
Contact bonding	2 mates	time window	time window	non-applicable	short under pressure
Solvent reactivation	1/2 mates	user defined*	user defined*	solvent	short under pressure
Heat reactivation	1/2 mates	user defined*	user defined*	heat	short under pressure

* Valid for 1-component application, otherwise the reaction time of the crosslinker has to be considered

Disadvantages

- exact compliance with defined procedures necessary
- only suitable for certain types of polymers
- where a crosslinker is used (2-component adhesive), the ease of activation decreases with time

Field of application

Heat reactivation is used in important industrial applications where large numbers of pieces require short, automated work cycles (e.g. automotive, shoe-making, and packaging industries). The various working techniques for application of solvent-based adhesives are shown schematically in Figure II-1.5 (adapted from [5]) and in Table II-1.5.

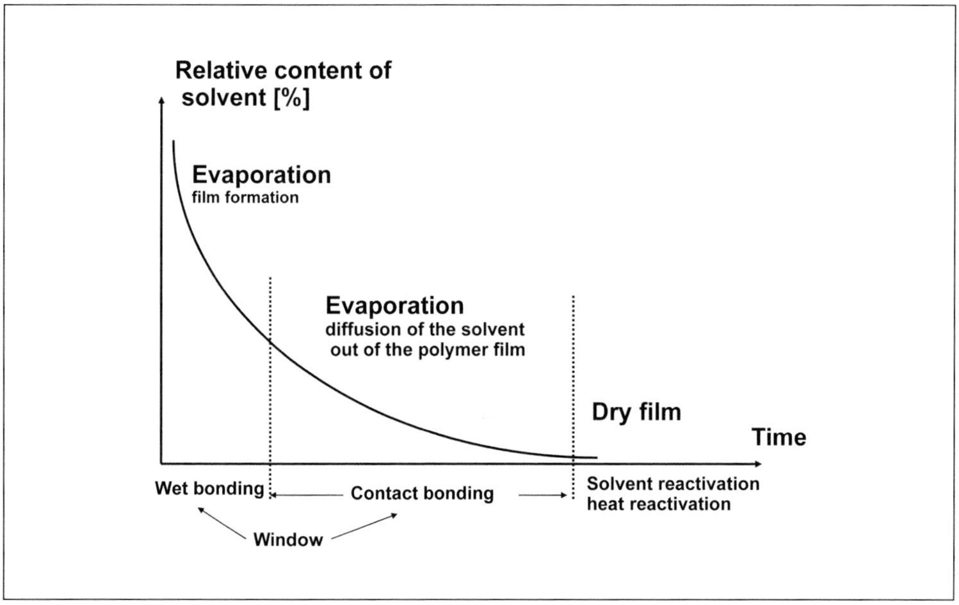

Figure II-1.5: Working technique for solvent based adhesives

1.1.4 Chemistry of solvent-based adhesives

1.1.4.1 Polyurethanes (PUR or PU)

Manufacture

There are basically two ways to manufacture solvent-based polyurethane adhesives. The easiest is to buy commercial polymers and to dissolve them, along with additives, in suitable solvents. The other way is to synthesise them in solution by reaction between polyisocyanates and polyols. For good final mechanical properties, it is necessary to build up a high molecular weight by meeting the exact stoichiometric conditions of the polyaddition reaction. For thermoplastic, i.e. soluble, polyurethanes, the monomers must have a defined functionality of two, as shown in Figure II-1.6.

The preferred isocyanates are aromatic, e.g. 4.4'-MDI or sometimes TDI, mainly due to their lower prices (see also Chapter II-2.2.1 on reactive polyurethane adhesives). MDI is symmetrical and yields polymers of high crystallinity while TDI is cheaper. Suitable polyols are mainly crystalline polyester polyols, because their tendency to crystallise is reflected in the resultant polyurethane polymers. In most cases, therefore, polyesters based on butanediol adipate (BD/AA), hexanediol adipate (HD/AA) or polycaprolactones having a molecular weight of 1000 to 6000 g/mol are used (see also Chapter II-2.2.1.10). The crystallinity of polyurethanes increases with the molecular weight of the constituent polyester. A polyurethane based on a polyester of molecular weight 1000 g/mol therefore tends to be less crystalline than a similar polyurethane based on a polyester of molecular weight of 3000 g/mol. HD/AA are more crystalline than BD/AA. On the other hand, polymers of less crystallinity are easier to heat-activate.

The resultant polyurethanes have crystalline melting points in the range 40 to 60 °C and are thus suitable for applications requiring heat-activated adhesives. Products of lower crystallinity can also be used as traditional contact adhesives.

The reaction between the polyols and isocyanates is usually performed in several steps. First, an isocyanate-terminated prepolymer is made from polyesters in the presence of excess isocyanate in solution. Second, in the reaction in which the polymer is formed, chain extenders, i.e. low-molecular diols, such as 1,4-butanediol, are added in an amount that will lead to a stoichiometric conversion rate (NCO/OH = 1). Sometimes a small excess of isocyanate is employed to compensate for small quantities of water and loss of isocyanate due to other side-reactions. The final polymer does not contain any isocyanate groups.

The influence of the NCO/OH ratio on the development of molecular weight is shown schematically in Figure II-1.7 (adapted from [6]). Any significant deviation from the stoichiometric NCO/OH ratio yields low-molecular products only.

The low-molecular chain extenders lead to domains of high urethane concentration within the polyurethane. These domains impart very good mechanical and thermal stability to the polymer due to their intermolecular hydrogen bonding. The quantity of chain extenders employed is typically in the same molar range as that of the polyester polyols. The reaction is performed in solution. Suitable solvents are ketones, such as butanone (MEK), acetone, esters, such as ethyl acetate

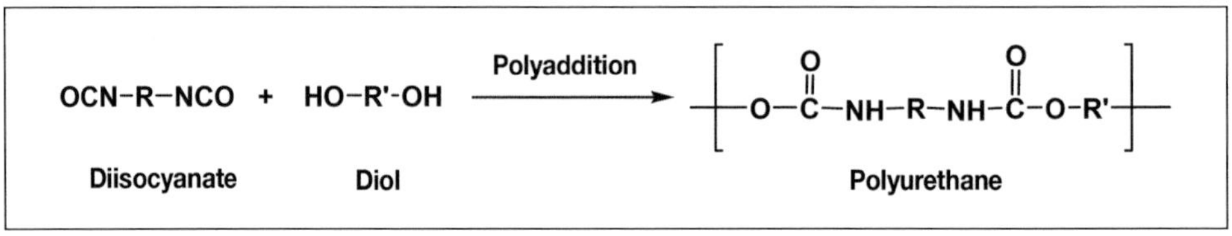

Figure II-1.6: Synthesis of linear polyurethanes

and methyl acetate and occasionally aromatic solvents, such as toluene and xylene or appropriate blends. Halogenated solvents are hardly used now.

Normally, the solvent is heated with the polyester and the isocyanate with slight stirring below the boiling point in the presence of small amounts of catalysts, such as DBTL (see Chapter II-2.2.1.2) to accelerate the reaction. The quality of the solvents as regards water content must be watched very carefully. Water reacts readily with isocyanates (see Chapter II-2.2.1.2) to form amines and ureas. The presence of water would affect the properties of the polymer substantially wherever an exact stoichiometric ratio was needed for achieving high molecular weights. If necessary, solvents must be dried over molecular sieves in advance. Alternatively, the water can be made to distil off azeotropically from the polyester and solvent before the isocyanate is added (ethyl acetate and water form an azeotropic mixture containing 9 % water). Note that solvents from solvent recovery systems, especially, may contain high amounts of water. Where esters are used, such as ethyl acetate, a very low concentration of ethanol is essential. Ethanol is a monofunctional reactant which acts as a chain-terminating agent and would yield a polymer of undesirable, low molecular weight.

For the most part, there are no modification stages in the case of solvent-based polyurethane adhesives. In special cases, carbodiimides can be added improve hydrolytic stability while fumed silica may be used to adjust the rheology. Resins can help to improve the tack. Where an adhesive bond must have good heat resistance, crosslinking isocyanates need to be additionally used (2-component system).

Guide formulations

The easiest way to formulate an adhesive is to use commercial polymers [7]. Table II-1.6 describes a slow-crystallising solvent-based PUR adhesive for the packaging industry. Besides the slightly crystalline polyurethane, this adhesive contains a liquid polyester which functions as a plasticizer by extending the open time and, in a two-component application, undergoing crosslinking. Table II-1.7 presents a fast-crystallising adhesive for the shoe industry based on a highly crystalline polyurethane. Both adhesives have a solids content of about 20 %. They can be optionally crosslinked by adding polyfunctional isocyanates (i.e. 3 % "Desmodur" RFE, Covestro) in a 2-component system.

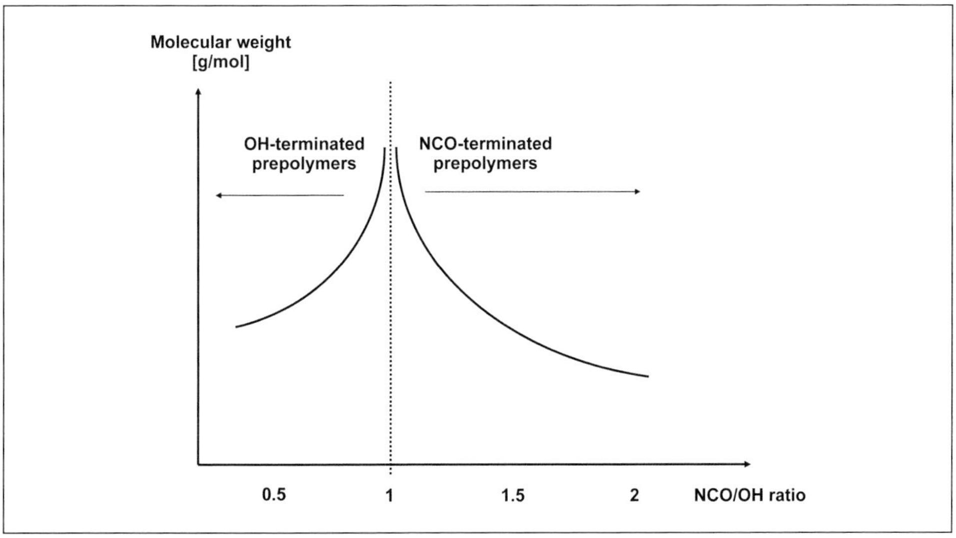

Figure II-1.7: Influence of the NCO/OH ratio on the molecular weight

Table II-1.6: Solvent-based polyurethane adhesive for the packaging industry

Item	Raw material	Parts by weight
1	Polyurethane (low crystallinity)	14
2	Polyester (liquid)	7
3	Solvent	79

Item 1: "Desmocoll 406", flexible polyurethane with a low crystallization rate. It has outstanding adhesion on flexibilised PVC and can be used to formulate adhesives with a long tack life, Covestro
Item 2: "Baycoll AD 2047", linear polyester polyol that is suitable for the formulation of solvent-borne adhesives and coatings, Covestro
Item 3: Ethyl acetate

Table II-1.7: Solvent-based polyurethane adhesive for the shoe industry according to DIN CEN/TR 15990

Item	Raw material	Parts by weight
1	Polyurethane (highly crystalline)	16
2	Rheological additive	1
3	Solvent 1	18.4
4	Solvent 2	65.6
5	Hardener (optional)	5 %

Item 1: "Desmocoll 540", Covestro
Item 2: "Aerosil 200", fumed silica, Evonik
Item 3: methyl ethyl ketone (MEK)
Item 4: Acetone
Item 5: "Desmodur RFE", solution of tris(p-isocyanatophenyl) thiophosphate in ethyl acetate, NCO content 7.2 ± 0.2 %, Covestro

Table II-1.8: Crystallising polyurethane adhesive

Item	Raw material	Parts by weight
1	Polyester (crystalline)	14.7
2	Chain extender	0.37
3	Diisocyanate (MDI)	2.02
4	Catalyst	0.012
5	Solvent 1	27.6
6	Solvent 2	55.2

Item 1: "Dynacoll 7345", partially crystalline, saturated polyester, hydroxyl number 27 – 34 mg KOH/g, molecular weight 3500 g/mol, Evonik
OH number = 30 g KOH/kg, M = 3730 g/mol, f = 2 val/mol
Item 2: 1.4-Butanediol, BASF SE, M = 90 g/mol, f = 2 val/mol
Item 3: "Desmodur 44 MS", Covestro
Item 4: DBTL, Biesterfeld
Item 5: Ethyl acetate
Item 6: Acetone

Table II-1.8 shows a guide formulation for making a crystalline polyurethane adhesive, including synthesis of the polymer [8]. The polyester is a crystalline type (BD/AA) with a molecular weight of approximately 3500 g/mol, while butanediol serves as chain extender. The chain extender and polyester are used in roughly the same molar quantities, with 4,4'-MDI being the preferred isocyanate. The total hydroxyl content and the isocyanate are used stoichiometrically, i.e. NCO/OH = 1 and the solids content is 17 %. There are no modification steps for this adhesive.

The calculation is based on the following figures:

- The polyester is characterised (manufacturer's figures) by an OH value of 30 mg KOH/g (30 g KOH/kg), corresponding to 536 mval OH/kg and, for a functionality of 2 val/mol, a molecular weight of 3730 g/mol (an OH value of 56 g KOH/kg ≙ 1000 mval/kg; see Chapter II-2.2.1.5).
- 3730 g Polyester thus corresponds to 1 mol and 2000 mval OH.
- 14.7 g Polyester corresponds to 14.7 · 2000/3730 = 7.88 mval OH.
- 1,4-Butanediol has a molecular weight of 90 g/mol; for a functionality of 2 val/mol this corresponds to 2000 mval OH for 90 g and 22,220 mval OH/kg.
- 0.37 g 1,4-Butanediol corresponds to 0.37 · 2000/90 = 8.22 mval OH.
- 4.4'-MDI has a molecular weight of 250 g/mol; for a functionality of 2 val/mol this corresponds to 2000 mval NCO for 250 g and 8000 mval NCO/kg.
- 2.02 g 4.4'-MDI corresponds to 2.02 · 2000/250 = 16.16 mval NCO
- →NCO/OH = 16.16/(7.88 + 8.22) = 1.0; i.e. equivalent quantities of polyols and isocyanates are converted.

The polyester is melted at 80 to 100 °C and carefully dissolved in approximately 90 % of the dry solvent blend consisting of ethyl acetate and acetone. The DBTL and 1,4-butanediol are added, and the MDI, dissolved in the remaining 10 % ethyl acetate, is added afterwards at approximately 60 °C. The reaction takes place at 60 °C over about three hours.

Evaluation

Solvent-based polyurethane adhesives provide a very good property profile. They show good adhesion to a variety of plastic substrates. Their "victory march" coincided with the introduction of soft PVC shoe soles, which could not be durably bonded with traditional polychloroprenes.

Advantages

+ variable structure due to availability of a multitude of raw materials
+ fast-setting products due to crystallisation and heat activation
+ good adhesion to various substrates, especially plastics
+ in some cases, adhesion to contaminated surfaces
+ very good resistance to plasticisers
+ can be used as 1- or 2-component systems

Disadvantages

- solvent-based
- in most cases, only a low-solids content is possible (often less than 20 %)
- limited heat resistance (without crosslinker: about 70 °C)
- limited hydrolytic stability without crosslinker
- on certain plastics, they can lead to stress corrosion

Field of application

Solvent-based polyurethane adhesives are still widespread in the **automotive industry**. Virtually the entire interior of a vehicle consists of laminated materials, one part being a dimensionally stable supporting part made of derived timber (i.e. Lignotok) or ABS, the other providing the aesthetic appeal and consisting of PVC soft-foam film (artificial leather) or textile. Examples of laminated parts are the headliners, door-panels, dashboards and the rear parcel shelf. These laminates are usually produced in automated bulk-manufacturing processes by sub-contractors. Bulk manufacturing requires short cycle times and therefore fast-setting adhesives. It may be achieved advantageously with a combination of crystallising adhesives in a heat-activation process. If the adhesives are formulated with crosslinkers as a 2-component system, adhesive joints can even withstand temperatures up to 100 °C, which can develop in a car roof exposed to intense sunlight. Recently, solvent-based polyurethanes have been supplanted by water-based PUR dispersions (see Chapter II-1.2.3.1), which are now capable of providing a similar product profile. In the **shoe industry**, the most demanding bonding operation is the attaching of shoe soles to the bootleg. Usually, the bootleg is brushed with a 2-component, solvent-based PUR adhesive. The sole is furnished with the same adhesive, often formulated as a 1-component system. The consequence is that the adhesive on the sole can be reactivated after an extended period of time. The soles are reactivated with an infrared heater, are manually attached to the bootleg in the right position and are then compressed for a few seconds. The fast-crystallising adhesive quickly develops enough green strength to withstand the strong restoring forces exerted by the complex geometries of fashionable ladies' shoes. In the **woodworking and furniture industry** as well as in the **packaging industry**, decorative and functional laminates of different materials are successfully produced with solvent-based polyurethane adhesives.

1.1.4.2 Polychloroprenes

Manufacture

Polychloroprene (neoprene, CR – chloroprene rubber) is produced by free-radical-initiated emulsion polymerization of 2-chloro-1,3-butadiene (chloroprene). Nowadays, the monomer is manufactured by a free-radical addition of chlorine to butadiene in the gas phase. The addition may be a 1,2- or 1,4-type (Figure II-1.8). The 1,4-adduct can be catalytically isomerised to the 1,2-adduct. This can be dehydrohalogenated in an elimination reaction in the presence of sodium hydroxide [9].

As shown in Figure II-1.9, the resulting polymer contains 1,2-, 3,4- as well as cis- and trans-1,4-linkages. The corresponding polymers differ in crystallinity and reactivity. Polymers with a high content (approx. 90 %) of trans-1,4 adducts confer a high degree of crystallinity and are very fast setting when used as contact adhesives [10, 11, 12]. The tendency to crystallise leads to high mechanical strength and good chemical resistance compared to other rubber polymers. 1,2-ad-

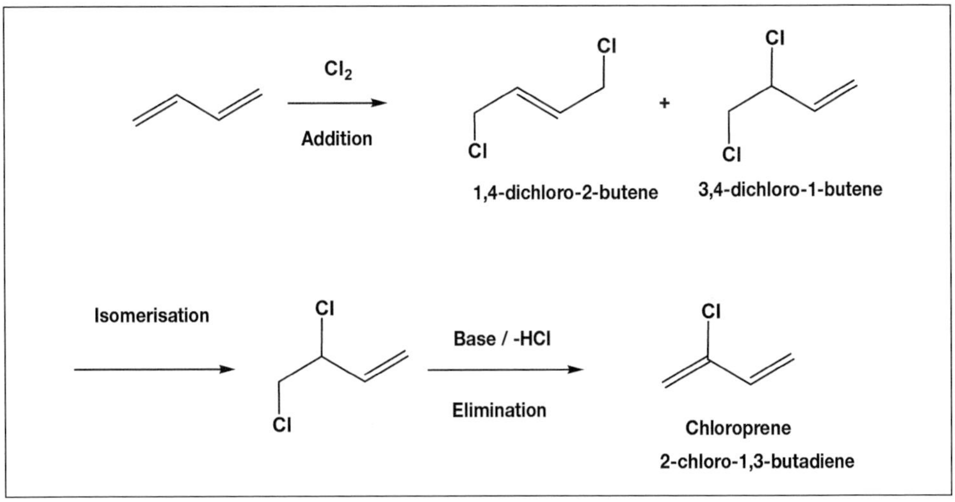

Figure II-1.8: Manufacture of chloroprene

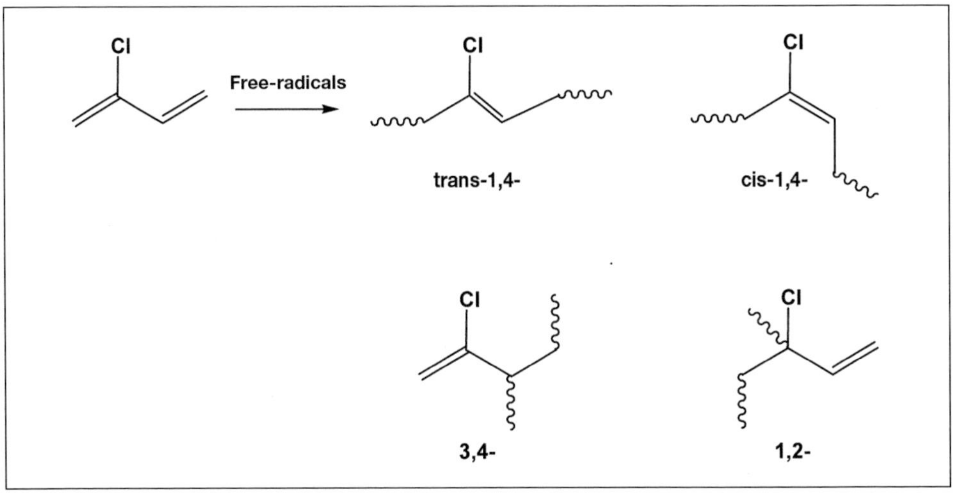

Figure II-1.9: Structure of polychloroprene

ducts can be crosslinked at the allyl chlorine. Substitution of the allyl chlorine by traces of water yields OH groups, which can be readily crosslinked by polyfunctional isocyanates.

A wide range of commercial **polychloroprene** chips is available for formulating adhesives, and the manufacturers provide assistance with selection and use [10, 11]. The typical quantities in a formulation range from 10 to 20 %. The individual polymers mainly differ in their tendency to crystallise. Fast-crystallising types, if used as contact adhesives, develop their bond strength very quickly, but have a limited open time. Polymers containing gel need to be masticated prior to formulation as solvent-based adhesives. Some types of polymers already contain thiuram disulphide (a free-radical scavenger; Figure II-1.10) added by the manufacturer. This improves the mastication process and the final storage stability of the adhesive. However, thiuram can lead to discolouration of the adhesive.

Phenolic resins (10 to 50 phr) are the most important class of **resins** for polychloroprenes [13]. Adding non-reactive terpene-phenolic resins (Figure II-1.11) mainly improves adhesion and tack.

The use of reactive alkylphenolic resins based on p-tert-butyphenols (Figure II-1.12) in combination with magnesium oxide (approx. 10 % of the resin quantity) boosts heat resistance by forming chelates (resinates) (Figure II-1.13). These chelates are normally prepared separately in the corresponding solvents. Small quantities of water can be added to achieve a catalytic effect.

Alternatively, colophony (rosin) or coumarone-indene resins can be used to modify the setting process. The use of various **additives** is unavoidable. Polychloroprenes have a tendency to lose hydrogen chloride, especially if exposed to heat. Contact with metallic substrates can trigger undesirable corrosion processes. Alkaline metal oxides, such as magnesium oxide (4 to

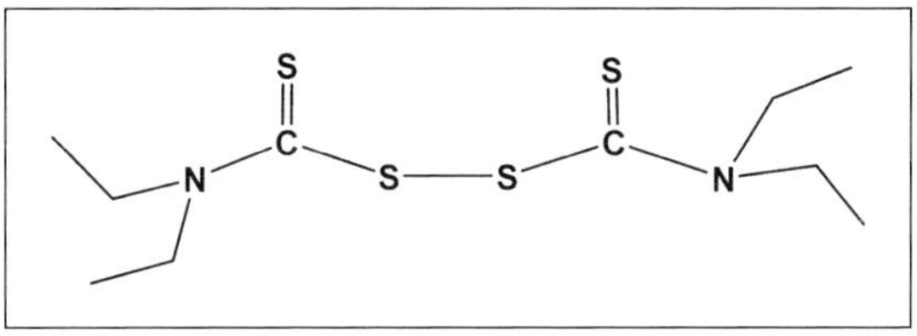

Figure II-1.10: Tetraethylthiuram disulphide

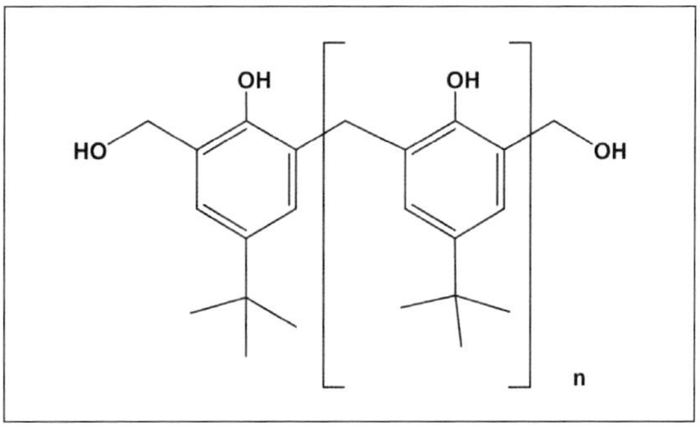

Figure II-1.12: Phenolic resins based on p-tert-butyl phenol

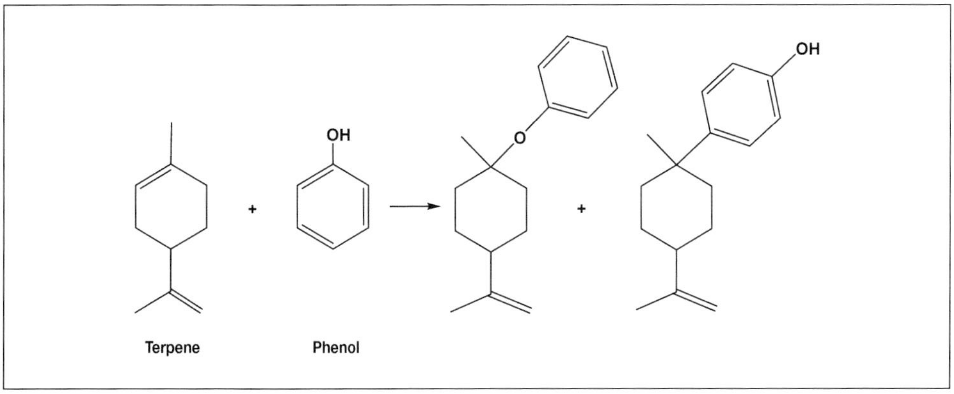

Figure II-1.11: Synthesis of terpene-phenol resins

Table II-1.9: Solvent-based polychloroprene adhesive of high green strength

Item	Raw material	Parts by weight
1	Polychloroprene (fast crystallising)	21.4
2	Stabiliser 1	0.85
3	Stabiliser 2	0.85
4	Chlorinated rubber	7.48
5	Solvent 1	15
6	Solvent 2	15
7	Solvent 3	30
8	Solvent 4	22.4

Item 1: "Baypren 320", Arlanxeo
Item 2: Mangnesium oxide
Item 3: "Bayoxid Z aktiv", tinoxide, Lanxess
Item 4: "Pergut S 40", Covestro
Item 5: Ethyl acetate
Item 6: Gasoline 65 to 95 °C boiling fraction
Item 7: Cyclohexane
Item 8: Butanone

8 phr) or tin oxide (approx. 2 phr) neutralise traces of released acid and stabilise the adhesive very efficiently. Magnesium oxide can be added during the mastication process, and will be homogenised very efficiently, reducing the tendency for sedimentation to occur in solution. Addition of antioxidants based on hindered amines or phenols (approx. 2 phr) inhibits oxidative degradation of unsaturated polymer chains and unsaturated resins. Hindered phenols have a lower tendency to discolour. Chlorinated rubber [10] can improve the green strength of the adhesive formulation. Fumed silica acts as a rheological additive. While **fillers**, such as chalk ($CaCO_3$) and barytes ($BaSO_4$), can theoretically be used, they are not recommended on account of sedimentation problems. **Solvents** are chosen from economical, technical and ecological aspects. Erstwhile aromatic hydrocarbon/aliphatic /ketone mixtures have now been replaced by petroleum ether/cycloaliphatic /ketone/ester blends. Small amounts of higher-boiling solvents, such as xylene, can extend the open time. For better durability of the adhesive joint, a 2-component system with isocyanate **crosslinkers** is recommended.

Guide formulations

Table II-1.9 shows a solvent-based polychloroprene adhesive with a high green strength achieved with a fast-crystallising polymer and chlorinated rubber [10].

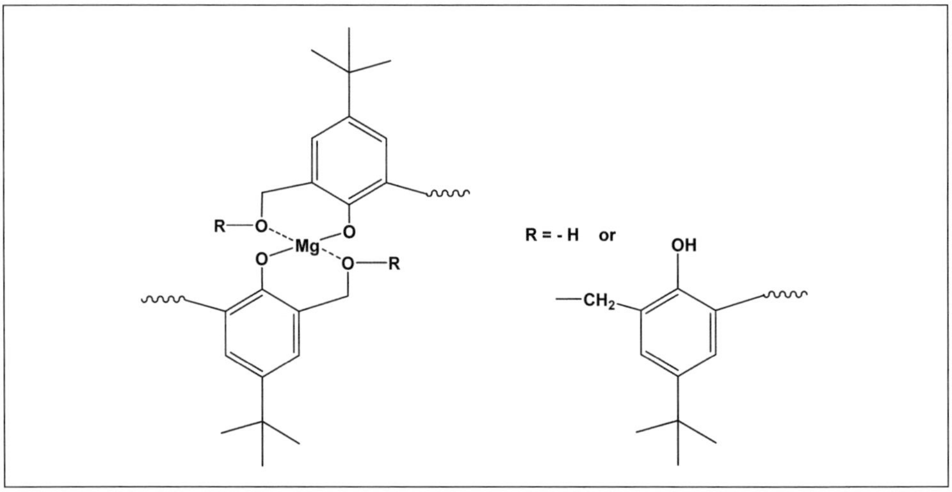

Figure II-1.13: Phenolic resin chelate

Table II-1.10 presents a solvent-based polychloroprene adhesive whose good heat resistance was achieved by adding a heat-activated alkylphenol resin [10].

Table II-1.11 shows a solvent-based polychloroprene adhesive of good heat resistance, which is additionally stabilised with antioxidants. The formulation is still based on traditional aromatic hydrocarbons. Water acts as the catalyst for forming the chelate complex [11].

Table II-1.10: Solvent-based polychloroprene adhesive of good heat resistance

Item	Raw material	Parts by weight
1	Polychloroprene (fast crystallising)	20.49
2	Stabiliser 1	0.82
3	Stabiliser 2	0.82
4	Heat reactive alkyl-phenolic resin	6.15
5	Solvent 1	7.2
6	Solvent 2	14.4
7	Solvent 3	28.7
8	Solvent 4	21.5

Item 1: "Baypren 320", Arlanxeo
Item 2: Magnesium oxide
Item 3: "Bayoxid Z aktiv", tinoxide, Lanxess
Item 4: "SP 126", Akrochem
Item 5: Ethyl acetate
Item 6: Gasoline, boiling fraction 60 to 95 °C
Item 7: Cyclohexane
Item 8: Butanone

Evaluation

Solvent-based polychloroprene adhesives are general-purpose types. As contact adhesives, they combine good setting with good, lasting adhesion to a variety of substrates. Their crystallinity confers both high strength and protects against ageing influences at room temperature.

Advantages

+ suitable for contact adhesives
+ fast-setting, brief fixturing
+ high green strength
+ high bond strength
+ good adhesion, even on less polar substrates
+ high resistance to water, oil and even ozone

Disadvantages

- solvent-based
- low resistance to plasticiser migration
- halogenated polymer
- can initiate corrosion on exposure to heat
- heat resistance of the polymer is limited

Field of application

Based on their good property spectrum, polychloroprene adhesives are regarded as general-purpose adhesives.

Table II-1.11: Solvent-based polychloroprene adhesive of good heat resistance, additional stabilisation, and water as catalyst

Item	Raw material	Parts by weight
1	Polychloroprene (fast crystallising)	15.2
2	Stabiliser 1	0.6
3	Stabiliser 2	0.8
4	Stabiliser 3	0.3
5	Reactive alkylphenol resin	7
6	Water	0.15
7	Solvent 1	15.2
8	Solvent 2	60.75

Item 1: "Neopren AD 20" (mastication recommended), DuPont
Item 2: Magnesium oxide
Item 3: Tin oxide
Item 4: Antioxidant, "Wingstay L", Synthomer
Item 5: "SP 154", Akrochem
Item 6: Water as catalyst
Item 7: Hexane
Item 8: Toluene

They are used in the **shoe industry** to attach soles to bootlegs. In the **furniture industry**, they serve in the manufacture of decorative laminates. In the **automotive industry,** they are used for interior applications. They are still employed for bonding carpets to floors in the construction industry. In many **do-it-yourself**-applications, they serve as a general-purpose adhesive for various adhesive joints.

1.1.4.3 Nitrile rubbers

Manufacture

As shown in Figure II-1.14, nitrile rubber is manufactured by free-radical-initiated hot-emulsion polymerization of butadiene and acrylonitrile to yield a random copolymer [11]. Stabilisation is effected with anionic surfactants (see also Chapter II-1.2.2). Sulphur compounds are added to control the molecular weights. Acrylonitrile is made by oxidising propylene in the presence of ammonia [12].

The emulsion is coagulated, dried and milled. Trade products are offered as bales or chips. They are surface-treated with talcum powder or the like to prevent caking. The polymers are high molecular or are crosslinked to a greater or lesser degree.

Commercial copolymers have an acrylonitrile content of 15 to 50 %, which greatly influences the polymer properties. The glass transition temperature of the copolymer is composed additively of the mass fractions and the glass transition temperatures of the homopolymers ($T_{g\ polybutadiene}$ -90 °C, $T_{g\ polyacrylonitrile}$ 90 °C).

The T_g of the copolymer can be calculated from Fox's equation [11]:

$$1/T_g = w_1/T_{g1} + w_2/T_{g2} \ldots$$

where

w_i is the mass fraction of component i and

T_{gi} is the glass transition temperature for the homopolymer of component i

In addition, the polarity and, consequently, the compatibility with other polymers varies with the composition. Polybutadiene is non-polar whereas polyacrylonitrile is polar. High contents of acrylonitrile increase resistance to non-polar solvents and fuels (solubility parameters: $\delta_{Butadiene} = 8.8$; $\delta_{Acrylonitrile} = 12.8$) [14]. Other monomers, such as acrylates and styrene, can be copolymerised, too. Hot polymerization (25 to 50 °C) affords high-molecular, branched and crosslinked products with a high content of cis-1,4- structures. This yields very good mechanical properties but, as a consequence, necessitates mastication before the dissolution step. Copolymers with a ni-

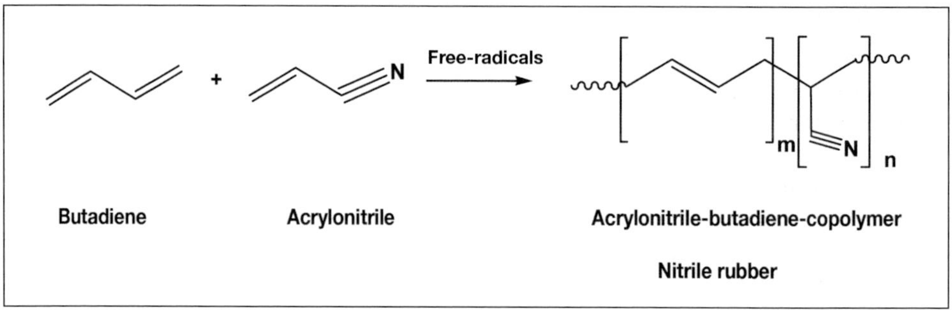

Figure II-1.14: Manufacture of nitrile rubber

trile content of about 40 % are normally used for adhesive applications. Typical **solvents** are ketones (such as acetone and butanone), esters (ethyl acetate and butyl acetate), aromatic hydrocarbons, and binary or tertiary mixtures. Compounding with other polymers is possible if they are compatible. **Fillers** can be carbon black, fumed silica (20 to 60 phr, primarily to increase cohesion), ferric oxide or silicates. Tack can be increased with **resins** based on colophony, chlorinated rubber or phenol. Commercial polymers already contain **additives**, such as antioxidants. Rheological properties can be adjusted with corresponding additives.

Table II-1.12: Leather adhesive based on nitrile rubber

Item	Raw material	Parts by weight
1	Nitrile rubber	9.9
2	PVC copolymer	9.9
3	Tin oxide	0.5
4	Stearic acid	0.15
5	Calcium silicate	0.2
6	Solvent	79.35

Item 1: "Perbunan", ARLANXEO or "Nipol", Zeon Chemicals L.P., with approx. 30 to 40 % AN content
Item 6: Butanone

Guide formulation

Table II-1.12 shows the starting formulation for a leather adhesive. The nitrile rubber is masticated and then dissolved along with a PVC copolymer. The solids content is about 20 % [11].

Evaluation

Nitrile rubber adhesives yield good chemical, hydrolytic and thermal stability. Their good compatibility with polar components facilitates formulation with other polymers. Nitrile rubber adhesives adhere well to substrates, such as glass, metals, and wood, and to plastics, such as PVC, PVAc and PA, and are thus suitable for a variety of applications.

Field of application

Nitrile rubber adhesives are used in many technical **lamination processes**. For example, PVC and polyester film have been joined successfully to make film packaging. They can be used as **flocking adhesives**. In combination with phenolic resins (see Chapter II-2.4.1), they serve to flexibilise highly resistant **friction linings**. Generally, the importance of nitrile rubbers in the adhesives industry lags behind that of polyurethanes and polychloroprenes.

1.1.4.4 Polyacrylates

Manufacture

As shown in Figure II-1.15, polyacrylates are manufactured by free-radical polymerization in solution or emulsion. They can also be bulk-polymerised.

The monomers employed are mainly esters of acrylic or methacrylic acid [11]. As shown in Figure II-1.16, acrylic acid can be produced by allyl oxidation of propylene [12].

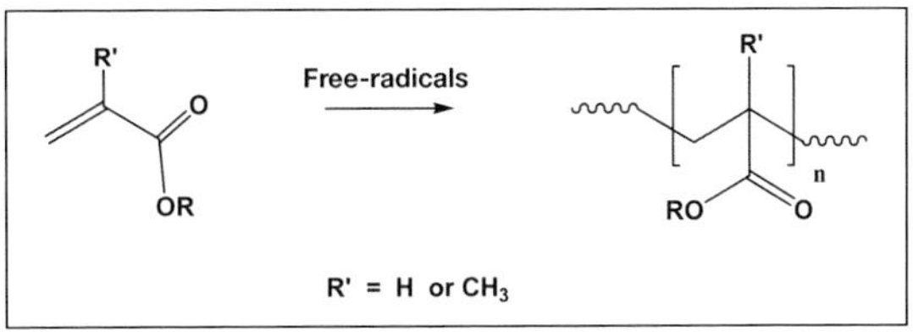

Figure II-1.15: Manufacture of polyacrylates

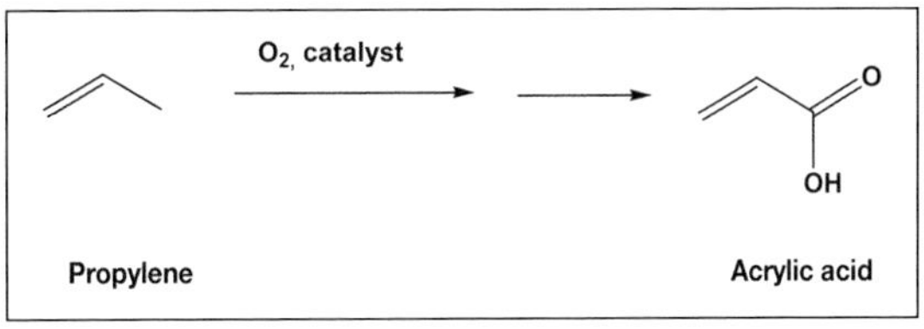

Figure II-1.16: Manufacture of acrylic acid

Table II-1.13: Glass transition temperatures of acrylic polymers [18]

Polymer	**T_g [°C]**	**Polymer**	**T_g [°C]**
Poly(acrylate)		**poly(methacrylate)**	
Acrylic acid	130	methacrylic acid	162
Methyl	22	methyl	105
Ethyl	-8	ethyl	67
Propyl	-25	propyl	33
Butyl	-43	butyl	32
Isobutyl	-17	isobutyl	64
tert-Butyl	55	tert-butyl	102
Hexyl	-51	hexyl	-14
Acrylonitrile	105	**Methacrylonitrile**	112

Further important comonomers:	polystyrene	107 °C
	polyvinyl acetate	42 °C
	polyvinyl chloride	77 °C

Methacrylic acid can be produced, as shown in Figure II-1.17, by making acetone and cyanide react to form cyanohydrin, with subsequent hydrolysis or alcoholysis and elimination [12].

The polymerization process can be performed without the application of pressure, and thus is feasible for small adhesives manufacturers, too. On the other hand, economic considerations often militate against in-house polymerization, as large monomer suppliers also offer polymers.

It is mainly the composition of the **base polymer** which determines the properties of the adhesive. The monomers can be drawn from a large number of acrylic or methacrylic esters and their acids, and from nitriles, amides, vinyl acetate, vinyl chloride, styrene and functionalised monomers, such as hydroxyethyl acrylate (see Figure II-1.18).

Compared with acrylates, methacrylate polymers exhibit reduced chain mobility. The esters' side-chains, formed from the corresponding alcohols, act as internal plasticisers, whose effect is proportional to the number of carbon atoms in their chain, up to a number of eight. Branched side-chains reduce the plasticising effect [11]. The molecular weight and consequently the mechanical properties of the polymers can be adjusted via the manufacturing conditions. Polyacrylates are amorphous substances with a characteristic glass transition temperature. Table II-1.13 shows how the glass transition temperature varies with structure [18]. The glass transition temperature of copolymers can be calculated from Fox's equation [11] (see also Chapter II-1.1.4.3).

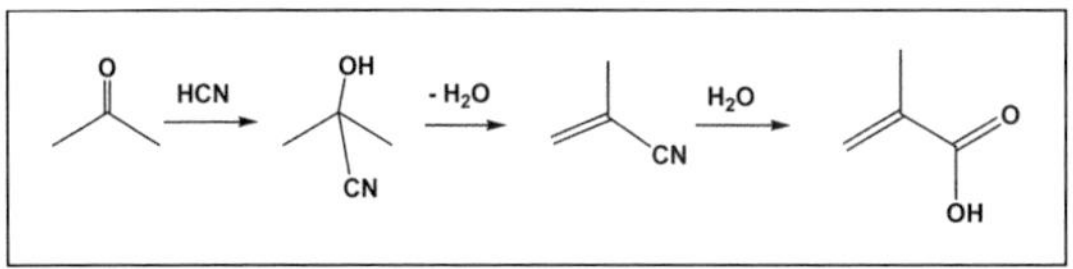

Figure II-1.17: Manufacture of methacrylic acid

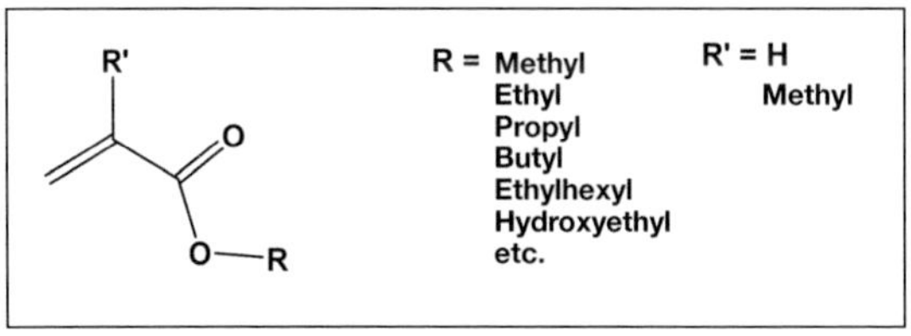

Figure II-1.18: Monomers for acrylic adhesives

The effect of structure on the glass transition temperature is also reflected in the mechanical properties. As shown in Table II-1.14, methacrylic polymers possess high mechanical strength and low elongation, as is typical of adhesives. Polyacrylates based on long-chain alcohols have high elongation and low mechanical strength. Consequently, the latter are preferred in pressure-sensitive adhesives and sealants.

Copolymerization with functional monomers, such as hydroxyethyl acrylate, can be performed if crosslinking in respect of a 2-component application is desired.

Table II-1.14: Mechanical properties and glass transition temperatures of poly(meth)acrylates

	Tensile strength [MPa]	Elongation [%]	T_g [°C]
Polymethacrylate			
Methyl	62	4	105
Ethyl	34	7	67
Butyl	6.9	230	32
Polyacrylate			
Methyl	6.9	750	22
Ethyl	0.2	1800	-8
Butyl	0.02	2000	-43

Guide formulations

A glossy film laminating adhesive can be formulated on "Acronal 500 L" (BASF SE), a 40 % solution of an acrylic copolymer in ethyl acetate with a glass transition temperature of about -21 °C. This solution can be further tailored to the application with resins based on colophony of terpene-phenol, plasticisers based on phthalates, and fillers [16].

A solvent-based pressure-sensitive adhesive (see Chapter II-3) can be prepared by copolymerising butyl acrylate (low glass transition temperature) with some acrylic acid, as shown in Table II-1.15 [17]. For laboratory purposes, a simple flask with reflux condenser is sufficient. The monomers are dissolved in the solvent under nitrogen, the mixture is heated to 60 °C and the free-radical initiator added. After a few hours, an additional amount of initiator is added to minimise the residual monomer content.

Table II-1.15: Solvent-based acrylic adhesive

Item	Raw material	Parts by weight
1	Butyl acrylate	52.8
2	Acrylic acid	2.2
3	Dibenzoyl peroxide	0.05
4	Acetone	45

Evaluation

Acrylic adhesives can be tailored to a variety of applications via the choice of monomer system. They adhere well to numerous substrates, have good resistance to UV radiation and are of low toxicity. They can be used as wet bonding adhesives, contact adhesives, heat-reactivation adhesives and even pressure-sensitive adhesives, with the type of application depending on the monomer system (glass transition temperature).

Advantages

+ broad spectrum of formulation possibilities
+ large number of bonding technologies possible
+ good adhesion to a large number of substrates
+ good hydrolytic stability
+ good light and UV stability
+ transparency
+ low toxicity

Table II-1.16: PVC adhesive

Item	Raw material	Parts by weight
1	Polyvinyl chloride	25
2	Rheological additive	2
3	Solvent 1	65.6
4	Solvent 2	9.4

Item 1: "Vinnol H 11/59", K-value = 59, Wacker
Item 2: "Aerosil 200", Evonik
Item 3: Butanone
Item 4: Cyclohexanone

Disadvantages

- solvent-based
- residual monomers
- limited molecular weight due to manufacturing process

Field of application

Applications for acrylic adhesives are determined largely by the glass transition temperature and the intended processing method. Polymers of low glass transition temperature serve as **pressure-sensitive adhesives** for labelling and adhesive tapes (see Chapter II-3). For use as contact adhesives, they need greater mechanical strength and consequently a higher glass transition temperature. Acrylic adhesives are used to produce decorative laminates in the **wood and furniture industry**. They are also employed in the manufacture of toys and sports equipment. Heat activation requires polymers of high glass transition temperature which are tack-free at room temperature. Acrylic adhesives serve as **laminating adhesives** for food packaging, glossy film lamination or in **automotive interior applications**.

1.1.4.5 Polyvinyl chloride

Manufacture

Polyvinyl chloride (PVC) is manufactured by free-radical emulsion polymerization of vinyl chloride, as shown in Figure II-1.19 and Figure II-1.20.

In the past, vinyl chloride was produced by adding HCl to acetylene. Nowadays, it is mainly produced by oxychlorinating ethylene with hydrogen chloride and oxygen [12]. The monomer is a putative carcinogen.

The emulsion is spray-dried. Commercial polymers differ mainly in their molecular weight. Copolymers containing 10 to 40 % vinyl acetate also exist.

Guide formulation

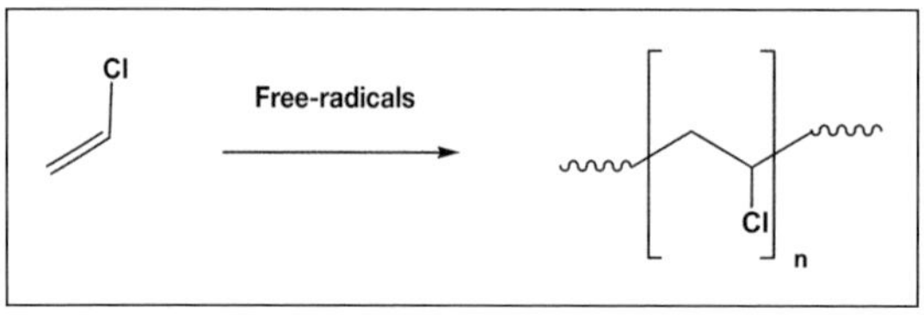

Figure II-1.19: Manufacture of polyvinyl chloride

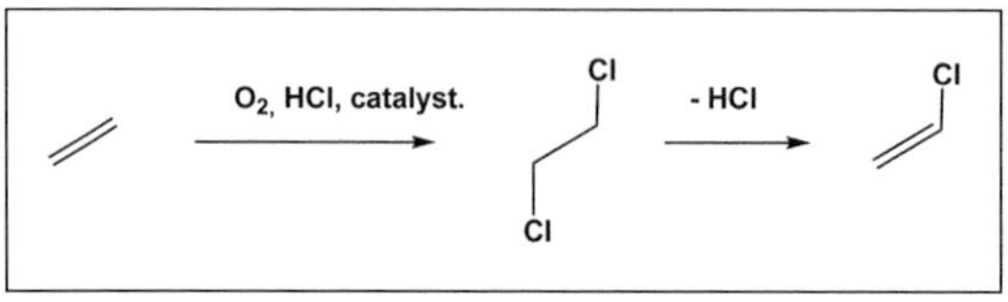

Figure II-1.20: Manufacture of vinyl chloride

Formulating a solvent-based PVC adhesive is simple. A commercial PVC polymer (15 to 30 %) is dissolved in a suitable solvent, such as a ketone or tetrahydrofuran (THF). The rheology is adjusted with additives.

Table II-1.16 shows a starting formulation for a typical PVC adhesive used to join PVC pipes in the construction industry. The solids content is about 25 % [18].

Evaluation

Solvent-based PVC adhesives are of no major importance. Their property profile is identical with that of PVC because there are usually no modification steps. PVC is resistant to hydrolysis and light. It is a typical thermoplastic polymer and so is readily attacked by polar solvents. Its heat resistance is limited as well.

Advantages

+ good adhesion to PVC due to diffusion bonding
+ good hydrolytic stability
+ good resistance to dilute acids bases, and salt solutions
+ resistance to light
+ good resistance to oil and grease

Disadvantages

- solvent-based
- the need to handle THF
- readily soluble in polar solvents
- moderate heat resistance

Field of application

PVC adhesives are mainly used in conjunction with PVC substrates. The latter are readily swollen by the solvent present in the adhesive. As the adhesive itself contains dissolved PVC, an ideal bond is created by mutual interdiffusion of adhesive and substrate molecules upon contact. Compatibility is ideal because the polymers are identical and interdiffusion is supported by the presence of solvent. PVC adhesives are used in the construction industry to join PVC pipes and to bond PVC sheeting for garden ponds.

1.1.4.6 Polyvinyl acetate

Manufacture

Polyvinyl acetate (PVAc) is manufactured by free-radical emulsion polymerization of vinyl acetate as shown in Figure II-1.21.

In the past, vinyl acetate was manufactured by adding acetic acid to acetylene in the gas phase. Nowadays, ethylene is made to react with sodium acetate in the presence of palladium chloride, copper salts and oxygen [12], as shown in Figure II-1.22.

The polymerization does not require pressure and so can be performed by small adhesives manufacturers. The dispersion is spray-dried. Again, as in the case of acrylics, major monomer suppliers also offer polymer dispersions or powders for applications in solvent-based adhesives.

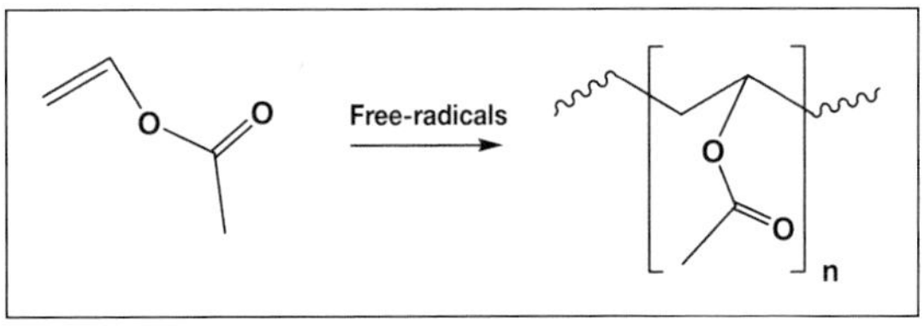

Figure II-1.21: Manufacture of polyvinyl acetate

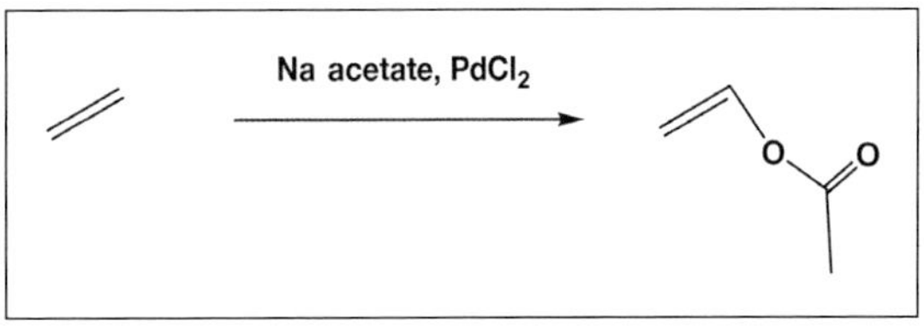

Figure II-1.22: Synthesis of vinyl acetate from ethylene

Table II-1.17: PVAc adhesive

Item	Raw material	Parts by weight
1	Polyvinyl acetate type 1	18.7
2	Polyvinyl acetate type 2	12.5
3	Solvent 1	28.3
4	Solvent 2	28.3
5	Water	6.1
6	Fumed silica	6.1

Item 1: "Vinnapas UW 10 fs", Wacker
Item 2: "Vinnapas B 17 sp", Wacker
Item 3: Methyl acetate
Item 4: Ethanol, 99 %
Item 6: "HDK N 20 P", Wacker

Guide formulation

Formulating solvent-based PVAc adhesives is simple. A commercial polymer or a blend of polymers of different molecular weights (15 to 30 %) is dissolved in a solvent, such as an ester, a ketone, aromatic hydrocarbon or an ether. The higher polymer imbues the adhesive with heat resistance and good mechanical properties. Additives may be used to adjust the rheology.

Table II-1.17 shows a starting formulation for a typical solvent-based PVAc adhesive for general applications. The solids content is about 40 % [19]. Two PVAc polymers of different molecular weights are used. Fumed silica is added to modify the rheological properties. The polymers are admixed to the solvent, and the fumed silica is then added.

Evaluation

PVAc adhesives are sometimes called general-purpose adhesives. But this name does not reflect the true situation. In a non-industrial environment, such as households and offices, it is usually only paper, cardboard or wood which needs to be joined. The adhesive does perform well in such applications. For other applications, though, the bond may prove to be brittle (the glass transition temperature of polyvinyl acetate is approx. 30 °C) and adhesion to many types of plastic is only moderate.

Advantages

+ low cost
+ general-purpose
+ good adhesion to wood and paper
+ non-toxic (except for the solvent)

Disadvantages

- solvent-based
- only moderate adhesion to many plastics
- moderate impact resistance
- moderate peel strength
- moderate resistance to solvents

Field of application

Polyvinyl acetate adhesives are mainly used in the **DIY sector**. They are inexpensive, not dangerous (except for the solvents) and bond well to most substrates in the home.

1.1.5 References

[1] H. G. Elias, Makromoleküle, 6th ed., Wiley-VCH, 1999
[2] J. M. G. Cowie, Chemie und Physik der synthetischen Polymeren, Vieweg,1997
[3] Ullmann's Enzyklopädie der Technischen Chemie, 4th ed., vol. 16, 283 ff
[4] B. Müller, U. Poth, Coatings Formulation, Vincentz Network, 2006
[5] G. Habenicht, Kleben, 3rd ed., Springer Verlag 1997
[6] M. Szycher, Szycher's Handbook of Polyurethanes, CRC Press, 1999
[7] solutions.covestro.com/en/materials/adhesives (Desmocoll)
[8] Internal papers, FH-Aachen, W. Rath, unpublished
[9] H.-J. Arpe, Industrielle Organische Chemie, 6th ed., VCH, 2007
[10] arlanxeo.com/en/brands/baypren
[11] I. Skeist, Handbook of Adhesives, 3rd ed., Chapman & Hall, 1990
[12] ROBBINS, J. Polychloroprene contact adhesive. Adhes Sealants Ind, 2003, 10. Jg., Nr. 2, S. 30-7.
[13] R. Jordan, R. Hinterwaldner, Klebharze, R. Hinterwaldner Verlag, 1994
[14] W. Hofmann, Rubber Technology, Hanser Publishers, 1996
[15] D. Satas, Handbook of Pressure Sensitive Adhesive Technology, 3rd ed., Satas & Associates, 1999
[16] "Acronal 500 L" BASF SE, Technical Information Bulletin TI/ED 1792 d, April 2000
[17] tonchemie.de/Klebstoffe.htm
[18] polybrands.com/sf.html
[19] Information of the Technical Service, Wacker Polymer Systems GmbH, 2004

1.2 Water-based adhesives and sealants

1.2.1 Basic principles sensitive

Water-based dispersion adhesives are by far the largest class of adhesives in terms of volume. Their market share ranges from 40 to 70 %, the exact percentage depending on which products are classified as adhesives. Water-based phenolic resin binders for manufacturing chipboards are not considered to be adhesives in this book.

The history of water-based dispersion adhesives stretches back a long way. Traditional wood glues made from animal bones have been around for a long time. Water is the most important solvent and by far the cheapest. It is readily available almost everywhere, non-toxic and easy to dispose of. However, most polymers are not polar enough to be dissolved at the molecular level (on the principle that "like dissolves like"). Before hydrophobic polymers can be applied from water, they must be dispersed in water as small particles.

Overview articles on the subject of water-based adhesives, including a focus on the use of bio-based raw materials, can be found in [22–27].

1.2.2 Dispersions – theory and stabilisation

Water-based dispersions under the microscope form a multi-phase or colloidal system. This system consists of the continuous aqueous phase and the dispersed particles of the adhesive polymer. Heterogeneous multi-phase systems, such as the dispersion of small particles in water, are thermodynamically unstable (to disperse the polymer, surface work has to be performed). Unless suitable measures were taken, Brownian motion would cause the small particles to continually collide. Due to their similar polarity, the particles would adhere to each other and grow into agglomerates. These agglomerates would either float or settle, in accordance with their relative density.

The adhesive particles of a water-based dispersion have diameters ranging from 0.5 to 2 μm (by comparison, the wavelength of visible light is 0.4 to 0.8 μm). To prevent coagulation of the dispersed particles, a suitable means of stabilisation is required [1, 2].

1.2.2.1 Ionic stabilisation

If charges in the form of ionic groups are generated on the surface of a dispersed particle, the counter-ions in the aqueous phase form a "diffuse ion cloud" around the dispersed particle. When two particles stabilised in such a way approach each other, the identically charged ion clouds repel each other as a result of electrostatic forces (see Figure II-1.23).Coagulation or flocculation is thereby prevented [1, 2].

This repulsion is described mathematically by the DLVO theory (see literature on colloid chemistry) [1, 3, 4]. For two particles stabilised in this way, there is a distance at which repulsion by electrostatic forces and attraction by van der Waals forces lead to a potential minimum (secondary minimum). If this distance is breached to any great extent, the attractive forces, which are very strong at short distances, will dominate and coagulation will occur (primary minimum). Ionically stabilised dispersions therefore cannot be redispersed after coagulation (irreversible film formation). The change in electric potential with increase in distance is shown schematically in Figure II-1.24.

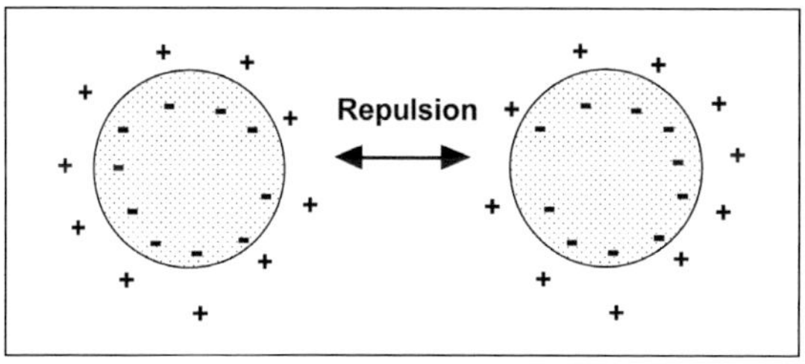

Figure II-1.23: Repulsion of ion clouds (electrostatic stabilisation)

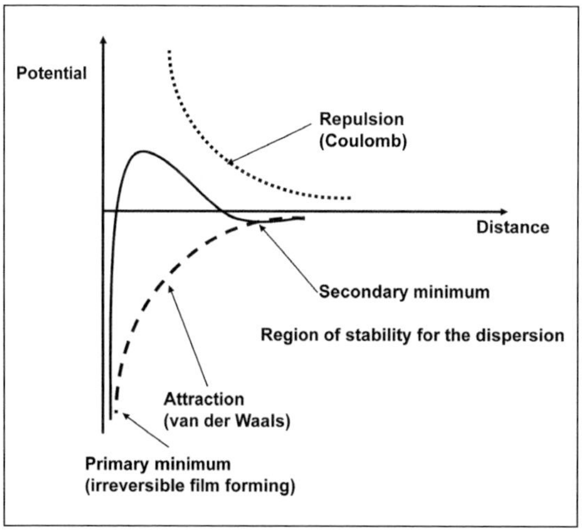

Figure II-1.24: Electric potential as a function of distance between two ionically stabilised particles

The particle surfaces can be covered with ionic groups in different ways. In the case of **ionomers** (or polyelectrolytes), the polymers already contain covalently bound ionic groups (Figure II-1.25). In a polar aqueous environment, these groups are oriented towards the water and the counter-ions form the stabilising ion cloud. Covalently bound groups can withstand shear forces, which can be substantial in the case of spray application. The presence of electrolytes containing polyvalent counter-ions causes bridging between the dispersion particles and coagulation occurs. Polyurethane and acrylic dispersions are examples of ionomers.

Advantages

+ irreversible film formation
+ high shear stability
+ functional groups for crosslinking reactions in the polymer backbone

Disadvantages

– sensitivity to electrolytes

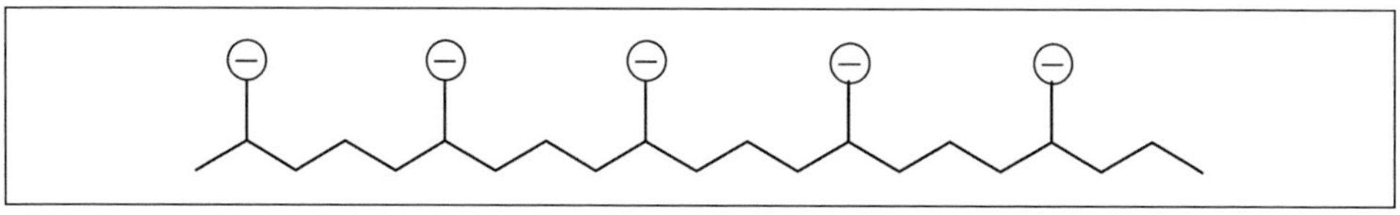

Figure II-1.25: Ionomers

1.2.2.2 Surfactants

Surfactants (surface active agents, emulsifiers, soaps, wetting agents, tensides) are amphiphilic molecules, i.e. they contain a hydrophilic and a hydrophobic moiety. The hydrophilic moiety can be either anionic or cationic (or even non-ionic) while the hydrophobic moiety is often a hydrocarbon chain. Figure II-1.26 shows sodium lauryl sulphonate, a typical anionic surfactant.

Amphiphilic molecules of this kind have the ability to lower surface tension along a phase boundary. In aqueous milieu, the hydrophilic, charged moiety can orient itself towards the polar water, while the hydrophobic moiety, as shown in Figure II-1.27, aligns itself with the air or a hydrophobic polymer particle. As a result, the surface tension is reduced and dispersed particles are stabilised against coagulation [1, 2].

Surfactants can be characterised by their chemical structure, i.e. the nature of the hydrophilic group (e.g. carboxylate, sulphate, sulphonate, ammonium ions), the hydrophobic group (usually a hydrocarbon) and by their solubility in water or non-polar solvents. Surfactants are held on the polymer particles by weak secondary valence forces only and so can be removed by strong shearing – this can lead to coagulation.

Advantages

+ the surfactant system can be tailored to the application

Disadvantages

- shear stability more limited than that of ionomers
- ionic surfactants are sensitive to electrolytes
- surfactant remains in the dry adhesive film

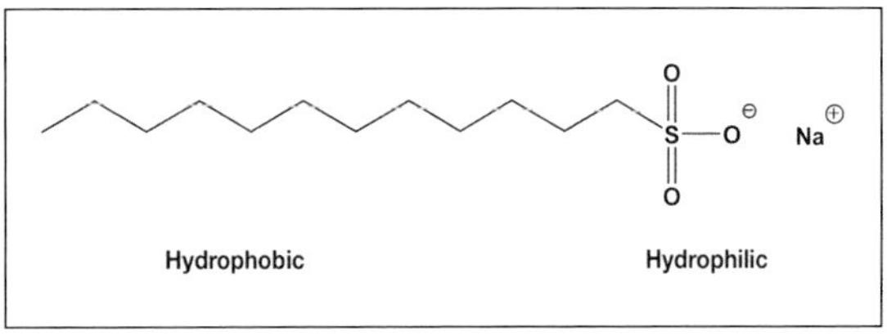

Figure II-1.26: Sodium lauryl sulphonate

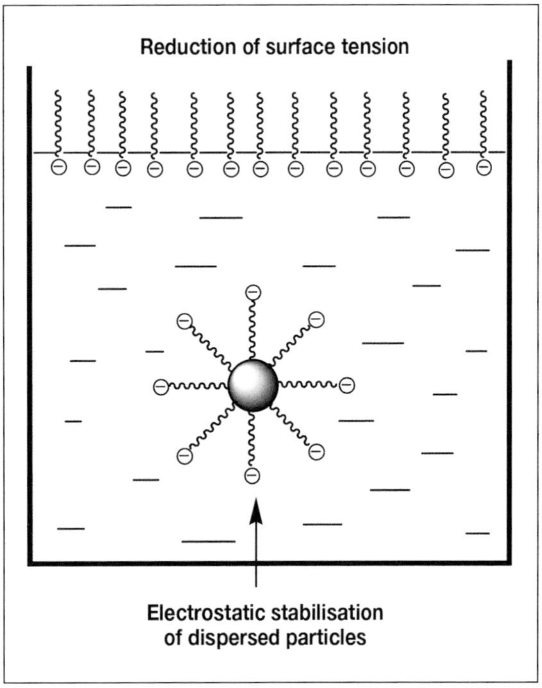

Figure II-1.27: Effect of surfactants

1.2.2.3 Steric stabilisation

Certain polymers, particularly hydrophilic-hydrophobic copolymers, are also suitable for stabilising dispersed polymer particles (**protective colloids**). Protective colloids derive their effect from adsorption of their hydrophobic moieties (in the form of "trains") onto the dispersed polymer particle, while the hydrophilic moieties in the form of "loops" or "tails" are oriented towards the water. Where two such stabilised particles approach each other, there is a certain distance at which the mobility of the hydrophilic polymer chains becomes limited (volume restriction). This restriction of mobility corresponds to enforced reduction in entropy, because mutual hindering reduces the number of possible orientations of the hydrophilic chains. This state is less favourable because it is less probable and is shown in Figure II-1.28 (not to scale) [1, 2].

An alternative interpretation of steric stabilisation is based on osmosis. Where two sterically stabilised particles approach each other, the concentration increases as the chains begin to overlap. In this region of increased concentration of hydrophilic protective colloid particles, the polar medium (water) starts diffusing in to dilute the system again. This dilution drives the approaching particles apart. Since stabilisation is not affected by ionic forces, such systems are able to resist electrolytes. Owing to the hydrophilic nature of the protective colloids in the remaining film of adhesive, the film can often be redissolved or redispersed again in water. This is a desirable property for many applications, such as powder adhesives for ceramic tiles. This type of stabilisation is unsuitable where a genuinely water-resistant bond is desired. Shear stability is greater than that of surfactant-stabilised particles. Steric stabilisation can also be employed in non-aqueous systems.

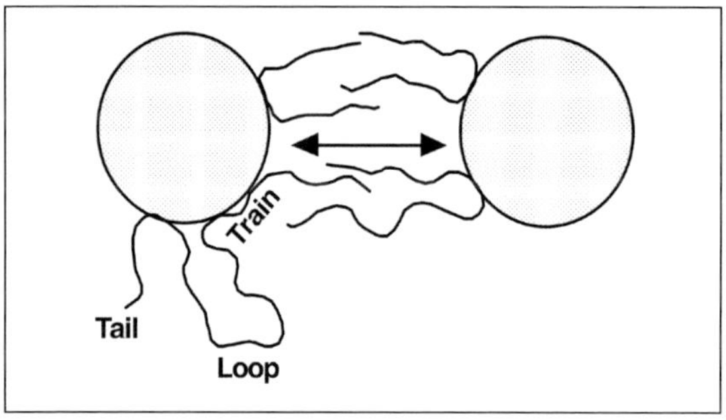

Figure II-1.28: Steric stabilisation

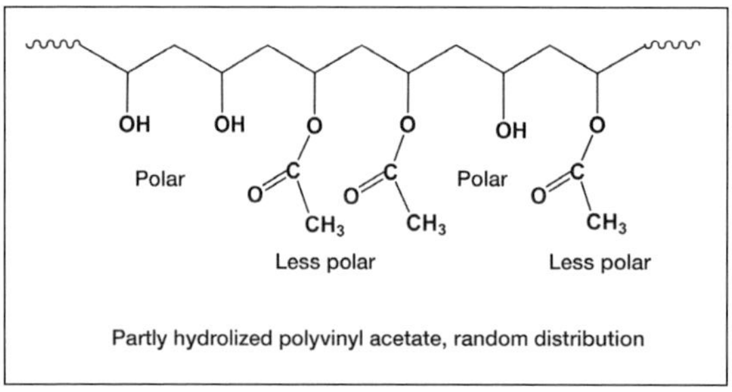

Figure II-1.29: Polyvinyl alcohol as protective colloid

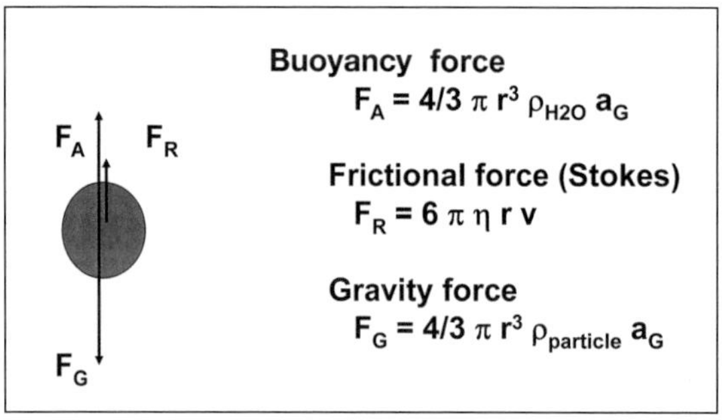

Figure II-1.30: Rheological stabilisation where $\rho = m/V$ and $V_{ball} = 4/3 \cdot \pi \cdot r^3$

Advantages

- + high stability to electrolytes
- + also adaptable to non-aqueous systems
- + redispersibility
- + high shear stability

Disadvantages

- – dry film is sensitive to water
- – stability is temperature-dependent

Typical protective colloids are polyvinyl pyrrolidone, polyvinyl alcohol (see Figure II-1.29) and certain polyurethanes [1, 2].

1.2.2.4 Rheological stabilisation

Flocculation occurs when thermal motion causes particles to approach each other until they eventually experience mutual attraction due to van der Waals-forces, and form agglomerates. The density dictates whether the agglomerates float or settle. High viscosities can inhibit this process due to the substantial restriction on particle mobility. A simple balance of forces (gravitational force = friction force plus buoyancy force) can be used to calculate the rate of floating or settling (sedimentation). This principle is utilised extensively in sealant formulations for the purpose of filler stabilisation. Figure II-1.30 shows the physical principles behind rheological stabilisation.

Buoyancy force	$F_A = m_{water\ displaced} \cdot a_G$	=	$4/3 \cdot \pi \cdot r^3 \cdot \rho_{water} \cdot a_G$
Friction force (Stokes)	F_R	=	$6\pi \cdot \eta \cdot r \cdot v$
Gravitational force	$F_G = m_{particle} \cdot a_G$	=	$4/3 \cdot \pi \cdot r^3 \cdot \rho_{particle} \cdot a_G$

where

- a_G is the acceleration due to gravity
- η is the viscosity of the dispersant
- r is the radius of the dispersed particle (spherical)
- v is the rate of sedimentation or floating
- ρ is the density
- V is the volume
- m is the mass

The rate of sedimentation (or floating) is: v

$$F_R = F_G - F_A$$
$$6\pi \cdot \eta \cdot r \cdot v = 4/3 \cdot \pi \cdot r^3 \cdot \rho_{particle} \cdot a_G - 4/3 \cdot \pi \cdot r^3 \cdot \rho_{water} \cdot a_G$$
$$v = 2\, a_G \cdot \Delta\rho \cdot r^2/9\, \eta$$
$$\rightarrow \quad v \sim 1/\eta \sim \Delta\rho \sim r^2$$

In other words, high viscosities, low differences in density between dispersed particle and dispersant and, especially, small particle sizes favour rheological stabilisation. Rheological stabilisation is mostly employed for (viscous) sealants and trowellable adhesives.

1.2.2.5 Formulation of water-based dispersion adhesives

Formulation of water-based adhesives is more complex than that of solvent-based types. The following principles need to be considered: The **base polymer dispersion** (20 to 100 % with a typical solids content of 30 to 70 %) determines the final properties via its chemical structure. For modification of the adhesive, a knowledge of the stabilisation principle employed for the dispersion is essential. Often, **blended dispersions** (up to 40 %) are added for modification purposes. In contrast to polymer solutions, dispersions of different polymers can be combined easily, provided that the same stabilisation system is employed. This can have significant consequences as regards attainable physical properties and cost reductions. As in the case of solvent-based adhesives, **resins** (up to approx. 30 %; nearly all types) can be used to increase tack and adhesion or to modify setting characteristics. Resins usually have to be added as res-

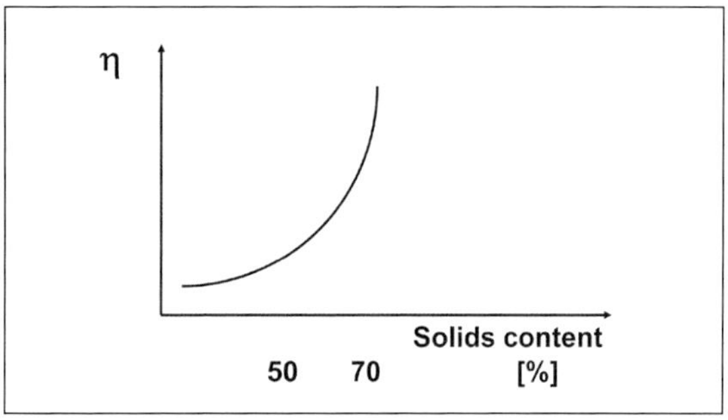

Figure II-1.31: Viscosity of dispersions

in dispersions. **Fillers** (up to 60 %, primarily chalk) can serve to reduce costs and adjust application properties. Trowellable adhesives or sealants, especially, have a high solids content. Sprayable dispersions normally are free of fillers. **Plasticizer** addition is mostly reserved for sealants. **Additives** (below about 5 %) are optional and dictated by the application. Many commercial dispersions already contain important additives that may influence the properties of the final bond. Flow characteristics can be adjusted with rheological additives (inorganic or organic). In the absence of rheological additives, water-based dispersions would have virtually the same viscosity as water over a wide range of solids content (see Figure II-1.31).

Defoaming agents help to reduce undesirable foaming during manufacture and application of the adhesive. Wetting agents boost the wetting of substrates of low surface tension and of fillers [2]. Water has a high surface tension and so is unable to wet most substrates during the bonding process. Coupling agents can be used to great effect with certain substrates. Coalescents help to improve film formation as the dispersion dries. Biocides are crucial for water-based systems. Without them, the polymer would degrade within a short time due to growth of mould and bacteria. Colorants are optional and sometimes help to make an adhesive applied to a substrate visible (quality control).

1.2.2.6 Application of water-based dispersion adhesives

Water-based dispersion adhesives are applied in the same way as solvent-based types. However, some special aspects need to be considered.

Components

Dispersion adhesives may be 1-component or 2-component systems. For 2-component systems, the correct mixing ratio and pot life must be observed. The crosslinking agents must be soluble or dispersible in water and they should react only slowly with the water, if at all. Many commercial crosslinkers are based on aliphatic isocyanates. These will have been modified to render them easy to disperse.

Application

Water-based adhesives can be applied by spraying techniques, by roller or brush, either to one substrate (one way) or to both substrates (two way). During application, it is important to bear in mind that, unlike ionomers, surfactant-stabilised water-based dispersions are sensitive to shear forces. These can be substantial in the case of spray application, and lead to undesirable coagulation at the nozzles. With ionically stabilised systems, too, water is sometimes inadequate for cleaning the application tools, due to irreversible film formation.

Procedure and working technique

The individual steps involved in achieving a bond are:

- preparation of the substrates (i.e. cleaning) and the adhesive (1-component/2-component)
- wetting of the substrate by the water present in the dispersion
- evaporation of the water
- coagulation of the dispersion, and film formation
- development cohesion
- wetting of the substrates by the adhesive polymer and development of adhesion

Wetting of substrates with water-based dispersions is much more difficult than with solvent-based types. Due to its high surface tension, water is very poor at wetting non-polar surfaces – which is typically the case for most plastics. Wetting agents (surfactants) therefore frequently have to be used or, alternatively, the substrate surfaces have to be pretreated properly in advance. Water evaporates much more slowly than most solvents. The evaporation number for water is 80. This means that water takes 80 times as long as diethyl ether to evaporate (see Chapter II-1.1.1). The outcome is longer (undesirable) cycle times or greater and costly energy input. Additionally, the process by which water evaporates can cause corrosion in industrial exhaust equipment. Film formation [1, 2, 5] by water-based dispersions is more complex than by their solvent-based counterparts. Let us take a simplified look at the individual steps in film formation of an ionically stabilised dispersion. During evaporation, the particles of the dispersion approach each other until a critical distance is achieved (see Figure II-1.24). At this distance, the van der Waals forces of attraction (proportional to $1/r^6$) begin to dominate the forces of repulsion (proportional to e^{-r}) and the particles stick together on contact. The residual dispersant (water) slowly continues to evaporate, with the spherical particles being deformed to a honeycomb-like structure and adhering to each other by intermingling of the polymer chains due to their identical composition. At this stage, cohesion and adhesion start developing between adhesive and substrate. The following parameters are important for film formation:

- temperature of the environment
- minimum film-forming temperature of the dispersion
- relative humidity of the ambient air
- rate of evaporation
- particle diameter and distribution
- water uptake by the substrates

Evaporation of the dispersant and mutual bonding of the dispersion particles due to "autohesion" are heavily influenced by the ambient temperature. For each system there exists a minimum film-forming temperature (MFFT) which is often similar to the glass transition temperature of the formulated polymer system. Where a dispersion is dried below the MFFT, there will be no smooth film and no development of adhesion to the substrates since the polymer chains are incapable of movement below the glass transition temperature. However, film formation of high-melting polymers can be boosted by adding high-boiling solvents (film-forming agents). Typical high boilers are glycol ethers and glycol esters (Figure II-1.32). N-ethylpyrrolidone (NEP) is also used today as a film former and as a replacement for N-methylpyrrolidone (NMP). This is an aprotic and highly polar organic solvent that is completely miscible with water.

The relative humidity and the weather conditions influence the rate of evaporation. If the water evaporates too rapidly, this can be hinder the formation of a homogeneous film, because the particles need to undergo deformation during film formation. For viscoelastic materials, such as dispersion particles, deformation will be more difficult at high evaporation rates, because the elastic component will dominate the viscous component. Finally, the particle diameter

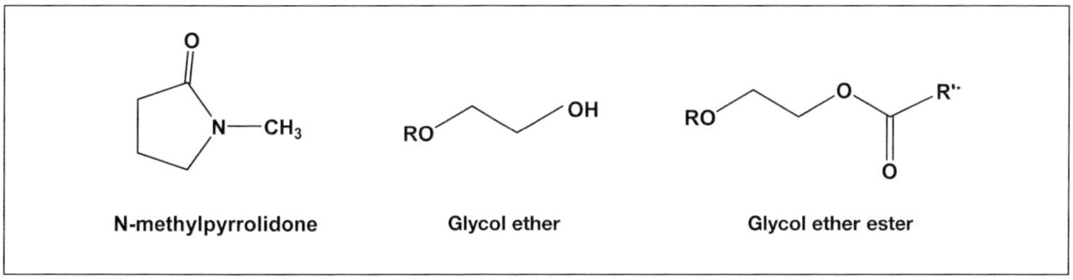

Figure II-1.32: Film-forming agents

is important if the film is to be homogeneous. Small particles form a smoother film than coarse particles, because film formation can only proceed from the surfaces of the individual particles. At the microscopic level, films formed from water-based dispersions are not identical with films formed from solution since the individual particles of the dispersion are not mixed fully homogeneously.

Working techniques

The working techniques for water-based dispersions are similar to those already discussed for solvent-based adhesives (see also Chapter II-1.1.3).

Wet bonding is only possible where moisture is able to permeate the substrates, e.g. wood, paper and textiles. The substrates must be fixtured until the adhesive has formed a sufficiently dry film. Many water-based adhesives, unlike solvent-based adhesives, have a tendency to corrugate paper due to its high polarity.

Contact bonding can be performed in the same manner as for solvent-based adhesives if the drying process provides a time window during which the polymers are not yet fully solidified (i.e. by crystallisation).

Anyone who has ever licked a stamp or an envelope is familiar with **solvent reactivation** by means of which an adhesive is moistened with water. This process is suitable for highly polar polymers, such as polyvinyl alcohol. Even though it is not commonplace, thermoplastic polymers can be reactivated with organic solvents.

Thermoplastic polymers can be **thermally reactivated** in the same way as solvent-based systems, especially in industrial processes. It should be noted, though, that measures taken to stabilise the dispersions (protective colloids, surfactants, ionomers) can influence the ease of reactivation.

Evaluation

There are many good reasons why water-based dispersions dominate the market for adhesives. In the current era, ecology and work safety are foremost among them. Water is non-toxic, non-flammable, easy to dispose of and not harmful to the environment. But the many ostensible advantages and disadvantages call for a certain degree of circumspection. Workers spraying water-based dispersions need appropriate protection, because adhesive aerosols are not harmless, especially where a 2-component system containing reactive crosslinkers is used. Even the assessment that water-based dispersions are environmentally friendly products needs to be put in perspective. When wash-water from cleaned application utensils or leftover adhesive is poured down the drain (illegally) under the false assumption that the product is environmentally friendly, the stabilised dispersion particles will nonetheless pollute the water and are very hard to degrade in the wastewater treatment plant. Where high-boiling solvents have served as film-forming agents, they will evaporate slowly after application. The term "solventless" in these cases can be very misleading.

The technical advantages accruing from the use of water-based dispersions are a wide scope for formulation, the higher solids content, and the possibility of using crosslinked or partly crosslinked polymers (microgels). The higher price resulting from the more complex manufacturing method becomes less important given the higher solids content. And yet salesmen have a very hard job convincing customers that the higher price vis-à-vis solvent-based adhesives is really justified. Technically disadvantageous are the poorer adhesion properties due to the poorer wetting power of water and the absence of "surface bite." Many dispersions are sensitive to frost and corresponding precautions need to be taken during transport and storage.

Table II-1.18: Comparison of water-based and solvent-based adhesives

Advantages of water-based dispersions	Advantages of solvent-based adhesives
Ecology	wetting
Work safety	adhesion
Viscosity independent of solids content	manufacturing effort
High solid content	fast evaporation
Polymer blends possible	(price per volume)

Advantages

+ environmentally friendly
+ solventless (?)
+ low toxicity
+ non-flammable
+ high solid content
+ viscosity virtually independent of solids content
+ viscosity independent of molecular weight
+ high molecular weight polymers can be used
+ crosslinked polymers can be used
+ irreversible film formation
+ large number of possible formulations
+ even incompatible polymers can be mixed using dispersion technology

Disadvantages

– higher manufacturing outlay
– higher price
– high evaporation number of water
– can initiate corrosion
– sensitive to frost
– difficult wetting of hydrophobic surfaces
– (necessary) ingredients can influence adhesion properties
– difficult cleaning of application tools
– coagulation at nozzles during spray application
– special crosslinkers necessary for 2-component application

Although water-based dispersion adhesives have now superseded solvent-based adhesives in many applications, the latter are still used in some applications. Dispersions undoubtedly offer an ecological advantage, but solvent-based systems are technically superior. The two systems are compared in Table II-1.18.

1.2.3 Chemistry of water-based dispersion adhesives

1.2.3.1 Polyurethanes (PUR or PU)

Manufacture and formulation

Historically speaking, PUR dispersions are newcomers to the field of dispersion adhesives. They are one of the fastest-growing groups, not least because a way has been found to transfer the extremely customisable properties of polyurethane technology to that of dispersions. PUR dispersions are not low-cost, mass-produced products, but rather are specialities. They shall therefore be treated in more detail at this point. PUR dispersions for adhesives are available from many suppliers nowadays. Naturally, appropriately formulated PUR dispersions can be applied by every standard bonding technique, of which thermal reactivation is the most important. It can be used for dispersions based on crystalline polyols of suitable crystalline melting point or on amorphous polyols of suitable glass transition temperature. Suppliers state a minimum reactivation temperature in their technical data sheets. For spray applications, ionic stabilisation with ionomers is best. For 2-component adhesives, the pot life must be taken into consideration. For formulation with, say, a further dispersion, the stabilisation mechanism is important. Most commercial PUR dispersions are anionic ionomers.

Commercial or self-made dispersions can be modified by using a blend (up to 50 %) based on acrylic, EVA or PVAc dispersions to reduce the costs significantly. Rheological additives, such as thickeners, wetting agents and colorants, are added as required to suit the application. For self-made dispersions, it is essential to add biocides.

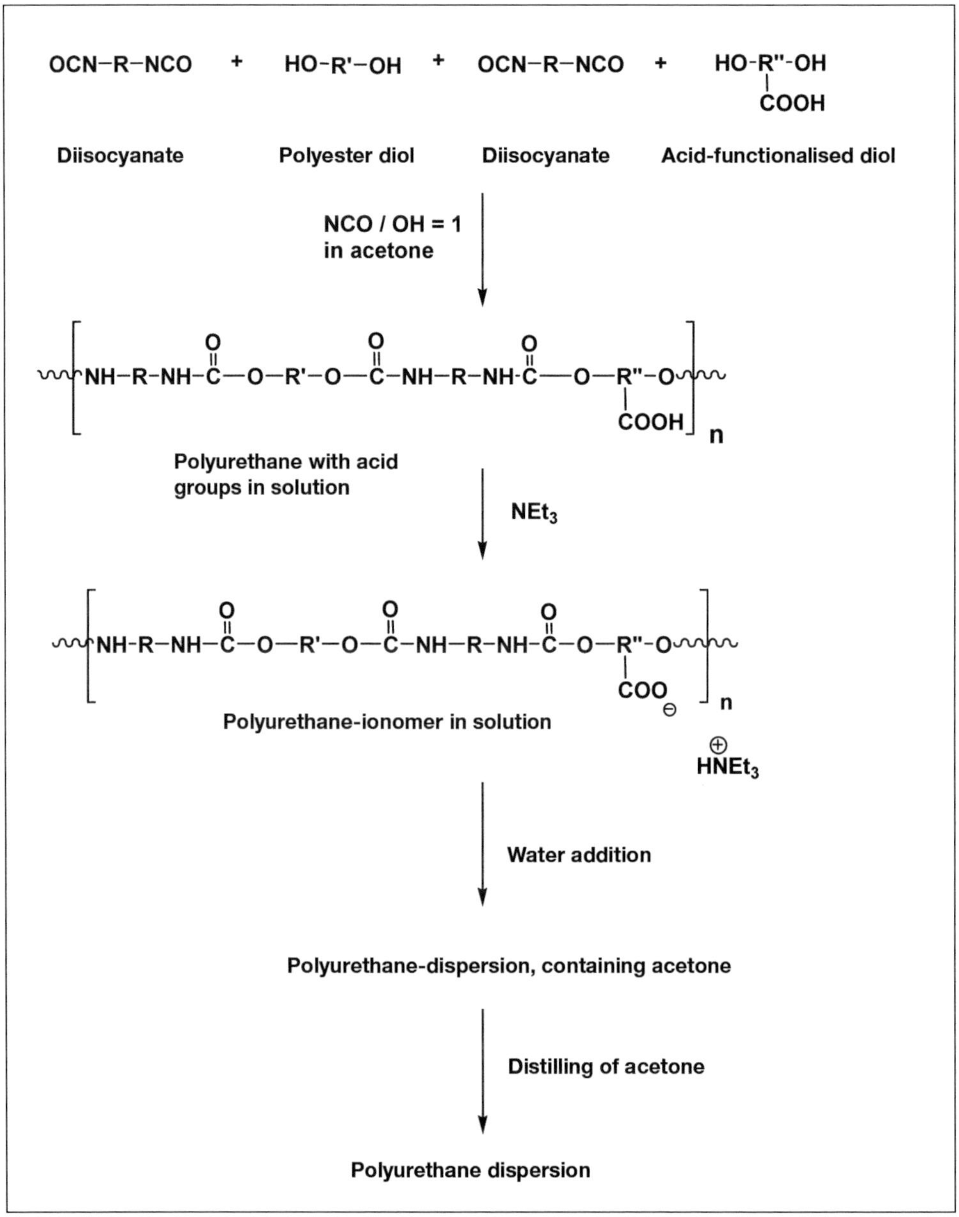

Figure II-1.33: Acetone process

For in-house manufacturing, different processes are available. The most important are the acetone process (also known as the ketone or solvent process) and the prepolymer process [6–8]. A brief treatment of both follows.

In the acetone process, as in the manufacture of solvent-based polyurethane adhesives, a high molecular weight polyurethane polymer is made by stoichiometric reaction between diols and diisocyanates. The characteristic feature of this polymer is that, in addition to the oligomeric polyols, ionic groups are incorporated into the polymer chain by means of a low molecular weight diol or diamine which bears, for example, a carboxyl group. The acidic groups can be converted into ionic

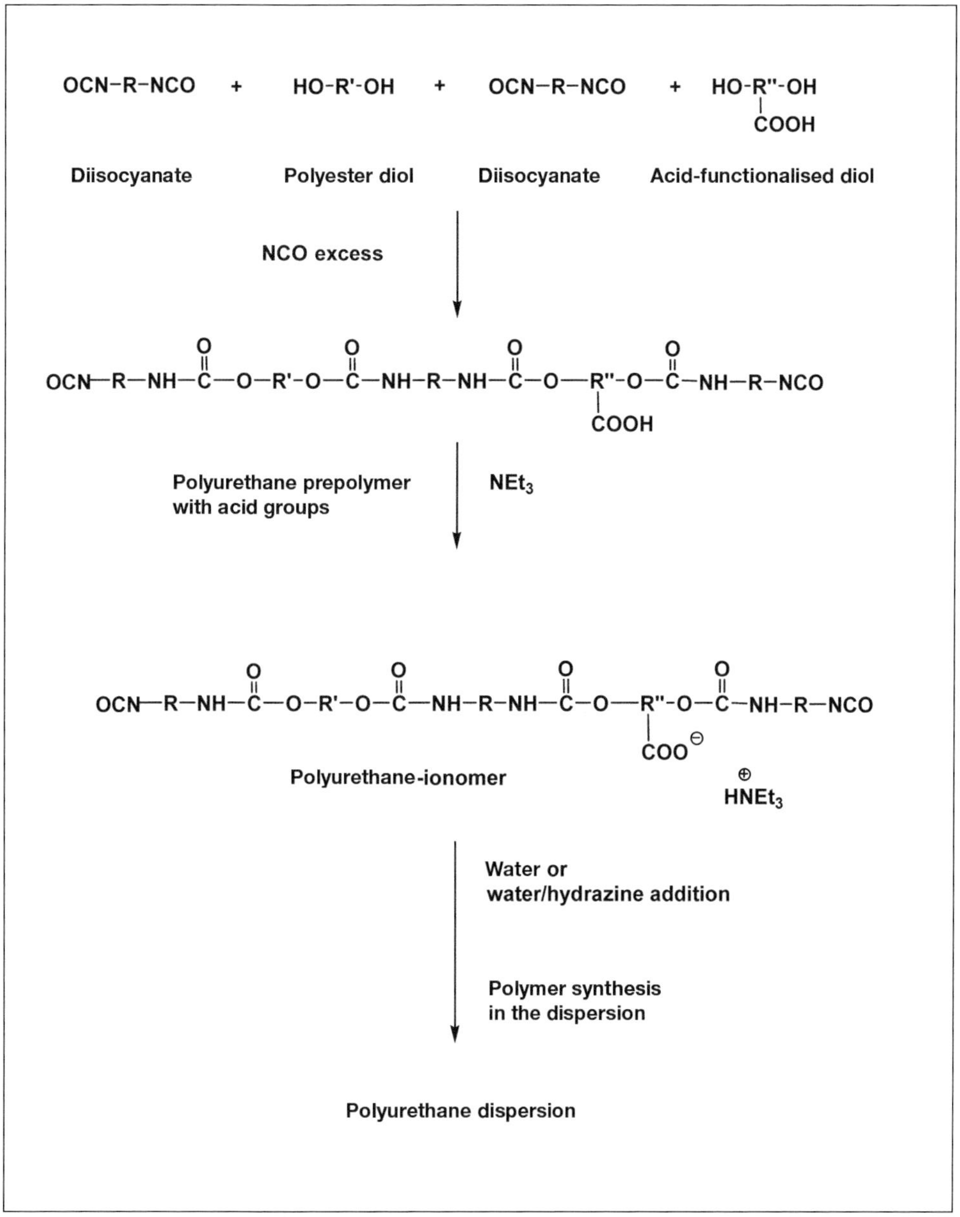

Figure II-1.34: Prepolymer process

groups by simple neutralisation. Water is added to the polymer solution in acetone until a dispersion of a polymer in water containing still acetone is obtained. The solvent is removed by distillation to yield the PUR dispersion. The resultant dispersion is also called a secondary dispersion, because the dispersion is generated in the second step. This process offers great flexibility and the polymers are made under very controlled conditions. Its disadvantage is the use of solvents in the manufacturing process. Figure II-1.33 shows the main features of the acetone process.

In the prepolymer process, a PUR prepolymer is made by making excess diisocyanate react with suitable diols (see also Chapter II-2.2.1). As in the acetone process, ionisable groups are also incorporated into this prepolymer. Unlike the acetone process, these prepolymers are still low-viscosity liquids at the reaction temperature, a fact which simplifies handling. These functionalised prepolymers are converted into ionic form by neutralisation and water is admixed.

This initially leads to a water-in-oil emulsion and then, as water is added, to an oil-in-water emulsion of isocyanate-terminated prepolymers. The ionic groups orient themselves towards the particle surface. Water can readily diffuse into the dispersion particles where it reacts, just as it does in 1-component PUR systems, by joining two isocyanate end groups via a urea linkage, with liberation of carbon dioxide. The reaction which yields the actual high molecular polymer therefore occurs inside the dispersion droplet and is a diffusion-controlled process. For viscosity reduction, high-boiling solvents, such as NEP, are added in low concentrations. These solvents remain in the dispersion, where they can support film formation during drying. The advantage of this process is that there are no low-boiling solvents involved that would have to be distilled off at the end. The disadvantages are that the synthesis reaction of the polymer is less easily controlled, and that high-boiling solvent remains in the dispersion. Figure II-1.34 shows the basic principles behind the prepolymer process.

The prepolymer process is ideal for in-house production by small companies, because it works without any low-boiling solvents. The preferred isocyanates are aliphatic types, such as IPDI, HDI and hydrogenated MDI (see Chapter II-2.2.1.2 for structures). IPDI is advantageous because the different reactivity of the two isocyanate groups leads to a lower molecular weight and thus to lower-viscosity prepolymers. However, its angled structure somewhat reduces the degree of crystallinity of the final product. Aromatic isocyanates, such as MDI, are usually too reactive with water, and so a simple dispersion step cannot be employed. Preferred polyols for thermally reactivatable PUR adhesives are crystalline polyesters based on hexanediol adipate (HD/AA), butanediol adipate (BD/AA) and polycaprolactones having molecular weights between 200 and 6000 g/mol. Ionic groups are usually generated with dimethylol propionic acid (DMPA; Figure II-1.35), because the carboxyl groups are less reactive with isocyanate. Consequently, DMPA enters into a bifunctional reaction with isocyanates.

For the prepolymer synthesis, the NCO/OH ratio is set to 1.5; at a reaction temperature of 80 to 100 °C, low-viscosity prepolymers are obtained. A high NCO/OH ratio would be more costly and yield a higher concentration of urea groups in the final product, which would impact the ease of reactivation. A low NCO/OH ratio leads to high molecular prepolymers of high viscosity, and poorer dispersibility. The particle size declines with increase in acid content, because smaller particles have a larger relative surface area, towards which the ionic groups in the water orient themselves. The smaller the dispersed particles, the more transparent the dis-

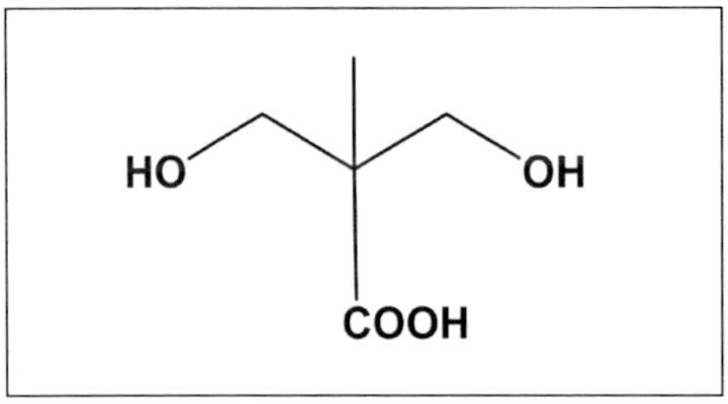

Figure II-1.35: Dimethylol propionic acid (DMPA)

persions become. A certain minimum concentration of ionic groups (acid number 150 to 250 mval COOH/kg polymer) is usually sufficient. Due to the low molecular weight of DMPA, however, higher concentrations of acid lead to a higher concentration of urethane bonds in the polymer, poorer ease of reactivation and higher costs. Conversion into ionic groups is performed immediately before the dispersion step, by neutralisation with bases, such as triethylamine (TEA). NMP serves as a high-boiling solvent for dissolving the DMPA in advance (5 to 20 % NMP, expressed in terms of the solids content), and remains in the final dispersion. For the synthesis, it is recommended first that the isocyanate be made to react with the crystalline polyester diol (2 h, 80 to 100 °C, addition of 200 ppm DBTL) and that the DMPA, dissolved in NMP, be added and allowed to react for another 1 to 2 hours at about 80 °C. Then, TEA should be added, followed immediately by deionised water, slowly and with stirring. As the water is being added, the viscosity increases briefly and then drops significantly when the oil-in-water state is reached. The dispersions are stable to sedimentation and, to an extent depending on the acid concentration, either are transparent or exhibit Tyndall scattering and have a blue shimmer.

Table II-1.19: PUR dispersion adhesive for footwear applications

Item	**Raw material**	**Parts by weight**
1	PUR dispersion	99
2	PUR thickener	0.5
3	Water	0.5
4	Optional crosslinker	3 weight-%

Item 1: "Dispercoll U54", Covestro
Item 2: "Borchigel L75 N", Borchers
Items 2 and 3 are prepared separately and admixed to item 1
Item 4: "Desmodur DN", Covestro

For toxicological reasons, NMP should be replaced by less harmful solvents, such as glycol ether or NEP, wherever possible. However, it is important to note that the use of NEP or other solvents depends on the specific requirements of the manufacturing process and the desired properties of the final product. Therefore, the suitability of alternative high boilers for a specific application should be carefully evaluated.

Guide formulations

Formulating a water-based PUR dispersion with commercial products is easy [9]. Table II-1.19 shows the ingredients for a heat-activated adhesive based on an anionic PUR dispersion. The flow properties are adjusted with a rheological additive (thinned 1 : 3 with water). The system can be thermally reactivated at 75 °C and has a high green strength after setting. If used as a two-component system, its pot life is about 8 hours. The addition of a crosslinker (3 to 5 % dispersible aliphatic isocyanate, "Desmodur DN", Covestro) raises the heat resistance from about 60 °C to more than 100 °C, which is sufficient for most automotive interior applications.

Evaluation

PUR dispersions possess adhesion properties comparable to those of the solvent-based polyurethanes. This also applies to many industrially important plastics, such as PVC, PC and ABS. Usually, polyester polyurethanes are the preferred adhesives for bonding soft PVC due to their good plasticiser resistance. If the joint will be exposed to thermal or hydrolytic conditions, a 2-component application is recommended. As is typical of PUR chemistry, PUR dispersions can be modified to the specific requirements of an application via the choice of isocyanates, polyols and adjustments of the manufacturing process. Heat-activated PUR dispersions rank among the most important types in industrial applications, and so are given prominence here.

Advantages

- + solventless (virtually)
- + high solids content (40 to 60 %)
- + blends with other dispersions possible
- + high shear stability
- + good sprayability
- + good heat reactivatability
- + high green strength
- + good adhesion to a variety of substrates
- + good resistance to plasticisers
- + can be used as 2-component systems
- + long pot life
- + good hydrolytic stability (as 2-component system)
- + good heat resistance (as 2-component system)

Disadvantages

- – price
- – frost sensitivity
- – poorer wetting of substrates than solvent-based adhesives
- – somewhat poorer reactivatability
- – somewhat poorer adhesion

Field of application

Nowadays, PUR dispersions are mainly used instead of traditional solvent-based PUR adhesives (see Chapter II-1.1.4.1). This includes, for example, nearly the entire **automotive interior**. Prominent examples here are headliners, door panels and rear parcel shelves. In all cases, high numbers of pieces have to be produced in short cycle times to high quality standards. 2-component application guarantees good heat resistance (exposure to the sun can lead to temperatures above 100 °C in the roof area). In the **shoe industry**, replacement of solvent-based products by water-based PUR dispersions for attaching shoe soles to the bootleg has begun but has so far failed to find widespread acceptance. This is astonishing, considering that technically suitable PUR dispersions have been commercially available since the mid-1980s.

1.2.3.2 Polychloroprenes

Manufacture and formulation

Polychloroprene (neoprene, (CR)) is produced by emulsion polymerization of 2-chloro-1,3-butadiene, as described in Chapter II-1.1.4.2. Adhesives manufacturers can obtain these dispersions commercially [10, 11]. Individual types differ mainly in the degree of crystallinity and gel fraction. While the crystallinity is responsible for the setting, the gel fraction determines the degree of cohesion and heat resistance. Copolymerization with 2,3-dichloro-1,3-butadiene affords very fast-crystallising products. In contrast to solvent-based polychloroprene adhesives, partly crosslinked polymers (microgels, gel fraction up to 70 %) can be used for dispersions. **Commercial dispersions** are usually stabilised with anionic surfactants or protective colloids, such as polyvinyl alcohol, and they have a solids content in the range of 35 to 60 %. Addition of further surfactants is possible, but if fast coagulation is needed, such as in the manufacture of mattresses, this can lead to undesirable side effects. For adhesive applications, the dispersions need fur-

Table II-1.20: Water-based polychloroprene adhesive

Item	Raw material	Parts by weight
1	Polychloroprene dispersion, fast crystallising	65
2	Anti-ageing agent	1.3
3	Emulsifier	1.3
4	Stabiliser	2.6
5	Resin dispersion, tackifier	approx. 30

Item 1: "Dispercoll C 84", Covestro
Item 2: "Rhenofit DDA-50 EM", amine-basis, Lanxess
Item 3: "Emulvin WA", non ionic surfactant, Lanxess
Item 4: "Akro-Zinc BAR 85", ZnO, Akrochem
Item 5: "Bremar SK 9700", neutralised tackifier solution that can be added directly to the formulation without prior emulsification or neutralisation processes, Robert Kraemer

ther formulation. The setting characteristics can be readily adjusted by blending different types of dispersion. Partly crosslinked dispersions increase the heat resistance. As with solvent-based polychloroprenes, **resins** (30 to 60 phr) in the form of dispersions are used to increase adhesion and tack. A large number of resin dispersions are commercially available, including alkylphenol resins and terpene-phenol resins. These resin dispersions can also be made with the aid of suitable emulsifiers. Starting materials include terpene-phenol resins, colophony esters and coumarone-indene resins. It is important for the stabilising system of the resin dispersion to be compatible with that of the polychloroprene dispersion. Alkylphenol resin/metal oxide chelates, of the kind used to improve the heat resistance of solvent-based adhesives, cannot be employed for water-based systems. In special cases, **fillers** (up to 200 phr), such as chalk and silicates, can be added. Typically, rheological **additives** based on polyacrylates, methylcellulose or polyurethanes can be used. Tin oxide or magnesium oxide (2 to 6 phr) is added to neutralise eliminated hydrogen chloride. Antioxidants (1 to 4 phr) based on hindered amines or phenols boost the ageing resistance of the polymer and the resins. For high-performance requirements, 2-component application is possible. Suitable crosslinkers include dispersible isocyanates, which are able to react with allyl OH groups produced from allyl chlorine.

Guide formulations

Table II-1.20 describes a water-based polychloroprene adhesive [11] without prior emulsification or neutralisation processes of the tackifier.

Evaluation

Due to their crystalline melting point of about 50 °C, polychloroprene dispersions exhibit good contact-adhesive properties. They combine a long open time with high green strength. They adhere well to a variety of substrates and have good hydrolytic stability. Like all halogenated adhesives, their heat resistance is limited. Their heat resistance is generally worse than that of solvent-based products, because chelates cannot be used. This disadvantage can be compensated to a certain extent by the use of gel-containing polymers. When used in a 2-component system with isocyanate crosslinkers, the adhesives offer a suitable property profile for most applications.

Advantages

+ solventless
+ high green strength
+ possibility of contact bonding

Table II-1.21: Water-based acrylic assembly adhesive with high initial tack [12]

Item	Raw material	Parts by weight
1	Acrylate dispersion, binder	625
2	Emulsifier	7
3	Dispersing agent	13
4	Filler	290
5	Thixotropic agent	20
6	Water	20
7	Thickening agent	25

Item 1: "Acronal ECO 6270 pH 7,8 with NH_3 (25 %)", BASF SE
Item 2: "Lutensol AT 18", BASF SE
Item 3: "Dispex AA 4135", BASF SE
Item 4: "Omyacarb 5 GU" chalk, Omya
Item 5: "Aerosil 300", Evonik
Item 7: "Rheovis AS 1180" (alternative: Rheovis AS 1187), BASF SE

Disadvantages

- halogenated product
- limited resistance to plasticisers
- limited heat resistance

Field of application

Polychloroprene dispersions are used wherever traditional solvent-based adhesives have been used in the past. Besides contact bonding, they lend themselves to wet bonding. Major applications are in the **shoe industry**, where shoe caps are bonded to the inner soles, in the **construction industry** to bond all kinds of flooring, in the **furniture industry** for decorative lamination of wood, in **mattress manufacture,** and in the **automotive industry** to make interior laminates.

1.2.3.3 Polyacrylates

Manufacture and formulation

Water-based acrylic dispersions are made directly from the monomer by emulsion polymerization. For this reason, they are also called primary dispersions. The polymerization process can be performed in the absence of pressure. Consequently, as with PUR dispersions, small companies can use commercial products or do their own in-house polymerization (see Chapter II-1.1.4.4). The properties of polymeric acrylates depend on the molecular weight (typically quite high for emulsion polymerization) and mainly on the composition of monomers employed. Besides the various acrylic esters, comonomers include styrene and other vinyl compounds. The fundamental structure-property relationships are briefly described again below. Polymeric acrylates are amorphous polymers. Their glass transition temperature is influenced by the type of acid (methacrylates > acrylates) and the alcohol component of the corresponding ester. Within certain limits, the glass transition temperature of the polymer decreases in proportion to the increase in length of the hydrocarbon chain of the alcohol in the ester. Thus, it is possible to make acrylates that exhibit pressure-sensitive properties by virtue of their low glass transition temperature, and acrylates whose high glass transition temperature confers high cohesive strength to the point of brittleness. For applications in water-based dispersions, there are certain upper limits imposed by the minimum film-forming temperature.

Formulation proceeds from the choice of suitable **base dispersion** (20 to 40 %). In accordance with the intended application, this can be modified with **resins** (10 to 30 %) and **fillers**, usually chalk (up to 60 %). **Additives** (only a few percent) are typically surfactants, defoamers, and wetting agents [10].

Guide formulations

Table II-1.21 presents a low-emissions, water-based acrylic adhesive with high initial tack [12]. The product is formulated with a high quantity of filler (to reduce costs and for ease of processing).

Table II-1.22 presents a starting formulation for a water-based acrylic contact adhesive [12, 13]. The product is modified with a phenolic resin dispersion.

Table II-1.23 presents a starting formulation for a water-based acrylic adhesive for ceramic tiles [14]. Besides the polymer and the resins, the formulation contains a number of additives typically employed in water-based dispersions.

Table II-1.22: Water-based acrylic-PUR contact adhesive [12]

Item	Raw material	Parts by weight
1	Polyester-polyurethane elastomer dispersion	50
2	Acrylate dispersion	30
3	Resin dispersion, tackifier	20
4	Acrylic rheology modifier	4
5	Optional crosslinker	5

Item 1: "Luphen D DS 3548", BASF SE
Item 2: "Acronal A 380", BASF SE
Item 3: "Bremar SK 9700", Robert Kraemer
Item 4: "Rheovis AS 1125 (8%, pH 8)", BASF SE
Item 5: "Basonat F 200 WD", solvent-free polyisocyanate to be added shortly before application, Improves heat resistance significantly, BASF SE

Table II-1.23: Water-based acrylic adhesive for ceramic tiles

Item	Raw material	Parts by weight
1	Water	69
2	Thickening agent	5
3	Defoamer	1
4	Acrylate dispersion, binder	200
5	Filler	435
6	Filler	290

Item 2: "Walocel MW 40000 PFV", Dow Wolff Cellulosics GmbH
Item 3: "FoamStar PB 2706", BASF SE
Item 4: "Acronal S 533 pH 8 with NaOH (20%)", BASF SE
Item 5: "Omyacarb 40 GU", Omya GmbH
Item 6: "Omyacarb® 130 GU", Omya GmbH

Evaluation

The numerous building blocks available for acrylates means that they offer more variety of properties than most other classes of polymer. Acrylates can be made in strengths ranging from "soft as butter" to high strength – whatever is feasibly is done. Acrylates are inexpensive, are highly tolerant of modification and can easily be tailored to the requirements of any application. However, copolymerization is usually a random process. Consequently, segmentation yielding properties akin to a multi-phase polymer of the kind found in polyurethanes, is not possible.

Advantages

+ flexible formulation due to diversity of available copolymers
+ good resistance to hydrolysis
+ good resistance to light and weathering
+ transparency (without fillers)

Disadvantages

– limited heat resistance

Field of application

Pressure-sensitive adhesives represent a very large field of application for water-based acrylic dispersions (see also Chapter II-3). Only polymers of low glass transition temperature are used here. Applications in the packaging industry include glossy film lamination. Major applications in the construction industry are flooring adhesives and tile adhesives. In the latter case, polymers of high glass transition temperatures are also in use.

Use as a sealant

In contrast to all other sealants, polyacrylate sealants contain a dispersion medium (water). The polyacrylates described in this chapter can contain a proportional amount of monomers other than (meth)acrylic acid and esters thereof (e.g. acrylonitrile, vinyl esters).

Solidification of latexes

Latexes (aqueous polymer primary dispersions) are manufactured by emulsion polymerization [15]. Sealants based on aqueous polyacrylic latexes dry physically and are comparable with latex paints as regards solidification (film formation) [15]. The solidification rate of latex based sealants depends heavily on ambient temperature and humidity.

As the water evaporates, the disperse polymer particles approach each other increasingly until spherical packing takes place (Figure II-1.36). Further evaporation of water causes the capillary forces of the very thin water lamellas to deform the spherical polymer particles, yielding rhombic dodecahedra. These fuse together (coalesce) at the particle boundaries. Intermingling of the polymer chains across the particle boundaries may occur. Primary dispersions (latexes) are thermodynamically unstable with regard to coagulation; i.e. solidification leads to a thermodynamically irreversible, stable state. All sealants made from latexes are non-crosslinked polymers and so are thermoplastic and swellable by solvents.

Coalescence (solidification or film formation) of primary dispersion particles only take place above the minimum film forming temperature (MFFT). Below the MFFT, the polymer is in the glassy state and unable to form a film. The MFFT of polymer dispersions for sealants is significantly below 0 °C as a rule and therefore distinctly lower than that of dispersions for paints. A low MFFT means that the polymer is soft and has a low glass transition temperature (T_g). Glass transition temperatures of polymers for latex based sealants lie between -25 and -30 °C [16]. Nevertheless, permanent plasticisers (e.g. phthalates) are added to latexes for sealants; sometimes coalescing agents (low-volatility organic solvents) serve as temporary plasticisers [15]. In contrast to permanent plasticisers, coalescing agents leave the polymer after a certain time and do not cause lasting plasticisation.

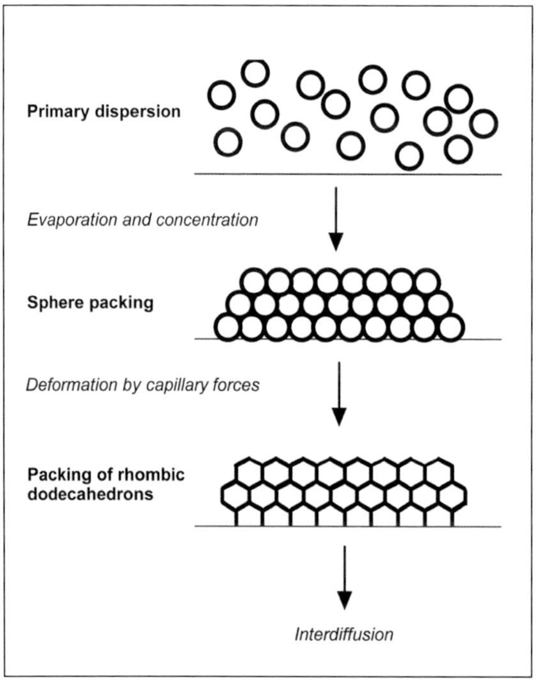

Figure II-1.36: Solidification of polymer latexes

Properties of polyacrylate sealants

Compared with earlier sealants based on dissolved polyacrylates, aqueous polyacrylate latexes permit the currently employed disperse polymer to have much higher molecular weights. Generally, higher molecular weights increase the elasticity at the expense of plasticity (which dominates at low molecular weights). In other words, the change from solvent-borne to water-based systems has brought not only improvements in environmental performance, but also improved quality.

Aqueous polyacrylate sealants have the following advantages and disadvantages.

Advantages

- \+ 1-component
- \+ inexpensive
- \+ good adhesion to damp substrates
- \+ good ageing stability
- +/- up to 15 % permissible total deformation (25 % is required for expansion joints, however)

Disadvantages

- high shrinkage of about 20 % during drying (evaporation of water)
- slow skinning
- sensitive to frost and rain when freshly applied
- limited suitability for exterior applications
- unsuitable for flooring and underwater joints

Table II-1.24: Formulation for a high elastic latex-based construction sealant [12]

Item	Raw material	Parts by weight
1	Acrylate dispersion, binder	310
2	Plasticizer	100
3	Surfactant	2
4	Dispersant	8
5	Pigment	30
6	Filler	545
7	Thickener	5

Item 1: "Acronal V 278 pH 8 with NH_3 (25 %)", BASF SE
Item 2: "Hexamoll DINCH ", BASF SE
Item 3: "Disponil FES 77", BASF SE
Item 4: "Dispex CX 4231", BASF SE
Item 5: "TiO_2 Kronos 2056", Kronos Titan GmbH, 51373 Leverkusen
Item 6: "Omyacarb 5 GU", Omya GmbH, 50968 Köln
Item 7: "Rheovis PU 1191", BASF SE

Latex-based sealants can be applied to all absorbent substrates, such as concrete, cellular concrete, plaster, gypsum and wood.

The performance of a solidified sealant is determined substantially by its elastic recovery (resilience) [14]. Ideally, elastic recovery should be 100 %, but in practice it is lower. In contrast to all other sealants described here, sealants based on polyacrylate latexes are not crosslinked. Consequently, elastic recovery at 50 % extension is 45 to 65 % and hence relatively low [14], i.e. the plastic component is relatively high.

Formulation of latex-based sealants

The choice of polymer latex plays an important role in formulation [17]. Fillers such as chalk and calcite increase the volume of the sealant (cost reduction), but can also improve technological properties. As the particle size of the filler increases, elastic recovery, stress (50 % extension) and elongation at break decrease. Furthermore, lamellar talc (Figure II-1.37) lowers the surface tack of the solidified sealant.

Rheologically active nanoparticles, such as fumed silica, improve the non-sag properties of freshly applied seal-

Figure II-1.37: Scanning electron micrograph of talc

Table II-1.25: Formulation for a latex-based construction sealant [12]

Item	Raw material	Parts by weight
1	Polymer latex (65 % solids)	31.5
2	Wetting agent	0.4
3	Dispersing agent	0.1
4	Titanium dioxide, rutile	4.0
5	Propylene glycol alkylphenyl ether	4.0
6	Ethylene glycol monophenyl ether	3.0
7	Chalk (filler)	56.5
8	Fumed silica	0.5
Total		100

Item 1: aqueous polymer latex based on acrylic acid esters, vinyl ester and acrylonitrile, adjusted to pH 8 with sodium hydroxide, glass temperature about -40 °C, MFFT <1 °C, e.g. "Acronal V 271", BASF SE
Item 2: , fatty alcohol ethoxylate
Item 3: , sodium salt of a polyacrylic acid
Item 5: propylene glycol alkylphenyl ether, plasticiser, e.g. "Plastilit 3060", BASF SE
Item 6: plasticizer, evaporation number > 10,000, BASF SE
Item 7: chalk, average particle diameter 5 µm
Item 8: , hydrophobic, BET surface area 90 m^2/g (rheological additive)

Table II-1.26: Formulation for a low cost latex-based construction sealant [12]

Item	Raw material	Parts by weight
1	Acrylate dispersion, binder	300
2	Plasticizer	90
3	Surfactant	3
4	Dispersant	6
5	Pigment	15
6	Filler	586

Item 1: "Acronal S 790", BASF SE
Item 2: "Hexamoll DINCH", BASF SE
Item 3: "Hydropalat WE 3437", BASF SE
Item 4: "Dispex AA 4135", BASF SE
Item 5: "TiO_2 Kronos 2056", Kronos Titan GmbH, 51373 Leverkusen
Item 6: "Omyacarb 5 GU", Omya GmbH, 50968 Köln

ants. Sealants can be coloured with pigments. Plasticisers and coalescing agents lower the glass transition temperatures of polymers; foremost among these are phthalates, polyisobutenes and propylene glycol phenyl ethers. The use of propylene glycol phenyl ethers as plasticisers yields low-temperature elastic sealants of low hardness and fast skinning after application. Dispersing agents, such as polycarboxylates, make it easier to incorporate fillers and pigments, and increase the stability of the sealants. Glycols can be added as antifreeze. Sealants based on latexes need to be treated with fungicides to prevent microbial colonisation.

Guide formulations

Guide formulations are presented in Table II-1.24, Table II-1.25 and Table II-1.26 [12].

Sealants based on aqueous polymer latexes suffer from a problem mentioned before, namely early rain resistance, i.e. the resistance of a freshly applied sealant to being washed out. Early rain resistance can be greatly improved by small additions of water-soluble or water-swellable polysaccharides [18] or polyvinyl alcohols [19].

1.2.3.4 Polyvinyl acetate

Manufacture and formulation

Polyvinyl acetate (PVAc) is manufactured by free-radical emulsion polymerization of vinyl acetate (see Chapter II-1.1.4.6). As with the acrylates, commercial dispersions (or powders that can be dispersed in water) are available. The polymerization can be performed in the absence of pressure. However, copolymerization with ethylene to yield the more flexible ethylene-vinyl acetate copolymers (EVA) requires pressurised conditions. Copolymerization with maleic esters or acrylates can be used to modify properties. The solids content of commercial products ranges from 50 to 60 %. The dispersions are stabilised by means of protective colloids (i.e. polyvinyl alcohol), surfactants or a combination of both. Where the dispersions have been stabilised with high quantities of protective colloids only, such films are water-sensitive because the polar stabilising system remains within the dry adhesive. On the other hand, they are usually redispersible. This redispersibility may be desirable if a powder-like adhesive (i.e. ceramic tile adhesive) can be formulated with water immediately before use. Conversely, dispersions stabilised only with ionic surfactants form films that are highly resistant to water, because they are not redispersible after coagulation. These dispersions are better suited to thermal reactivation, because no high-melting protective colloids are present. Some commercial dispersions already contain film-forming agents.

For many applications, modification is unnecessary, but may be effected with fillers (chalk or silicates), resins (colophony, coumarone-indene or phenol based), plasticisers (phthalates or other esters) and additives, such as rheological additives, wetting agents, defoamers and biocides.

Guide formulation

PVAc dispersions such as "Resyn 5763" and "Vinac DPN890" (D3 grade, Celanese) can be used directly as a wood adhesive (wood glue) [20]. The dispersion already contains a film-forming agent. For bonded joints at risk of hydrolysis or exposure to heat (wood for exterior applications), dispersible isocyanates ("Desmodur DN", Covestro) can serve as crosslinkers.

Evaluation

PVAc dispersions are low-cost products. They adhere strongly to cellulosic substrates, such as wood, cardboard and paper. Emulsion polymerization affords high molecular weight polymers which, in combination with a high glass transition temperature, guarantee high mechanical strength. High solid dispersion adhesives dry and set rapidly. Copolymerization with ethylene yields highly flexible products, which can serve in flexible laminated packaging.

Advantages

+ low cost
+ good adhesion to wood and paper
+ high solids content (up to 60 %)
+ rapid setting
+ good water resistance and optional redispersibility
+ good mechanical properties due to high molecular weight
+ can be used with crosslinkers as 2-component system
+ non-toxic

Disadvantages

– moderate adhesion to many plastics
– moderate peel strength
– moderate impact resistance

Fields of application

The **woodworking** and **furniture industry** are key application areas for PVAc dispersions due to their high strength and good adhesion to wood. Various adhesive applications are also found in the **paper** and **packaging industry**. EVA copolymers are easy to thermally reactivate. PVAc dispersions are also used in the **construction industry** for flooring and mortars.

1.2.3.5 Polyvinyl alcohol

Monomeric vinyl alcohol (PVAl) does not exist in the pure form. As shown in Figure II-1.38, it tautomerises to ethanal.

Polyvinyl alcohol is therefore made from polyvinyl acetate. The vinyl acetate function is chemically an ester, which in the presence of base catalysts, can be transesterified with methanol to methyl acetate and polyvinyl alcohol. However, like polyacrylates, polyvinyl acetate is not ranked as a polyester, as the ester is in the side group only and is not a part of the polymer chain. To an extent depending on the degree of saponification, polyvinyl acetate yields variously saponified products, extending as far as pure polyvinyl alcohol. The polarity of the polymer increases with increase in saponification. Consequently, highly saponified polymers (>70 %) are readily soluble in water. However, pure polyvinyl alcohol is less soluble in water than the highly saponified grades due to its high crystallinity. Polyvinyl alcohol has a glass transition temperature of 85 °C and a crystalline melting point of 230 °C. It is therefore very brittle in the pure state [10, 21].

Guide formulation

A water-reactivatable adhesive can be obtained from a 30 % solution of a highly saponified polyvinyl acetate, such as "Mowiol 4-88" (Merck) [21].

Fields of application

Polyvinyl alcohol can serve as **adhesive,** but it can also be used as an additive for other adhesives, such as polyvinyl acetates. Due to its polar structure, PVAl adheres strongly to such substrates as wood and paper. It has various uses as an **additive**. The effect of partly saponified polyvinyl acetates as protective colloid has already been described. The pronounced polarity can extend the open time of water-based dispersions by keeping the water in the polar environment for longer. PVAl can also act as a rheological additive. For paper bonding or **water-reactivatable adhesives** (stamps, labels, envelopes), PVAl is used as an adhesive directly.

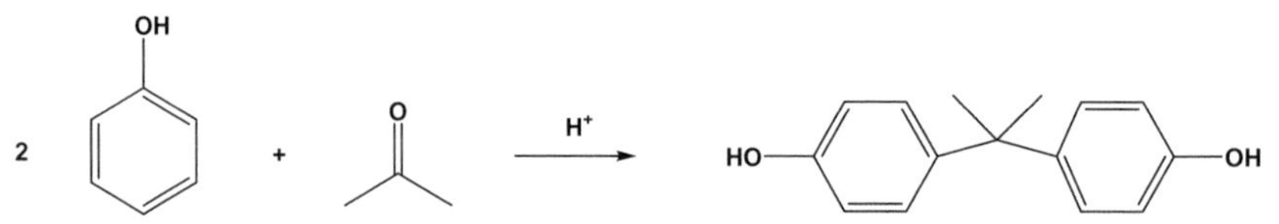

Figure II-1.38: Tautomerism of ethanal

1.2.4 References

[1] H. D. Dörfler, Grenzflächen und Kolloidchemie, VCH, 1994
[2] B. Müller, U. Poth, Coatings Formulation, Vincentz Network, 2006
[3] E. J. W. Verwey, et al., Theory of the Stability of Lyophobic Colloids, Elsevier Publishing Company, INC, 1948
[4] H. G. Elias, Makromoleküle, 6th ed., Wiley-VCH, 1999
[5] K. Weinmann, FARBE&LACK, 93 6 (1987), 447
[6] D. Dieterich, Adv. Org. Coat. Sci., Technol. Ser.1 (1979), 55 ff
[7] D. Dieterich, Angew. Makromol. Chem, 98 (1981), 133 ff
[8] D. Dieterich, Progr. Org. Coat., 9 (1981), 281 ff
[9] "Dispercoll U", brochure Bayer AG, edition 03.01, new: solutions.covestro.com/brands/dispercoll
[10] I. Skeist, Handbook of Adhesives, 3rd ed., Chapman & Hall, 1990
[11] "Dispercoll C", brochure Bayer AG, edition 02.01, neu: solutions.covestro.com/brands/dispercoll
[12] BASF SE, Technical Sales Polymer Dispersions for Architectural Coatings & Construction, Europe
[13] Formulation Suggestion K-2226 at dow.com/ucarlatex/tech/start.htm - 1

[14] Formulation Suggestion TA-1504 at dow.com/ucarlatex/tech/start.htm
[15] B. Müller, U. Poth, Coatings Formulation, Vincentz Network (2006) Chapters III-1 and III-2
[16] J. Krobb, E. Wistuba, Adhäsion (1992) No. 5, p. 14–20
[17] D. Distler, Wässrige Polymerdispersionen (in German), Wiley-VCH (1999) Chapter 7.3.3
[18] DE-OS 42 24 044 A1
[19] DE-OS 42 05 358 A1
[20] Celanese, "Mowilith LDL 2555 W", Technical Data Sheet, Issue 03/2003
[21] "Kuraray", "Kuraray Poval", Mowiol, Technical Data Sheets
[22] Magalhães, Solange, et al. "Brief overview on bio-based adhesives and sealants." Polymers 11.10 (2019): 1685
[23] Islam, Md Nazrul, et al. "An overview of different types and potential of bio-based adhesives used for wood products." International Journal of Adhesion and Adhesives 112 (2022): 102992
[24] Heinrich, Lydia Alexandra. "Future opportunities for bio-based adhesives–advantages beyond renewability." Green chemistry 21.8 (2019): 1866-1888
[25] Presti, Marco Lo, et al. "Bioinspired Biomaterial Composite for All-Water-Based High-Performance Adhesives (Adv. Sci. 16/2021)." Advanced Science 8.16 (2021)
[26] Pletcher, W. A., and Edmund J. Yaroch. "Water-Based Adhesives." Adhesives in Manufacturing. Routledge, 2018. 407-423
[27] Pletcher, W. A., and Edmund J. Yaroch. "Water-Based Adhesives." Adhesives in Manufacturing. Routledge, 2018. 407-423

1.3 Hot-melt adhesives

This section deals with hot-melt adhesives whose hardening takes place solely by the physical process of solidification from the melt and subsequently do not exhibit any surface tack. The group of permanently tacky hot-melt adhesives are dealt with in Chapter II-3, and the so-called reactive hot-melts, i.e. those that additionally crosslink by means of a chemical reaction, are dealt with e.g. in Chapter II-2.2.1.12. and Chapter II-2.2.2.

1.3.1 Basic principles

Hot-melts or hot-melt adhesives (the terms are equivalent) so called 100 % solids adhesives which means they do not contain solvents or water, are based on thermoplastic polymers. These polymers are solid at room temperature and soften when heated to yield a viscous liquid. This process is reversible and can be repeated several times, theoretically infinitely. They are applied as a melt allowing a good wetting of the adherend. After joining, which, too requires a sufficiently low viscosity to allow again a good wetting, this time of the second adherend the adhesive solidifies upon cooling to room temperature and develops cohesive strength (Figure II-1.39).

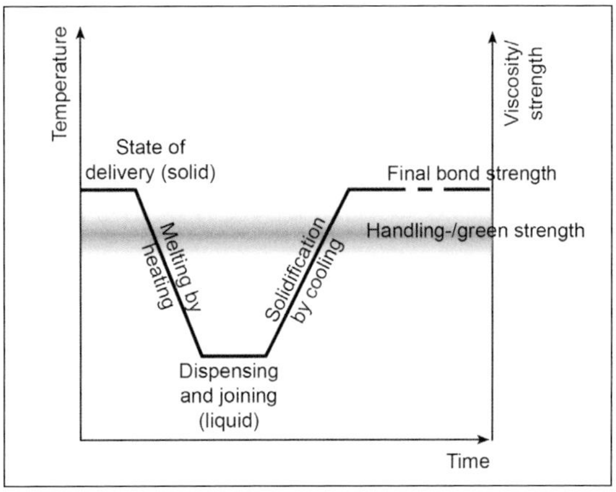

Figure II-1.39: Hot-melt processing principle

According to the spatial arrangement of the polymer chains in a plastic, a distinction is made between amorphous and crystalline orientation. If the polymer chains are present in a disordered state, in some cases also with interpenetration, this is referred to as an amorphous structure.

Even if it is difficult for the long, filamentous polymer chains to crystallize in the solid state, such structures of regularly arranged chains are possible. Although it would be energetically favourable for the polymer chains to align parallel to each other, such alignment is hindered by entanglement. Therefore, the polymer chains within the crystalline regions are both aligned and folded. These domains are therefore neither crystalline nor amorphous and are classified as semi-crystalline. The ability of polymers to crystallize when cooled depends on their chain structure. The presence of straight, unbranched, only slightly branched chains or chains with small, evenly arranged side chains facilitates the formation of semi-crystalline regions. Crystal growth is obtained by a further accumulation of folded polymer chains, but this can only happen at temperatures below the melting temperature (T_m) and above the glass transition temperature T_g. Above T_m, the molecular motions are so strong that the molecular arrangement is destroyed. Below the glass transition temperature, the movement of the molecular chains is frozen, so that reorientation can only take place extremely slowly in the scale of several months or years, if at all.

With an increase in the crystalline areas (crystallinity), the strength and stiffness as well as the melting temperature ranges and the heat resistance increase. Within certain limits, the crystallinity is influenced by the cooling conditions, in particular their speed, so that the crystallinity cannot be regarded as a material constant.

1.3.1.1 Hot-melt specific key parameters

As a result of the increasing chain mobility the viscosity of hot-melts decreases with increasing temperature. The transition from the solid to the liquid state, as well as the reverse process of solidification, usually occurs within a certain temperature range. The following parameters are relevant, especially for the processing of hot-melt adhesives.

Melt viscosity

Temperature dependent viscosity of the adhesive in a molten state. It influences the dispense rate, spread of the applied adhesive, and the wetting of the adherends' surfaces. The melt viscosity is typically measured according to EN ISO 2555 (Chapter IV-4.1) using a single cylinder type rotational viscometer method (e.g. Brookfield "Thermosel").

Melt flow index (MFI)/melt mass flow rate (MFR)/melt volume flow rate (MVR)

The melt flow index (MFI) is a measure of how much hot-melts can be forced through a defined capillary at a given temperature and pressure. The MFI is therefore more or less a measure of viscosity and provides valuable information regarding the dimensioning of the application system. The MFI can be determined both as melt mass-flow rate (MFR) with the unit [g/min] and as melt volume-flow rate (MVR) in [ml/min] according to EN ISO 1133 (Chapter IV-4.1).

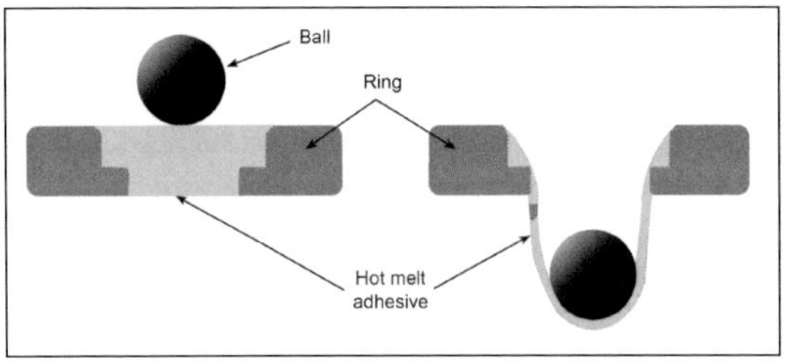

Figure II-1.40: Ring-and-ball softening point, left at start of the test, right at end of the test

Melting and softening point:

For determining the softening point according to EN 1238 [12] (Chapter IV-4.1) a metal ring of defined dimensions is filled with hot-melt adhesive, the ring is weighted with a ball and placed in an inert liquid bath by means of a specimen holder. This is then heated at a specified heating rate. The softening point is defined as the temperature at which the adhesive, deformed by the ball weight, first reaches a certain position below the ring. (Figure II-1.40)

Alternatively, the melting point can be determined by using the so called Kofler heating bar. This is a metal strip with an approximately linear temperature gradient (between room temperature and 300 °C). The hot-melt to be characterized is placed on a section of the strip corresponding to its approximately expected melting range and the exact softening point is then determined.

Heat resistance

In the case of hot-melts, the heat resistance of a bond is related to the softening point of the adhesive. Just as the result of determining the softening point by the ring-ball method depends essentially on the mass of the ball, the heat resistance of an adhesive bond is also determined by the force acting on the bond.

To determine this, test specimens are usually loaded with a defined weight in a heating oven at a constant temperature and the time after which the bond fails is determined.

Alternatively, the temperature is gradually increased until the bond fails. The highest temperature that the bond has withstood without damage is then specified as the heat resistance.

Open time

The open time is the period after the start of adhesive dispensing during which the adhesive can still sufficiently wet the second adherend. For hot-melt adhesives, the materials of the adherends, i.e. their thermal conductivity and the form of the adhesive application (ratio of quantity to surface) have a major influence.

To determine the open time, the hot-melt is applied to the adherend in accordance with the type (bead, film, etc.) and quantity intended for the application. Subsequently, paper or other defined test strips are inserted into the adhesive at regular intervals adapted to the expected open time. After the adhesive has cooled down to such an extent that the surface is tack-free, the test strips are peeled off. The open time is then the insertion time of the first test strip that separates adhesively from the adhesive.

Set time/setting time

The setting time is the time required to achieve a certain strength of the bond. Like the open time, this is also depending on the material of the adherends and the type of application of the hot-melt.

Two test specimens are bonded together in a T-shape specimen, loaded with a weight (e.g. 1 kg) after a certain time. The setting time is the time after which the test specimen withstands the load without deformation of the bond.

Green strength

This term refers to the ability of the hot-melt to withstand (low) forces in the period immediately after joining. The determination is similar to that of the setting time, but instead of the time after which the bond is loaded, the weight is varied.

Reactivation capability

Conditions under which the already applied and (cooled) adhesive can be converted back into a 'tacky' state by heating.

Thermal stability

To determine the influence of the adhesive's residence time in the melting tank, the hot-melt adhesive is heated under specified conditions. Samples are obtained at regular intervals and analyzed with regard to viscosity (acc. to ISO 2555, Brookfield [29]) and softening point (acc. to DIN EN 1238, ring-ball method [12]).

1.3.1.2 Classification by setting mechanism

Based on the processes that take place during the solidification, hot-melt adhesives can be classified into the following three groups.

Class A

Amorphous polymers with a glass transition temperature above room temperature possess high strength (high elastic modulus) at room temperature. However, they are quite brittle at room temperature and exhibit very slight elongation. Examples of such amorphous polymers are PVC, PS, PC and PMMA. For formulation as hot-melts, either they are compounded with another polymer of low glass transition temperature, or block copolymers (i.e. S-I-S, Chapter II-3.3.3.1) are used. On being heated, these polymers lose their strength as soon as the glass transition temperature is reached. Above the glass transition temperature, the polymer gradually liquefies. Close to the glass transition temperature, the viscosities of such polymers are quite high. Application, therefore, calls for high temperatures, significantly above the glass transition temperature, but care must be taken to avoid thermal damage. [1, 2]

Figure II-1.41 shows the viscosity-temperature dependency for a Class A polymer (amorphous). The glass transition temperature T_g is above room temperature. In order to achieve the viscosity reduction required for quality processing, temperatures significantly, i.e. approx. 50 to 100 °C above the glass transition temperature, are required.

Class B

Semi-crystalline polymers having a crystalline melting point T_m above room temperature are not brittle if the molecular weight is sufficiently high and if the glass transition temperature T_g is significantly below room temperature. When modifying such polymers for use as hot-melt adhesives, it should be considered that the crystallization process may be influenced or disturbed by additional ingredients. Examples of such crystalline or semi-crystalline polymers are EVA copolymers, and many polyesters and polyamides. When heated, these products retain their mechanical strength virtually unchanged until the crystalline melting point is reached. Above the melting point, the products liquefy. The viscosities of these polymers above the melting point are usually significantly lower, because the distance from the glass transition temperature is much greater.

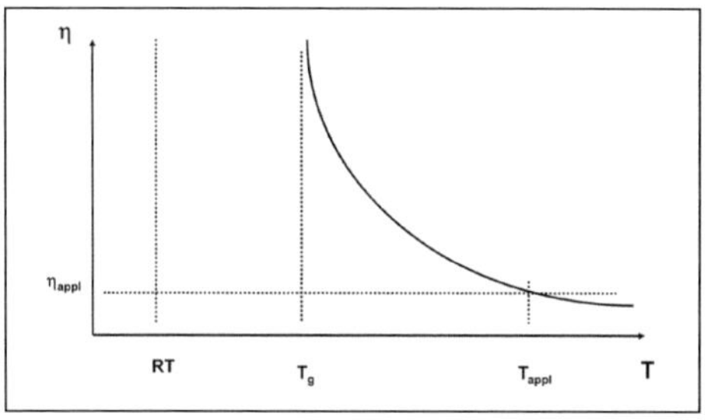

Figure II-1.41: Viscosity-temperature dependency for class A hot-melt

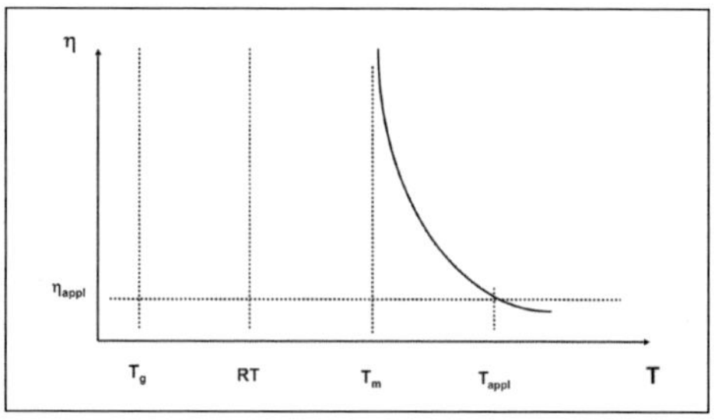

Figure II-1.42: Viscosity-temperature dependency for class B hot-melt

Figure II-1.42 shows the viscosity characteristics of a Class B hot-melt. The crystalline melting point T_m is above, the glass transition temperature T_g below room temperature (semi-crystalline). The viscosity of the melt (besides the molecular weight) is mainly determined by the temperature difference between the glass transition temperature T_g and the actual temperature. Since the glass transition temperature is far below the crystalline melting point T_m, temperatures which are with approx. 50 °C just moderately above the crystalline melting point T_m already allow a sufficiently low application viscosity.

Class C

High molecular weight polymers having a glass transition temperature T_g below room temperature are, strictly speaking, very high-viscosity liquids at room temperature, but they exhibit many signs of solids, such as a defined shape. An example of such polymer is butyl rubber (e.g. present in chewing gum). Due to their low cohesion, hot-melts based on these polymers can generally only serve as sealants and, and only where no really large loads have to be transferred (see Chapter II-3.3.3.2). Depending on the temperature, such polymers show a more or less pronounced cold flow, i.e. a tendency to irreversible deformation with or without load applied. When heated, they gradually lose their strength, which is already low, and behave more and more like liquids. In order to provide at least a certain strength at room temperature, hot-melt adhesives based on such polymers must have a high molar mass. This in turn requires an application temperature well above room temperature to achieve an adequate viscosity reduction.

Since they are characterized by a certain degree of permanent tackiness even in the solid state, they actually belong as pressure-sensitive adhesives to the group of adhesives without solidification. Even though they are dealt with in more detail in Chapter II-3.3.3.2, they should not remain unmentioned here.

Figure II-1.43 shows the situation for a Class C hot-melt. The amorphous polymer has a glass transition temperature T_g far below room temperature. To provide sufficient strength at room temperature, it must be a highly viscous liquid, which can be achieved by a correspondingly high molecular weight. According to the viscosity-temperature dependence shown in Figure II-1.44 for polymers of different molecular weight, a comparatively high application temperature is required.

1.3.1.3 Formulating principles

As already pointed out, hot-melts are so called 100 % adhesives, i.e. they don't contain processing auxiliaries, such as water or solvent, which would have to be removed during solidification. The following components with the functions listed are typically used for formulation where the percentage ranges given in parentheses are to be understood as reference values.

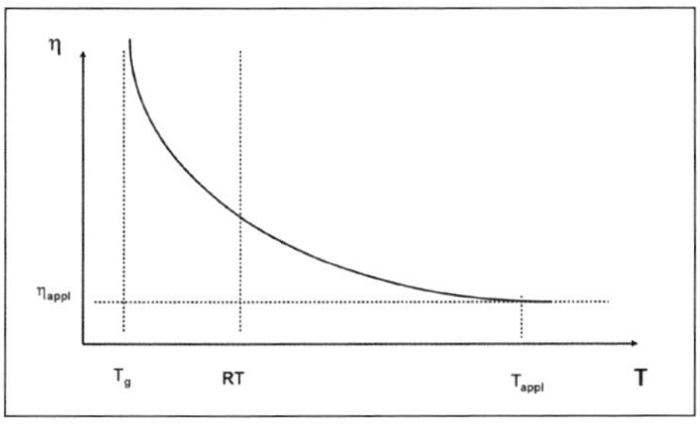

Figure II-1.43: Viscosity-temperature dependency for class C hot-melt

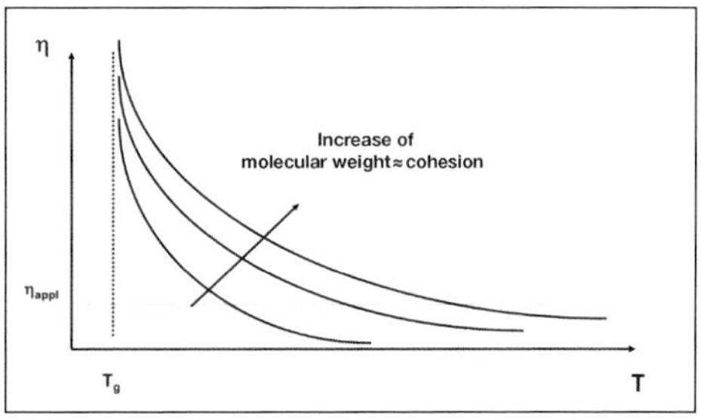

Figure II-1.44: Viscosity as a function of molecular weight for a class C hot-melt

- The **base polymer** (30 to 100 %) defines the main properties of the adhesive, such as its application conditions, adhesion, cohesion bond strength, chemical and hydrolytic stability, softening characteristics, setting mechanism and, of course, price. Polymer blends can be used, too.
- **Resins** (up to 40 %) are used to improve tack, adhesion, and heat resistance, as well as to lower the application viscosity. Compatibility with the polymer and any influence on the setting characteristics (T_g or T_m) must be considered.
- **Plasticizers and waxes** (up to 50 %) are added to lower the application viscosity, influence the setting behaviour, improve flexibility, and to reduce raw material costs. Besides liquid plasticizers, solid plasticizers are used as well, although the distinction between the latter and resins is often difficult to draw. When plasticizers are added, in addition to the desired effect of a certain improvement in elasticity, it must be taken into account that they can certainly have a negative effect on the glass transition temperature of amorphous polymers and on the crystallization properties of crystalline polymers.
- The purpose of **fillers** (up to 50 %) employed in large quantities is in most cases to lower costs. Many hot-melt adhesives are formulated without added fillers.
- **Pigments** are used in small percentages for colouring.
- **Additives** (usually <5 %), especially antioxidants to improve the stability of the adhesive against thermoxidative degradation under application conditions.
- **Rheological additives** and **adhesion promoters** may also be used.

1.3.1.4 Manufacture

Hot-melts based on commercially available polymers are usually manufactured in a discontinuous process by simply mixing the polymer with the other ingredients in the melt. If ingredients, such as resins, fillers or additives, are to be added, mixing should be performed according to the "high to low viscose" rule. For this purpose, only a partial quantity of the polymer is initially charged, and a highly viscous mixture is generated by adding the solid ingredients. Due to the high viscosity, these can be well dispersed avoiding agglomeration, provided that suitable mixers are used. The remaining polymer and the other viscosity-reducing ingredients like plasticizers, resins and waxes are then incorporated. For most types of hot-melt adhesives (e.g. EVA polymers), a heatable mixer with a simple stirrer is sufficient. Highly filled systems are often mixed in kneaders or internal mixers. A continuous manufacturing process by mixing the raw materials in an extruder is certainly feasible but is limited to high-volume products.

Some hot-melts based on polycondensation products, such as polyesters and polyamides where modification by additional ingredients is of secondary importance, are polymerized directly from the monomers under conditions of heat and vacuum.

1.3.1.5 Delivery form

With the exception of pressure-sensitive hot-melt adhesives (see Chapter II-3) and the so called reactive hot-melt adhesives that crosslink by a chemical reaction (see Chapter II-2.2.1.14), hot-melt adhesives do not necessarily require special packaging.

The delivery form is determined by the product properties but also by the subsequent processing method and the requirements of the application.

Depending on the product, they are supplied in cartridges, pails, kegs or drums, but also as granules or powders packed in bags. In addition, there are products in the form of rods, sticks, films, or nonwovens. Often anti-caking agents are added to prevent agglomeration.

1.3.1.6 Bonding process

The hot-melt adhesive must first be melted in order to allow dispensing, but also in order to be able to wet the adherends. Dispensing on one and joining with the second adherend must be completed before the hot-melt has cooled down and its viscosity has reached a state where a sufficient wetting of the second adherend is no longer possible. This time frame is relatively short and named "open time". The adhesive layer cools further, with the thermoplastic polymer of the adhesive solidifying at a rate dependent on the adhesive, the bond line thickness and the thermal conductivity of the adherends whereby adhesion and cohesion are developed. Typically, the bond reaches the strength required for further processing of the adhesively bonded part (initial (green-) strength, handling strength) within a very short time. Although it may possibly take several hours to reach the final bond strength, but the difference between initial (green-) and final strength is usually small (Figure II-1.39).

Dispensing and application

The dispensing and application system for hot-melt adhesive systems consist of different components. The main components include the melter. Here, the hot-melt adhesive, which is solid when delivered, is melted by applying heat and kept at application temperature, and then delivered to the application unit (e.g. nozzle, spray gun, roller, etc.) by means of a pump and via heated hoses.

When selecting the application system, various criteria must be taken into account, these are in particular:

- the size of the bonding area
- the number of parts to be bonded per unit of time and, as a result, the required degree of automation
- the specific properties and delivery form of the hot-melt adhesive used
- the type of application, e.g. spot, line or bead application, two-dimensional application, e.g. by spraying or coating in a wafer application or by means of a wide slot nozzle

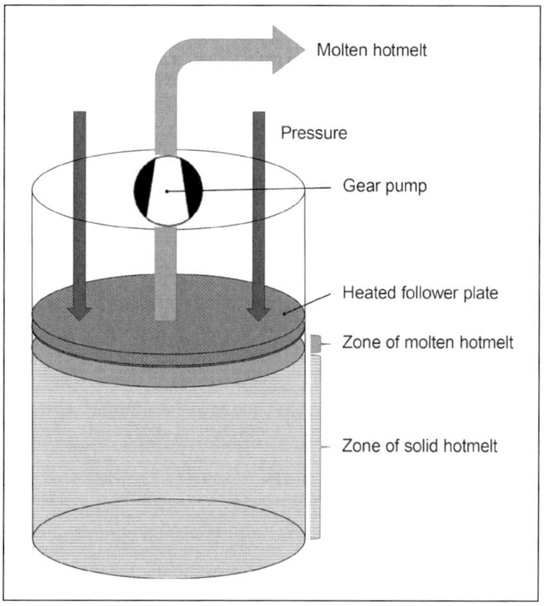

Figure II-1.45: Drum melter with heated follower plate

Hot-melt adhesives supplied in pails, drums, in the form of blocks or pellets are melted in tank-melters and stored in the molten state at application temperature. Obviously, volume and the heating capacity of the melters must be adjusted to the designated hot-melt consumption. The melters are designed to prevent or limit access of oxygen to the hot adhesive so that oxidation of the molten polymer can be avoided or reduced to a minimum. Closed tanks also reduce the environmental impact by odours from emitted fumes. A particularly gentle variant is the processing of the adhesive from drums or pails using a heated follower plate. Figure II-1.45 schematically shows a drum melter with heated follower plate. Here, only the top layer of the adhesive close to the heated follower plate is exposed to the high application temperature and melted. The remaining adhesive in the drum is not subjected to thermal stress.

The most common application methods for hot-melt adhesives are spot, bead or line application, spray application and flat application via slot nozzles or rollers.

In spot, bead or line application, the adhesive is dispensed for a specific time via a nozzle fitted with a valve. By moving the nozzle accordingly, the adhesive can be positioned on the areas of the component intended for bonding.

By means of slot nozzles, it is possible to apply precisely defined, wide and usually very thin films of hot-melt, so that this process can be used, in addition to the use of rollers for flat application, e.g. on web-shaped substrates (paper, foil, fabrics, etc.) but also on piece goods (glass, metal plates, etc.).

By spraying a continuous layer of hot-melt adhesive can be applied. Typically, areas with no hot-melt applied must be avoided to avoid defective areas. Areas with too much adhesive can, depending on the adherend, lead to the adhesive seeping, e.g. through a textile, onto the surface underneath. In the production of functional clothing, the presence of voids between the individual very small adhesive dots is deliberately utilized to maintain the water vapor permeability of the bonded substrates.

To achieve a constant and reproducible adhesive application, temperature control in the entire application system from the melting tank to the nozzle is of particular importance. Temperature fluctuations can lead to a change in the adhesive viscosity and thus to a change in the application pattern. Excessive temperatures, even if they occur only temporarily or locally, e.g. on the walls of the melting tank or the piping, can lead to thermal damage to the adhesive and even to degradation with the formation of coking. These particles can clog the nozzles or, if dispensed with the adhesive, will lead to visual deterioration of the bond.

The application temperature is of critical imprtance because it directly affects how long the hot-melt remains "open" after it has been applied to the adherend. If the temperature of the hot-melt is too low, the open time will be reduced, and the substrates might fail to adhere properly. Hot-melts usually have, depending on the polymer base and the specific formulation a short open time (in the order of several seconds or just a very few minutes). Joining the second adherend must therefore best be performed immediately after adhesive application. If, on the other hand, temperature is too high, the viscosity of the adhesive is reduced, which can lead to a penetration through the substrate, thermal decomposition of the hot-melt, a thermal damage of the adherends and a number of other problems.

A common alternative is thermal reactivation. One or in some cases also both adherends are coated with the adhesive in advance at an earlier time e.g. by the substrate manufacturer. Bonding can be performed at a later time simply by thermal reactivation. Thermal reactivation is normally used if hot-melt application is intended for bonding large flat substrates. The open time would be too short to fully cover one face of a large area with adhesive. Another alternative used e.g., for textile lamination is to use the already mention hot-melt adhesives delivered as powder, non-woven or film by using heated presses.

For low-volume applications, e.g. in handicraft or DIY, hot-melt adhesives either packaged in aluminium cartridges or available in the form of sticks or rods can be processed using heated hand-held dispense guns.

Solidification

The solidification of hot-melt adhesives, i.e. setting, occurs quite quickly, typically within seconds or a few minutes, depending on the thermal conductivity of the adherends, the quantity of adhesive applied, the polymer base and the respective formulation. As long as the effect of abrupt cooling of the hot-melt, e.g., on a cold metallic adherend, does not prevent wetting of the adherend, the adhesive will bond to the adherends within a very short time, resulting in a fast strength build and the substrates can often be further processed shortly after joining without the need for additional fixation.

When a melt of an amorphous adhesive **Class A (T_g > RT)** is cooled to below the glass transition temperature, solidification occurs almost immediately, without any delay. Such adhesives have only a very short open time but very quickly develop a high initial (green-) strength. Solidification at glass transition temperature occurs spontaneously.

Semi-crystalline **Class B** adhesives **(T_m > RT)** show similar cooling characteristics under similar conditions; however, the solidification process, namely crystallization, is exothermic and requires some time, because the polymer chains have to orient themselves in order to build up the ordered crystalline structure. Semi-crystalline hot-melts therefore need some time to solidify. Because the crystallization process does not always occur in the same way, industrial production lines with crystalline hot-melt adhesives must have a greater tolerance with regard to the solidification speed. On the other hand, crystalline adhesives with a low melting point offer the advantage of a longer open time compared to a Class A product. Due to the delayed crystallization, the adhesive can still be liquid even after cooling below its crystalline melting point

Amorphous, amorphous **Class C** adhesives **(T_g << RT, high molecular weight)** show only a viscosity change upon cooling, which is proportional to the temperature change. This viscosity change occurs concurrently with cooling, but the mechanical strength always remains low.

1.3.1.7 Evaluation – bond properties

Like every adhesive technology and, for that matter, every joining process, hot-melt adhesives also have specific advantages and disadvantages. Tables II-1.27 and Table II-1.28 summarizing the general advantages and disadvantages of hot-melts. However, it should be kept in mind that the properties can differ significantly from one type of polymer to another, and even within a single polymer type. This can be seen, for example, both in the softening temperatures that determine the temperature resistance of the bond and in the required application temperatures (Table II-1.29).

The disadvantages resulting from the thermoplastic state were largely eliminated by reactive hot-melts developed in the early 1980s. In these systems, discussed in Chapter II-2.2.1.14 of this book, rapid solidification by cooling is followed by a slow chemical crosslinking under the influence of humidity.

Figure II-1.46 summarizes, without claiming to be complete, the main criteria to be considered when selecting or developing a hot-melt adhesive for a specific application. In many cases, properties can be adapted to the requirements by appropriate selection of the polymer type and intelligent formulation. In some cases, however, requirements from the bonding process contradict those from the bonding requirements. In such cases, it is necessary to make compromises.

1.3.1.8 Fields of application

The group of hot-melt adhesives accounts currently for a market share of 15 – 20 %. It is expected that the steady increase observed over the last years will continue in the years to come. (Figure I-1.11) They are used in many industries, often due to their fast solidification and thus the possi-

Table II-1.27: Advantages of hot-melt adhesives

Fast strength build High initial strength	– setting times in the range of seconds – short cycle times – in many cases, temporary fixation is not required – can be used as a fixation for bonding, especially with 2-component adhesives
100 % solids system	– hardening does not require the evaporation of a solvent or water, – no need for space-intensive drying sections – not classified as hazardous material
Single component	– no dosing and mixing of components
Suitable for both automated and manual application	– delivery form and application technology must match
Gap filling	– depending on viscosity at application temperature
Thermoplastic	– debondability Bonded parts ready for circular economy – bonding of block-free precoated parts by the heat-sealing process
Almost unlimited shelf life	– however, the rules of quality assurance require the limitation of the shell life
Typically, not classified as hazardous material (e.g. solvents, monomers)	– only minimal environmental and occupational safety requirements – nevertheless, harmful vapours can be emitted from the molten adhesive. – products for direct and indirect food contact possible
Material costs	– depending on the base polymer

Table II-1.28: Disadvantages of hot-melt adhesives

Thermoplastic	– limited cohesive strength – limited heat resistance – limited resistance against solvents – tendency to creep under continuous static load
Application in a molten form	– risk of damaging temperature sensitive adherends – heated application system with temperature control required – risk of thermal degradation of the adhesive – risk of burns on contact with molten adhesive – harmful fumes can be emitted from the molten adhesive
Short open time	– requires a correspondingly adapted process
Limited strength level	– depending on polymer type and formulation
moderate adhesion	– poor wetting of materials, especial to those with high thermal conductivity like metals due to rapid cooling of the adhesive surface on first contact with the adherend – adhesion can be increased by pre-warming the adherend which will also extend open time but on the other hand will slow down strength build
Shrinkage upon solidification	– formation of tension in the bond

bility to realize short cycle times. Furthermore, they are generally not subject to labelling requirements, so that occupational safety measures are basically limited to the avoidance of burns and, if necessary, local exhaust ventilation of vapours.

In the **packaging industry,** paper, cardboard and similar substrates of low cohesion often need to be joined quickly with low-cost products in automated processes. The need to withstand high loads under extreme conditions is mostly of minor importance. On the other hand, fast bonding is essential to allow very short cycle times. These are ideal conditions for hot-melt adhesives. Hot-melts can often even be used for food packaging due to the absence of harmful additives in the adhesive formulation. The most important hot-melts for packaging applications are based on copolymers of ethylene and vinyl acetate, of propylene, and of styrene.

*Table II-1.29: Typical processing parameters for selected **hot-melts** [10, 19]*

Base polymer	Softening range [°C]	Processing temperature [°C]
Ethylene vinyl acetate copolymers	45 – 90	130 – 200
Polyolefin	85 – 160	120 – 200
Polyester	50 – 230	150 – 240
Polyamide	95 – 195	120 – 240
Polyurethane	65 – 185	90 – 200

In the **hygiene industry**, the manufacture of disposable diapers, sanitary napkins and similar products involves the use of hot-melts. They also make laminates and resealable closures possible. The vast quantities involved are an ideal field of application for hot-melts, which are usually tolerated well by the skin. Hot-melt pressure-sensitive adhesives based on styrene block-copolymers are very common (see Chapter II-2.2.1.14).

In the **book-binding industry,** bonding the backs of paperback books is a major application of hot-melts. Again, the high volumes and fast strength build requirements are the determining factor in the selection of hot-melts. Standard hot-melts for book-binding applications are usually based on EVA copolymers. However, these show only limited heat resistance, which is why reactive hot-melts, which crosslink by reacting with atmospheric moisture (see Chapter II-2.2.1), are increasingly being used for high-quality books.

In the **textiles industry,** many garments (e.g. suits) are lined with non-wovens to improve the appearance (broad shoulders) and wearing comfort. Specific applications are to be found in male and female outer-door wear, collars for shirts, wristbands, adhesively bonded labels, stitching etc.

Delivery forms

- Volume of application
- Geometry of application
- Type of applicator
- Packaging
- Price

Application of the melt

- Temperature
- Viscosity
- Stability under application conditions
- Geometry of application
- Damage of substrates
- Open time
- Wetting

Setting characteristics

- Setting time
- Setting mechanism

Bonding properties

- Bond strength
- Adhesion
- Heat resistance
- Resistance to hydrolysis etc.

Figure II-1.46: Criteria to be considered when selecting or developing a hot-melt

These laminates are manufactured with high-performance hot-melts (polyesters or polyamides). Application is normally a continuous process, either on large gravure rolls or entailing hot-melt powders with subsequent thermal reactivation leading to point-shaped bonds. Full-face bonding is less desirable because it would impair the wearing comfort of the corresponding garment. The fast-setting process again renders hot-melts ideal for use in automated processes. High-performance hot-melts are also used to make multi-layer laminates for car-seat covers. In special instances, an adhesive non-woven or an adhesive film can be incorporated directly between the substrates using a thermal reactivation process. In domestic households, hot-melt applications can be found in the form of iron-on pictures or patches – this is a typical heat activation process.

The solubility of thermoplastics, e.g. in the solvents used in dry cleaning, can have a negative effect. Here, too, the reactive hot-melts discussed in Chapter II-2.2.1.14 offer a good alternative.

In the **automotive industry,** hot-melts are used for example in interior trim applications were the requirements for thermal resistance are not too high. Where the heat resistance of conventional hot-melts is not sufficient, the relatively new group of reactive hot-melts (see Chapter II-2.2.1.14) have partially replaced water-based dispersion or solvent-based systems that were most commonly used in the past.

Polyamide based hot-melts are used due to their high heat resistance for so called under the hood applications, like encapsulating of electrical connectors as well as for endcap mounting of air and oil filters.

Class C hot-melts (T_g > RT<T_{appl}) typically based on butyl rubber are used both in the form of prefabricated moulded parts and in the molten state for various sealing applications, e.g. in the car body. Another application for class C hot-melts is the mounting of a polyethylene film to the inner side of the doors to protect the backside of the decorative inner door panel from water that inevitably penetrates into the door interior through the side windows seal.

Hot-melts are also used in the **woodworking and furniture industries**. Even though their bond strength may be inferior to that of water-based wood adhesives, hot-melts based on polyamides, EVA or PP are used for edge banding. If the edge veneer is already precoated with the adhesive, it can be mounted at any desired time by the combination of heat and pressure (e.g. iron or industrial ironing machine). Hot-melts are also used as a fixturing aid making use of the rapid-setting.

In the **shoe industry** for example, hot-melts based on polyester or polyamide are used for the permanent attachment of the upper shaft to the insole of shoes. Polyesters usually set faster and yield a higher initial (green-) strength, whereas polyamides are somewhat more flexible. Again, fast setting and high initial (green-) strength enables the process to be automated for rapid mass production. Laminates of non-woven with leather or polyamide (sports shoes) are also made with the help of hot-melt powders based on polyester, polyamide or EVA.

In the **construction industry**, the use of hot-melt adhesives directly on the construction site is limited to simple fixing tasks using hot-melt sticks and light handguns because of the usually inconvenient use of appropriate melting equipment. In addition, class C hot-melts are used in the form of prefabricated adhesive tapes, e.g. for sealing roofing membranes.

In the manufacture of insulating glass panes, class C hot-melts based on butyl are used as moisture-impermeable primary seals between the spacer and the glass panes.

In the **DIY sector**, the use of hot-melt sticks, mostly EVA-based, in combination with simple electrically heated glue guns is quite common for a number of different applications. Hardware stores usually offer a pretty good selection of different glue sticks and the guns to go with them.

1.3.2 Chemistry and formulation of hot-melt adhesives

It is obvious that when formulating a hot-melt adhesive, a base polymer should be chosen that provides good adhesion to the substrates being bonded. A low melt viscosity, good thermal stability of the melted adhesive, a solidification speed adapted to the respective bonding process, a softening temperature adapted to the respective application and resistance to moisture are important criteria to be considered. A wide range of thermoplastic polymers can be used for the manufacturing of hot-melt adhesives. Each polymer has specific properties and advantages which are together with their market relevance briefly summarized below [11]. A direct comparison of the different base polymers is difficult due to the many formulation options. Properties that are important for the suitability of an adhesive for its intended use, such as melt stability, melt viscosity, adhesion, shear and peel strength, are not solely determined by the base polymer, but can also be at least partly influenced by the formulation.

More detailed information on the formulation is provided in the following subchapters.

Hot-melts based on **copolymers of ethylene and vinyl acetate (EVA)** offer a favourable price-performance ratio. Therefore, it is not a surprise, that they are with a market share of about 50 % the most widely used group of hot-melts. They are mainly used for packaging application like carton sealing, for bookbinding, edge banding in the furniture production, in the shoe industry and as heat seal coatings.

Hot-melts based on **polyolefins**, accounting for around 10 to 15 % of the market are not as important as EVA hot-melt adhesives but are becoming increasingly popular due to their interesting properties.

Hot-melts based on **polyesters**, typically based on terephthalic acid or isophthalic acid and various diols like hexane diol are characterised by a high chemical resistance which is why they are used besides others for manufacture of fuel filters in the car industry and for textile lamination.

Hot-melts based on **polyamides** are polycondensation product of dicarboxylic acids or dimer fatty acids with diamines. In many cases the unformulated polyamide already offers excellent hot-melt properties at a narrow molecular weight distribution and show compared to EVA hot-melts a much wider adhesion range. The very good temperature resistance, some grades allowing bonds exposed to up to 140 °C and at the same time outstanding cold flexibility, down to -40 °C makes them ideal for under the hood applications in the automotive industry. Due to the good chemical resistance of special polyamides, they are also used for bonding of fabrics exposed to oily environments and for bonding of plasticized plastics. Additionally, the flame-retarding properties can be utilised.

On the other hand, the high water absorption capacity of polyamides and the high processing temperatures require appropriate measures. Foaming in the melting tank can be prevented by appropriate packaging or storage at very low humidity. In order to minimize oxidation and, in extreme cases, coking of the hot-melt in the melting tank, covering with dry nitrogen or carbon dioxide is recommended.

Hot-melts based on **thermoplastic, non-reactive polyurethanes** are only of secondary importance. This is related to the high required melt temperatures, at which the polymer structure is already undergoing some thermal degradation reactions. In converted form, they are used as films, e.g. for flame or hot lamination. The by far more important group of reactive polyurethane hot-melts, discussed in Chapter II-2.2.1.14.

100 % solids adhesives based on **butyl rubber** (copolymers of isobutene and perhaps isoprene), **synthetic rubber** (block copolymers of styrene and butadiene or isoprene) and **acrylates** are typically pressure-sensitive adhesives and are covered in Chapter II.3. Nevertheless they

should not remain unmentioned here, since individual products are heated for application in certain applications.

Hot-melts based on **butyls** are usually plastic, permanently tacky sealants of low cohesion and poor heat resistance. They are used in the insulating glass industry and the automotive industries primarily as sealants. (see Chapter II-3).

Hot-melts based on **synthetic rubber** SBS, SIS (block copolymers of styrene and butadiene or isoprene) are permanently tacky adhesives (pressure-sensitives). Their primary applications are in the field of labels, adhesive tapes and packaging and are covered in Chapter II-3.3.3.2.

Hot-melts based on **acrylates** have limited applications. The radiation-curing products for pressure-sensitive applications are discusses in Chapter II-3.3.4.2.

1.3.2.1 EVA hot-melt adhesives

The main raw materials used in the formulation of EVA hot-melts besides the copolymer of ethylene and vinyl acetate are tackifier resins, waxes, fillers, and additives, especially antioxidants.

EVA Polymer

The copolymers of ethylene and vinyl acetate (Figure II-1.47), which give this group of hot-melts its name, are made by a free-radical polymerization under pressure of ethylene and vinyl acetate and are available from several suppliers like Dow ("Elvax"), Celanese ("Ateva") and ExxonMobil Chemical ("Escorene"), typically supplied as pellets. In order to avoid thermally induced degradation reactions during formulation, a small amount of an antioxidant, e.g. butoxylated hydroxytoluene (BHT), is added to commercially available EVA copolymers during their manufacture.

The properties of the polymer are primarily determined by its molecular weight, including the degree of branching and the ratio in which the two monomers are contained in the monomer and are mainly characterized by the following three parameters that influence the processing and bonding properties of the hot-melt.

Vinyl acetate content

The polymers are usually divided into three categories. Up to approx. 20 % vinyl acetate (VA) is referred to as low VA content, up to approx. 30 % as medium VA content and above this as high VA content. As the VA content in the copolymer increases, the ability of the polyethylene fractions to crystallise is hampered. This increases the flexibility, transparency, and solubility of the polymers. At the same time, the polar vinyl acetate increases the polarity of the polymer, which generally leads to better wetting and thus better adhesion of polar substrates. For formulating hot-melt adhesives copolymers with a VA content of 15 % – 30 % are preferred. To achieve special properties, EVA grades with lower or higher VA contents can and are used as well.

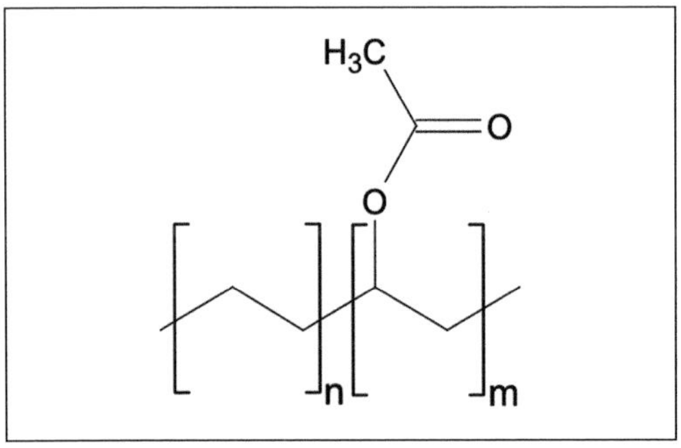

Figure II-1.47: EVA Copolymer

Melting and softening point/melt flow index

While the melting or softening point of the EVA type used essentially determines the temperature application range of the resulting bonded joint, the MFI has a direct influence on the application conditions of the hot-melt adhesive. Both values are proportional to the molecular weight of the EVA copolymer (see Figure II-1.42).

Table II-1.30 lists some of the EVA grades from the DOW Chemical Company product range [8] that are suitable for the production of hot-melt adhesives. The table shows, illustrated by Figure II-1.48, the correlation between VA content and melting point. The different melting points occurring for products with the same VA content are due to correspondingly different molecular weights.

With regard to both melting point and melt flow index, it must be considered that the addition of tackifier resins and/or wax leads to a decrease in the melting or softening point and an increase in the melt flow index (corresponding to a decrease in viscosity). As a guide, it can be assumed that adhesive joints made with EVA hot-melts will not withstand temperatures above 60 °C where a mechanical load acts simultaneously.

Table II-1.30: Selected commercial EVA copolymers [8]

EVA Grade	**VA-content [wt %]**	**MFI [g/10 min]**	**DSC melt point [°C]**
"ELVAX 750"	9	7	98
"ELVAX 660"	12	2.5	96
"ELVAX 550"	15	8	85
" ELAVAX 410"	18	500	73
"ELAVAX 420"	18	150	73
"ELAVAX 450"	18	8	86
"ELAVAX 460"	18	2.5	88
"ELAVAX 470"	18	0.7	89
"ELAVAX 350"	25	19	74
"ELAVAX 360"	25	2	78
"ELAVAX 210W"	28	400	60
"ELAVAX 220W"	28	150	70
"ELAVAX 240W"	28	43	74
"ELAVAX 250"	28	25	70
"ELAVAX 260"	28	6	75
"ELAVAX 265"	28	3	73
"ELAVAX 150"	32	43	63

In many cases, commercially available EVA copolymers are formulated for their use as hot-melt adhesives with tackifier resins, waxes, fillers, and additives, especially antioxidants. Here, as in the selection of the EVA type, a detailed knowledge of the intended application is required. It

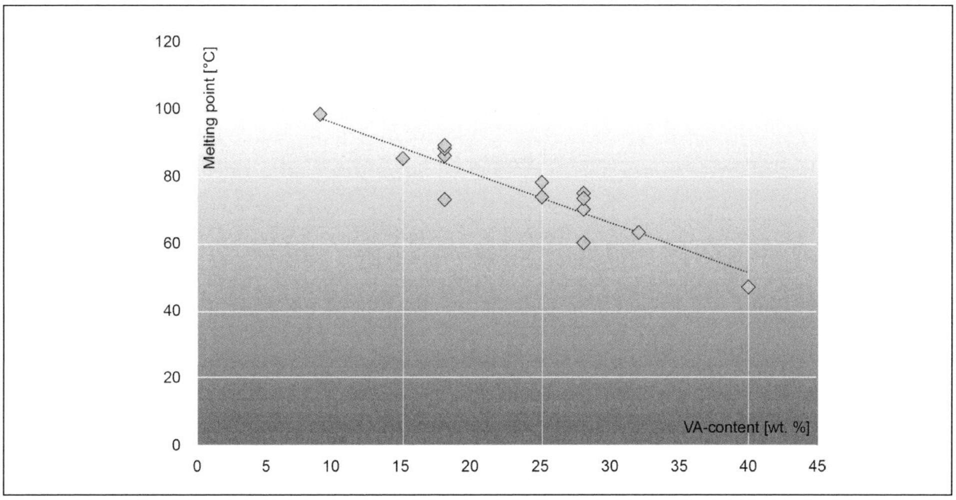

Figure II-1.48: Melting point vs. VA-content for the Dow EVA grades listed in Table II-1.30

should not go unmentioned that blends of two EVA polymers, can offer processing advantages. Blending a high-molecular and a low-molecular EVA type can lead to good mechanical properties combined with low melt viscosity for ease of processing.

This includes the intended processing method, a definition of the substrates and the thermal and mechanical requirements imposed on the joint. This information is used to establish the main criteria concerning the VA content, MFI and melting point of the optimum EVA copolymer as well the selection of additional ingredients.

Resins

Resins [3, 26, 27] are usually added to improve adhesion and tack (tackifier resin), especially during the bonding process when joining the substrates and the adhesive is in a molten state (hot tack). Due to their low molecular weight, the resins cause a reduction in viscosity, thus facilitating the application of the adhesive and leading to better wetting and adhesion. Typical addition rates are 30 – 40 % based on the total formulation.

Chemically, the most important resins are petrochemical products based either on aliphatic C_5 hydrocarbons (e.g. oligomers of isoprene, cyclopentadiene, isobutylene, 1,3-pentadiene etc.) or on aromatic C_9 fractions (e.g. oligomers of α-methylstyrene, coumarone-indene, vinyl toluene). Resins based on natural raw materials, such as colophony (rosin, chemically: mainly abietic acid) and their derivatives (hydrogenated, oligomerised, esters) or polyterpenes are used, too.

The main criteria governing the choice of resins are compatibility with the EVA copolymer, the influence on adhesion, melting or softening point, processing viscosity, open time, colour, smell, temperature stability, sensitivity to oxygen (due to carbon-carbon double bonds) and obviously also price.

Waxes

Waxes [4, 27] are typically added in a proportion of 20 to 30 %. They are mainly used to reduce costs and viscosity. Chemically, waxes are hydrocarbons of low molecular weight.

They are classified as follows:

Paraffin waxes: Distillates obtained from the petroleum industry which are mainly linear aliphatic hydrocarbons with melting points in the range of 30 to 70 °C and molecular weights between 200 and 500 g/mol.

Microcrystalline waxes: They are obtained from the residues from petrochemical distillation. Compared with paraffin waxes, they have a higher molecular weight and are more strongly crosslinked. Their melting points are usually in the range of 60 – 95 °C.

From a chemical viewpoint, **synthetic waxes** or **polyethylene waxes** are just low-molecular polyethylenes with melting points ranging from 90 to 115 °C.

Fillers

The primary intention behind adding fillers e.g. chalk, barytes, gypsum, talk and silicates is to reduce the raw material cost. This cost advantage is partially counterbalanced due to the increase in volume-based price caused by the increase in density. It must also be taken into account that fillers can affect the rheology of the melt and the mechanical properties of the adhesive. Nevertheless, fillers are used up to 50 % based on the total formulation. For more information about filler please refer to Chapter IV-1.4.

Additives

Additives are usually functional raw materials added in small quantities, such as antioxidants in the case of hot-melt adhesives. Their task is to protect the polymer against thermo-oxidative degradation due to exposure to high temperature especially during the processing step and thus to increase the lifetime in the melting tank. Sterically hindered primary phenolic stabilizers are used in particular.

In some cases, organo-functional silanes are added to improve adhesion to certain substrates [6] or fumed silica is added to modify the rheology and reduce stringing [7].

Table II-1.31: General purpose EVA hot-melt [13]

Item	Raw material	Parts by weight
1	EVA Copolymer (28 % VA, MFI 150)	30.0
2	Resin, thermoplastic pentaerythritol (PE) rosin ester	34.5
3	Paraffin wax (60 °C m.p.)	20.0
4	Polyethylene homopolymer, softening point 108 °C	15.0
5	Sterically hindered primary phenolic antioxidant	0.5

Item 1: "Elvax 220W", DOW
Item 2: "Permalyn 3100", Neville Chemical
Item 3: "HYMAX 4117", Hymax GmbH
Item 4: "Epolene N-14", Westlake Chemical
Item 5: "Irganox 1010" BASF SE

Processing

EVA polymers and their additives are compounded at elevated temperature typically in a melt mixing process using an anchor stirrer or similar. Air (oxygen) should be excluded to avoid degradation effects. The mixing time and temperature are critical and should be kept to a minimum. In no case should the melt temperature be allowed to rise above 220 °C as the polymer will degrade very fast.

Table II-1.32: EVA hot-melt for book-binding [13]

Item	Raw material	Parts by weight
1	EVA Copolymer (28 % VA, MFI 25)	30.0
2	EVA Copolymer (28 % VA, MFI 6)	15.0
3	Tackifier, aromatic modified aliphatic hydrocarbon resin, softening point 124 °C	25.0
4	Aromatic modified aliphatic hydrocarbon resin, softening point 92 °C	15.0
5	Plasticizer, phthalate-free	5.0
6	Microcrystalline wax (84 °C m.p.)	10.0
7	Sterically hindered primary phenolic antioxidant	0.5

Item 1: "Elvax 250", DOW
Item 2: "Elvax 260", DOW
Item 3: "Escorez 2101", ExxonMobile
Item 4: "Escorez 5320", ExxonMobile
Item 5: "Mesamoll", Lanxess
Item 7: "Irganox 1010" BASF SE

Guide formulations

Quantities of about one-third polymer, resin and wax each as used for the starting point formulation for a **general-purpose hot-melt** provided with Table II-1.31 [13] are typical for EVA hot-melt formulations. The adhesive is characterized by a melting temperature of 97 °C resulting in a viscosity of 1400 cps at 150 °C and a shear adhesion failure temperature of 65 °C. The open time is, approximately 3 minutes.

Table II-1.32 presents an EVA hot-melt for **book-binding** applications [13]. The hot-melt adhesive is characterized by a viscosity of approx. 12,000 cps at 175 °C and, in addition by a good adhesion to paper. The elongation at break of approx. 500 % with an embrittlement tempera-

Table II-1.33: EVA hot-melt for carton sealing [13, 15]

Item	Raw material	Parts by weight
1	EVA Copolymer (28 % VA, MFI 25)	44.5
2	Microcrystalline wax (84 °C m.p.)	30
3	Tackifier, glycerol ester of partially hydrogenated wood rosin	25
4	Sterically hindered primary phenolic antioxidant	0.5

Item 1: "Elvax 250", DOW
Item 3: "Staybelite Ester 10", Les Dérives Résiniques Et Terpéniques (DRT)
Item 4: "Irganox 1010" BASF SE

Table II-1.34: EVA hot-melt for wood and plastic veneer [13, 16]

Item	Raw material	Parts by weight
1	EVA Copolymer (32 % VA, MFI 43)	15.0
2	EVA Copolymer (28 % VA, MFI 400)	10.0
3	EVA Copolymer (28 % VA, MFI 25)	14.0
4	Tackifier, hydrogenated hydrocarbon resin,	5.5
5	Tackifier, thermoplastic aromatic resin based on styrene, coumarone and indene, softening point 98 °C	5.0
6	Tackifier, glycerol ester of highly hydrogenated wood rosin	8.0
7	Filler, barite ($BaSO_4$)	41.5
8	Titanium dioxide	0.8
9	Sterically hindered primary phenolic antioxidant	0.2

Item 1: "Elvax 150", DOW
Item 2: "Elvax 210W", DOW
Item 3: "Elvax 250", DOW
Item 4: "NOVARES TN 100 - TN 160" (products with graduated softening point up to 160 °C), Rain carbon Inc.
Item 5: "Hikotack V90", Kolon Industries
Item 6: "Foral 85" Les Dérives Résiniques Et Terpéniques (DRT)
Item 7: "Barytmehl N", Sachtleben Minerals
Item 8: "Kronos 2220" Kronos B.V.
Item 9: "Irganox 1010" BASF SE

ture of below 0 °C, offers the high flexibility over a wide temperature range that is required for bookbinding applications. When a book is bound using an adhesive, the spine of the prepared paper block is coated with the adhesive and then bonded to the book cover. In this process, the adhesive must have a sufficiently long open time but should then solidify as fast as possible to allow the shortest possible fixing times.

Table II-1.33 provides an example for a hot-melt for industrial **packaging applications** [13, 15]. This application requires a low melt viscosity, a high tack, a very fast setting time, good adhesion to the relevant materials at a low material price. Therefore a PE-wax is used to reduce viscosity and regulate the setting time and softening characteristics. This formulation is also recommended for the use in a heat-sealing process where the hot-melt adhesive is precoated to the substrates and thermally reactivated during the joining process. The adhesive should be easy to reactivate and not be tacky in the cold state.

Edge veneering and **edge banding** constitute a primary market for hot-melt adhesives based on EVA. Table II-1.34 provides a starting point formulation for an EVA hot-melt suitable to bond both wood and plastic veneer [13, 16].

Table II-1.35 is providing a starting point formulation for an **EVA glue stick hot-melt** [13, 14]. The required properties can also be achieved, as shown by the formulation provided with Table II-1.36 by the addition of a rheology modifier like fumed silica [7]. This hot-melt does not exhibit any stringing and will not run-down vertical surfaces.

1.3.2.2 Polyolefin hot-melts adhesives

While olefinic hot-melt adhesives have long had the reputation of being limited to applications with very low performance requirements, they now represent an important class of hot-melt adhesives with a market share of about 10 – 15 %. A comparative study conducted by Westlake Chemical [18] of hot-melt formulations based on its "Epolene" polyethylene polymers with two EVA polymers shows, they can compete with EVA hot-melt adhesives shows (see Table II-1.37).

Table II-1.35: EVA hot-melt glue stick [13, 14]

Item	Raw material	Parts by weight
1	EVA Copolymer (28 % VA, MFI 400)	26.5
2	EVA Copolymer (25 % VA, MFI 19)	23.4
3	Tackifier, Styrenated terpene resin	35.0
4	Plasticizer, cyclohexane dimethanol dibenzoate	15.0
5	Sterically hindered primary phenolic antioxidant	0.1

Item 1: "Elvax 210W", DOW
Item 2: "Elvax 350", DOW
Item 3: "Styrenated terpene resin", Foreverest Resources Ltd.
Item 4: "Benzoflex 352" Eastman
Item 5: "Irganox 1010" BASF SE

Table II-1.36: Shear thinning EVA hot-melt glue stick [7]

Item	Raw material	Parts by weight
1	EVA Copolymer (28 % VA, MFI 400)	56.4
2	EVA Copolymer (32 % VA, MFI 43)	9.4
3	Tackifier, hydrocarbon resin, softening point 90 °C	28.2
4	Rheology modifier	5.9
5	Sterically hindered primary phenolic antioxidant	0.1

Item 1: "Elvax 210W", DOW
Item 2: "Elvax 150", DOW
Item 3: "Piccotex LC", Synthomer
Item 4: "Aersil 200" Evonik
Item 5: "Irganox 1010" BASF SE

Raw materials

Polyethylene like the highly crosslinked wax or the maleic anhydride modified form both mentioned, in Table II-1.37 allow the formulation of hot-melt adhesives with properties similar or beneficial to EVA based hot-melt adhesives.

Polypropylene can only be used for hot-melt adhesives in its atactic, irregular form (APP). APP is a low-cost by-product from the synthesis of isotactic, crystalline polypropylene (IPP). It is mainly amorphous, tacky, soft and has a low molecular weight of about 20,000 g/mol. Due to these properties, hot-melt applications based on APP are limited to low-end demands, where high strength is not required. Polypropylene and corresponding formulations have a tendency to embrittle at low temperatures (T_g approx. -15 °C).

Commonly used polyolefin raw materials used for the formulation of hot-melt adhesives are **conventional copolymers or terpolymers** polymerized from ethene, propene and other 1-alkenes polymerized by Ziegler-Natta catalysis or the more modern so called **metallocene polyolefin** ones made by metallocene catalysis. Metallocene catalysis (available e.g. from Clariant, ExxonMobile and Lyondellbasell) allows a better control of the molecular weight, branching and polymer architecture, especially the tacticity. A direct consequence of the narrow molecular weight distribution is a significantly reduced content of low-molecular-weight components, which to some extend are volatile and extractable resulting in good organoleptic properties. Compared to EVA-based hot-melt adhesives. The main advantages over conventional polyolefin and also EVA hot-melt adhe-

Table II-1.37: Comparison of hot-melt adhesive formulations based on polyethylene polymers versus two EVA polymers [18]

Item	Raw material	Parts by weight			
1	Highly branched polyethylene wax	72	–	–	–
2	Low-density polyethylene wax modified with maleic anhydride	–	72	–	–
3	EVA (18 % VA, 150 melt index) %	–	–	35	–
4	EVA (28 % VA, 150 melt index) %	–	–	–	35
5	C_5-hydrogenated hydrocarbon tackifier resin, m.p. 130 °C	22	22	45	45
6	medium density polyethylene	6	6		
7	Synthetic Fischer-Tropsch wax, drop melting point 112 °C	–	–	10	10
8	Paraffin wax (m.p. 152 F	–	–	10	10
9	Sterically hindered primary phenolic antioxidant	0.45	0.45	0.45	0.45

Item 1: "Epolene C-15 Polymer", Westlake Chemical
Item 2: "Epolene C-18 Polymer", Westlake Chemical
Item 3: grade and supplier not disclosed
Item 4: grade and supplier not disclosed
Item 5: "Eastotac H130 E" resin, Sythomer
Item 6: "Epolene N-21 Polymer", Westlake Chemical
Item 7: "Paraflint H-1", Sasol
Item 8: grade and supplier not disclosed
Item 9: "Irganox 1010" BASF SE

Resulting properties				
Viscosity @ 177 °C [cp]	1200	1200	1500	1550
Softening point [°C]	106	107	105	108
Set time [sec]	3.8	3.3	3.6	4.4
T_g [°C]	-23	-23	-12	-16
Viscosity change (100 h @ 177 °C) [%]	7	15	30	40

sives is an improved resistance to high application temperatures which allows longer resident times in the melting tank.

Also, **amorphous poly-α-olefins (APAO or APO)** (available e.g. from Evonik, Rextac, Synthomer) are used. They are atactic co- and terpolymers specifically produced from ethene, propene and 1-butene in a Ziegler low-pressure polymerization.

In the formulation of polyolefin hot-melt adhesives, similar to EVA hot-melt adhesives, tackifying resins, viscosity modifiers, antioxidants and, in some formulations, small portions of EVA and fillers are added. To achieve the desired properties, it is quite common to use blends of different polyolefins.

Processing

Same as for EVA hot-melt adhesives the liquid ingredients are compounded at elevated temperature typically in a melt mixing process. The solid tackfying resins and other additives are then added and mixed until completely blended.

Air (oxygen) should be excluded to avoid degradation effects. The mixing time and temperature are critical and should be kept to a minimum.

Guide formulations

Table II-1.38 provides a guide formulation for a hot-melt based on LDPE for packaging applications.

The formulation provided with Table II-1.39 gives the formulation of a hot-melt adhesive based on a metallocene catalyzed polyolefin [17] with compared to other such hot-melt adhesives containing besides a low level of viscosity modifiers which are liquid at room temperature a special polyethylene wax, that is partially incompatible with the metallocene polymer. Compared to other polyolefin hot-melt adhesives it does not show the very negative phenomenon known as bleed-through when used to bond fibrous or perforated substrates as used e.g. for the manufacture of absorbent hygienic articles.

Table II-1.40 shows a hot-melt adhesive based on propylene-based olefin polymers and a hydrocarbon tackifying resin for the manufacturing of mattresses. It offers compared to common EVA-based hot-melt adhesives a reduced odour, shows very good adhesion to the materials involved and short open times allowing short cycle times.

The so-called APAO hot-melt adhesives based on **amorphous poly(α) olefins** have gained market share in recent years. This is to some extend related to their special affinity to polyolefin plastics like polyethylene and poly-

Table II-1.38: Hot-melt based on LDPE for packaging industry [28]

Item	Raw material	Parts by weight
1	Branched LDPE	70
2	C_5-hydrogenated hydrocarbon tackifier resin, m.p. 130 °C	30
3	Sterically hindered primary phenolic antioxidant	0.15
4	Thioester-based antioxidant,	0.15

Item 1: "Epolene C-15 Polymer", Westlake Chemical
Item 2: "Eastotac H130 E" resin, Synthomer
Item 3: "Irganox 1010", BASF
Item 4: "Cyanox 1212", Solvay

Table II-1.39: Bleed-through-free hot-melt adhesive based on metallocene catalyzed polyolefin for bonding fibrous or perforated substrates [17]

Item	Raw material	Parts by weight
1	isotactic metallocene butene-1 polymer	49.0
2	Hydrogenated hydrocarbon tackifier resin	44.2
3	Paraffinic mineral oil, viscosity modifier	3.8
4	Polyethylene homopolymeric wax	1.0
5	Sterically hindered primary phenolic antioxidant	2.0

Item 1: "Koattro PB M 1500M", LyondellBasell
Item 2: "Regalite R1100". Eastman
Item 3: "Primol 352", ExxonMobil
Item 4: "Polywax 850", Baler Hughes
Item 5: "Irganox 1010", BASF

Table II-1.40: Polyolefin based hot-melt adhesive for mattress manufacturing [24]

Item	Raw material	Parts by weight
1	Amorphous polyolefin	45
2	Low molecular weight propylene-based olefin polymer	12
3	hydrocarbon tackifier resin	30
4	White mineral oil	10
Total		100

Item 1: "Aerafin 18 Polymer" Synthomer
Item 2: "Aerafin 17 Polymer" Synthomer
Item 3: "Picotac 1095", ExxonMobil

Properties
Viscosity @ 180 °C: 800 mPas•s
Ring and ball softening point: 115 °C
Glass transition temperature: -28 °C
Open time: 3 min

Table II-1.41: Multipurpose APAO hot-melt adhesive with a medium open time [13]

Item	Raw material	Parts by weight
1	Poly-α-olefin, low ethylene copolymer	70
2	Tackifier resin, aliphatic hydrocarbon resin	30

Item 1: "RT 22980", Rexene Products
Item 2: "Escorez 1102" resin, ExxonMobil

Table II-1.42: APAO hot-melt packaging adhesive – effect of replacing part of the APAO by a wax [13]

Item	Raw material	Parts by weight	
1	Saturated, low molecular weight, amorphous poly-α-olefin, propylene homopolymer	59.7	49.7
2	Tackifier, hydrogenated hydrocarbon resin,	30.0	30
3	Branched low density polyethylene homopolymer	10.0	10
4	High melting point Fischer-Tropsch wax	0	10
5	Sterically hindered primary phenolic antioxidant	0.3	0.3

Item 1: "Eastoflex P1023 ", Synthomer
Item 2: "Eastotac H-130E", Synthomer
Item 3: "Epolene C-13" Westlake Chemical
Item 4: "Sasolwax H-1" Sasol
Item 5: "Irganox 1010", BASF

Resulting properties		
Viscosity at 190 °C [cps]	2100	1190
Ring and ball softening point [°C]	141	144
Set time [sec]	4.5	1.9
Tensile strength [MPa]	6.0	13.2
Elongation at break [%]	23.7	27.1

propylene and their delayed setting behaviour. APAO hot-melt adhesives are used e.g. in packaging applications, in the hygiene sector, in book-binding, filter bonding, in woodworking and in heat seal applications [4, 11, 29].

The polymers can be formulated with hydrocarbon resins to lower viscosity and improve tack, adhesion and heat resistance. Wax is added to reduce raw material costs, lower the melt viscosity and adjust open time and setting time. The processing usually takes place in a melt extruder.

Table II-1.41 shows a guide formulation for a multipurpose APAO hot-melt adhesive with a medium open time of 60 sec, a softening point of 103 °C and a melt viscosity of 3275 mPas.

Based on two variants of a guide formulation of an APAO hot-melt adhesive for the packaging sector, the influence of replacing part of the APAO by a wax on the adhesive properties is shown in Table II-1.42. Adhesive bonds made with both formulations are able to withstand high peel and shear stresses, even at elevated temperatures, for extended periods of time.

The formulation presented with Table II-1.43 shows an APAO hot-melt adhesive for general product assembly applications. The two used APAO polymers differ in the softness indicated by their penetration hardness (according to ASTM D 5) and their softening point. Item 1 is compared to item 2 somewhat softer indicated by a higher penetration value (40 dmm vs 18 dmm) and a lower softening point (135 °C vs 155 °C). By varying the ratio of the APAO polymers the tensile strength can be adjusted.

Table II-1.43: APAO hot-melt adhesive for product assembly [13]

Item	Raw material	Parts by weight
1	Low molecular weight amorphous poly-α-olefin, copolymer of propylene and ethylene	59.7
2	Saturated, low molecular weight, amorphous poly-α-olefin, propylene homopolymer	30.0
3	Tackifier, hydrogenated hydrocarbon resin,	10.0
4	Low molecular weight, medium density polyethylene	
5	Sterically hindered primary phenolic antioxidant	0
6	Thioester-based antioxidant	0.3

Item 1: "Eastoflex E1060 ", Synthomer
Item 2: "Eastoflex P1023 ", Synthomer
Item 3: "Eastotac H-130E", Synthomer
Item 4: "Epoxlene N-21"
Item 5: "Irganox 1010", BASF
Item 6: "Cyanox 1212" Solvay

Table II-1.44: Main areas of application for polyester hot-melt adhesives

– Packaging
  – food and beverage
  – pharmaceutical packaging
  – flexible packaging
	– Automotive
	  – filters
	  – upholstery
	  – laminates
	  – airbag

– Foorwear/leather – lasting cement	
– Electrical and electronics	– Textiles and fabrics – wash and dry-cleaning resistant

1.3.2.3 Polyester hot-melt adhesives

Polyester hot-melts are high-performance products. Compared to the EVA and polyolefin-based products mentioned earlier, they offer the following significant advantages, albeit at the expense of higher raw material costs.

- Good adhesion to various polar substrates like
  - wood
  - paper
  - leather
  - fabrics, textile
  - polyurethan foam
  - plasticized plastics (vinyl)
- High thermal resistance, maintaining a strong bond even when exposed to high temperature
- High chemical resistance, e.g., mineral oil, solvents and diesel vapours
- Good UV-resistance
- Fast setting
- Compliant with various food contact regulations

It is therefore not surprising that polyester hot-melt adhesives, despite their higher price, are used for demanding bonding tasks in various industries (Table II-1.44). Their market share is therefore with approx. 5 % low compared to that of the low performance EVA and olefinic hot-melt adhesives.

Raw materials

Since the properties of the hot-melt adhesive can already be adjusted by the composition of the **polyester resin**, a formulation of the polymer by the adhesive manufacturer is rather unusual in contrast to EVA and polyolefin hot-melt adhesives, where the final properties are largely determined by other formulation components (tackifier resins, viscosity-reducing waxes).
To improve the hydrolysis resistance of polyester hot-melt adhesives, **stabilizers** such as
- sterically hindered primary phenolic antioxidant ("Irganox 1010", BASF) or
- phosphite-based colour and molecular weight stabilizer ("Weston 618"; SI Group)
- carbodiimide, calcium oxide or certain anhydrides to improve resistance against hydrolysis

are used.

Tack can be increased by adding small amounts of **tackifier resins**, such as terpene phenol resins (SP-560 RESIN, Akrochem Corporation), if necessary, without greatly reducing temperature resistance.

Small amounts of **powder additives**, such as fumed silica (e.g. "Aerosil", Evonik) act as crystallization nuclei and allow a controlled acceleration of crystallization, i.e. shortening of the setting time.

It is also possible to add **fillers** and, if necessary, **colour pigments**. Modification of polyesters by adding waxes and the like, is not usually recommended as these ingredients could have an undesirable effect on the crystallization process needed for setting and on the melting point.

Formulation

Chemically, **polyesters** are polycondensates of polycarboxylic acids and polyols but also synthesis from hydroxycarboxylic acids which have both a carboxyl group and a hydroxyl group is an option.

The formation reactions for the synthesis of polyesters are shown in Figure II-1.49 using the example of the reaction of a dicarboxylic acid with a diol to form linear polyesters. Branched polyesters are formed with the addition of polyfunctional carboxylic acids or alcohols. In a first reaction step, the diester of the carboxylic acid is formed in a condensation reaction from the dicarboxylic acid and the excess diol with liberation of water. The reaction takes place at approx. 200 – 240 °C under a catalyst, e.g. zinc acetate or titanium or tin(IV) alcoholates, with elimination of water.

In the industrial scale, instead of dicarboxylic acids, the corresponding methyl esters is used in a transesterification reaction. The advantage is that the more volatile methanol (boiling point

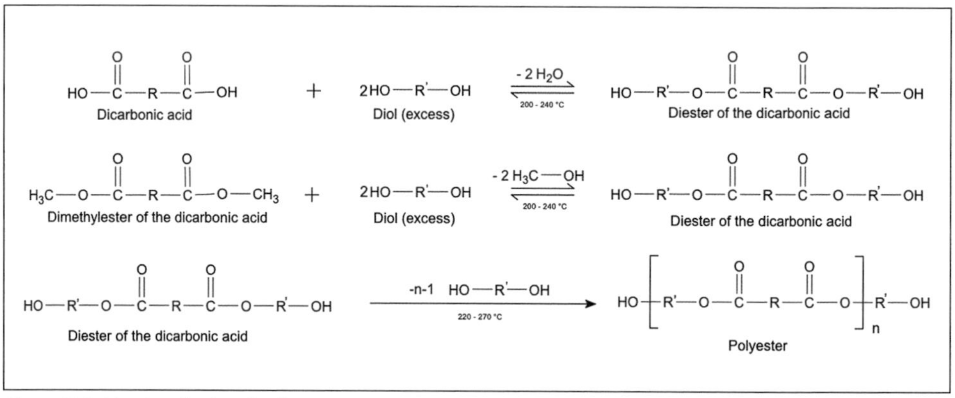

Figure II-1.49: Synthesis of polyester

65 °C compared to water with 100 °C) has to be distilled off instead of water. Transesterification reactions also play an important role wherever recycled polyester is to be incorporated.

In a second step, in the presence of a catalyst (e.g. tetra isopropyl titanate or cobalt acetate), polycondensation by transesterification to the diester of the desired molecular weight occurs in vacuum at 220 – 270 °C with liberation of the diol till the aimed for molecular weight (typically in range of 20,000 – 30,000 g/mol) is reached. The progress of the reaction can be followed, for example, by measuring the hydroxyl value or acid value or by viscosity measurements on samples of the reaction mass (see Chapter IV-4.1). Tracking the power consumption of the stirrer motor as a measure of viscosity can at least give good indications for when to perform the first measurement of the hydroxyl- or acid value. The change in viscosity and molecular weight as a function of conversion rate, shown schematically in Figure II-1.50, show that high molecular weights are achieved only at high conversion rates.

The relatively high reaction temperatures of the synthesis can lead to undesirable side-reactions, such as the formation of ethers (butanediol yields tetrahydrofuran), ester pyrolysis with the formation of olefins and ring formation by dicarboxylic acids to yield ketones, with decarboxylation.

To accelerate cooling to stop the polycondensation reaction, the finished polyester reaction mass is often discharged into flat trays and pelletized or ground, where necessary in the presence of liquid nitrogen. For special applications, such as the laminating of textiles, the product can be spun into a nonwoven. In any case, water contact must be avoided, as this can lead to damage of the polymer due to hydrolysis effects.

The synthesis of polyesters, both during development in laboratory and especially in production scale, requires a certain degree of experience. The resulting polyester resin and thus its properties are affected by the equipment used, especially size and geometry of the reactor, as well as the prevailing agitation conditions, the heating rates and the quality of the vacuum. Accordingly, experience is also required for upscaling from the laboratory to the production scale.

For the formulation of polyester hot-melt adhesives, if the expertise and equipment required for synthesis are not available, commercially available polyester products (e.g. provided by Mitsubishi Chemical under the trade name "Nichigo-Polyester") can be used. These can be formulated by mixing different grades to achieve the properties required for the intended application and if necessary further additives (see before) can be added as well. In any case, it should be checked whether the polyester types considered are already sufficiently stabilized. If not, suitable stabilizers (see before) should be used to improve the properties.

For use as hot-melt adhesives, crystalline grades in particular are used, whether in the form of commercially available products or those developed and manufactured in-house.

With regard to the product properties, adhesion to certain substrates, the softening characteristics upon exposure to heat, and stability to hydrolytic degradation and solvents are primary concerns. The crystallization properties resulting from the monomer composition have a significant influence on

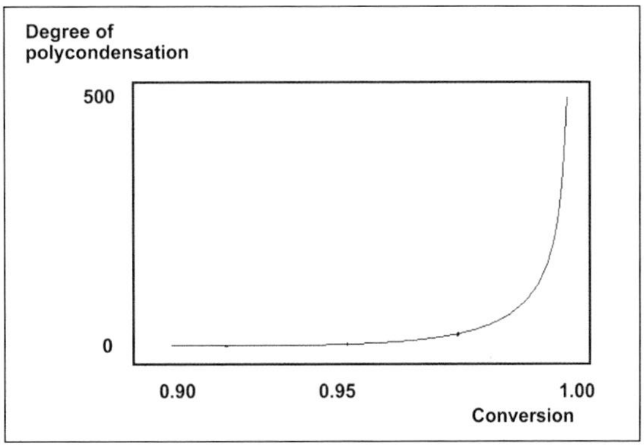

Figure II-1.50: Schematic diagram showing the dependence of viscosity and molecular weight from the conversion rate

the setting behaviour, i.e. the open time and the setting time of the hot-melt adhesive. However, an excessive tendency to crystallization can lead to adhesion problems during bonding on high-modulus substrates due to the volume shrinkage associated with crystallization.

Polyesters composed of terephthalic acid and simple diols such as ethylene glycol, widely used for the manufacture of plastic parts e.g. one-way beverage bottles have very high melting points, which prevent their use as hot-melt adhesives. Combination with other acids and diols lowers the melting point and leads to good processability. At the same time, the addition of further components reduces the degree of crystallinity and increases flexibility. However, less crystallinity means less resistance to hydrolysis and solvents. In most cases, therefore, mixtures of different dicarboxylic acids and different polyols are used to achieve the desired product profile. To counteract problematic gel formation, bifunctional monomers are used almost exclusively. Figure II-1.51 shows some of the most often used monomers in polyester synthesis.

Influence of dicarboxylic acids on polyester properties

Due to the higher symmetry of terephthalic acid compared to isophthalic acid, polyesters based on terephthalic acid have a higher degree of crystallization and a higher melting point than those based on isophthalic acid [1]. The degree of crystallinity can be adjusted by the controlled use of isophthaletes such as:

- Polyethylene terephthalate m.p. 256 °C (crystalline)
- Polyethylene isophthalate m.p. 103 °C (amorphous)
- Polyethylene terephthalate with 20 % isophthalate m.p. 210 °C

Polyesters from aromatic carboxylic acids melt at higher temperatures than those from aliphatic carboxylic acids due to their lower chain mobility [1]. Examples are:

- Polyethylene terephthalate m.p. 256 °C
- Polyethylene adipate m.p. 45 °C

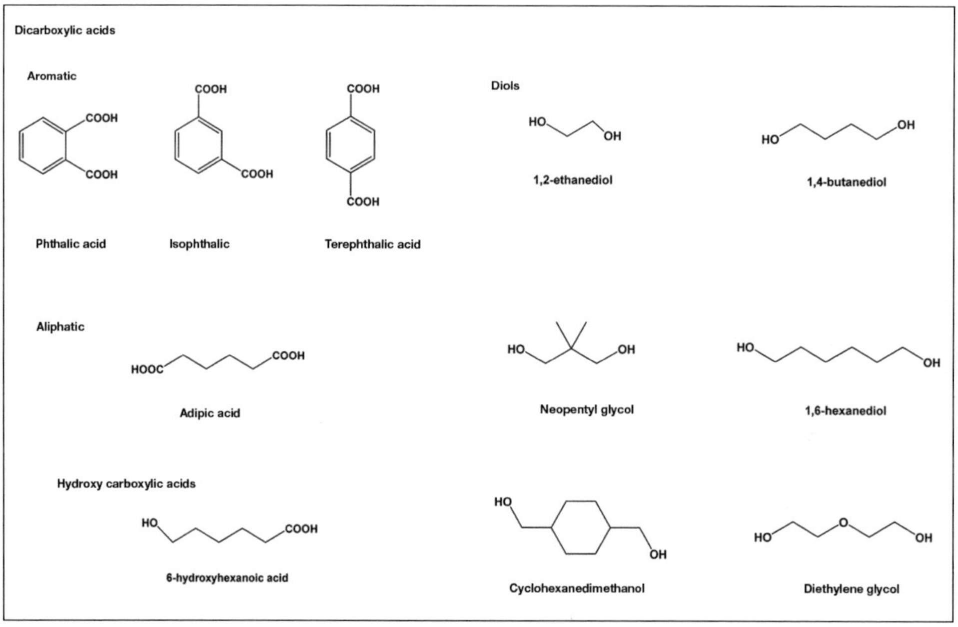

Figure II-1.51: Typical monomers used for the synthesis of polyesters

The addition of long-chain dicarboxylic acids, such as sebacic acid (C_{10} dicarboxylic acid), to terephthalates lowers the melting point but can lead to faster crystallization on cooling due to better chain mobility.

Table II-1.45: Polyester hot-melt adhesive for the use as lasting adhesive in the shoe industry [9]

Item	Raw material	Parts by weight	[mval]
1	Dimethyl terephthalate	60.9	628
2	Dimethyl isophthalate	3.4	35.1
3	Dimetyl phthalate	3.4	35.1
4	1,2-ethandiol	32.4	1045
5	Co-acetate (catalyst)	0.013	–

Influence of diol on polyester properties

The shorter the diol chain, the higher is the melting point of the corresponding polyester [1]:

- Polyethylene terephthalate m.p. 256 °C
- Polybutylene terephthalate m.p. 222 °C
- Polyhexamethylene terephthalat m.p. 148 °C
- Poly(diethylene glycol) terephthalate m.p. 65 °C

Branched diols, such as neopentyl glycol, reduce crystallinity.

Stoichiometric considerations

Using the example of a relatively simple polyester hot-melt adhesive for the use as lasting cement in shoe manufacturing (Table II-1.45), the calculation of the stoichiometrically required quantities of the monomers is explained below. A hot-melt suitable for this application needs to have a sufficient open time and a very short setting time (may. fixturing time of 15 s). The initial (green-) strength should be very high because the restoring forces are strong. The polyester described by Table II-1.45 where crystallinity has been reduced by small proportion of isophthalic acid and phthalic acid meets these requirements. [9].

The three monomers Dimethyl terephthalate, dimethyl isophthalate and dimethyl phthalate have the formula $C_{10}H_{10}O_4$ and thus a molecular weight of 194 g/mol. Their functionality is 2, so that 194 g of the corresponding me-

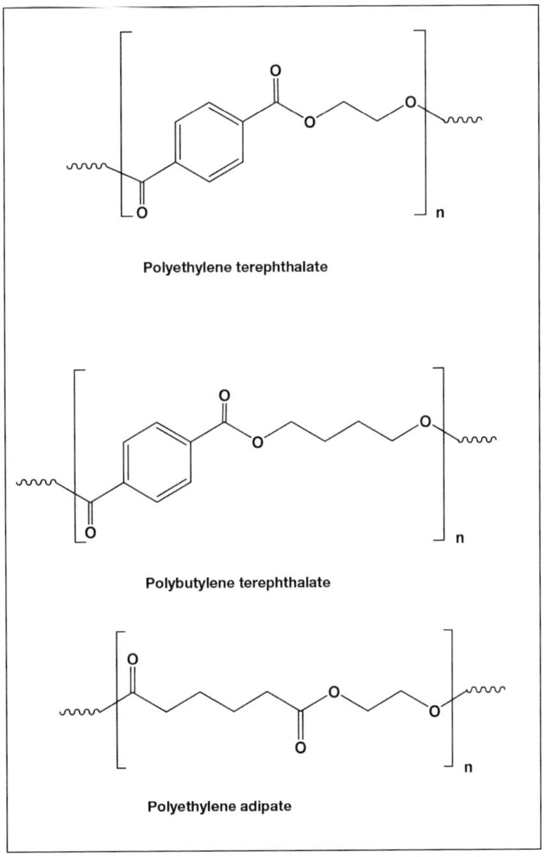

Figure II-1.52: Structural formulas of exemplary polyesters

Table II-1.46: Polyester-based hot-melt adhesive for edge banding of chip- or particle board [20]

Item	Raw material	Parts by weight
1	Copolyester obtained from – 35 % by weight of dimethyl terephthalate (DMT), – 42 % by weight of diglycol adipate (DGA), – 23 % by weight of ethylene glycol Viscosity at 180 °C: 250,000 cps Melting point of about 145 °C	37.5
2	Copolyester obtained from – 31.5 % dimethyl terephthalate (DMT) – 15.5 % by weight of butanediol (BD) – 1.5 % by weight of isophthalic acid – 51.5 % by weight of polyoxytetra-methylene glycol, molecular weight of 1,000 Viscosity at 180 °C: 220,000 cps melting point of about 155 °C	13.9
3	Phenolic resin	16.1
4	Barite	32.1
5	Sterically hindered primary phenolic antioxidant	0.4

Item 3: "SP560 RESIN", Akrochem Corporation
Item 4: "Barytmehl N", Sachtleben Minerals
Item 5: "Irganox 1010", BASF

Resulting properties
Viscosity at 180 °C [cps] 220,000
Ring and ball softening point [°C] 146

thyl ester has 2 val (2000 mval) of latent acid functionality. Thus, 1000 g of the ester corresponds to 10309 mval COOH (latent).

Using the weight percentages for the three dimethyl esters given in Table II-1.41, the mval equivalents given in the right column can be calculated:

- 60.9 g dimethyl terephthalate corresponds to 60.88 • 2000/194 = 628 mval (latent) carboxylic acid.
- 3.4 g dimethyl isophthalate or dimethyl phthalate corresponds to 3.4 · 2000/194 = 35.1 mval (latent) carboxylic acid.

This results in the reaction mixture containing 698.2 mval of latent COOH groups.

The diol 1,2-ethanediol has the formula $C_2H_6O_2$, and thus a molecular weight of 62 g/mol. The functionality is 2, 62 g ethanediol corresponds to 2000 mval OH. Hence, the 32.4 g in the example corresponds to 32.4 · 2000/62 = 1045 mval OH.

Therefore, the diol is present in a 1.5-fold excess (1045 mval/698.2 mval = 1.5) in the reaction mixture, so that a polyester containing hydroxyl groups is obtained.

For the synthesis, the components 1 –- 5 are heated for 4 – to 8 h under a temperature gradient of 150 – 225 °C (transesterification). The reactor is then placed under vacuum (approx. 0.1 kPa) and heating is continued for another 6 h at 225 – 250 °C for the polycondensation process. The resulting hot-melt has a crystalline melting point of about 220 °C.

Guide formulations

Table II-1.46 gives a starting point formulation offering a thermal resistance greater than the required 75 °C for **edge banding of chip- or particle board panels**. It also shows a sufficient resistance against solvents used in painting operations.

The adhesive described by Table II-1.47 is particularly suitable for processes requiring very short setting times like **industrial bookbinding**. Moreover, the adhesive has the advantage of

Table II-1.47: Polyester-based hot-melt adhesive for industrial book binding [20]

Item	Raw material	Parts by weight
1	Copolyester obtained from – 35 % by weight of dimethyl terephthalate (DMT) – 23 % by weight of 1,2-ethanediol – 42 % by weight of diglycol-adipate DGA Viscosity at 180 °C: 15,000 cps Melting point of about 145 °C	69.7
2	Copolyester obtained from – 31.5 % dimethyl terephthalate (DMT) – 15.5 % by weight of butanediol (BD), – 1.5 % by weight of isophthalic acid – 51.5 % by weight of poly-oxy-tetra-methylene-glycol, molecular weight of 1,000 Viscosity at 180 °C: 200,000 cps Melting point of about 155 °C	9.9
3	Copolyester obtained from – 40.6 % dimethyl terephthalate (DMT) – 20.0 % by weight of butanediol (BD) – 39.4 % by weight of polyoxytetra-methylene glycol, molecular weight of 1,000 Viscosity at 180 °C: 10,000 cps Melting point of about 170 °C	9.9
4	Natural esterified rosin resin	9.9
5	Sterically hindered primary phenolic antioxidant	0.4
6	Phosphite-based colour and molecular weight stabilizer	0.2

Item 5: "Irganox 1010", BASF
Item 6: "Weston 618"; SI Group

Resulting properties
Viscosity at 180 °C [cps]: 8000
Ring and ball softening point [°C]: 132
Setting time [sec]: 5 – 6
Open time [sec]: 15
Resistance to repeated bending at -10 °C: good

Table II-1.48: Polyester hot-melt for the textile industry [6, 28]

Item	Raw material	Parts by weight [g]	[mval]
1	Adipic acid	23.8	326
2	1,4-butanediol	29.2	649
3	Tetraisopropyl titanate (catalyst)	0.02	–
4	Recycled polyethylene terephthalate	47	490

being more easily separated from cellulosic fibres during paper recycling without formation of sticky agglomerates of the adhesive, the so called „stickies".

In addition to commercially available polyesters, **recycled PET**, e.g. from the reprocessing of disposable beverage bottles, can also be used for the production of polyester hot-melts. Table II-1.48 describes a polyester hot-melt for the textile industry where about half of the polyester comes from PET recycling. [6]. The manufacturing process shows the typical two steps of esterification. The use of the recycled PET in the second step is an interesting alternative from both an ecological and an economic point of view but requires the recycled PET to be available in both the required quantity and in a defined quality.

In the first reaction step, adipic acid reacts with a twofold molar excess of butanediol. For this purpose, components 1 to 3 are heated for 90 minutes to distil off the reaction water and some butanediol, and the temperature of the batch is raised to 260 °C.

In the second step the recycled PET (item 4) is added in an amount sufficient to obtain a 60/40 mol-percent-ratio of terephthalic acid to adipic acid and the transesterification and polycondensation are carried out for 60 minutes at 260 °C under vacuum (10 Pa). The resulting polyester has a melting point of about 135 °C.

1.3.2.4 Polyamides

Like their polyester counterparts, polyamide hot-melts adheives are high-performance products offering a range of beneficial properties:

- Good adhesion to various substrates including low surface energy materials, e.g.:
  - wood
  - paper
  - leather
  - fabrics, textile
  - many plastics incl. plasticized plastics (vinyl)
  - pretreated polyethylene and polypropylene
  - metals
- High cohesive strength due to strong hydrogen bonds between the polymer chains
- High thermal resistance, maintaining a strong in high temperature applications
- Good creep resistance under load
- Good impact and vibration resistance even at low temperatures
- Good UV-resistance
- High chemical resistance, e.g., against mineral oil, diesel vapours, solvents
- Fast setting, open time from a few seconds up to about 1 minute allowing short cycle times
- Flame-retarding properties

On the other hand, some special requirements must be met when processing polyamide hot-melt adhesives. Owing to the good heat stability of the bonds, the processing viscosity of polyamide hot-melt adhesives offering the often desired high strength even at high temperature is usually higher than with other hot-melt adhesives, so that the use of gear pumps is advisable if there are high requirements for dispensing accuracy. Due to the relatively high processing temperatures of 120 – 240 °C, depending on the adhesive type, and the high oxidation sensitivity, it is necessary to blanket the melt in the tank with an inert gas (e.g. Ar, N_2, CO_2). It must also be considered that the rather high application temperature limits the application to substrates with a corresponding temperature resistance. Finally, polyamides absorb moisture relatively quickly (0.1 – 2 % per 24 h) which requires a special water-vapor-tight packaging. Opened packs should be stored in special containers to prevent the access of atmospheric moisture. If these measures are not observed, the absorbed water will evaporate in the melting tank causing foaming of the adhesive melt.

As mentioned above, polyamide hot-melt adhesives are specialty adhesives, often developed for a specific application or even a specific application process. It is therefore not surprising that their market share of around 10 % is lower than that of hot-melts used in particular for mass packaging. Table II-1.49 lists some typical applications for polyamide hot-melts.

Table II-1.49: Main areas of application for polyamide hot-melt adhesives

– Packaging – hot filling process	– Automotive – various under-the-hood applications
– Footwear/leather	– Electrical and electronics
– Textiles and fabrics – wash and dry-cleaning resistant	– Battery production – electronics encapsulation – heat shrinkable articles
– Filter industry – oil- and gas filter	

Raw materials

Similar to polyester hot-melt adhesives, the properties of the polyamide hot-melt adhesive can already largely be adjusted by the composition of the **polyamide polymer**, so that formulating the polymer by adding further ingredients (tackifier resins, plasticizer, viscosity-reducing waxes) is rather uncommon.

The use of **stabilisers,** such as antioxidants (e.g. sterically hindered primary phenolic antioxidant ("Irganox 1010" BASF) or phosphite-based colour and molecular weight stabilizer (e.g. "Weston 618"; SI Group), is recommended due to the sensitivity of molten polyamides to thermal oxidation.

Within certain limits, the properties can be modified by adding resins. The tack can be increased by adding small amounts of tackifier resins, such as glycerol ester of partially hydrogenated wood rosin ("Stabilite Ester 10", Les Dérives Résiniques Et Terpéniques, DRT), sometimes also **plasticizer** like ethyltoulene sulfonamide ("Santicizer 8", Valtris Specialty Chemicals Company) are used. It is also possible to add **fillers** and, if necessary, **colour pigments**.

Formulation

Polyamides synthesized from diamines and dicarboxylic acids in analogy to polyesters are called diamine-dicarboxylic acid type polyamides (AA-SS). In contrast, those synthesized from aminocarbox-

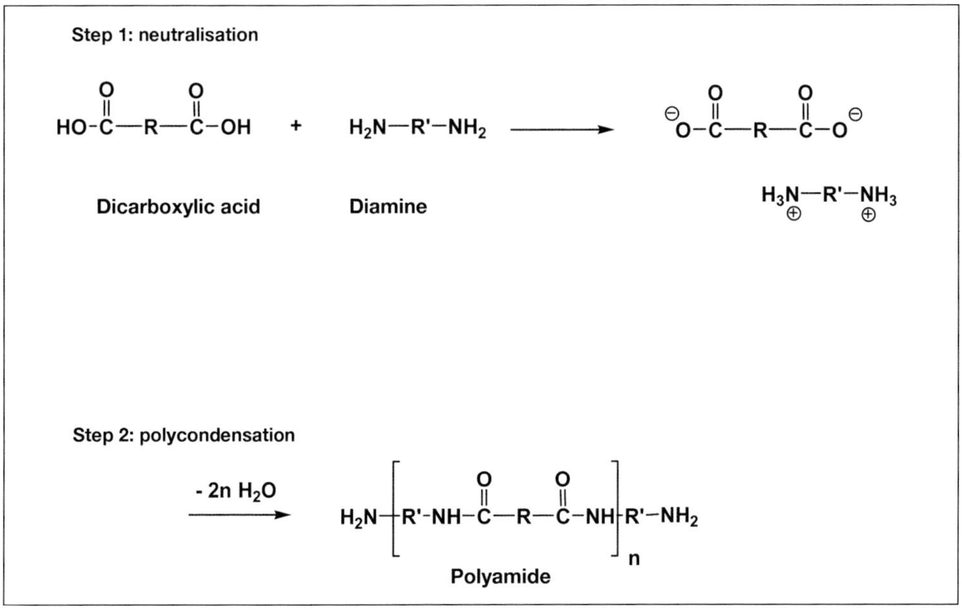

Figure II-1.53: Formation reactions of a polyamide from dicarboxylic acids and diamines (AA-SS-Type)

ylic acids, or their lactams are referred to as aminocarboxylic acid type polyamides (AS). In each case, A stands for an amino group and S for a carboxy group. In both cases linear polyamides are obtained

The formation reactions for the synthesis of AA-SS-type polyamides are shown in Figure II-1.53. Making the polyamide polymer consists in mixing the monomers (salt formation occurs initially by neutralization) and heating at 220 to 230 °C (in special cases, up to 300 °C) for 2 to 3 h. To avoid oxidative degradation, it is recommended to start the reaction under inert gas. Heating is continued under vacuum (approx. 2 kPa) for approx. another 2 h at 220 to 240 °C until the reaction is complete. The resulting polyamides for hot-melt applications have a molecular weight in the range of 15,000 to 50,000 g/mol.

The formation of AS-type polyamides by polycondensation of aminocarboxylic acids or after a ring-opening reaction from their lactams, which requires the presence of water as shown in Figure II-1.54 using the example of ε-aminocaproic acid and ε-capropactam respectively.

Figure II-1.55 shows a selection of important monomers suitable for the synthesis of polyamide melts.

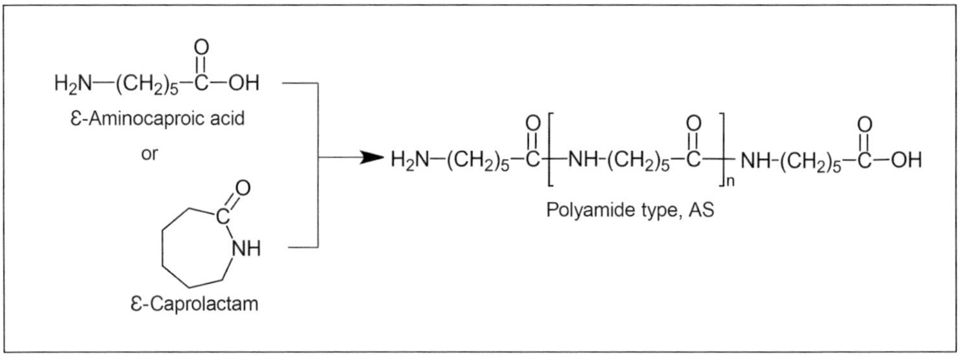

Figure II-1.54: Formation reactions of Polyamides from aminocarboxylic acids (AS-type)

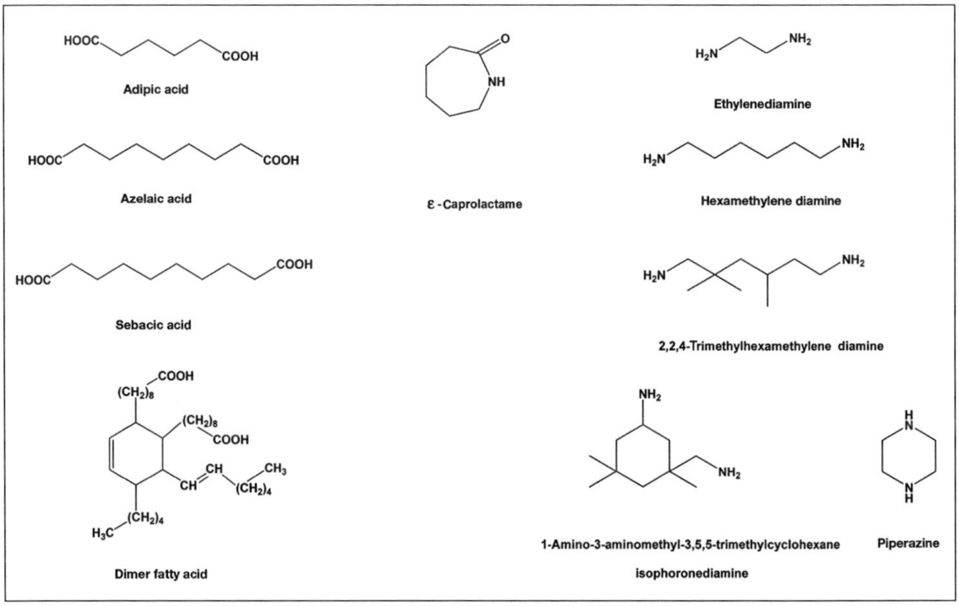

Figure II-1.55: Typical monomers used for the synthesis of polyamides

Table II-1.50: Melting point of polyamides [21,22]

Type	Abbreviation	Base monomers	Melting point [°C]
AS-Type	PA 6	ε-capropactam	210 – 220
AS-Type	PA 11	Aminoundecanoic acid	180 – 190
AA-SS Type	PA 6.6	Hexamethylenediamine/adipic acid	264
AA-SS Type	PA 6,10	hexamethylenediamine/sebaceic acid	240

The melting behaviour (melt temperature, melt viscosity, open time and setting time), the mechanical properties and degree of water absorption of polyamides can be adjusted by an appropriate selection of the monomers used for synthesis. In general, the tendency of water absorption can be reduced by using monomers with long hydrocarbon chains.

Instead of naming the monomers used for synthesis, polyamides are often described by the letters 'PA' followed by one or two numbers. Here 'PA' stands for polyamide and the numbers indicate the number of carbon atoms in the diamine and the dicarboxylic acid used or respectively in the amino acid. For example, 'PA-6.10' designates a polyamide made from hexamethylenediamine (6 C-atoms) and sebacic acid (10 C-atoms), and 'PA-12' a polyamide made from lauric lactam.

Pure polyamides, like 'PA-6.6' made from hexamethylenediamine and adipic acid used as synthetic fibres in textile industry, are not suitable as hot-melts due to their high melting points and melt viscosities. Table II-1.50 shows some melting points of nylon-type polyamides [1].

The melting behaviour, an important factor for the use of polyamides as hot-melt adhesives, can be influenced in the synthesis in various ways.

1. **Dimer fatty acid** serving as acid component, in combination with short-chain diamines, leads to amorphous, low-melting polyamides. For instance, the polycondensate of dimer fatty acid and ethylenediamine yields a polymer whose melting point ranges from 105 to 110 °C. Heat resistance is improved by replacing defined amounts of dimer fatty acid by adipic acid, azelaic acid or sebacic acid. Due to the hydrocarbon chains dimer fatty acid yields polyamides of low water uptake.

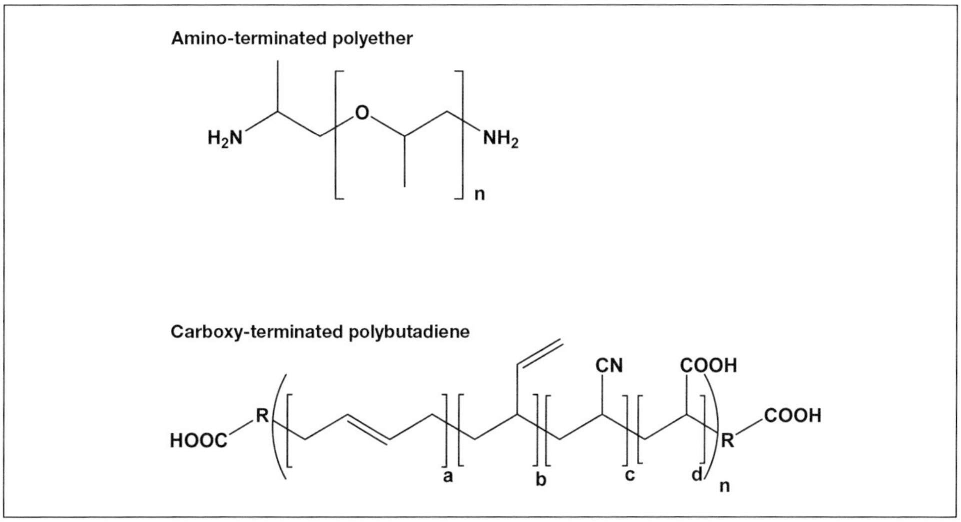

Figure II-1.56:Amino-terminated polyether and carboxy-terminated polybutadiene as flexibilizer for polyamide hot-melt adhesives

Table II-1.51: Polyamide hot-melt use as long lasting adhesive in the footwear industry [5]

Item	Raw material	Parts per weight [g]	[mval]
1	Dimer fatty acid, acid number 192 mg KOH/g	87.1	299
2	Adipic acid	1.7	18.5
3	Ethylene diamine	9.4	313
4	Triphenyl phosphate	0.9	–
5	Stabiliser	0.9	–

2. **Coplymerisation:** Compared with homopolymers, **copolymers** (di- or terpolymers) have much reduced melting points because the irregular configuration reduces crystallinity. While pure PA-6.6 melts at 264 °C and pure PA-6.10 melts at 240 °C, a 70/30 mixture has a melting point of just 200 °C. A terpolymer consisting of 50 parts polycaprolactam (m.p. 225 °C), 10 parts 'PA-6.6' (m.p. 264 °C) and 10 parts 'PA-6.10' (m.p. 210 °C) melts at just 140 °C.
3. **Secondary diamines**, such as piperazine, reduce the ability to form hydrogen bonds between the different polymer chains. The corresponding polyamides thus have less crystallinity and lower melting points. The use of piperazine can also boost adhesion to certain plastics.
4. Incorporation of reactive **flexibilizers** into the polymer chain. Polyamides can be flexibilized by incorporating small amounts of long-chain diamines, such as hexamethylene diamine, amino-terminated polyethers ("Jeffamine", Huntsman) or carboxyl-terminated polybutadienes ("Structol Polydis", Schill+Seilacher "Struktol" GmbH) (see Figure II-1.56) [4, 6].

Stoichiometry

The determination of the monomer composition required for a complete conversion is carried out in the same way as described for polyester hot-melt adhesives. For the formulation of a polyamide hot-melt adhesive for footwear production shown in Table II-1.51, the val COOH contained per kilogram is first determined from the respective functionalities and molar weights.

The functionality of dimer fatty acid ($C_{36}H_{64}O_4$, M = 560 g/mol) is 2, so that 1 Mol (= 560 g) has 2 val (2000 mval) of COOH functionality. Thus, 1000 g of the fatty acid corresponds to 3571 mval COOH.

In data sheets, the acid number determined according to EN 1241 (see Chapter IV-4.1) is often given in mg KOH/g polymer (or g KOH/kg). Using the molecular weight for KOH of M = 56 g/mol, it follows that 56 g KOH (1 mol KOH) corresponds to 1 val. Thus, an acid number of 56 g KOH/kg would correspond to 1000 mval COOH/kg. The dimer fatty acid would therefore have a (theoretical) acid number of

$$56 \cdot \frac{3570}{1000} = 199.9 \text{ g KOH/kg}$$

The dimer fatty acid used in the example of Table II-1.51 has according to technical datasheets with [192 g KOH/kg compared to the theoretical value a somewhat lower acid number, so that 1000 g dimer fatty acid corresponds to just 3429 mval. The amount of 87.1 g of dimer fatty acid used in the formulation corresponds to only 299 mval. Therefore, the formulation of Table II-1.51 contains

$$299 \text{ mval} + 18.5 \text{ mval} = 317.5 \text{ mval COOH}$$

and theoretically, 317.2 mval NH_2 is required for a complete conversion.

The functionality of ethylenediamine is (M = 60 g/mol) is 2 so that 60 g contain 2 val (2000 mval) of NH_2 functionality and therefore

$$\frac{60}{2000} \cdot 317.2 = 9.5 \text{ g}$$

ethylenediamine are required. Thus, the amine used in the formulation is a slightly lower amount than required stoichiometrically.

Guide formulations

Table II-1.52 shows a guide formulation for a laminating adhesive for the textile industry [5]. With ε-caprolactam as the main ingredient, the formulation yields a fast-setting polyamide with good heat resistance. Acids and amines are used in an almost equal ratio. The resulting polyamide has a melting point of 130 °C.

Table II-1.53 shows a guide formulation for a flexibilized polyamide hot-melt adhesive [34]. The properties are mainly determined by the high fraction of the dimer fatty acid and the polyoxyalkylene diamine as flexibilizing agent.

Table II-1.52: Textile laminating hot-melt adhesive [5]

Item	Raw material	Parts per weight [g]	[mval]
1	Adipic acid	15.4	211
2	1-amino-3-amino-methyl-3,5,5-trimethyl-cyclohexan (isophorondiamine)	11.9	140
3	2,2,4-trimethyl-hexa-methylenediamine 	4.8	61
4	ε-Caprolactam	67.9	–

Table II-1.53: Flexibilized polyamide hot-melt adhesive [30]

Item	Raw material	Parts by weight [g]	[mval]
1	Dimer fatty acid, acid number 192 mgKOH/g	66.47	231
2	Stearic acid ($C_{17}H_{35}$-COOH)	0.4	<< 0.1
3	Azelaic acid (HOOC-C_7H_{14}-COOH)	3.6	41.5
4	Ethylene diamine	8.2	273
5	Polyoxyalkylene diamine MW 2000	20.8	20

Item 1: "UNIDYME 18", Kraton or "Empol 1028", BASF
Item 3: "Emerox 1144", Emery Oleochemicals
Item 5: "Jerffamine D 2000", Huntsman

For the formulation of polyamine hot-melt adhesives, if the expertise and equipment required for synthesis are not available, commercially available polyamide products (e.g. provided by BASF under the trade name "Versamid") can be used. These can be formulated by mixing different grades to achieve the properties required for the intended application and if necessary further additives (see before) can be added as well. In any case, it should be checked whether the polyamide grades considered are already sufficiently stabilized. If not, suitable stabilizers (see above) should be used to improve the properties.

Compounding of polyamide polymers and their additives is usually carried out by the melt mixing process. To avoid oxidative degradation of the polymers, oxygen should be kept out by covering with a dry protective gas and the mixing time and temperature should be kept to a minimum.

Table II-1.54, for example, shows a general-purpose polyamide hot-melt adhesive formulation based on a commercially available polyamide with a broad adhesion spectrum, good flexibility even at low temperature and good oil, grease, and water vapor resistance. However, the adhesive has a slight surface tack, i.e. it is not completely block-free.

Table II-1.54: General-purpose polyamide hot-melt adhesive [23]

Item	Raw material	Parts by weight
1	Polyamide resin	100
2	Paraffin wax (melting point 55 °C)	4
3	Tackifier resin (glycerol ester of partially hydrogenated wood rosin)	10
4	Plasticizer	10

Item 1: "Versamid 940", BASF SE
Item 3: "Staybelite Ester 10", Les Dérives Résiniques Et Terpéniques (D.R.T.)
Item 4: "N-ethyl-4-methylbenzenesulfonamide", Hefei TNJ Chemical Company Co., Ltd

Table II-1.55: Polyamide Hot-melt for metal bonding [13]

Item	Raw material	Parts by weight
1	Polyamide from condensation of polymerized linoleic acid polyamines	40
2	Reactive polyamide resin based on dimerized fatty acid and polyamines	60

Item 1: "Versamid 900", BASF SE
Item 2: „Versamid 100", Huntsman Corp.

Table II-1.56: Heat seal polyamide hot-melt adhesive [23]

Item	Raw material	Parts by weight
1	Polyamide from condensation of polymerized linoleic acid polyamines	90
2	Tackifier resin (glycerol ester of partially hydrogenated wood rosin)	10

Item 1: "Versamid 940", BASF SE
Item 2: "Staybelite Ester 10", Les Dérives Résiniques Et Terpéniques (D.R.T.)

Table II-1.55 shows the formulation for a polyamide hot-melt for metal bonding. Besides metal-to-metal bonding the adhesive effectively joins also many plastics (except low surface energy plastics like polyethylene and polypropylene), foils and paper. The resulting bonds are tough and feature a high impact resistance.

Table II-1.56 shows the formulation for a heat seal polyamide hot-melt adhesive where strength retention at low temperatures is required. Besides good flexibility the adhesives shows good adhesion to a wide variety of substrates.

1.3.2.5 Polyurethane

When talking about polyurethane adhesives, which account for a good 10 – 15 % of world sales, a distinction must be made between the much less important non-reactive thermoplastic polyurethanes suitable for the formulation of hot-melts and the reactive polyurethane adhesives, which are discussed in Chapter II-2.2.1.

The non-reactive polyurethane hot-melt adhesives are typically highly crystalline, thermoplastic, predominantly linear hydroxyl polyurethanes. Such products are prepared by polyaddition of diisocyanates and diols. Mostly 4,4' methylene diphenyl diisocyanate (MDI) is used here, and only if light-stable products are required aliphatic isocyanates such as isophorone diisocyanate (IPDI) or hexamethyldiisocyanate (HDI) are used. Linear diols used are e.g. butane diol-1,4 or hexane diol-1,4 but also long-chain polyester or polyether diols. As in the example of the formulation shown in Table II-1.57 polycaprolactone can also be used to prepare a non-reactive polyurethane hot-melt adhesive.

Compounding generally requires both high shear and vacuum processing. After melting the polycaprolactone at 100 °C the two diols are added, and moisture is eliminated by nitrogen at < 10 mbar and 120 °C before the MDI is added. After completion of the reaction, the product is filtered and bottled, as required by the intended application.

Replacing a small portion of the butanediol chain extender with 1-butanol reduces the molecular weight of the polymer by a factor of 2 to 3, thereby reducing the viscosity of the polymer melt by more than an order of magnitude. On the other hand, this slightly reduces the tensile strength of the polymer.

High shear mixing is required to blend the polymers while vacuum is required to remove moisture from the raw material to avoid unwanted reaction of the isocyanate with moisture. Low-speed, high-torque, high-shear mixers are required to draw the solid rubber into melt.

Table II-1.57: Nonreactive polyurethane hot-melt adhesive [13]

Item	Raw material	Parts by weight
1	Polycaprolactone (MW = 530)	29.5
2	1,4 butane diol	14.24
3	1-butanol	0.62
4	MDI	55.64

Item 4: "Desmodur VKS20F", Covestro

Resulting properties

Viscosity at 190 °C [cps]: 10000

T_g [°C]: 0

Elongation [%]: 500

Due to the high melt viscosities and the resulting high application temperatures at which the polyurethane already decomposes noticeably, the products are often used in the form of hot-melt adhesive films. Activation takes place at temperatures of up to 160 °C. Due to the short exposure time compared to processing from a melt tank, no significant decomposition occurs.

To avoid the chemical reaction of polyurethane production, ready-made hydroxyl polyurethanes (e.g. "Desmocoll 530" and "Desmocoll 540" from Covestro) can also be used for the formulation of polyurethane hot-melts.

1.3.2.6 Block copolymers based on polystyrene

Since these hot-melt adhesives are usually permanently tacky and therefore mainly used for the production of hot-melt pressure-sensitive adhesives, they are discussed in Chapter II-3.

1.3.3 References

[1] H. G. Elias, Makromoleküle, 6th ed., Wiley-VCH, 1999
[2] J. M. G. Cowie, Chemie und Physik der synthetischen Polymeren, Vieweg,1997
[3] I. Skeist, Handbook of Adhesives, 3rd ed., Chapman & Hall, 1990
[4] R. Jordan, Schmelzklebstoffe, Band 4a, Hinterwaldner Verlag, 1985
[5] R. Jordan, Schmelzklebstoffe, Band 4b, Hinterwaldner Verlag 1986
[6] R. Jordan, Schmelzklebstoffe, Band 4c, Hinterwaldner Verlag 1987
[7] EP 0 376 461
[8] Dow ELVAX EVA copolymer resin grade selection guide, https://www.dow.com/documents/en-us/catalog-selguide/914/914-375-01-elvax-eva-copolymer-resins-grade-selection-guide.pdf, accessed 2023-08-26
[9] M. J. Satriana, Hot-melt Adhesives, Manufacture and Applications, Noyes Data Corporation, 1974
[10] Fraunhofer IFAM Adhesive Bonding – the Right Way section 4.6.1, https://leitfaden.klebstoffe.com/en/4-6-1-adhesives-which-harden-on-cooling/ accessed 2023-08-26
[11] H. Onusseit, Adhesive Technology – Basis Principles, Beuth Verlag GmbH, Berlin 2012
[12] EN 1238:2011-07 Adhesives - Determination of the softening point of thermoplastic adhesives (ring and ball), Beuth Verlag Berlin
[13] https://adhesives.specialchem.com/formulations/ht-hot-melt-adhesives
[14] Streeter, B.E., "Plasticizers for Solvents and Hot-melt Adhesives", Velsicol Chemical Corp.)
[15] Adhesives Formulary Handbook, NIIR Board, Asia Pacific Business Press, Inc., 2010, p. 266
[16] A. Pizzi andK. L. Mittal, eds., Handbook of Adhesive Technology, Chapter 37, "Hot-melt Adhesives", A. Pizzi, Marcel Dekker, New York, 2003.
[17] EP 3 858 933 A1
[18] https://www.chempoint.com/insights/westlake-epolene-polyethylene-wax-for-hot-melt-adh

[19] Covestro: Bayfol®, Platilon®: TPU hot-melt adhesive films brochure, https://solutions.covestro.com/de/marken/platilon accessed 2023-08-26
[20] EP 0 261 669 B1
[21] https://de.wikipedia.org/wiki/Polyamide accessed 2023-08-26
[22] https://en.wikipedia.org/wiki/Nylon_66 accessed 2023-08-26
[23] J. Nelson, Adhesives Formulary Handbook, NIIR Board, Asia Pacific Business Press, Inc., 2010)
[24] https://adhesives.specialchem.com/formulation/low-viscosity-short-open-time-polyolefin-hot-melt-mattress-adhesive-by-eastman-chemical?li=100170309&lr=pas2410765&utm_campaign=AS-Product-NL&utm_medium=email&_hsenc=p2ANqtz–6Tkl7xcRN9v84rLDbMA1VDmJnUHLj1B4sLpXnTUx-xslT0grFTnYU4sTNawhhwBHWqU3WvtQhm-L1SMm2vshpgX8qTew&_hsmi=330430551&utm_content=330430551&utm_source=hs_email, accesses 2024/10
[25] ISO 2555 Plastics – Resins in the liquid state or as emulsions or dispersions – Determination of apparent viscosity using a single cylinder type rotational viscometer method
[26] R. Jordan, R. Hinterwaldner, Klebharze, R. Hinterwaldner Verlag, 1994
[27] Ullmann's Encyclopedia of Industrial Chemistry, 6th edition, release 2001
[28] R. A. Miller, Polyethylenes and amorphous Polyolefins in Adhesives and Sealants, www.eastman.com
[29] "Vestoplast", brochure, Degussa
[30] Cognis Deutschland, technical information 2004

1.4 Plastisols

Plastisols are colloidal systems in which small thermoplastic polymer particles are dispersed in a liquid plasticizer. Fillers are used to adjust the rheological properties, e.g. to obtain a flowable but to a certain extent sag-resistance paste. The addition of adhesion promoters is necessary to achieve the adhesion properties required for the respective application.

Solidification takes place at elevated temperature. The liquid plasticizer diffuses into the polymer particles, causing them to swell and fuse together forming a viscous gel. This step is practi-

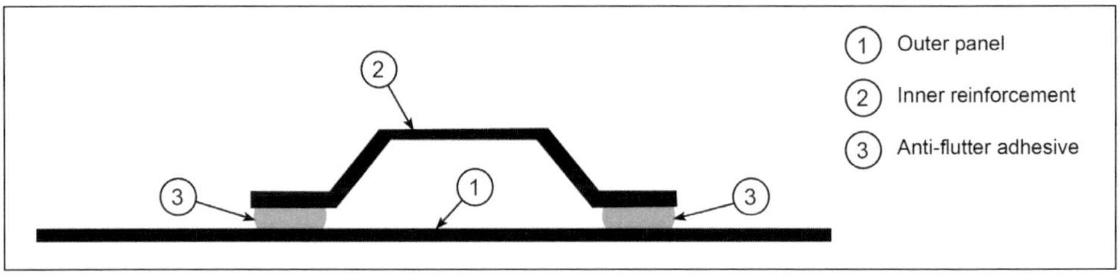

Figure II-1.57: Anti flutter bonding

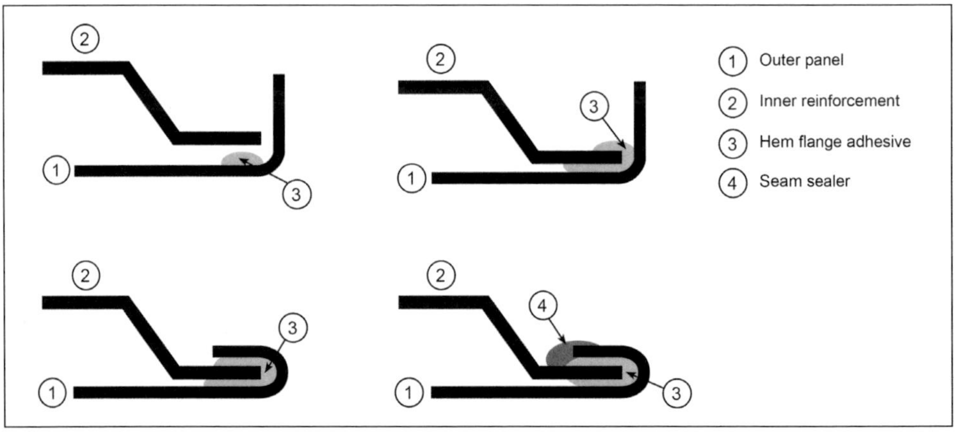

Figure II-1.58: Hem-flange bonding and seam sealing

cally irreversible. During subsequent cooling, the gel solidifies into an elastic and dimensionally stable material with a certain thermoplastic behaviour. Plastisols therefore show a non-negligible dependence of mechanical properties on temperature.

The main area of application for plastisols as adhesives or sealants is car body construction in the automotive industry with the so-called anti-flutter bonding (Figure II-1.57), hem-flange bonding and sealing (Figure II-1.58) as is common on hoods (Figure II-1.59) and doors and seam sealing in general. These plastisols are almost exclusively PVC-based. They are in direct competition with adhesives based on polyurethane (see Chapter II-2.2.1) and epoxies (see Chapter II-2.2.2), which cure by a chemical reaction when exposed to heat.

Excursus: Adhesives and sealants in car body construction

Automotive body construction is characterized by the necessity for the majority of bonding to be performed on oily surfaces. Removing the oil required for corrosion protection and to enable the deep-drawing processes before bonding or sealing is not practicable in series production, as the corrosion protection must be maintained until immediately before the painting process.

The painting process starts with cleaning the car body, usually by means of high-pressure cleaning (power wash), before the surface is further pre-treated in immersion baths, e.g. by phosphating, and the four-step paint build-up takes place. In the first step the cathodic dip coating , the so-called e-coat is applied and then cured at approx. 180 °C. Subsequently, the filler, the base coat and finally the clear coat is applied by spraying and cured at approx. 165 °C and 145 °C respectively. In addition to the adhesives and sealants already applied to the oiled components in

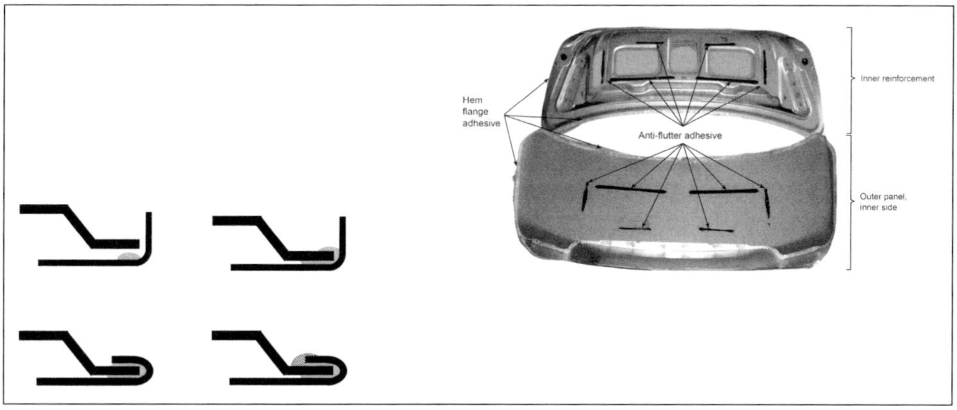

Figure II-1.59: Trunk lid consisting of the outer panel and inner reinforcement separated after bonding indicating the position of the hem flange adhesive and the anti-flutter adhesive

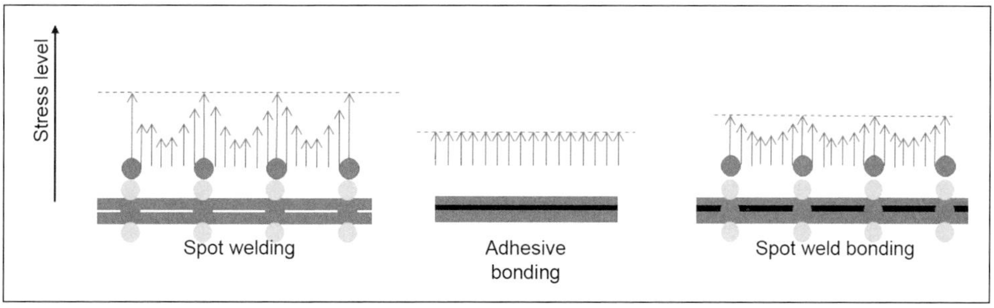

Figure II-1.60: Stress distribution comparison

the so-called body-in-white and cured in the e-coat oven, further bonding and sealing operations are carried out before the filler or base coat is applied, which then cure in the subsequent ovens.

Joints between the various metal parts are nowadays increasingly made by a combination of spot welding and adhesive bonding. Spot welding provides the basic strength and thus allows the body to be assembled, while the adhesive, cured during the paint process provides additional strength leading to a reduction in the stress peaks occurring in the area of the spot welds under load (see Figure II-1.60). The crash properties of components can be improved by using suitably toughened adhesives, usually based on epoxies (see Chapter II-2.2.2). In the event of an accident, the crash structure absorbs more energy, and the passenger compartment is stiffer, resulting in a reduced deformation of the passenger compartment. Figure IV-4.22 impressively shows that the energy absorption of crash structures can be optimised by means of a good crash-stable adhesive.

Another advantage of spot-weld-bonding is the option of reducing the thickness of the sheet metal used without adversely affecting the stiffness and stability of the car body; bonding enables lightweight construction and thus contributes to a reduction in fuel consumption.

The special characteristics of car body construction result in additional requirements for the adhesives and sealants used. In addition to the usual requirements to be met by adhesives, the adhesive or sealant must be adapted to the respective oven temperatures (see before) in terms of their solidification and resistance, regardless of whether the bonding is carried out before or after e-coat. In addition, adhesives or sealants applied before e-coat must not only be able to adhere to oily sheet metal but must also have a consistency that prevents them from being washed out during the cleaning of the car body or the subsequent immersion processes of pre-treatment and e-coat application. This can be achieved by a thermal pre-curing step at a reduced temperature compared to the e-coat process. However, as this requires an additional oven, which takes up space, consumes energy and makes it more difficult to remove the oil from the bodywork, this is avoided wherever possible.

1.4.1 PVC plastisol

Although in principle almost other thermoplastics than polyvinyl chloride (PVC) are suitable for the formulation of plastisols PVC is by far the most commonly used polymer. Besides homopolymers of vinyl chloride in some cases also copolymers containing besides vinyl chloride a small portion of e.g. vinyl acetate are used.

The industrial production of PVC is mainly carried out by suspension polymerization. The resulting polymers have a molecular weight in the range of 30,000 – 130,000 g/mol, which corresponds to K values of 45 – 80. During polymerization, small primary particles (0.1 – 2 µm) are initially formed, which coalesce into larger secondary particles (15 – 25 µm) as the monomer conversion rate increases. The reaction is usually stopped at monomer conversions of 75 – 90 %. [1]

The most important properties of the PVC powder used for the formulation of plastisols are the particle size and the molecular weight or K value. The particles must be sufficiently small so that the particles remain in suspension and do not tend to settle. This usually requires a PVC powder with a particle size of about 1 µ. As the K value increases (increasing molecular weight), the tendency to the formation of gels decreases.

For a long time, phthalate plasticizers such as diethylhexyl- or diisononyl phthalate (DEHP, DINP) were used almost exclusively for the production of PVC plastisols. Due to their potential to impair health by altering the endocrine system, e.g. affecting male fertility, and their very slow biodegradability once released into the environment, they are increasingly being replaced by al-

Table II-1.58: PVC plastisol adhesive/sealant [4]

Item	Raw material	Parts by weight
1	PVC homopolymer, dispersion grade	58
2	Non-phthalate plasticizer	25
3	Heat and light stabilizer	2
4	Additive	3
5	Calcium carbonate	12

Item 2: "Eastman 168", Eastman Chemical Company
Item 3: "Akrostab LT 4789", Valtris Specialty Chemicals
Item 4: "Eastman TIXB", Eastman Chemical Company
Item 5: "Hubercarb Q325", J.M. Huber Corporation
Resulting properties
Tensile strength [MPa]: 15
Tensile strength at 100 % elongation [MPa]: 9
Elongation [%]: 315
Shore A Hardness: 80
Specific gravity [g/ml]: 1.3

Table II-1.59: Model PVC plastisol formulation [3]

Item	Raw material	Parts by weight
1	Plasticizer, diisonyl phthalate	30.0
2	PVC homopolymer	7.5
3	PVC copolymer, containing vinyl acetate (1)	5.0
4	PVC-copolymer, containing vinyl acetate (2)	17.5
5	Coated calcium carbonate	16.0
6	Calcium carbonate	16.0
7	Fumed silica	0.5
8	Drying agent	2.0
9	Stabilizer	0.2
10	Solvent	4.3
11	Polyamide-based adhesion promoter	1.0

Item 2: "PVC Vestolite 1353 K", Vestolite GmbH & Co KG
Item 3: "Lacovyl PA 1384", Arkema
Item 4: "Vinnolit SA 1062/7", Westlake Vinnolie GmbH
Item 5: "Socal 312", Solvay Chemicals
Item 6: Ulmer Weiss XM, Eduard Merkle GmbH
Item 7: „Aerosil 200", Evonik Coporation
Item 8: „CaO Precal 305", Schaefer Kalk
Item 9: "Zinc Oxide", AppliChem GmbH
Item 10: "Exxsol D 80", Exxon Mobile Chemical
Item 11: "Nourybond", Evonik Coporation
Resulting properties
Different grades of "Nourybond" adhesion promoter have been tested at different loads of 0.5 – 1.5 %, cure temperature 30 min at 130 °C, 140 °C or 160 °C for detailed test results see [3]
Viscosity at 23 °C [mPas]: 1.200 – 2.400
Bake time and temperature: 30 min @ 130 – 160 °C
Tensile strength at break [MPa]: 1.1 – 3.1
Tensile elongation at break: 30 – 130
Lap shear strength on e-coated sheet metal [MPa]: 0.5 – 3.1

ternative products such terephthalate esters. Some grades offer a certain percentage of certified recycled content.

When selecting the plasticizer system, care must be taken to ensure that the diffusion process is sufficiently slow under transport and storage conditions to prevent gelling during transport and storage and thus ensure a sufficiently long shelf life. Such plastisols generally require gelling temperatures above 120 °C, often in the range of 160 – 180 °C.

Fillers such as calcium carbonate are added to achieve the desired rheological properties and for commercial reasons. Calcium carbonate provides a secondary benefit in its ability to remove hydrochloric acid (HCl) formed by any partial decomposition of the plastisol. Functional precipitated calcium carbonate is generally used in combination with ground calcium carbonate. If required sag resistance can be further improved by adding small quantities of fumed silica.

Stabilizers, e.g. metal soaps or zinc oxide are added in small quantities to improve temperature and light resitance. Adhesion promoters e.g. maleic acid modified vinyl terpolymer resin, blocked isocyanate or modified polyamide adhesion promoters are added to improve adhesion to certain substrates [2, 3].

PVC plastisols are produced by dispersing the PVC and fillers in the plasticizer mixture. Additives to improve theological behaviour, adhesion properties and stabilization are added in the final step after the dispersion process has been completed

Table II-1.58 provides a guide formulation based on a PVC-homopolymer and a non-phthalate plasticizer. With this formulation the additive (item 4) is a low viscose material compatible with PVC in all proportions reducing the viscosity and thus often allowing a higher filler content, resulting in cost savings.

Table II-1.59 determine the resulting properties as a function of the type and the proportion of the adhesion promoter used as well as the hardening conditions. Detailed results about the effect of the different parameters and ingredients used are published in [3].

1.4.2 References

[1] J. Falbe, M. Regitz, Römpp Chemie Lexikon 9. Auflage, Georg Thieme Verlag, Stuttgart 1995

[2] Bullman, A.R., "Plastisol Sealants for Automotive Applications", Journal Elastoplastics, Vol., 1, October 1969, pp. 256–265

[3] Evonik Corporation, Nourybond® Product literature, 2017, https://products.evonik.com/assets/89/71/Nourybond_adhesion_promoters_for_automotive_plastisol_applications_Americas_EN_Asset_818971.pdf, accessed 2023-11-15

[4] https://adhesives.specialchem.com/formulation/low-cost-pvc-plastisol-adhesive-or-sealant-a-by-eastman-chemical-company

2 Reactive adhesives

2.1 Basic principles

2.1.1 Resins

According to the classification of adhesives described in Chapter I-1.2, reactive adhesives are adhesives in which hardening occurs through chemical reactions. They consist essentially of reactive low-molecular monomers, oligomers or so-called prepolymers, often referred to as building blocks that react under suitable conditions with each other to solidify into a polymer. The polymer chains determining important properties of the cured adhesive are formed durting the bonding process. Depending on the mechanism of the polyreaction, a distinction is made between polyaddition (Figure II-2.1), polymerization (Figure II-2.2) and polycondensation (Figure II-2.3).

These reactive monomers or oligomers are also referred to as 'reactive resins' and should not be confused with non-reactive resins, which are used as tackifiers, e.g. in hot-melt adhesives. Reactive resins have functional groups through which the corresponding polymers are formed during cure of the adhesive.

There are three reasons why reactive building blocks are used for the formulation of adhesives instead of polymers.

Application

Low-molecular building blocks do not require processing aids such as solvents or water to enable the wetting of the adherends required for the formation of adhesion. Therefore 100 % solids

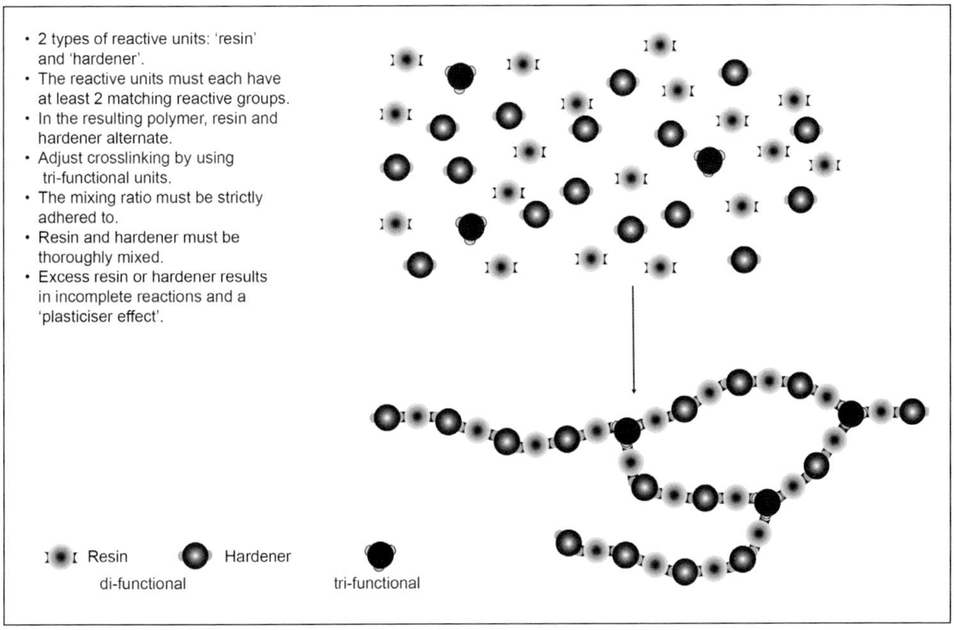

Figure II-2.1: Illustration of polyaddition reaction

adhesives that can be processed at room temperature not requiring the evaporation of solvents or water can be formulated. By virtue of their low molecular weight, they are of low viscosity and can be processed easily and wet the substrates very well.

Cohesion

Proceeding from low-molecular products, it is possible to synthesise the desired polymers on demand (e.g. polyurethanes). The functionality of the monomers offer an easy way of adjusting the crosslink density of the final polymer. The crosslink density defines the chemical and thermal resistance of a polymer as well as its cohesion.

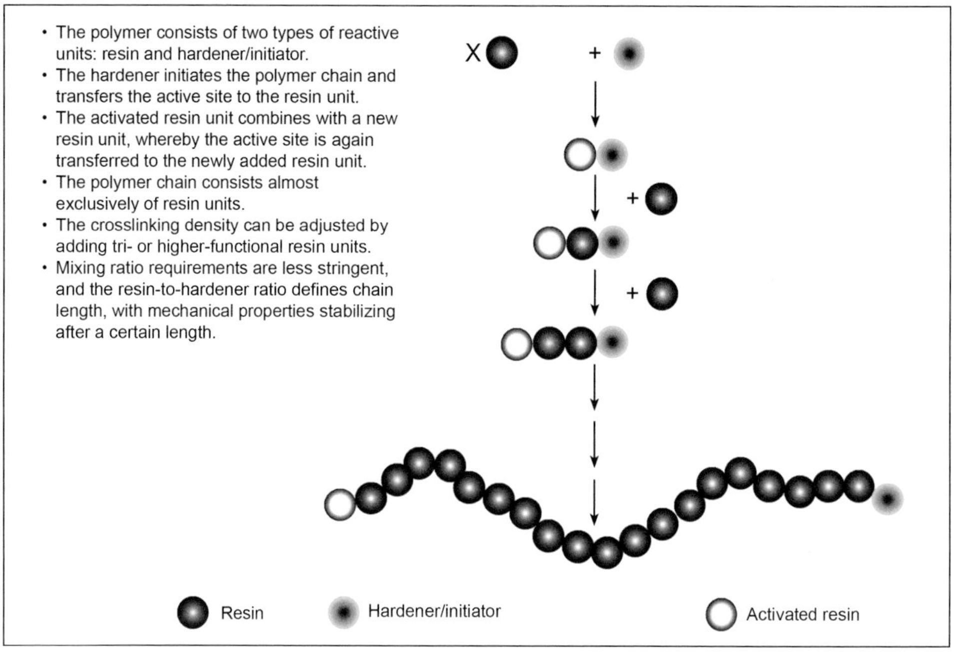

Figure II-2.2: Illustration of polymerization reaction

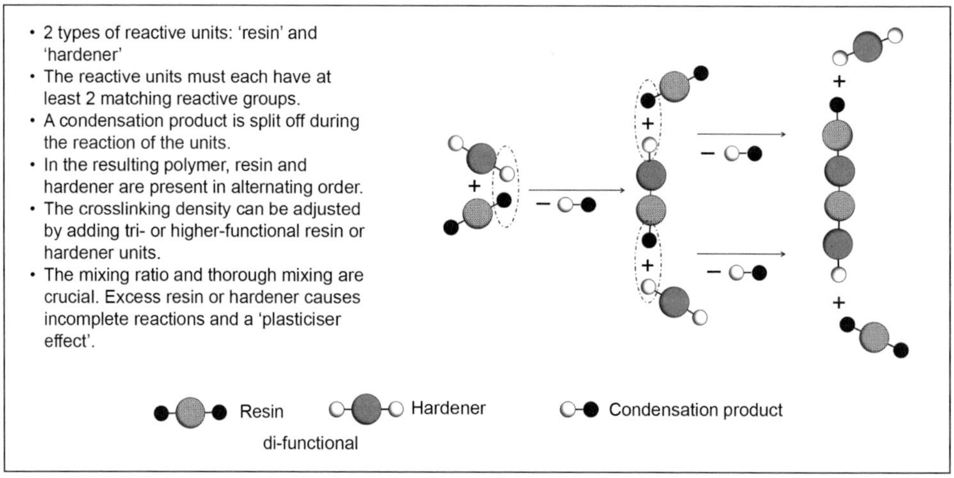

Figure II-2.3: Schematic illustration of polycondensation reaction

Adhesion

Reactive functional groups in the adhesive formulation can react with functional groups on the surface of adherends to yield highly resistant covalent bonds which confer excellent adhesion.

2.1.2 Fields of application

Reactive systems are found in both adhesives and sealants. The most important area of application for reactive adhesives is structural bonding. The goal here is to permanently join substrates together with a predictable bond strength. This requires not only good adhesion, but also good cohesion achieved through judicious monomer selection. Important fields of application lie in the automotive, aircraft, mechanical engineering and electronics industries.

Reactive sealants are widely used for different applications, especially but not limited to construction, transportation and electronics whenever a sealant with either certain chemical resistance and/or reliable mechanical properties is required.

2.1.3 Methods of curing

2.1.3.1 2-component systems

The two components are mixed homogeneously in the specified ratio immediately before application, whereby the curing reaction starts at the first contact of the two components with each other and progresses more or less rapidly depending on the formulation.

Each chemical system composed of resin and hardener has a specific optimum mixing ratio for optimum cure. Some systems, such as the polyaddition reaction of epoxies and polyurethanes, require a well-defined stoichiometric mixing ratio to ensure the desired adhesive properties. Not only a deviating mixing ratio, but also insufficient mixing quality can significantly impair the curing reaction resulting in a loss of bond performance. Other reactive systems, such as acrylates, which cure in a polymerization reaction, show a certain tolerance to mixing errors.

Depending on the amount consumed, the 2-component systems are either processed using so-called metering and mixing systems (Figure II-2.4) directly from the delivery containers or from supply containers. By means of metering pumps, the two components are fed according to the mix ration separately to a mixer where they are then mixed. Mixing can be performed by static or dynamic mixers. Static mixers

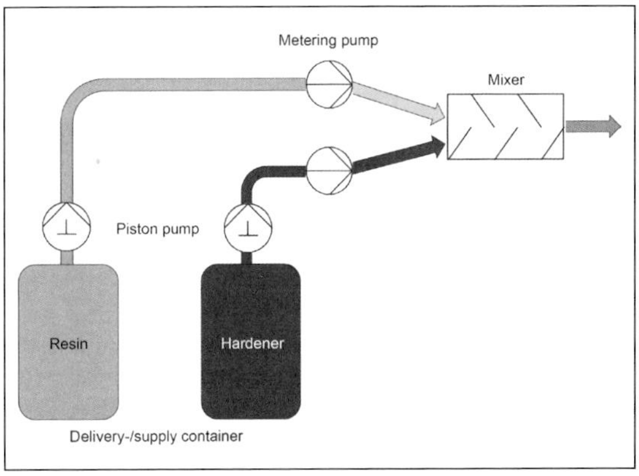

Figure II-2.4: Processing of 2-component adhesives by means of a metering and mixing unit

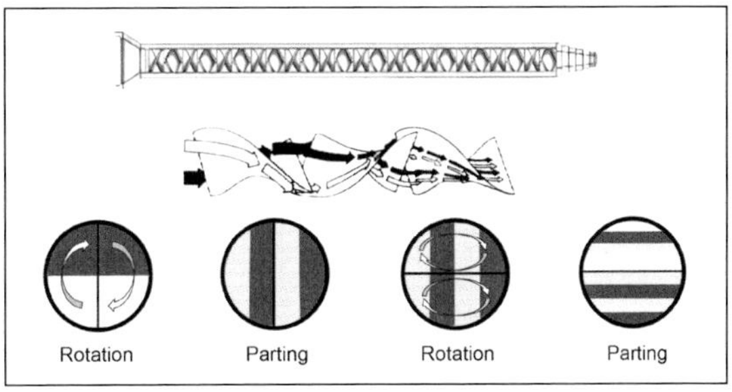

Figure II-2.5: Mixing principle of static mixers

(Figure II-2.5) have mixing elements that divide the volume flow into ever smaller lamellas. For truly homogeneous mixing, a multitude of mixing elements in succession is required. Dynamic mixers (Figure II-2.6) operate in a virtually continuous manner. As long as the two components are injected into the mixing chamber, which has a volume of only a few cubic centimetres and is equipped with a rapidly rotating stirrer, the mixture of the components is discharged through the outlet nozzle.

Today the use of disposable static mixers is very common, and dynamic mixers are only used in special cases if special requirements have to be fulfilled. The main advantage of the disposable static mixer is that it does not require any maintenance or cleaning, if the adhesive inside the mixer has been cured to a state that it cannot be further dispensed by fresh adhesive, or the mixing effect and quality is no longer sufficient the mixer is simply replaced by a new one.

As dynamic mixers are precision parts made of steel, they are relatively expensive and of course are not disposable. Whenever the adhesive in the mixer has reached a certain state of cure it needs to be purged out of the mixer by a suitable solvent or water following by a drying step. This is normally done by pressured air, generating aerosols which includes a certain EH&S risk. In addition, dynamic mixers require a high degree of maintenance to ensure a good and homogenous mix and are subject to wear. The main advantage of dynamic mixers is that due to the small volume much smaller quantities of adhesive are in the mixer and therefore the dynamic mixer should be preferred if just small quantities of adhesive, are dispensed per shot. Another advantage of the smaller volume is that the loss of adhesive by regular purging is much less as it would be with a static mixer. Dynamic mixers are typically used for mixing low-viscosity and fast-reacting products.

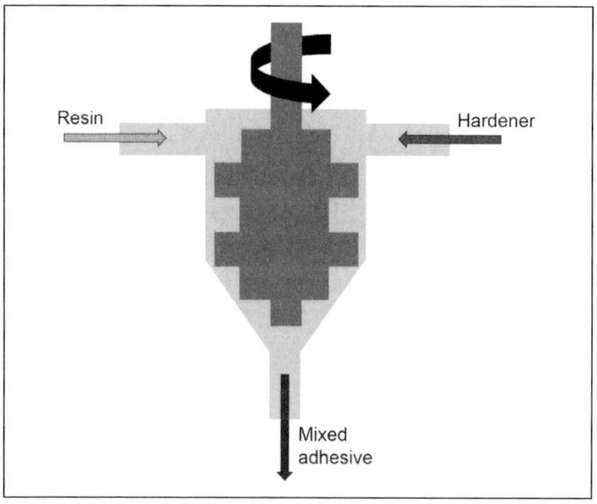

Figure II-2.6: Mixing principle of dynamic mixers

For the processing of small quantities, dispensing can also be carried out

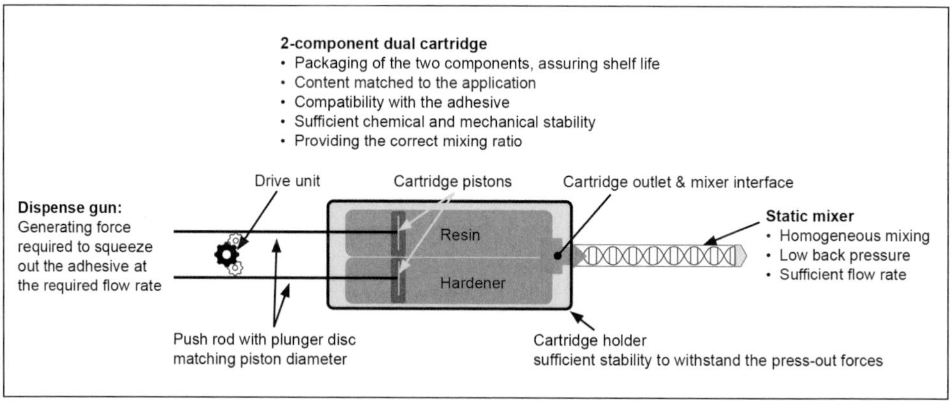

Figure II-2.7: Processing 2-component adhesives from dual cartridges

from 2-component dual cartridges (Figure II-2.7). The resin and hardener are contained in separate chambers of the cartridge, with the mixing ratio being controlled by the diameter of the respective chamber. The material is dispensed by simultaneously advancing the cartridge pistons using a manual, pneumatic or electric dispensing gun. Mixing is achieved by means of static mixers placed on the cartridge outlet.

Manual mixing by gravimetric metering the quantities of the two components and manual mixing with a spatula or a rotating stirrer is limited to slow curing adhesives having a sufficiently long pot life. When mixing, care must be taken to ensure that both components are fully mixed, i.e. no unmixed material remains in the corners of the mixing cup. In addition, the incorporation of air must be avoided.

The chemical reaction starts as soon as the two components come into contact resulting depending on the adhesive formulation in a more or less fast strength build (Figure II-2.8). The faster the system composed of resin and hardener reacts, the faster the increase in viscosity. The pot life is the period of time after mixing of resin and hardener within the mixed adhesive is still capable to wet the second adherend and forming a good and reliable bond. Therefore, a slow chemical reaction that offers a long pot life is desirable for good process reliability.

On the other hand, both adherends need to be held in the desired position for a certain time (clamp time) because the adhesive initially has no cohesive strength. As the chemical reaction progresses, cohesion and adhesion increase and at a certain time the substrates can be handled without mechanical fixturing; the so-called handling strength has reached. It has to be mentioned that the value of the handling strength cannot be considered as an adhesive specific property as it depends heavily on the loads applied to the bond during the downstream handling processes. For most industrial manufacturing processes, the curing reaction should be fast to ensure short cycle times. Clearly, a long pot life and fast curing are contradictory to each other. However, a good compromise can often be achieved via careful selection of the components for the reactive system.

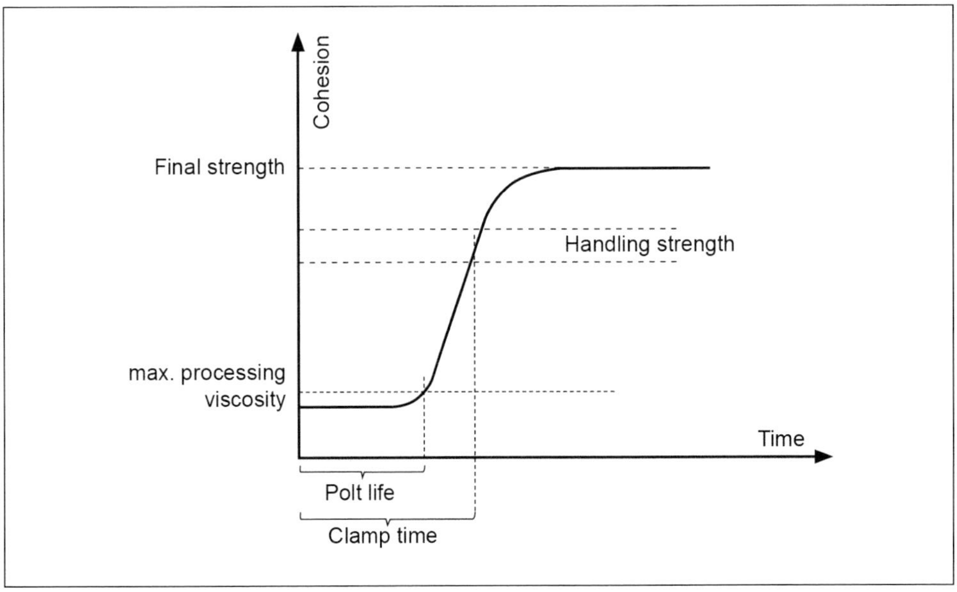

Figure II-2.8: Strength build of a 2-component adhesive

2.1.3.2 1-component systems

Single-component systems are often preferred to avoid the necessary metering and mixing, with the associated potential for errors and higher equipment costs compared to single-component systems. 1-component systems either already contain all ingredients required for the curing reaction in one component but one of the reactive components is blocked to avoid a premature start of the hardening reaction or the curing reaction is initiated e.g. by the addition of moisture from the adherends or the ambient air.

1-component systems work on different principles.

Moisture-curing systems

The curing reaction is initiated by ambient moisture. Ambient air and many substrates offer a permanent supply of moisture. Certain functional groups are capable of reacting with this moisture to form polymers. If an adhesive is formulated based on such functional groups and stored under exclusion of moisture (i.e. metal-foil packaging), no curing can occur. When the adhesive has been applied to the adherend(s), the moisture can gain access to the functional groups and trigger the curing reaction; cure is propagating from the outside to the inside (Figure II-2.9). Strictly speaking, such a 1-component adhesive is in reality a 2-component system, with the second component coming from the environment and not being part of the adhesive itself. These adhesives naturally have some limitations. They can only be used for large-area bonding where the water required for the cure reaction is either provided by the adherend, e.g. wood, or at least one of them is permeable to moisture. The diffusion of water through already cured adhesive is significantly slowed down with increasing thickness of the already cured layer and can even come to a complete stop. Climatic conditions play an important role as both a too low or a too high relative humidity can adversely affect the cure. Typical moisture curing adhesives or sealants are silicones, silane-modified polymers, cyanoacrylates, and polyurethanes based on isocyanates.

Heat-curing, thermosetting systems

The adhesive formulation already contains all the ingredients needed for the curing reaction. However, these are selected in such a way that they essentially do not react at ambient temperatures. This is achieved very effectively with latent curing agent (hardeners). These are unreactive at low temperatures but react with the reactive resin when heated above a certain temperature to form a polymer. Latent curing agents, often in the presence of resins, do not follow Arrhenius' law concerning the influence of temperature on the reaction rate (rule of thumb: a 10 °C increase in temperature doubles the reaction rate). This is achieved by various methods. Often, the curing agent is dispersed in the resin as a fine powder. If solubility is low, a reaction can occur only at the phase interface and therefore stable, coexisting phases are formed. Upon heating the solid hardener melts and solubility in the resin increases and the reaction rate increases as well – the system starts curing. Microencapsulated and thus passivated curing agents can behave in this way, too. Alternatively, the unreactive precursor of a curing agent can be chosen which, when heated, releases a molecule which reacts with the resin and triggers the curing reaction. In order to achieve a sufficient

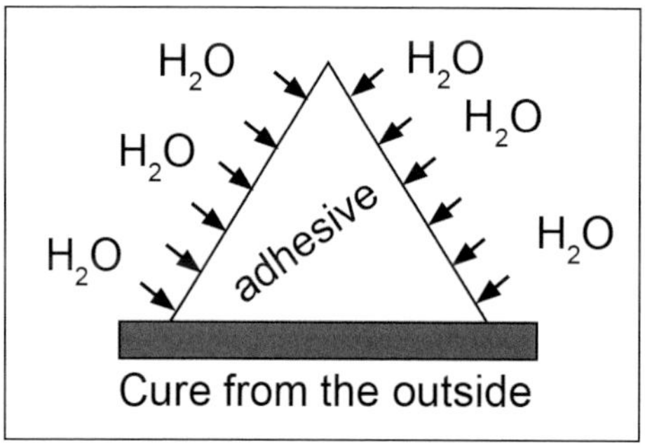

Figure II-2.9: 1-component moisture curing adhesive

shelf life even at elevated temperature, e.g. during transport typical curing temperatures are above 150 °C and thus rule out most plastics as material for the adherends. 1-component epoxies and adhesive based on phenolic resins are important classes of heat-curing adhesives.

Anaerobic systems (anaerobics)

Many free-radical reactions cannot be performed in the presence of oxygen because it acts as an inhibitor. Such an adhesive formulation cannot cure if oxygen is present. But if the adherend hinders oxygen to come in contact with the adhesive, curing can take place. Besides absence of oxygen the presence of metal ions provided from the adherend(s) (preferred are copper and iron) is required to catalyse the reaction. Such adhesives based on special acrylates are known as thread lockers work in this way.

Radiation-curing systems

Curing of a reactive adhesive can also be initiated by exposure to radiation. This may be UV (λ = 200 to 400 nm) or even an electron beam (high-energy electrons, EB curing). The adhesive is applied, and the adherend are mated and exposed to radiation. The high-energy radiation generates free-radicals or ions which initiate the polymerization reaction electron beam curing requires adequate radiation protection measures. One disadvantage of electron beam curing is the effort involved in complying with the legally prescribed radiation protection measures.

A precondition for using radiation curing is therefore that at least one of the substrates must be transparent to the radiation. This is a severely limiting factor, and so the technology is normally reserved for glass and some amorphous plastics, such as PC, PMMA and PVC. Chemical systems suitable for formulating radiation-curing adhesives include acrylates, whose UV-induced free-radical polymerization is triggered by a suitable photoinitiator. Special epoxies can be cured by a cationic mechanism in the presence of photosensitive curing agents.

2.1.4 Environmental, health and safety (EHS)

All reactive chemicals pose a certain health risk, with both acute and chronic impairments possible, so that they should only be handled in accordance with the instructions in the safety data sheets and after careful instruction of the employees about the risks involved to assure a safe handling. In some countries, there are specific legal training requirements for handling certain chemicals. The possibility of unwanted contact via the respiratory tract or skin should not be underestimated and must be avoided by using appropriate protective measures (room ventilation, local exhaust systems, if necessary, suitable protective clothing and gloves, safety goggles).

The information in the safety data sheets relating to environmental protection and waste disposal must also be observed. Waste from most reactive adhesive systems must be treated as hazardous waste if it is not or just partially cured. Fully cured adhesives can be disposed of as normal industrial waste, naturally in compliance with regional regulations.

An important task in the formulation of adhesives, in particular reactive adhesives, is to select raw materials and, if necessary, to reduce their hazard potential by means of chemical reactions during the adhesive production process, so that the formulated adhesive has the lowest possible hazard potential. Adhesives classified e.g. as toxic are not accepted by practically any user. Some users have so-called blacklists of substances that are not accepted as ingredients in formulations.

2.2 Polyaddition adhesives

2.2.1 Polyurethanes

2.2.1.1 Basic principles

Polyurethanes are by far the most versatile group of reactive adhesives. Comprehensive papers can be found in [1–10]. Urethanes are derivatives of carbonic acid, namely the semi-esters/semi-amides. The chemical and physical properties of urethanes are therefore closely related to the properties of esters and amides.

Urethanes can be synthesised simply by making an isocyanate react with an alcohol. As shown in Figure II-2.10, the reaction involves nucleophilic addition of the alcohol to the electrophilic isocyanate.

For the synthesis of high-molecular polyurethanes (abbreviated as PUR and often as PU), polyfunctional isocyanates (polyisocyanates) and polyols must therefore be forced to react in a stoichiometric ratio. By analogy with the reactions between carbonyl compounds and nucleophiles, it is possible to perform either electrophilic catalysis with special metal compounds (increasing the electrophilic nature of the isocyanates) or base catalysis with tertiary amines (increasing the nucleophilicity of the polyols).

Adhesives containing isocyanates require special care because they are carcinogenic and can trigger allergic asthma [11]. For the handling of polyurethanes or isocyanates, special training courses are prescribed within the EU today, which can be carried out online, for example [12]. Particularly in the case of warm- or hot-curing adhesives, attention must be paid to exhaust systems and other occupational safety equipment [13]. In the case of room-temperature curing adhesives, however, the risk of contamination via the respiratory tract is considered to be low [13].

2.2.1.2 Raw Materials

Isocyanates

Isocyanates are usually manufactured by two-step phosgenation of the corresponding amines (Figure II-2.11).

Phosgene-free manufacturing processes involving urea and alcohols have now been developed [4].

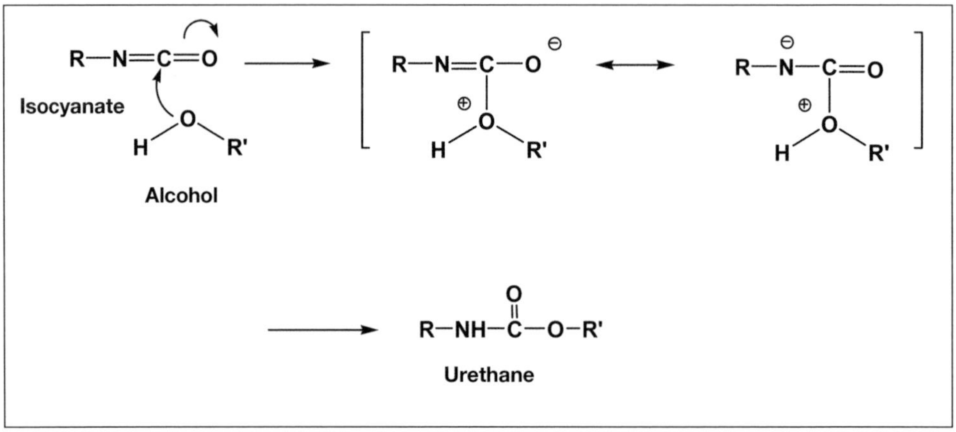

Figure II-2.10: Isocyanate/alcohol reaction

A good overview of current developments in the field of bio-based isocyanates can be found in [14]. It is also theoretically possible to substitute isocyanates as crosslinkers with cyclic carbonates, but these raw materials are not yet commercially available at the time of going to press [15].

A distinction needs to be drawn between aromatic and aliphatic isocyanates (the terms refer to the carbon chain directly linked to the isocyanate group). The former are more susceptible to nucleophilic attack and much cheaper, while the latter are more light-stable. Reactive PUR adhesives are mostly made with aromatic diisocyanates, because it is rare for adhesive joints to be exposed to direct light and the industry is always looking for low-cost solutions. The most important diisocyanates are shown in Figure II-2.12.

MDI or 4,4'-diisocyanatodiphenylmethane (sometimes also called **m**ethylene**d**iphenyl **d**iisocyanate) is now the most common isocyanate employed in PUR adhesives. Pure MDI consists of pale-yellow crystals (m.p. 40 °C) and the two isocyanate groups are nearly equally reactive. MDI is manufactured by reaction between aniline and formaldehyde (with subsequent phosgenation). This reaction is not homogeneous. Besides 4,4'-MDI, certain quantities of 2,4'-MDI are obtained (the amino group directs a second substituent to the o- and p-positions). In addition, aniline can react several times with formaldehyde (o- and p-) and so phosgenation yields isocyanates of higher functionality. The reaction mixture is separated by distillation. Depending on the intended application, products of suitable functionality and different raw material price can thus be used to formulate the adhesive. MDI has a high molecular weight relative to other isocyanates and so its vapour pressure is relatively low. It is therefore much safer to handle both during manufacture and application of the adhesive.

TDI (diisocyanatotoluene or **t**oluene **di**isocyanate) is a colourless liquid with a pungent smell. It is made by double-nitration of toluene followed by reduction and phosgenation. The reaction mixture contains certain amounts of 2,6-TDI in addition to 2,4-TDI. Typical trade products consist of a 80/20 mixture of the 2,4- and 2,6-isomers. The two isocyanate groups in 2,4-TDI exhibit significant differences in their reactivity towards nucleophiles (about 1 : 10), mainly for steric reasons. At curing temperatures > 100°C, the much higher vapor pressure of TDI compared to MDI must be taken into account, which can lead to appreciable monomer concentrations in the room air if ventilation is inadequate.

IPDI (**i**so**p**horone **di**isocyanate) and **HDI** (**h**examethylene **di**isocyanate) mostly play a secondary role in reactive adhesives due to their price and reactivity. Both are toxic, colourless liquids with a pungent smell. **H_{12}MDI** (4.4'-diisocyanatocyclohexylmethane) is a colourless liquid (melting point close to room temperature).

Commercially available isocyanates based on MDI are mainly polyisocyanate prepolymers or polyisocyanates and homologs.

Isocyanates can react in various ways. Besides reacting with alcohols to form urethanes, they can react with other nucleophiles bearing an acidic proton. The mechanism of the reaction with

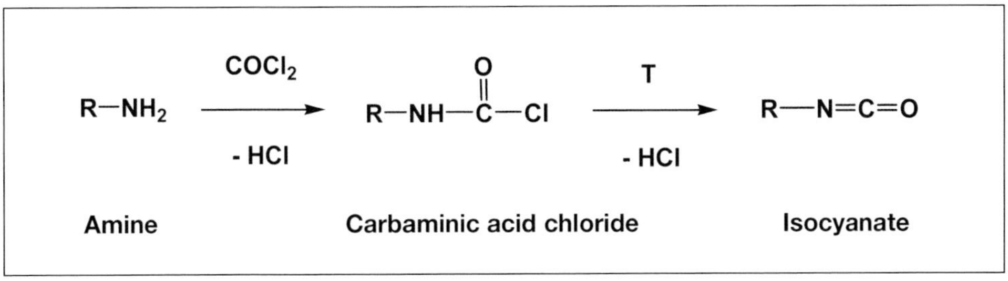

Figure II-2.11: Manufacture of isocyanates

amines (primary or secondary is similar to that with alcohols but is very fast (occurs within seconds) and exothermic. It is therefore only of indirect importance and not really useful for adhesives.

The Nolax company introduced such systems, which, using special mixing nozzles, enabled isocyanate-amine mixtures to be injected into cavities, resulting in a moulded joint within seconds. In the meantime, these formulations can also be used as flexible potting compounds [16, 17].

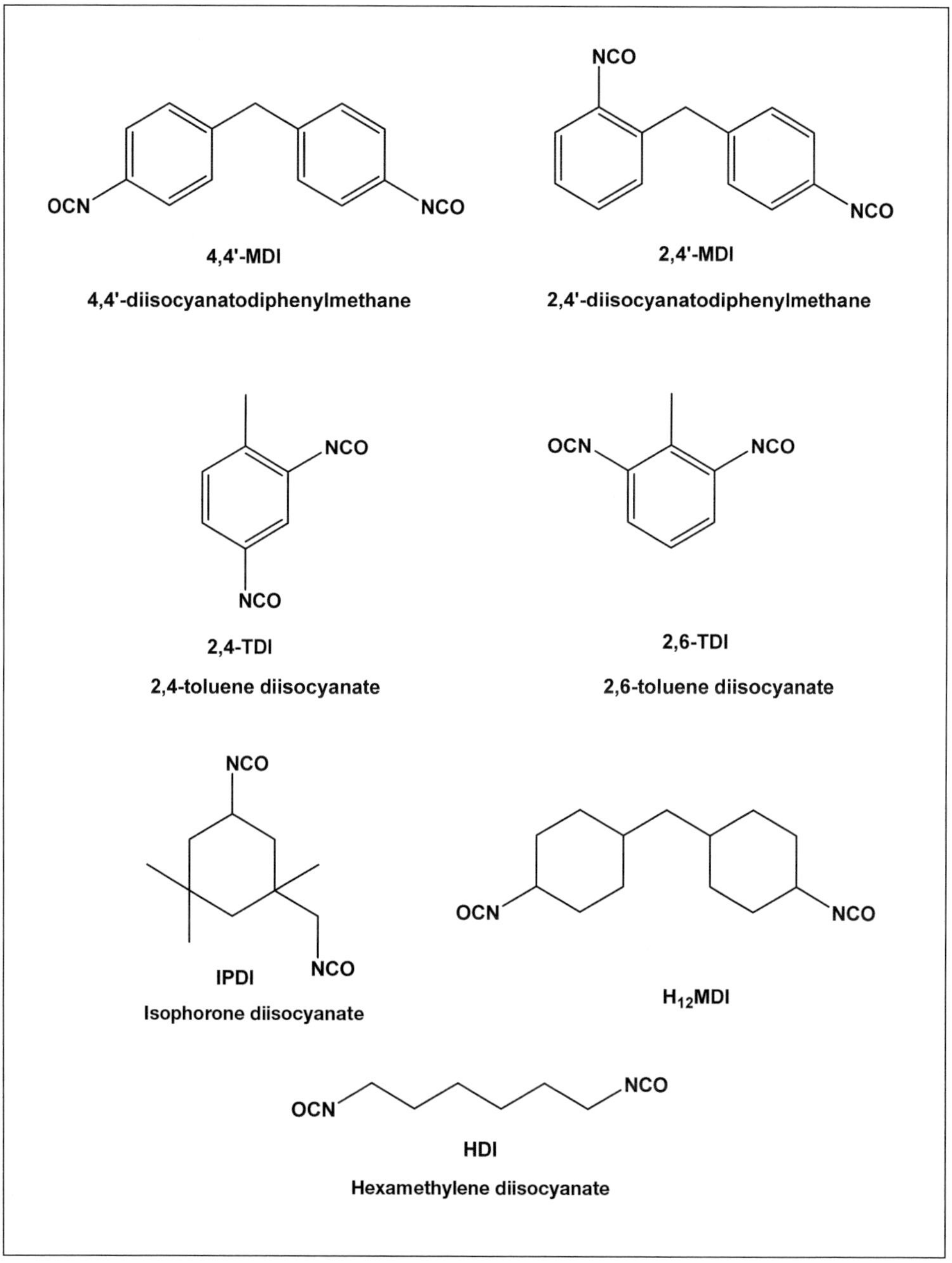

Figure II-2.12: Important diisocyanates

Figure II-2.13 shows the formation of urea from isocyanates and amines.

The reaction between isocyanates and water is of major importance for reactive adhesives (see also Figure II-2.14). The first step is nucleophilic addition of water. Carbamic acid, formed by way of intermediate, generates an amine by releasing carbon dioxide (this reaction can also be used to produce PUR foams, e.g. for sealings in control cabinets). If there is still isocyanate available, this amine can react rapidly by forming a substituted urea, as shown in Figure II-2.14.

However, if water is present in high excess compared to isocyanate, the reaction can be stopped at the amine step.

As shown in Figure II-2.15, isocyanates can react in other ways, too [1]. The reaction with urea proceeds mainly at elevated temperatures and leads to biurets. Reaction with urethanes leads in a similar way to allophanates. In the presence of alkaline catalysts, trimerisation to isocyanurates can occur, while, with other special catalysts, dimerisation leads to uretdiones.

Polyols

Polyols for synthesising polyurethanes are OH-functionalised oligomers with typical molecular weights in the range of 400 to 8000 g/mol and a functionality of 2 to 3 val/mol. The various polyols differ in their base chemical.

Polypropylene oxides are colourless, viscous liquids. In polyurethanes, they can act as flexible components. Due to side-reactions, the "true functionality" of polyethers is usually some-

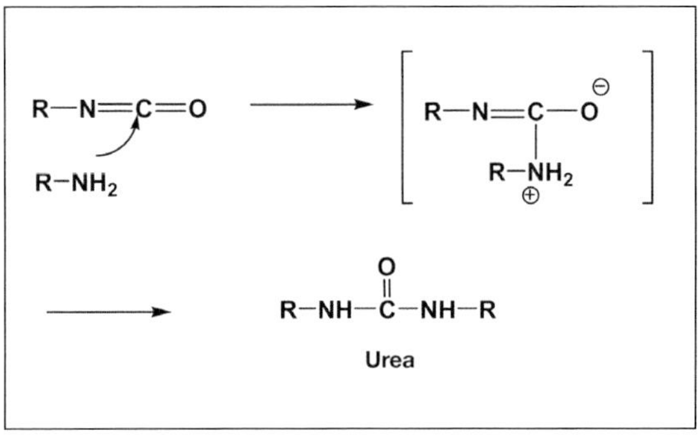

Figure II-2.13: Reaction between isocyanates and amines

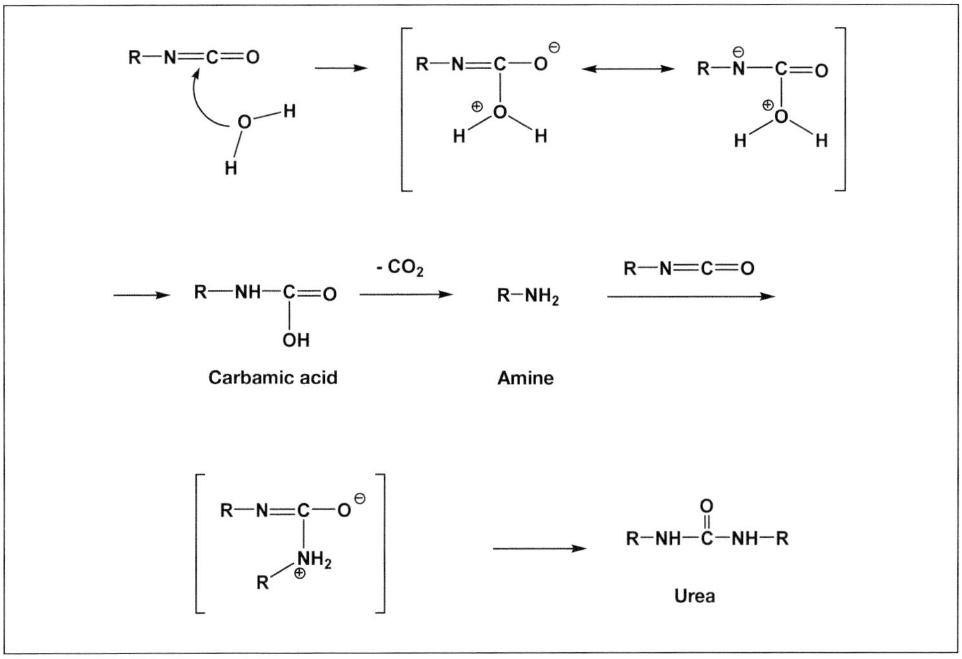

Figure II-2.14: Reaction between isocyanates and water

what lower than the theoretical value defined by the initiator molecule. New manufacturing processes (e.g. DMC) allow the synthesis of polyols that offer even more defined functionality.

Besides polyethers, **polyester polyols** are used to synthesise reactive polyurethanes. Commercial polyesters have molecular weights ranging from 1000 to 6000 g/mol and functionalities of 2 or 3 val/mol. They are synthesised by a polycondensation reaction in the same way as polyesters for hot-melts (see Chapter II-1.3) from polycarboxylic acids with excess polyols. A distinction is

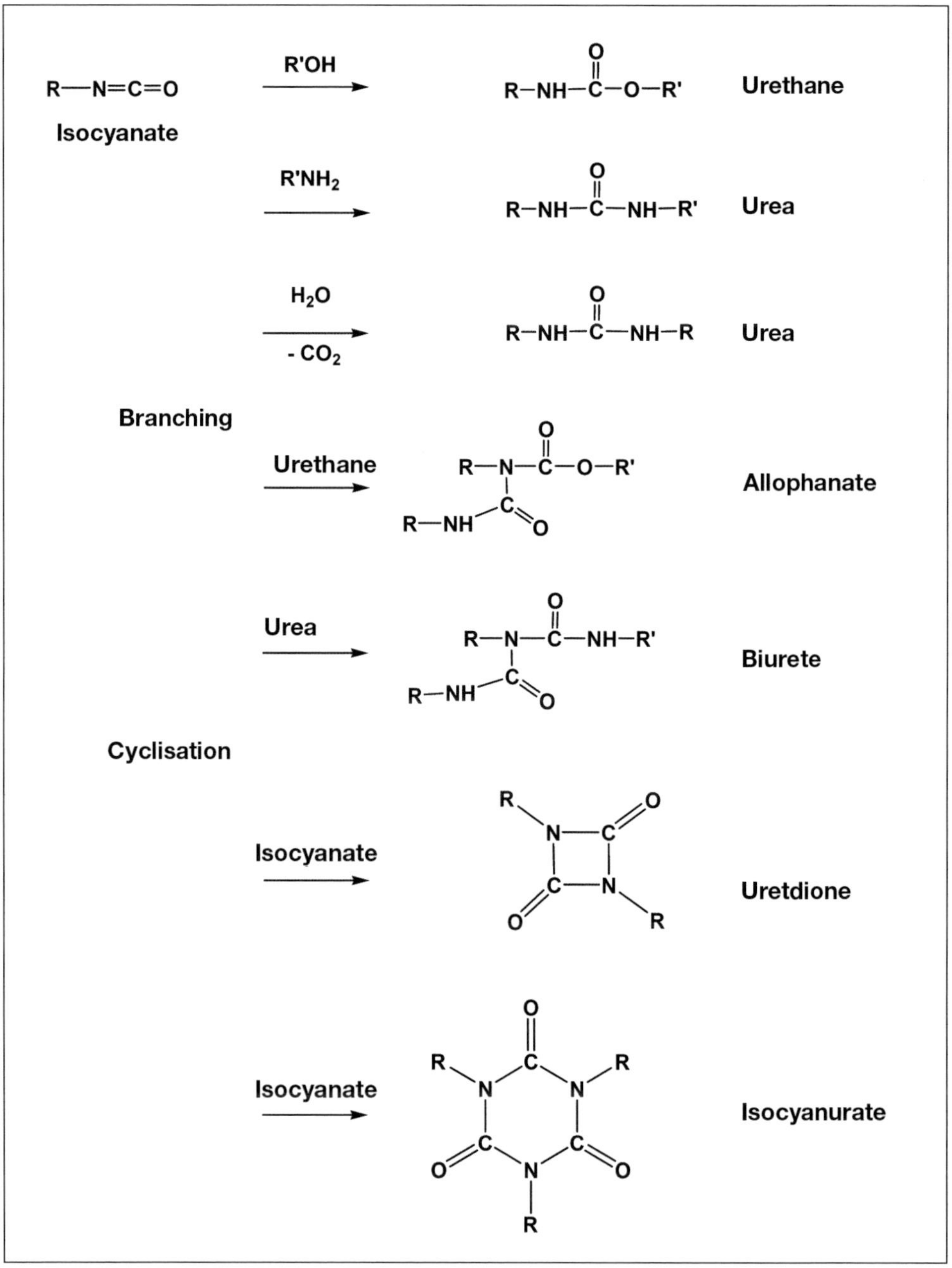

Figure II-2.15: Important reactions of isocyanates

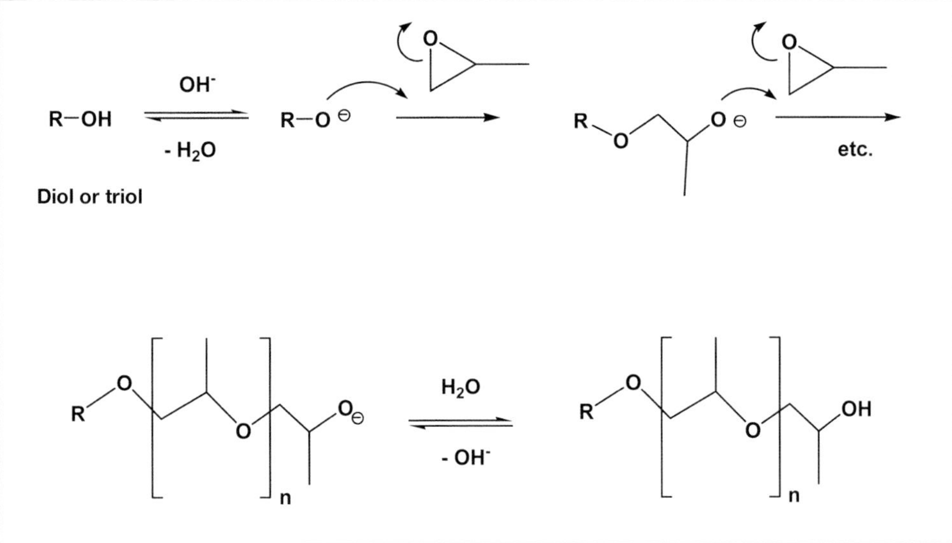

Figure II-2.16: Synthesis of polypropylene oxide

drawn between crystalline (usually adipates) and liquid or solid amorphous (mixed) polyesters. Polyesters are significantly more expensive than polyethers. Polycarbonates are special polyesters made by phosgenation of diols.

Hydroxyl-functionalised polybutadienes ("Polybd") have been commercially available since the late 1970s. These oligomers are made by free-radical polymerization of butadiene in the presence of hydrogen peroxide. They are supplied as viscous liquids of weak characteristic smell in molecular weights of 1000 to 2500 g/mol and functionalities of approx. 2.5 val/mol. Figure II-2.17 shows the idealised structure of hydroxyl-functionalised polybutadiene.

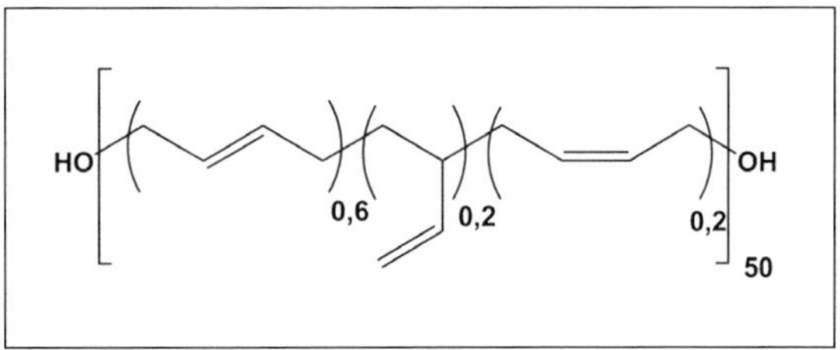

Figure II-2.17: Hydroxyl-functionalised polybutadiene

A polyol based on natural resources is **castor oil** (oil from the ricinus communis tree). It is a pale-yellow liquid with a molecular weight of about 1000 g/mol and a functionality of approx. 3 val/mol. As shown in Figure II-2.18, castor oil is the glycerol ester of ricinoleic acid (C_{18} carboxylic acid with an OH group and a double bond) [4]. As it is a derivative of a natural product, the exact structure of castor oil varies somewhat with the season and the weather. This must be borne in mind when secondary products are being formulated.

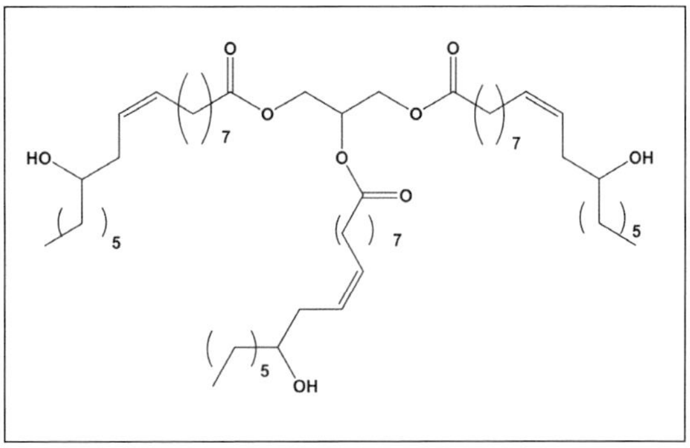

Figure II-2.18: Structure of castor oil

In addition to castor oil, there are now other possibilities for using biobased polyols for polyurethane formulations. These can be prepared, for example, by the alcoholysis of epoxidized fatty acids or oils (Figure II-2.19) [71].

Chain extenders

In polyurethane chemistry, chain extenders are short-chain, i.e. low-molecular (molecular weight below 200 g/mol) components (already familiar from polyester chemistry), which can react with isocyanates, and which have a defined functionality (mainly OH groups, f = 2 or 3 val/mol). Figure II-2.20 shows the most important chain extenders for formulating reactive PUR adhesives.

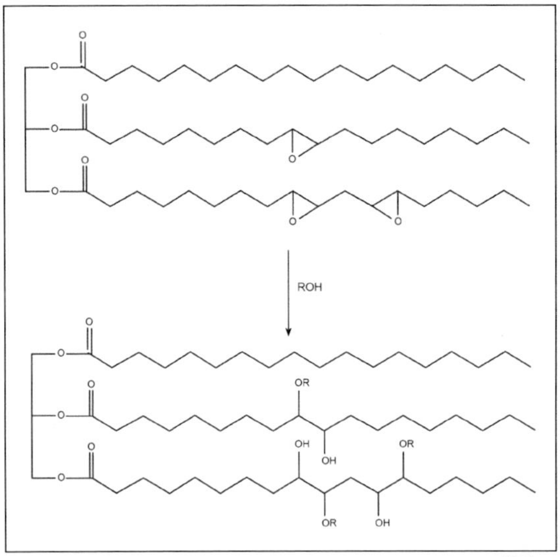

Figure II-2.19: Bio-based polyether glyceride polyols [71].

Additives

Reactive PUR adhesives are formulated with special additives. **Catalysts** are chosen for their ability to influence the reactivity and for their selectivity for specific reactions. Catalysts based on metal salts increase the electrophilic nature of isocyanates. Figure II-2.21 shows dibutyltin dilaureate, an important metal catalyst.

DBTL mainly catalyses the isocyanate -OH reaction, as well as the isocyanate-water reaction. Reactions by which isocyanates form dimers or trimers are not catalysed [2]. Alkaline cata-

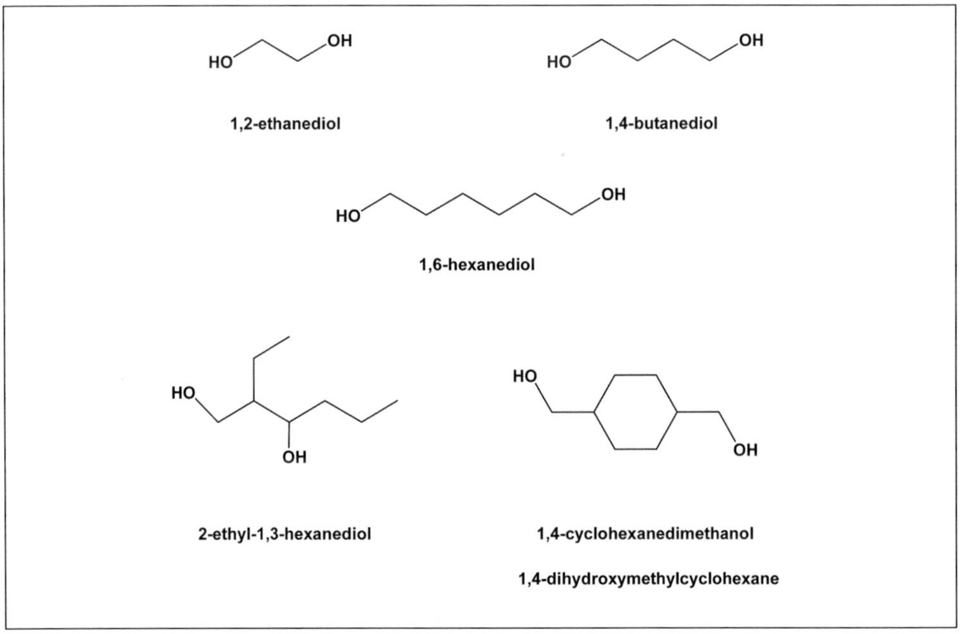

Figure II-2.20: Important chain extenders

lysts, such as 1,4-diaza[2.2.2]bicyclooctane (Figure II-2.22, DABCO) mainly catalyse the reaction between isocyanates and water, as well as the reactions of isocyanates with OH and of isocyanates to yield cyclic oligomers. For this reason, isocyanates or formulations containing isocyanates cannot be stored in the presence of these catalysts. The number of commercial metal- or amine-based catalysts is immense. Catalyst mixtures can exhibit substantial synergistic effects [3].
Please refer to the Chapter II-2.2.1.10 on new developments in the field of catalysis.

2.2.1.3 Curing technologies

2-component PUR adhesives consist of two separately packaged and stored components (polyol and isocyanate hardener) which, after mixing, react to yield a polymer under mild conditions. In the case of PUR adhesives, the A component is usually a polyol or a mix of polyols and the hardener is a liquid isocyanate or an isocyanate-terminated prepolymer. As shown in Figure II-2.23, after mixing, the components react to yield a polymer linked via urethane links.

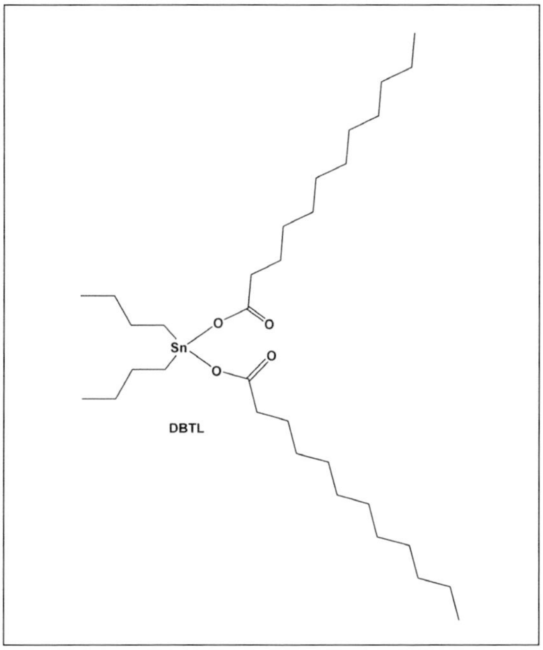

Figure II-2.21: Dibutyltin dilaureate (DBTL)

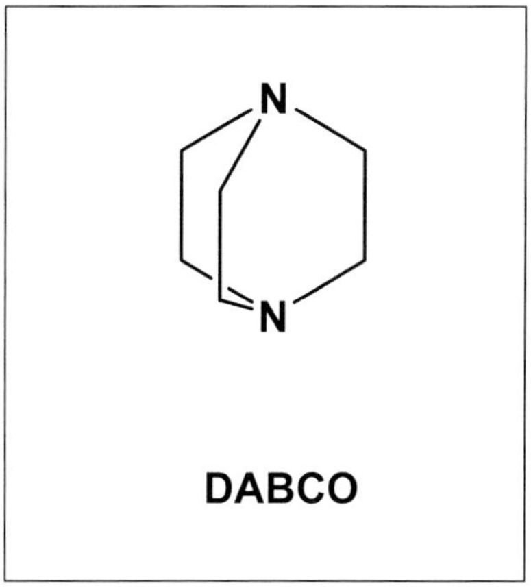

Figure II-2.22: 1,4-Diazabicyclo[2.2.2]-octane

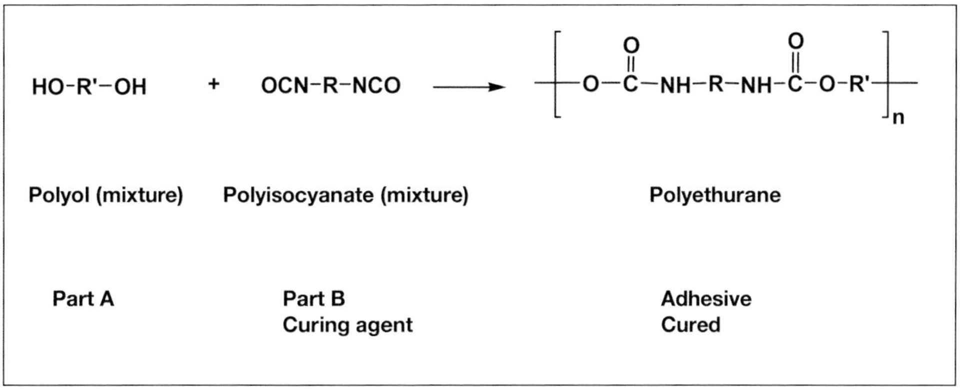

Figure II-2.23: Polymer formation (linear) in a 2-component polyurethane

1-component moisture-curing PUR adhesives consist of isocyanate-terminated prepolymers. They are manufactured and stored in the absence of moisture. After application, moisture from the environment triggers the reaction of isocyanate groups with water by forming urea links. If the prepolymers have two or more iscocyanate groups, as shown in Figure II-2.24, polymers are formed.

1-component heat-curing PUR adhesives can be obtained in various ways. Formulating isocyanate-terminated prepolymers with compounds which are unable to react at room temperature (latent hardeners), but which are capable of reacting with isocyanates at elevated temperatures affords a means of producing heat-curing systems similar to the epoxies. Alternatively blocked isocyanates, i.e. isocyanate adducts which decompose into isocyanates at elevated temperatures, can be mixed with polyols or polyamines. Heating initiates decomposition and curing.

2.2.1.4 Structure/property relationships

No other class of polymers provides such a broad scope for formulation as the polyurethanes. This also applies to adhesives and sealants. Polyurethanes can be formulated to afford very hard and brittle products as well as soft products which are highly extensible. The reaction rate for forming the polymer can be adjusted between seconds and days via the choice of raw materials and additives. Other properties, such as hydrolytic stability, water permeability, water uptake, thermal resistance, UV stability of polyurethanes, can be optimised for almost any application.

A look at the typical structure of polyurethanes reveals that the weight fraction of urethane groups is usually quite low. Where the polymer has been built up by the isocyanate/water reaction, it will contain urea groups in addition to the urethane groups. Typical polyisocyanates have a molecular weight of 150 to 250 g/mol, unlike the polyols, whose molecular weights often exceed 3000 g/mol. The properties of a polyurethane are thus mainly dominated by the polyols used. Most of the polyols are genuinely telechelic, i.e. they have a defined functionality, and the functional groups are located at the end of the chain. This guarantees a very controlled synthesis to high-molecular structures. **In the case of fatty acid-based polyols, endo-OH functions are also present.** The **mechanical properties** of polymers are mainly determined by the molecular weight, the crosslink density and the glass transition temperature.

The **molecular weight** of polyurethanes can be controlled via the stoichiometry and the functionality of the reaction partners. At this point, to keep things simple, only the situation for bifunctional reactants will be considered. When a polyol is made to react in excess, OH-terminated oli-

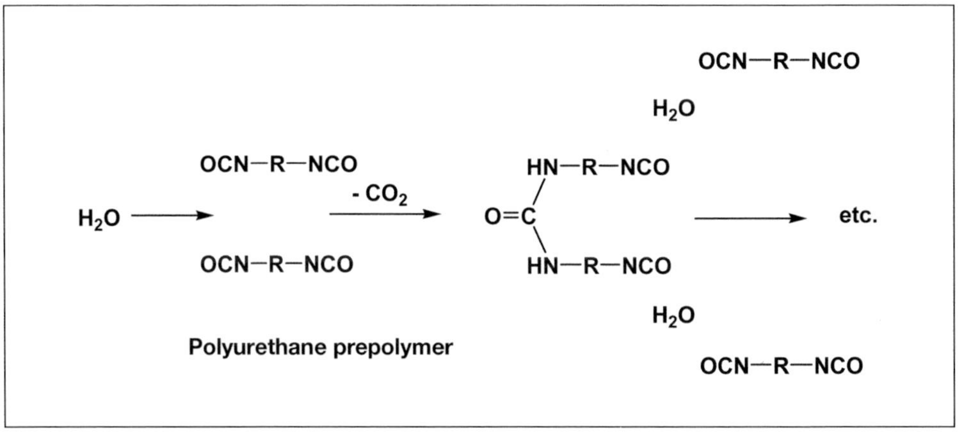

Figure II-2.24: Polymer formation in 1-component polyurethanes

gomers (prepolymers) are obtained, with the actual oligomer depending on the degree of excess. If the isocyanate is added in excess, isocyanate-terminated oligomers (reactive prepolymers) are obtained. Maximum molecular weight can be achieved (see Table II-2.4) only when isocyanate and polyol react stoichiometrically (NCO/OH = 1). A high molecular weight is needed for achieving good mechanical properties. **Usually, a small isocyanate excess (e.g. 1 : 1.1) is used.**

The selective use of triols or triisocyanates in the formulation or the catalysing of specific branching reactions during polymer formation can be used to modify the **crosslink density**. For example, elastomer properties (large network arc length) or thermosetting properties (short network arc length) can be obtained by judiciously blending diols and polyols of higher functionality. Crosslinks in the polymer can also be obtained by controlled reaction of isocyanate to yield allophanates or biurets (see Figure II-2.15). While the two crosslinking reactions mentioned primarily occur at elevated temperatures and lead to thermally unstable links, certain alkaline catalysts can cause trimerisation to isocyanurates, which are very stable.

As the polyols make the greatest contribution to synthesis of the polymer, their physical and chemical properties critically determine the end properties of the resulting polyurethanes. Polyols of low **glass transition temperature**, such as polypropylene oxides and hydroxyl-functionalised polybutadienes, yield polyurethanes of low glass transition temperature. Polyols that have a marked tendency to crystallise, such as some polyesters and poly-THF, lead to **crystalline polyurethanes**. Similarly, polyesters of high glass transition temperature provide glassy polymers.

Highly specific effects result from the use of chain extender s, i.e. short-chain polyols **or also amines, e.g. N,N-propoxylated aniline.**The reaction between stoichiometric quantities of pure short-chain polyols and polyisocyanates affords polar products which have a high concentration of urethane groups. Reaction between a mixture of short and long-chain polyols may suffer from local phase separation, because the highly polar (and usually crystallising) polymer moieties arising from the chain extenders (hard segments) are not compatible with the less polar polymer moieties arising from the high polyols (soft segments). The result is a polyurethane polymer with a block structure, called a **segmented polyurethane** (see Figure II-2.25).

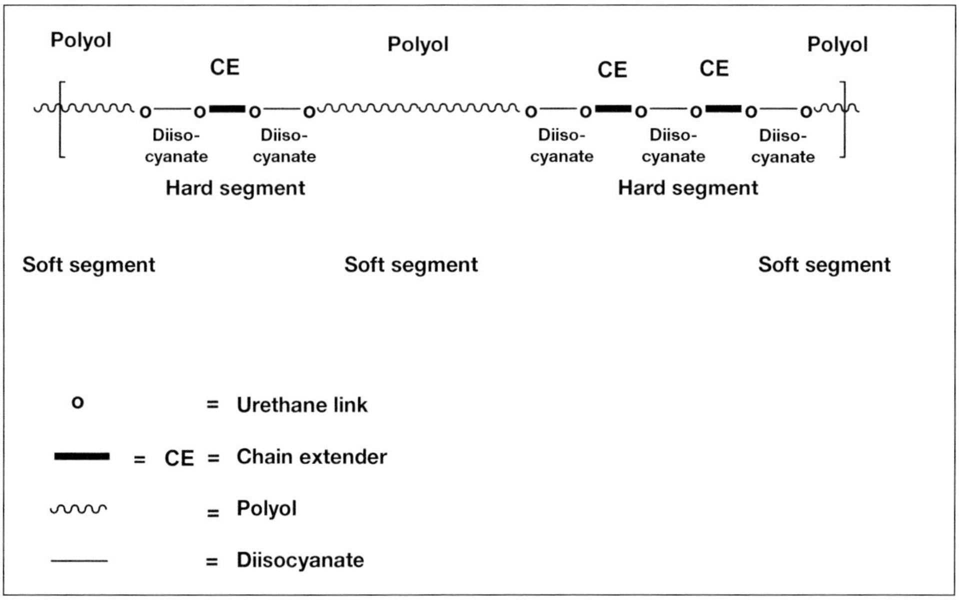

Figure II-2.25: Segmented polyurethane

Such segmented polyurethanes often exhibit incompatibility among the segments, a fact which leads to heterogeneous morphology (see Figure II-2.26). This morphology gives rise to special mechanical properties. Using just diols and diisocyanates produces thermoplastic polymers. These polymers are physically and thermoreversibly crosslinked via the incompatible, high-melting crystalline hard segments.

Such polymers are called thermoplastic elastomers, a term which refers to products which have the properties of an elastomer, such as reversible elongation at ambient temperatures, but which behave like a thermoplastic polymer at elevated temperatures by virtue of the fact that they melt.

As a general rule, the use of short-chain polyols leads to a proportional increase in the mechanical strength of the end polymer.

Compared to epoxies or phenolic resins, polyurethanes generally have inferior **thermal properties**. Above 150 °C the urethane linkage can be thermoreversibly opened to yield isocyanates and alcohols. The order of stability for various urethanes is shown in Figure II-2.27 [1].

Thermal stability of a urethane made from

Aliphatic isocyanate > Aromatic isocyanate

Primary alcohol > Secondary alcohol > Phenol

Figure II-2.27: Thermal stability of urethanes

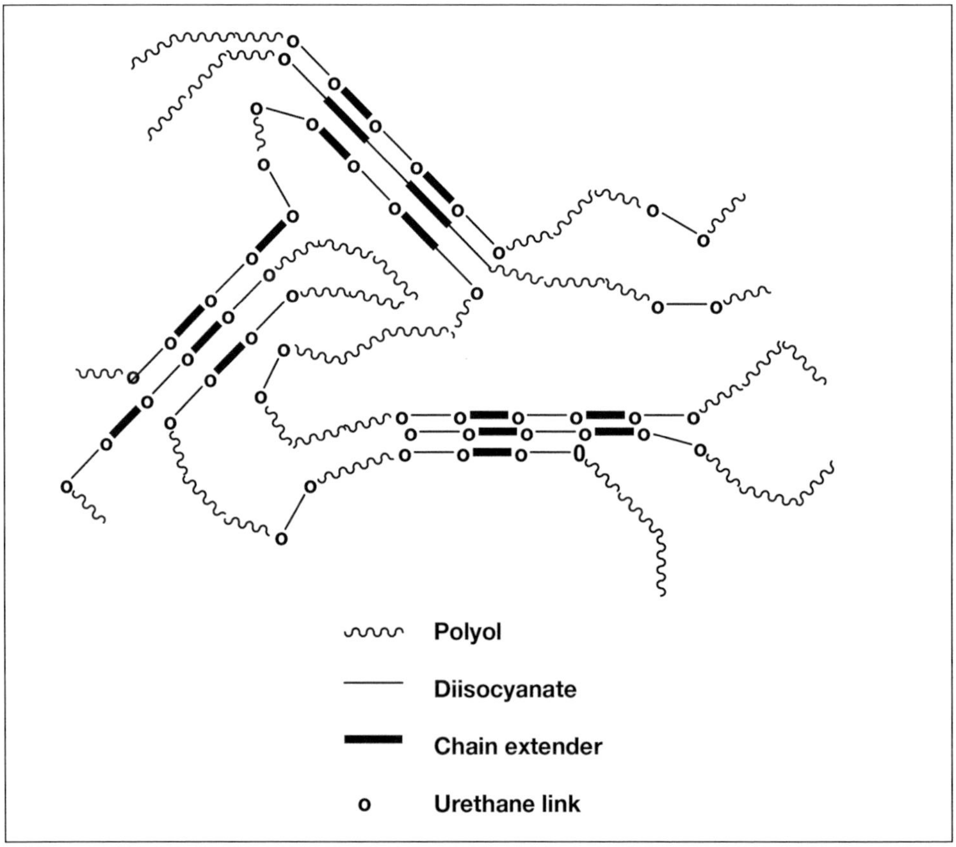

Figure II-2.26: Morphology of segmented polyurethanes

When the other functional groups obtainable from the isocyanates are considered, the order given in Figure II-2.28 is obtained [1].

Thermal stability data found in the literature are sometimes misleading. For instance, Table II-2.1, based on [1], shows the thermal stability limits for various structures.

Table II-2.1: Limits of thermal stability

Urethane	**T-limit [°C]**
n-Alkyl-NHCOO-n-alkyl	250
Aryl-NHCOO-n-alkyl	200
n-Alkyl-NHCOO-aryl	180
Aryl-NHCOO-aryl	120

However, the true heat resistance of polyurethane sealants and adhesives can differ substantially from these values. Thermal stability is also a function of catalytic influences and the presence of other reactants. Permanent exposure of polyurethanes made from aromatic isocyanates and polyethers to a temperature of about 120 °C can effect a marked change in mechanical properties in just one week due to irreversible degradation. In addition, polyurethane adhesives, like all other adhesives, lose their cohesive strength and adhesion at elevated temperatures, with the result that the cited thermal stability limits are of hardly any practical relevance.

Oxidative damage to polyurethanes can be initiated either thermally or under the influence of high-energy radiation (e.g. UV light). It is mainly observed in polyether-urethanes (tertiary C in PPO) or polybutadienes (allyl and double bonds). However, this damage proceeds very slowly in ambient conditions and can be further reduced by adding antioxidants and UV stabilisers. Adhesives are not normally exposed to direct radiation, whereas sealant surfaces are, and tend to become brittle and crack. Polyurethanes based on aromatic, not aliphatic, isocyanates have a tendency to yellow under the influence of light (which is why aromatic types are not frequently used in paints). The yellowing process is thought to be a consequence of free-radical formation and subsequent reaction to quinoid structures (similar to the yellowing of aniline). Figure II-2.29 shows the corresponding orders of stability.

T-limit [approx. in °C]

Isocyanurate	>	urea	>	urethane	>	biuret	>	allophanate
270		160		200 to 120		130		120

Figure II-2.28: Thermal stability of reaction products of isocyanates

Influence of polyol

Polybutadiene < polyether < polyester

Influence of isocyanate

Aromatic isocyanate < aliphatic isocyanate

Figure II-2.29: Oxidative stability of polyurethanes

Table II-2.2: Stoichiometric parametersof important isocyanates

Diisocyanate	Molecular weight [g/mol]	f [val/mol]	Equivalent mass [g/val]	NCO content [mval/kg]	NCO content [%]
MDI	250	2	125	8000	33.6
TDI	174	2	87	11,490	48.3
IPDI	222	2	111	9009	37.8
HDI	168	2	84	11,905	50.0
$H_{12}MDI$	262	2	131	7634	32.1

The **hydrolytic stability** of polyurethanes mainly depends on the polyol used. Polyester-polyurethanes are stable in a non-polar milieu but, unlike polyether- or polybutadiene-polyurethanes, are readily degraded in strongly hydrolytic conditions at elevated temperatures. Stabilisers, such as carbodiimides, can limit this damage. Figure II-2.30 shows the corresponding orders of stability. The hydrolytic stability of the urethane and urea linkages ranks between that of polyesters and polyethers.

The use of polyurethane adhesives is not generally recommended under very strongly hydrolytic conditions, such as in lead batteries containing sulphuric acid. However, polyester polyols are also used in polyurethane gaskets in sanitary applications, as the suitability of these polyols in humid environments is highly dependent on the overall formulation and prevailing conditions.

2.2.1.5 Stoichiometric considerations

Concentrations in PUR technology are stated in different ways by different authors and especially in technical data sheets, Table II-2.2.

Isocyanates

Usually, the isocyanate content is quoted in [%]. This is the quantity of NCO groups in [g] per 100 g of the formulation; hence, 10 g NCO groups in 100 g of the formulation (or 100 g NCO groups in 1 kg) corresponds to 10 %.

For stoichiometric calculations, the equivalents tend to be more important than the quantities. One NCO group has an equivalent mass of 42 g (N = 14, C = 12, O = 16), i.e. 1 equivalent of NCO corresponds to 1 val = 1000 mval ≙ 42 g.
So, 42 g (1000 mval) NCO per kg formulation corresponds to 4.2 % NCO.

4.2 % NCO ≙ 1000 mval/kg

Influence of polyol

Polyester < polyether < polybutadiene

Influence of isocyanate

Aromatic isocyanate < aliphatic isocyanate

Figure II-2.30: Hydrolytic stability of polyurethanes

Table II-2.3: Stoichiometric parameters of important polyols

Polyol molecular weight [g/mol]	f [val/mol]	Equivalent mass [g/val]	OH content [mval/kg]	OH content [%]	OH number [g KOH/kg]
500	2	250	4000	6.8	224
1000	2	500	2000	3.4	112
2000	2	1000	1000	1.7	56
3000	2	1500	667	1.13	37.4
1000	3	333	3000	5.1	168
3000	3	1000	1000	1.7	56
6000	3	2000	500	0.85	28
For comparison					
18 (water)	2	9	111,110	189	6220
90 (butanediol)	2	45	22,220	37.8	1244

Often, the equivalent mass of a compound (molecular weight M in [g/mol]/functionality f in [val/mol]) is stated. This can be calculated as follows:

NCO equivalent mass [g/val] = 42 · 100/% NCO = M/f

e.g. **MDI** (M = 250 g/mol, f = 2 val/mol)
The NCO equivalent massis given by: M/f = 250/2 = 125 g/val
→ 125 g ≙ 1 val ≙ 1000 mval
250 g ≙ 1 mol ≙ 2000 mval
1000 g ≙ 4 mol ≙ 8000 mval
→ 8000 mval/kg ≙ 125 g/val ≙ 8000 · 4.2/1000 = 33.6 % NCO

Polyols

The hydroxyl content is often quoted in [%] and is the quantity of OH groups per 100 g of polyol, i.e. 10 g of polyol in 100 g (or 100 g OH groups in 1 kg) formulation corresponds to 10 % OH, Table II-2.3.

For stoichiometric calculations, the equivalents again tend to be more important than the quantities. An OH group has an equivalent mass of 17 g (H = 1, O = 16),

i.e. 1 equivalent OH = 1 val = 1000 mval ≙ 17 g.

Hence, 17 g OH (≙ 1000 mval) per kg corresponds to 1.7 % OH in the formulation.

Often the OH concentration of polyols is stated as the hydroxyl number in equivalents to [mg KOH/g polyol] or in [g KOH/kg polyol].

1 val (1000 mval) OH corresponds to 56 g KOH (K = 39, O = 16, H = 1).

A hydroxyl number of 56 g KOH/kg polyol corresponds to 1 val (1000 mval) OH per kg polyol.

1.7 % OH ≙ 1000 mval/ kg ≙ 56 g KOH/kg polymer

The equivalent mass of OH groups in a polyol in [g/val] is calculated as follows:

OH equivalent mass [g/val] = 17 · 100/% OH = M/f

e.g. **Polyol** (where M = 1000 g/mol, f = 2 val/mol)
the equivalent mass is calculated as: M/f = 1000/2 = 500 g/val.
→ 500 g ≙ 1 val ≙ 1000 mval
1000 g ≙ 1 mol ≙ 2000 mval
where 1000 mval/kg ≙ 56 g KOH/kg ≙ 1.7 % OH
→ 2000 mval/kg ≙ 2000 · 56/1000 = 112 g KOH/kg polyol
≙ 1.7 · 2000/1000 = 3.4 % OH.

e.g. **Polyether "Desmophen 1380 BT"**
(supplier information: f = 3 val/mol, OH number = 385 g KOH/kg polyol)
where 1000 mval/kg ≙ 56 g KOH/kg ≙ 1.7 % OH
→ 385 g KOH/kg polyol ≙ 385 · 1000/56 = 6875 mval/kg
≙ 1000 · 1000/6875 = 145 g/val
≙ 6875 · 1.7/1000 = 11.7 % OH
→ M = f · equivalent mass = 3 · 145 = 435 g/mol

e.g. **Chain extender** 1,4-butanediol (M = 90 g/mol, f = 2 val/mol)
→ the equivalent mass is calculated as: M/f = 90/2 = 45 g/val
90 g ≙ 1 mol ≙ 2000 mval
1000 g ≙ 2000 · 1000/90 = 22,220 mval
where 1000 mval/kg ≙ 56 g KOH/kg ≙ 1.7 % OH
→ 22,220 mval/kg ≙ 22,220 · 1.7/1000 = 37.8 % OH
≙ 56 · 2,220/1000 = 1244 g KOH/kg polyol

e.g. **Water** (M = 18 g/mol, f = 2 val/mol)
→ equivalent mass = 19/2 = 9 g/val
18 g ≙ 1 mol ≙ 2000 mval
1000 g ≙ 111,10 mval
where 1000 mval/kg ≙ 56 g KOH/kg ≙ 1.7 % OH
→ 111,110 mval/kg ≙ 111,110 · 1.7/1000 = 189 % OH
≙ 111,110 · 56/1000 = 6220 g KOH/kg

The relevant stoichiometric parameters of important isocyanates and polyols are summarised in Table II-2.2 and Table II-2.3.

NCO/OH ratio

The stoichiometric conditions for a reaction between isocyanates and polyols are usually given by the NCO/OH ratio. This can be calculated as follows:

Method 1:
NCO/OH = Σ equivalents NCO/Σ equivalents OH

or Method 2:
NCO/OH = [Σ weight parts polyisocyanate 17 %NCO]/[Σ weight parts polyol 42 %OH]

Example

The NCO/OH ratio is to be calculated for the reaction between 40 g MDI and 100 g of a diol having a hydroxyl number of 112 g KOH/kg (suppliers often fail to state the correct units in their data sheets).

Method 1

The isocyanate equivalents are calculated as follows:
250 g (molecular weight of MDI) = 2000 mval (functionality of MDI = 2 val/mol); hence 40 g MDI corresponds to 320 mval NCO.
A hydroxyl number of 112 g KOH/kg corresponds to $112 \cdot \frac{1000}{56} = 2000$ mval/kg, i.e. the quantity of 100 g polyol corresponds to 200 mval OH.
The NCO/OH ratio is then $\frac{320}{200} = 1.6$.

Method 2

MDI has an NCO concentration of 33.6 %, and the polyol has a hydroxyl number of 112 g KOH/kg corresponding to a hydroxyl content of $112 \cdot \frac{1.7}{56} = 3.4$ %.
Thus, for the NCO/OH ratio:

$$40 \cdot 17 \cdot \frac{33.6}{100} \cdot 42 \cdot 3.4 = 1.6$$

This corresponds to a 60 % excess of isocyanate. An NCO-terminated prepolymer is obtained.

NCO/OH ratio and prepolymers

The properties of the products resulting from the reaction between polyisocyanates and polyols to yield polyurethanes depend critically on the stoichiometric ratios. As already shown, the reaction of bifunctional polyols with bifunctional isocyanates leads to a high molecular weight polymer only in the case of equivalent quantities. In case of excess polyol, the molecular weight decreases accordingly, and OH-terminated oligomers are obtained. Similarly, excess NCO leads to NCO-terminated, reactive prepolymers. The molecular weight of these prepolymers can be controlled via the stoichiometric ratio of NCO to OH. Such reactive prepolymers are used in 1- and 2-component PUR adhesives. As shown in Table II-2.4 for MDI (M = 250 g/mol, f = 2 val/mol) and a polyether polyol (M = 2000 g/mol, f = 2 val/mol), the molecular weight of these prepolymers can be estimated (assuming identical reactivity on the part of the functional groups) as a function of the stoichiometric ratio of the reactants.

If reactants with a functionality greater than two are made to react, a crosslinked polymer system can be obtained. Depending on the functionality and conversion rate, this crosslinked polymer can also be obtained under non-stoichiometric conditions. For structural adhesives, crosslinked polymers are advantageous because they are more resistant to mechanical, chemical and thermal influences. The **gel point** characterises the moment of a reaction when a quasi infinitely large molecule is generated by increasing branching and ultimately crosslinking. The corresponding theory, developed by Flory and extended by Stockmayer, can be applied to the reaction yielding crosslinked polyurethanes [18]. Where reactive, NCO-terminated prepolymers are to be synthesised from higher functional reactants, for the **critical NCO/OH ratio**, the Flory-Stockmayer equation, can be used to calculate where gelling of the batch will occur. On the other hand, a knowledge of the critical NCO/OH ratio provides information on the extent to which crosslinked PUR adhesiv e s are obtained from 2-component PUR adhesives when, for instance, the mixing ratio is not as it should be.

Table II-2.4: Molecular weight as a function of the NCO/OH ratio

NCO/OH ratio	Molecular ratio	Most probable molecule	Molecular weight [g/mol]
2	2/1	M-P-M	2500
1.5	3/2	M-P-M-P-M	4750
1.33	4/3	M-P-M-P-M-P-M	7000
1.25	5/4	M-P-M-P-M-P-M-P-M	9250
1.2	6/5	M-P-M-P-M-P-M-P-M-P-M	11,500
1.17	7/6	M-P M-P-M-P-M-P-M-P-M-P-M	13,750
1.0	1/1	... M-P-M-P-M-P-M-P-M-P-...	endless

M = MDI, molecular weight = 250 [g/mol],
P = polyether, molecular weight = 2000 [g/mol] = urethane linkage

According to [19], the **gel point** is given by Equation (1):

$$(P_A \cdot P_B)_{gel} = (f_{OH} - 1)^{-1} \cdot (f_{NCO} - 1)^{-1} \qquad \text{Equation (1)}$$

where f_i is the functionality of component i [val/mol]
P_i is the conversion of component i [%/100] ($P = 1 \triangleq 100$ % conversion)

For mixtures of various polyols or isocyanates, an average functionality f_{av} can be calculated as follows from Equation (2):

$$f_{av} = \frac{\sum N_i \cdot f_i}{\sum N_i} \qquad \text{Equation (2)}$$

where N_i is the amount of substance i (in [mol] of component i)
f_i is the functionality of this monomer in [val/mol]

Example 1

A triol (f_{OH} = 3 val/mol) and a diisocyanate (f_{NCO} = 2 val/mol) react to completion. At which NCO/OH ratio can gelling be expected if isocyanate is used in excess?

$P_{OH} = 1$ (due to NCO excess and full conversion)

From equation (1): $(3 - 1)^{-1} \cdot (2 - 1)^{-1} = P_{NCO} = 0.5$ (or 50 %).

At the gel point, 50 % of the isocyanate has reacted if full conversion is achieved. This means that, at the end of the reaction, 50 % of the isocyanate must still be present. Therefore, compared to the polyol at the beginning of the reaction, twice the stoichiometric quantity of isocyanate has to be used (reciprocal value of the conversion) if gelling is to be avoided. This means that gelling occurs at a critical NCO/OH ratio< 2. At this excess of NCO, 50 % of the isocyanate still remains at the end of the reaction ($P_{NCO} = 0.5$).

Example 2

For a reaction with an isocyanate, a mixture of a triol (f = 3 val/mol) and a diol (f = 2 val/mol) is used. What is the average functionality of the polyol mixture if the molar ratio of triol/diol is 2/1?

From equation (2) $f_{av} = \frac{2 \cdot 3 + 1 \cdot 2}{(2 + 1)} = 2.67$ val/mol

Example 3

Commercial hydroxyl-functionalised polybutadiene ($f_{OH} = 2.5$ val/mol) and a diisocyanate, such as MDI, are to be converted into a prepolymer. What is the critical NCO/OH ratio for gelling?

Given an isocyanate excess and full conversion $P_{OH} = 1$.

From Equation (1), the gel point is given by: $(2.5 - 1)^{-1} \cdot (2 - 1)^{-1} = P_{NCO} = 0.67$

This means that 67 % of the original quantity of the isocyanate has reacted with the polyol.

So, during the reaction, from 100 parts of the isocyanate, 67 parts have reacted with the polyol. Consequently, at the beginning of the reaction, the excess of isocyanate must have been $\frac{1}{0.67} = 1.5$ ($NCO/OH_{critical} = \frac{1}{0.67} = 1.5$). Therefore, it is necessary to choose an NCO/OH ratio greater than 1.5 if gelling is to be avoided in the synthesis of a prepolymer.

Generally, simplified Equation (3) holds for diisocyanates, isocyanate excess and full conversion:

$$NCO/OH_{crit.} = f_{OH} - 1 \qquad \text{Equation (3)}$$

Example 4

A diol/triol mixture is to be used to make a 2-component PUR adhesive. What is the minimum mixing tolerance of polyol system and diisocyanate needed that will still produce a crosslinked polyurethane?

Assumption: Full conversion and no side-reactions.

If the molar ratio of diol and triol is 1 : 1; according to Equation (2), the average functionalityis 2.5 val/mol.

From Eequation (1), the gel point is: $(2.5 - 1)^{-1} \cdot (2 - 1)^{-1} = P_{NCO}$ (or P_{OH}) = 0.67

As in Example 3, when the isocyanate is in excess, the polyol system has reacted completely ($P_{OH} = 1$). The conversion rate for the isocyanate is 67 %, and so the critical NCO/OH ratio is 1/0.67 = 1.5.

At full conversion, gelling occurs below an NCO/OH ratio of 1.5.

When the polyol is in excess, the critical OH/NCO ratio for full isocyanate conversion is again $\frac{1}{0.67} = 1.5$, corresponding to an NCO/OH ratio of 0.67 (the NCO/OH ratio is the reciprocal of the OH/NCO ratio).

Therefore, within an NCO/OH ratio of 0.67 to 1.5 and given full conversion, gelling occurs in any event. The system has a relatively broad mixing tolerance.

Table II-2.5: Functionalities of polyethers

Molecular weight of PPO and apparent functionality	Actual functionality KOH process [val/mol]	Actual functionality Impact process [val/mol]
2000 diol	1.94	1.99
4000 diol	1.67	1.98
6000 triol	2.14	2.94

Synthesis of prepolymers and content of free monomers

For safety labelling and establishing the toxicology of formulations containing isocyanate, it is essential to know the content of free reactive monomers. Monomeric isocyanates are determined by liquid and gas chromatography after reaction of the reactive isocyanate groups with chromophores or inhibitors. The standardized measurement method is described in ASTM D7252-17. For MDI-based formulations, it can be calculated with relatively good accuracy from equation (4) [20]:

$$m_{res}\,[\%] = [(NCO/OH - 1)/(NCO/OH)]^2 \cdot m_{orig}\,[\%] \qquad \text{Equation (4)}$$

where:
m_{res} is the content of residual monomer, in [%]
m_{orig} is the amount of monomer originally used, in [%]
NCO/OH is the molar stoichiometric ratio

Example

An MDI -based prepolymer was made with 10 % MDI in the original formulation at an NCO/OH ratio of 2.
From Equation (4):

$$m_{res}\,[\%] = [(2 - 1)/2]^2 \cdot 10\,[\%] \qquad \text{Equation (4)}$$

The residual monomer content is therefore 2.5 %.

If, in the same example, the NCO/OH ratio is lowered to 1.5 and the 10 % MDI in the original formulation is retained, the residual monomer content similarly computes to 1.1 %:

$$m_{res}\,[\%] = [(1.5 - 1)/1.5]^2 \cdot 10\,[\%] = 1.1\ \%.$$

These figures are reduced, of course, by modification with fillers, plasticisers etc. The use of polyfunctional reactants bearing different reactive groups also reduces the monomer content.

True functionality of raw materials

Many polyethers, especially those made by the traditional KOH process, do not have an exact functionality of 2 or 3 val/mol, as is ideally assumed. With growing molecular weight during synthesis, side-reactions become more prevalent and no functional OH group is formed at the end of the chain. As shown in Figure II-2.31, propylene oxide isomerises in a rearrangement reaction to yield allyl alcohol. The true functionalities of polyols are therefore always somewhat lower [21].

The use of special catalysts in conjunction with recently developed manufacturing processes (Impact process, ARCO/Bayer) affords high-molecular polyols of defined functionality (Acclaim polyols). Table II-2.5 presents an overview of the functionalities of polyethers [21].

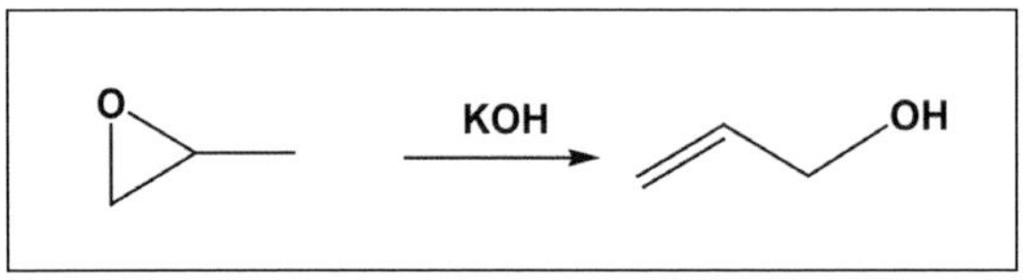

Figure II-2.31: Rearrangement of propylene oxide

2.2.1.6 Reactivity

Statistical considerations regarding the content of free monomers proceed from the assumption that the isocyanate groups in the monomer and in the growing polymer are of identical reactivity. This assumption is just a rough estimation, even for the symmetrical MDI. The relative reactivities can even be influenced by the catalyst system. Table II-2.6 gives an overview of the relative reactivities of various isocyanates [2, 3, 22].

Table II-2.6: Reactivity of isocyanate groups with alcohols

Isocyanate	k_1	k_2	k_1/k_2
2,4-TDI	400	33	12
4,4'-MDI	320	110	2.9
2,4'-MDI	50	2	25
HDI	1	0.5	2
IPDI	0.62	0.23	2.7

k_1, k_2 = relative rate constants of the two isocyanate groups in reaction with alcohols

If the reacting molecules contain functional groups of different reactivity, a smaller molecular weight distribution and a lower content of residual monomers result automatically. For instance, if a diol is made to react with a diisocyanate having two isocyanate groups of different reactivity at an NCO/OH ratio of 2, a significantly lower content of residual monomer is obtained. In this case, the more reactive isocyanate groups react quickly with the polyol, so that the isocyanate end groups in the resulting prepolymers are less reactive. These end groups no longer compete with the more reactive monomers. Were both isocyanate groups of the same reactivity, residual polyols would be equally likely to react with the isocyanate monomers and with the prepolymers already formed. The reaction with a prepolymer would lead to an increase in molecular weight, so that ultimately monomeric isocyanate would still remain.

The same holds for estimation of the gel point. For reactants bearing functional groups of different reactivity, the gel point is reached at a lower NCO/OH ratio than is predicted by the Flory-Stockmayer equation. This renders the manufacturing process more reliable, and the end products have a lower viscosity due to a smaller molecular weight distribution.

2.2.1.7 Side-reactions

The reaction between isocyanates and polyols does not proceed without side-reactions. Although it is common practice to use catalysts for the desired reaction, side-products are obtained at elevated temperatures, especially. This includes the formation of allophanates from urethanes, the formation of biurets from ureas and even the trimerisation of isocyanates to isocyanurates (see Figure II-2.15).

At 100 °C, for example, the rate at which biurets are formed from ureas can be as much as 10 % of the rate of urethane formation [3]. All these reactions lead to branching of linear molecules and therefore to an increase in functionality. This leads to unexpected gelling. The synthesis of prepolymers based on KOH-catalysed polyethers, especially, should not be performed without acidic stabilisers.

2.2.1.8 Water content

Water reacts readily with isocyanates to form urea (see Figure II-2.14). The small water molecule has an OH content of 111,110 mval OH/kg. Even tiny amounts of water (weight fraction) may provide huge equivalents for the reaction. Water is mostly introduced into the formulation via the polyols and the fillers. Most polyols contain about 0.05 to 0.1 % of water, while the water content of fillers (usually below 0.1 %) can fluctuate extensively with storage conditions. It takes comparatively little isocyanate to transform high-molecular polyols into prepolymers. If too much water is introduced with the raw materials, this can influence the true stoichiometry significantly. In critical cases, it is always advisable to perform a quantitative analysis for water in the raw materials (Karl-Fischer titration), with subsequent drying. On a laboratory scale, the polyol component is typically dried with 5 to 10 wt.% freshly activated molecular sieve (3 Ångström) over the weekend on a roller stand. The amount of molecular sieve required correlates with the polarity of the polyol component.

Example

A prepolymer is to be made from MDI (125 g/val) and a triol of M = 6000 g/mol (2000 g/val) at an NCO/OH ratio of 2.3 (from Equation (3), the critical NCO/OH ratio for gelling is 2.0).

Calculation: $125 \cdot 2.3 = 288$ g (2300 mval) MDI and
2000 g (1000 mval) polyol.

If we assume a water content of 0.1 % in the polyol (for 2000 g, this corresponds to 2 g), the true stoichiometry changes as follows:

2 g water corresponds to 222 mval OH, i.e. the originally calculated 2300 mval NCO is reduced to just 2078 mval. Thus, the true NCO/OH ratio is just 2.08. If just a small amount of isocyanate were to undergo other side-reactions or be consumed by water from the filler, the batch could undergo gelling in reality.

2.2.1.9 Formulation of reactive PUR adhesives

Reactive PUR adhesives are formulated with **reactive monomers and oligomers** (typically 10 to 100 % of the formulation) based on polyisocyanates and polyols. **Fillers** (up to 60 %), such as carbonates, silicates etc. can lower the price, while active fillers, such as carbon black and fumed silica (up to approx. 20 %) can influence rheology and improve the mechanical characteristics. **Plasticisers** (up to approx. 30 %), usually phthalates, reduce the processing viscosity and can lead to swelling of the end product. Fillers, plasticisers and polyols have to be tested for their water content and dried as necessary. Therefore, formulations often contain drying agents based on molecular sieves. Typical **additives** are catalysts, stabilisers, rheological additives, flame retardants [23–25].

2.2.1.10 2-component PUR adhesives and sealants

Formulation

2-component PUR adhesives usually cure by means of a polyaddition between an isocyanate and a polyol component. For controlled polymer synthesis, polyaddition especially requires the correct stoichiometric ratio. This imposes high demands on mixing quality and mixing tolerance. Often, systems are formulated such that a certain excess of isocyanate is present. This can have positive effects regarding adhesion to certain substrates, e.g., if isocyanate groups can react with functional groups on the substrate surface. In addition, after application to the substrates, the excess of isocyanate eventually reacts like a 1-component system by curing with moisture to yield polymeric ureas. As polyols typically have a much higher molecular weight than monomeric isocyanates, it is often helpful to replace some of the monomeric isocyanates with isocyanate-terminated prepolymers for the purpose of achieving a more favourable mixing ratio. The pot life and setting time are mainly regulated via the catalyst system. The mechanical strength (cohesion) can be adjusted via the crosslink density, i.e. the functionality and the molecular weights of the monomers.

The **A-component** (polyol component) consists of the polyol, and any necessary chain extenders, fillers, plasticisers, stabilisers, catalysts; in other words, most ingredients are put into the non-reactive polyol part. Mostly, mixtures of different polyols are used, but mutual compatibility must be tested in advance.

Where **polyesters** are used, liquid grades are necessary. These are typically mixed polyesters with molecular weights ranging from 500 to 3000 g/mol and a functionality of 2 to 3 val/mol (or even higher). As a result of their structure, polyesters are quite polar. This polarity renders the corresponding adhesives highly adhesion, especially to many polar substrates. The limited hydrolytic stability is a disadvantage, but they do offer resistance to light and oxygen. Polycarbonates are a special class of polyesters that feature improved hydrolytic stability.

A-components based on **polyethers** yield low-cost products (the price of polyethers is about half that of polyesters) boasting very good hydrolytic stability. The corresponding adhesives often feature poor adhesion, especially to plastic substrates, and so polyethers are frequently mixed with polyesters or castor oil. Polyethers are low in viscosity, a fact which can be advantageous for processing. Low-molecular polyether triols yield polymers with a high crosslink density and therefore high mechanical strength. The use of high molecular weight polyethers leads to highly flexible polyurethanes that have excellent low-temperature performance.

Furthermore, both biobased polyester and polyether polyols are now commercially available. Examples are the polyether polyols based on carbon dioxide from Covestro (trade name cardyon) and polyester-polyols made from renewable raw materials marketed under the trade name "Dynacoll Terra" from Evonik. These bio-based polyesters contain between >30 and 100 ppw of renewable monomers. According to the building block system amorphous, liquid and crystalline grades are offered. An overview of current developments can be found in [26].

In combination with compatible low-molecular chain extenders, **hydroxyl-functionalised polybutadiene** (Polybd) can be used to formulate highly elastic 2-component PUR adhesives offering excellent low temperature resistance, good hydrolytic stability and low water vapour permeability. Without the use of chain extenders, products would be too soft for adhesives applications (see Chapter II-2.2.1.2).

Due to the relatively low molecular weight of about 1000 g/mol and a functionality of about 3 val/mol, **castor oil** (often in combination with low-molecular weight polyethers) yield polymers of high cohesive strength, but also significant brittleness. Castor oil is a very inexpensive polyol.

Where the application does not call for any special properties, one of the preferred **fillers** is chalk. Milled chalk can be obtained for a very low price, while precipitated, coated chalks sometimes show suitable rheological benefits. Coarse fillers can convey the impression of better adhesion, because they promote cohesive failure of the bond. Talcum (Mg silicate), barytes ($BaSO_4$) and silica (SiO_2) are other important fillers. Abrasive fillers, such as silica, are sometimes unsuitable for machine processing.

Among **additives**, catalysts play the most important role in the processing of adhesives. **Catalysts** based on metals like i.e. DBTL catalyse mainly the NCO/OH reaction, and to a lesser extent the NCO/water reaction. Especially, amine catalysts (e.g. DABCO) catalyse not only the NCO/water reaction, but also the oligomerisation of isocyanates. Where these catalysts are used, therefore, the moisture content in the raw materials as well as in the substrates should be minimised, as otherwise bubbling (CO_2) may occur. The common use of metal and amine catalysts can lead to marked synergistic effects [3]. Catalysts which catalyse the NCO/OH reaction and also to a certain extent the trimerisation of isocyanates provide desirable long pot lifes and, due to increasing branching during the reaction, rapid curing (favourable pot life /curing ratio). Proper use of cata-

Table II-2.7: Current metal catalysts from Evonik [65]

Product	Description
Kosmos T 12 N	strong gelation catalyst, Dibutyltindilaurate
DABCO T 820	strong gelation catalyst with good hydrolytic stability
Kosmos T 9	stannous octoate, strong gel catalyst, low hydrolytic stability
Kosmos T 900	stannous neodecanoate, strong gel catalyst, low hydrolytic stability
Kosmos T 100	strongest gel catalyst particularly suitable for fast curing applications such as spray coatings and RIM, suitable in particular for use with aliphatic isocyanates
Kosmos 16	delayed-action gel catalyst with good hydrolytic stability and improved EHS profile compared to standard DBTL
DABCO T 131	strong gel catalyst with good hydrolytic stability which provides a more delayed action than DBTDL

lysts allows the pot lives to be varied from a matter of minutes up to hours, to suit the requirements of the application.

Although the risks of tin-based catalysts have long been known, metal catalysts still play an important role in the synthesis of prepolymers and the curing of coatings, adhesives, sealants and elastomers. The main **advantages** of these catalysts are:

- low consumption values
- high catalytic activity in the curing of aliphatic systems
- delayed curing profiles
- low effect on colour

These advantages are offset by some **disadvantages**:

- classified as CMR (carcinogenic, mutagenic, toxic to reproduction).

For DBTDL, for example, this means the classifications "H360FD Germ cell mutagenicity category 1B" and "H341 Toxic to reproduction category 2".
For DOTL in Europe, the following labels are planned: Change from 0 CMR to "H360FD Germ Cell Mutagenicity Category 1B", which will appear in SDS from 0.1 % by weight.
Tin ethyl hexanoate will be labeled as toxic to reproduction 1B from 2023.

Possibilities for substituting metal catalysts with catalysts with lower toxicological problems and current metal catalysts from Evonik are shown in Figure II-2.32 and Table II-2.7 [65].

Stabilisers are used for different functions. Antioxidants and light stabilisers (e.g. sterically hindered phenols) are mainly employed for products that contain polyether or polybutadiene in anticipation of oxidative damage by exposure to heat and light. The hydrolytic stability of polyester-polyurethanes can be improved by adding carbodiimides (2 to 4 %). For in situ drying of the raw materials, a molecular sieve, sometimes as a paste in polyols, is usually added in a proportion of 2 to 5 %. Fumed silica (approx. 5 %) can serve to adjust the **rheology**, but it can also help to

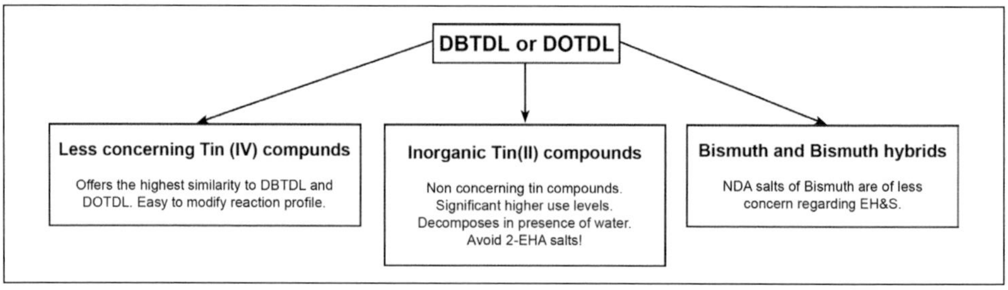

Figure II-2.32: Alternative metal catalysts with fewer toxicological risks (non-CMR metal catalyst) from Evonik [65]

avoid settling of fillers. Alternatively, small amounts of amines (approx. 0.5 %) can be used. Amines react very rapidly with the isocyanate hardener to yield ureas which also exert a pronounced rheological influence.

The **B-component** (hardener, isocyanate component) consists of monomeric isocyanates or mixtures of prepolymers terminated with oligomeric isocyanate. Usually, liquid isocyanates are preferred. For a more favourable mixing ratio, modification with prepolymers is possible. In addition, fillers, plasticisers and stabilisers might be added. The B-component has to be stored in the absence of moisture and, if warranted by the content of monomeric isocyanate, must be labelled as hazardous goods.

Aliphatic isocanates are now also available biobased, e.g. by fermentation of glucose to pentamethylene diamine and subsequent phosgenation to pentamethylene diisocyanate. The trimer of this molecule is marketed under the product name DESMODUR eco N 7300 by Covestro (Figure II-2.33). [66]

Alternative 2-component PUR concepts employ isocyanate-terminated prepolymers as base component and either water or an amine-based hardener which is liberated by water and which reacts with isocyanates (see also Chapter II-2.2.1.11).

Evaluation

2-component PUR adhesives are characterised by a product profile which is highly adaptable to the requirements of an application. It is a simple matter to formulate products of high mechanical strength and very high extensibility. Adhesion to most substrates can be classified as being very good, and hydrolytic stability and thermal resistance as good. The fact that a 2-component system and dangerous isocyanates are used is disadvantageous.

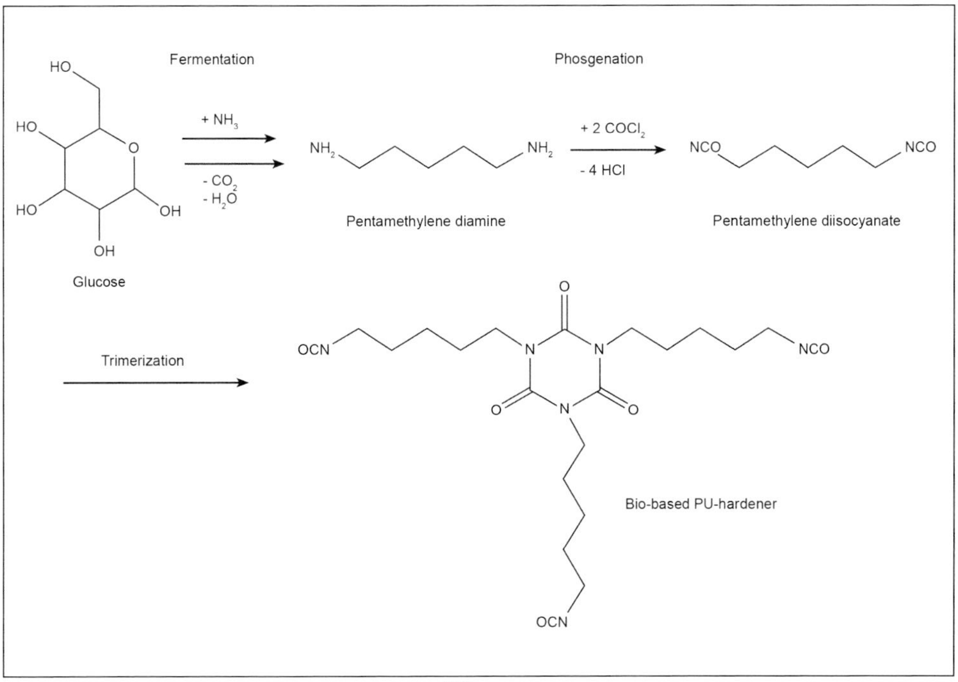

Figure II-2.33: Aliphatic bio-based PU hardener [66]

Advantages

+ broad formulation latitude
+ gap filling
+ very good adhesion, even to plastics and metals
+ good heat resistance
+ good impact strength
+ very good low-temperature properties
+ good hydrolytic stability

Disadvantages

- 2-component system
- isocyanates are dangerous goods

Fields of application

2-component PUR adhesives are used wherever 1-component systems cannot be used due to substrate or process-related circumstances, such as lamination of metal sheets. In addition, they are usually cheaper because there is no need for prepolymer synthesis in the manufacturing process.

Typical applications are to be found in the manufacture of sandwich element s for **utility vehicles** (caravans), in the **woodworking industry** as flooring adhesive for parquet floors, and the manufacture of metal-laminated flooring elements for computer rooms.

In the **automotive industry**, they are used for attaching plastic fasteners to the metal body and for bonding different plastic parts to each other. Another major application is the manufacture of headlamps, where the lens of glass or polycarbonate is bonded to the plastic body.

Finally, it should be noted that due to their chemical and morphological versatility, polyurethanes could be susceptible to manufacturing conditions (curing process, humidity). The industry reports "maturation phenomena", i.e. unstable mechanical properties of PU elastomers and PU-based adhesive bonds immediately after manufacture. Research showed that the properties of a fully cross-linked PU stored dry remain stable during a three-week observation, suggesting that maturation can be prevented by full cure and dry conditions. In contrast, chemical crosslinking after insufficient curing is likely to lead to maturation phenomena [27, 28].

Examples

Polyether as base material

Table II-2.8 lists a 2-component PUR adhesive based on polyether and castor oil [29].
The NCO/OH ratio can be calculated as follows:

$$\text{NCO/OH} = \frac{[\Sigma \text{ pbw polyisocyanate} \cdot 17 \cdot \%\text{NCO}]}{[\Sigma \text{ pbw polyol} \cdot 42 \cdot \%\text{OH}]}$$

$$\text{NCO/OH} = \frac{(38 \cdot 17 \cdot 31.5)}{(23.4 \cdot 42 \cdot 11.7 + 9.3 \cdot 42 \cdot 3.4 + 16.4 \cdot 42 \cdot 5)} = 1.25$$

i.e. a defined excess of isocyanate is used. However, this is not critical, as the gel point is easily reached due to the high functionality of the rea ctants.

As shown in Chapter II-2.2.1.5, the average functionality computes to $f_{OH} = 2.88$ for the polyols and to $f_{NCO} = 2.3$ for the isocyanates.

Table II-2.8: 2-component PUR adhesive based on polyether polyol and castor oil

Component A

Item	Raw material	Parts by weight*	Functional group [mmol]
1	Polyethertriol	50	342.5
2	Polyetherdiol	20	40.0
3	Castor oil	30	85.7
4	Rheological additive	4	
5	Filler	100	
6	Drying agent **	10	

* Parts by weight in [g]

** Contains polyol

Component B

7	polyisocyanate	81.3	609.0

Item 1: Polypropylene oxide, "Desmophen 1380 BT", hydroxyl number = 385 g KOH/kg, OH equivalent weight = 146 g/mol, Covestro

Item 2: Polypropylene oxide, "Desmophen 1112 BD", hydroxyl number = 107.4-115.4 g KOH/kg, OH equivalent weight = 500 g/mol, Covestro

Item 3: „Albodry" Rizinusöl PU Qualität, hydroxyl number – 160 g KOH/kg, Alberdingk Boley

Item 4: Fumed silica, "Aerosil 200", Evonik

Item 5: Milled chalk, "Omya BSH-OM", Omya

Item 6: Molecular sieve approx. 50 % in castor oil, paste, "UOP (Baylith) L-paste", Covestro

Item 7: Oligomeric, liquid MDI, "Desmodur VK 10", f = 2.3 val/mol, NCO = 31.5 % ≙ 7500 mval/kg, isocyanate equivalent weight = 133 g/mol ≙ 133 g/val, Covestro

From the figures:

$N_{Item1} = 0.053$ mol, $f_{Item1} = 3$ val/mol,
$N_{Item2} = 0.0095$ mol, $f_{Item2} = 2$ val/mol,
$N_{Item3+6} = 0.00163$ mol, $f_{Item3} = 3$ val/mol

we can compute the average functionality of the polyols:

$$f_{av} = \frac{\Sigma\, Ni \cdot fi}{\Sigma\, Ni} = \frac{0.053 \cdot 3 + 0.0095 \cdot 2 + 0.0163 \cdot 3}{0.0788} = 2.88 \text{ val/mol}$$

For the critical NCO/OH ratio, with $P_{OH} = 1$ and $f_{NCO} = 2.3$, it follows

$$(P_{OH} \cdot P_{NCO})_{gel} = (f_{OH} - 1)^{-1} \cdot (f_{NCO} - 1)^{-1} \rightarrow P_{NCO} = \frac{1}{1.88} \cdot 1.3 = 0.41$$

In other words, the critical NCO/OH ratio at which gelling sets in is 1/0.41 = 2.44; the NCO/OH ratio in the formulation is just 1.25, so crosslinking occurs in any case at full conversion.

The mixture has a pot life of about 2 h and, after 24 h on beech wood, a tensile shear strength of 9 MPa can be achieved. The adhesive can be used for applications in the furniture industry, in the construction industry, and for packaging.

Dollhausen developed a simple formula for calculating the mixing ratio for **100% conversion** of polyol and isocyanate, which has proved to be particularly practical [70]:

$$\mathbf{D} = 2.47 \cdot \frac{\%[OH]}{\%[NCO]} \cdot \mathbf{B} \qquad \text{Equation (5)}$$

Table II-2.9: 2-component PUR adhesive based on hydroxyl-functionalised polybutadiene

Component A

Item	Raw material	Parts by weight*	Functional group [mmol]
1	polyol	45.7	38.4
2	chain extender	7.8	73.6
3	filler	45.7	
4	coupling agent	0.5	
5	defoaming agent	0.05	

* Parts by weight in [g]

Component B

6	polyfunctional isocyanate	17	117.6

Item 1: OH-functionalised polybutadiene, "Polybd PolyBD R45, hydroxyl number = 47.1 g KOH/kg, OH equivalent weight = 1190 g/mol, Cray Valley

Item 2: Highly reactive amine initiated polyether polyol, "Voranol 220-530", hydroxyl number = 530 g KOH/kg, OH equivalent weight = 106 g/mol, Dow Chemical

Item 3: Hydrated aluminum silicate, "Dixie Clay", Vanderbilt

Item 4: Epoxy silane, "Silquest A-187", Momentive

Item 5: Solution of foam-destroying polymers, silicone-free, Byk-A 535, Byk-Chemie GmbH

Item 6: MDI oligomer, liquid, "Isonate M 143", M = 332 g/mol, f = 2.3 val/mol, NCO = 29.2 % ≙ 6952 mval/kg, isocyanate equivalent weight = 144.5 g/mol ≙ 144.5 g/val, Dow Chemical

Mixing ratio (weight): A/B = 100/17

Table II-2.10: Characteristic data for hydroxyl-terminated polybutadienes

Characteristic	"Poly bd R-45 HTLO" (Cray Valley)	"Liquiflex H" (Petroflex)
Molecular weight	2800	2700
Hydroxyl content [meq/g]	0.84	0.79
Hydroxyl content [% OH]	1.4	1.3
Hydroxyl functionality	2.4 to 2.6	2.1
Glass temperature [°C]	-75	

where

D	Desmodur by weight (isocyanate)
B	Baycoll by weight (polyol)
%[OH]	Hydroxyl content in % (Table II-2.3)
%[NCO]	Isocyanate content in % (Table II-2.2)

Hydroxyl-functionalised polybutadiene as base material

Table II-2.9 lists the ingredients of a 2-component PUR adhesive based on hydroxyl-functionalised polybutadiene. To increase cohesion, a short-chain diol based on propoxylated aniline is used. As this is a tertiary amine, it also provides catalytic activity [29].

Characteristic data for different hydroxy-terminated polybutadienes are summarised in Table II-2.10.

Table II-2.11: Biobased 2-component PUR adhesive PUR adhesive based on Cashew nut shell liquid

Component A

Item	Raw material	Parts by weight*	Functional group [mmol]
1	CNSL-Diol	100	204.7
2	Polyethertriol	17.15	77.6
3	Filler	100	
4	Coupling agent	1.086	
5	Defoaming agent	0.1	

* Parts by weight in [g]

Component B

6	Modified MDI	61.9	428.4

Item 1: Polyol based on cashew nut shell liquid, "NX-9203LP", hydroxyl number = 100-135 g KOH/kg, OH equivalent weight = 416-561 g/mol, Cardolite
Item 2: Polyether triol, "Voranol 260 CP", hydroxyl number = 647-676 g KOH/kg, OH equivalent weight = approx. 221 g/mol, Dow Chemical
Item 3: Milled chalk, "Omya BSH - OM", Omya
Item 4: Epoxy silane, "Silquest A-187", Momentive
Item 5: Solution of foam-destroying polymers, silicone-free, Byk-A 535, Byk-Chemie GmbH
Item 6: polycarbodiimide-modified diphenylmethane diisocyanate, liquid, "Isonate 143LP", M = 332 g/mol, f = 2.3 val/mol, NCO = 29.2 % ≙ 6952 mval/kg, isocyanate equivalent weight = 144.5 g/mol ≙ 144.5 g/val, Dow Chemical

The NCO/OH ratio is calculated as follows:

$$NCO/OH = \frac{45.7 \cdot 42 \cdot 1.43 + 7.8 \cdot 42 \cdot 16}{17 \cdot 17 \cdot 29.2 = 1.05}$$

The gel point is reached easily:

$$f_{av} = \frac{\Sigma Ni \cdot fi}{\Sigma Ni} = \frac{0.0154 \cdot 2.5 + 0.037 \cdot 2}{0.0524} = 2.14 \text{ val/mol}$$

Given an NCO functionality of 2.3 val/mol for the isocyanate, the gel point for an NCO/OH ratio of 1.5 ($P_{OH} = 1$) computes to:

$$(P_{OH} \cdot P_{NCO})_{gel} = (f_{OH} - 1)^{-1} \cdot (f_{NCO} - 1)^{-1} \rightarrow P_{NCO} = \frac{1}{1.4} \cdot 1.3 = 0.68$$

This corresponds (reciprocally) to a critical NCO/OH ratio of 1.5.

The formulation contains a silane adhesion promoter and, as necessary, a defoamer. For the manufacturing process, a degassing step is recommended. The pot life is 20 to 30 minutes, and full cure takes nearly 16 hours.

The adhesive shows a tensile shear strength of 9.1 MPa on galvanized steel which does not change significantly even after exposure to temperature and humidity (e.g. 2 weeks at 40 °C, 98 % r.h.). The adhesive is excellent for metal bonding, and even oil-contaminated surfaces can be bonded reliably. Especially worthy of attention are the good low-temperature performance, the impact resistance and the hydrolytic stability.

Bio-based alternatives

An alternative to polybutadiene diols (PolyBD) are bio-based polyols based on cashew nut oil, such as those marketed by Cardolite (NX-9203 and NX-9203LP). Similar to PolyBD diols, these prod-

Table II-2.12: Formulation for a 2-component polyurethane construction sealant [3]

Item	Raw material	Parts by weight
Component A		
1	Polyol	20
2	Plasticiser	22
3	Desiccant	2.0
4	Titanium dioxide, rutile	2.0
5	Rheological additive	3.0
6	Catalyst	0.3
7	Blocking agent for the catalyst	0.3
8	Filler 1	10
9	Filler 2	20.4
Total		80
Component B		
10	Polyisocyanate	4.7
11	Polyol	8.0
12	Plasticiser	7.3
Total		20

Item 1 and 11: Polyol: linear, aliphatic polycarbonate-polyester, solvent-less, 1.7 % OH, e.g. "Desmophen C 1200", Covestro
Item 2: Plasticizer, alkylsulphonic acid ester of phenols, e.g. "Mesamoll", Lanxess
Item 3: Dessicant: molecular sieve paste 3Å, e.g. "Finma-Sorb 430 PR", Finma or "UOP L-PASTE", Azelis
Item 5: Rheological additive, hydrogenated castor oil, e.g. "Efka RM 1920", BASF SE
Item 6: Catalyst: diazabicyclooctane 33 % in dipropylene glycol, e.g. "Dabco 33 LV", Evonik
Item 7: Blocking agent for the catalyst: oleic acid, for extending the processing time
Item 8: Filler 1, quartz-kaolin (Neuburg Siliceous Earth), e.g. "Sillitin Z 86", Hoffmann-Mineral
Item 9: Filler 2: heavy spar, e.g. "Schwerspat (heavy spar) EWO", Sachtleben
Item 10: Polyisocyanate, oligomeric MDI, solvent -less, 32.5 % NCO, e.g. "Desmodur VL 50", Covestro
Item 12: Plasticizer, alkylsulphonic acid ester of phenols, e.g. "Mesamoll", Lanxess

ucts offer excellent hydrophobicity and hydrolytic stability. These systems also have very good mechanical and adhesive properties. An alternative recipe using cashew nut shell liquid can be found in Table II-2.11.

Flexible construction sealants

In addition to adhesives, 2-component polyurethanes can also be used to formulate flexible construction sealants and secondary window sealants. Example formulations can be found in Table II-2.12 and Table II-2.13.

2.2.1.11 Moisture-curing, 1-component PUR adhesives and sealants

Formulation

Moisture-curing 1-component PUR adhesives are based on **isocyanate-terminated prepolymers** or prepolymer mixtures (20 to 100 %). These prepolymers can be made by making commercial polyols react with a corresponding excess of isocyanates. Due to their relatively low vapour pressure, the most important isocyanates are 4,4'-MDI and 2,4'-MDI. When aliphatic isocyanates are specified, IPDI is used in most cases. TDI or HDI are less common, but can be employed, too. For prepolymer synthesis, polyols and polyisocyanates are made to react in the absence of moisture in a nitrogen atmosphere and in the presence of small amounts of catalyst (e.g. DBTL, approx. 200 ppm) for 1 to 3 h at 80 to 100 °C. The only catalysts suitable are those which do not boost oligomerisation reactions by isocyanates. Therefore, basic catalysts, such as tertiary amines, should not be used. Higher-functional raw materials can suffer from the problem of gelling.

Table II-2.13: Guide formulation for a 2-component polyurethane secondary seal

Item	Raw material	Parts by weight
1	"Poly bd R-45 HTLO"	100
2	Short-chain diol	3
3	Calcium carbonate (0.5 µm)	85
4	Calcium carbonate (0.07 µm)	40
5	Carbon black	2
6	Fumed silica (rheology modifier)	7.5
7	Plasticiser	75
8	Antioxidant	2
9	Dibutyltin dilaurate (DBTL, catalyst)	0.045
10	Adhesion promoter	2
11	Oligomeric MDI (curing agent)	28.3

Item 2: Short-chain diol: 2-ethyl-1,3-hexanediol or phenyl diisopropanolamine
Item 7: Plasticiser: phthalates, e.g. "Elatur DPT", the plasticiser can be added completely or partially to the polyisocyanate as a curing agent formulation, Evonik
Item 10: Adhesion promoter: glycidylpropyl trimethoxysilane
Item 11: Oligomeric MDI: e.g. "Desmodur E21" equivalent weight = 262.5, Evonik, (NCO/OH = 0.95)
Desiccants such as molecular sieves (in component A) or tosyl isocyanate (in the curing agent) can be added

Polyether-PUR prepolymers are characterised by their great versatility. Long-chain diols in combination with short-chain triols lead to typical elastomers, as used in sealants. For typical adhesives, greater cohesion and therefore a higher crosslink density are necessary. This can be achieved with low-molecular weight polyols of higher functionality. Adhesives based on polyethers usually have a low viscosity, a low price and good hydrolytic stability. Synthesis of the prepolymers necessitates stabilisation. The use of low-molecular weight and high-functionality polyols affords end products with a high content of isocyanate. This can be disadvantageous due to bubbling in the subsequent curing reaction.

Liquid **polyester-PUR** prepolymers are only obtained with liquid polyesters, unless large amounts of plasticisers or solvents are used. Polyester-PUR prepolymers typically have a high viscosity but exhibit better adhesion properties in adhesives. Prepolymers based on hydroxyl-functionalised polybutadiene can be made with aromatic or aliphatic isocyanates, although an adequate excess of isocyanate and stabilisation are necessary. These prepolymers are of minor importance for moisture-curing systems, because the low water-vapour permeability of such low-polarity polymers would create very long curing times.

Castor oil is not very suitable for prepolymer synthesis due to its low molecular weight and its high functionality.

Prepolymers can be further formulated with the typical auxiliaries. These include **fillers** (up to approx. 50 %) based on chalk, silicates or carbon black. In most cases, filler addition is useful for price reduction. Typically, fillers are predried to minimise the entrained water. In higher concentrations (approx. 20 %), carbon black acts as a reinforcing filler. In any case, adding filler raises the viscosity. The viscosity can be adjusted with plasticisers (up to 30 %), usually based on phthalates.

Furthermore, typical **additives** may be used. For the synthesis of polyether-PUR prepolymers, it is best from the start of the reaction to use (latent) acidic stabilisers (approx. 0.5 %), such as toluenesulphonyl isocyanate ("Additive TI", Borchers Americas Inc, reacts in the mixture to yield a sulphonyl urethane or a sulphonamide), p-toluenesulphonamide (pk_a approx. 10), malonic ester (pk_a approx. 13), benzoyl chloride (produces traces of HCl) or phosphoric acid [30, 31]; see also Figure II-2.34.

Table II-2.14: Starting formulation for 1-component PUR adhesive

Item	Raw material	Parts by weight
1	Isocyanate-terminated prepolymer	49.3
2	Filler	49.3
3	Polyol	1
4	Rheological additive	0.27
5	High-boiling solvent	0.13

Item 1: "Desmodur E 23", NCO = 15.4 % ≙ 3667 mval/kg, ≙ /val isocyanate equivalent weight = approx. 275 g/mol ≙ approx. 275 g/val; Covestro
Item 2: "Quarz Flour W 3", Quarzwerke GmbH
Item 3: "Desmophen 5035 BT", hydroxyl number = 35 g KOH/kg, OH equivalent weight = 1600 g/mol, Covestro
Item 4: aromatic polyamine, "Addolink® 1701", Lanxess
Item 5: propylene carbonate
Preperation
Items 1 and 2 are homogenised in a high-speed stirrer. Items 3, 4 and 5 are mixed and added with vigorous stirring to the mixture of 1 and 2.

Additive TI is sometimes said to have water-binding properties, but in view of the concentrations present, this is not very likely. The following example shows how drying properties can be estimated:

Example

A prepolymer is to be made from 100 g of a polyether diol with a molecular weight of 2000 g/mol and a water content of 0.05 %.

1000 g polyether therefore contains 2000 mval OH, i.e. 100 g contains 200 mval OH and 0.05 g water. Given that the molecular weight of water is 18 g/mol, this amount of water corresponds to just 5.6 mval. In order to be really effective, toluenesulphonyl isocyanate needs to react more than 30 times as readily with water than with OH. This is definitely not the case. Just 0.5 % TI, corresponding to 2.5 mval, is enough to effect stabilisation. This quantity would be insufficient to bind the water in the polyol (not to mention the water in the fillers). Hence, stabilisation must be a consequence of the acidic properties of the corresponding TI urethane or the sulphonamide.

Besides stabilisers, adhesion promoters, such as silanes can be employed. Amino-functionalised silanes cannot be used, but epoxysilanes can, and confer good adhesion on metals and glass. Reactivity with moisture can be increased to a certain extent by using additional catalysts (up to 1000 ppm in the formulation). For 1-component systems, catalysts based on tin or bismuth are

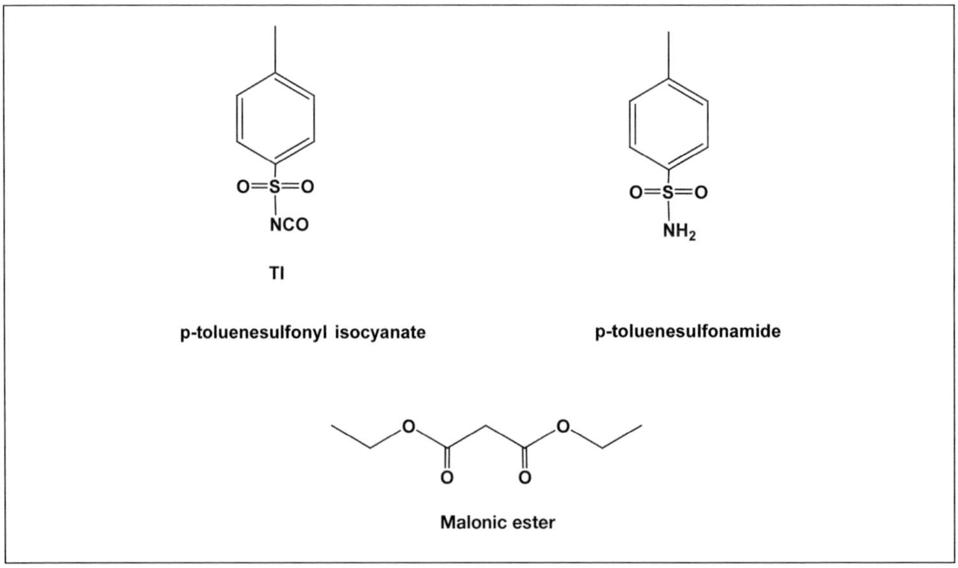

Figure II-2.34: Stabilisers

Table II-2.15: Starting formulation of a moisture-curing, 1-component, white polyurethane sealant [33]

1. Isocyanate-terminated prepolymer	**Parts by weight**
Polypropylene glycol (molecular weight: 2000)	41.4
EO-PO-EO-blockcopolymer (molecular weight 2000, 10 % EO, Figure II-2.35)	41.4
Trimethylolpropane	0.4
Diphenylmethane-4,4'-diisocyanate (MDI,33.6 % NCO)	16.8
Total	100
Synthesis at 70 to 80 °C until NCO content is 1.7 %; dilution with diurethane (see Figure II-2.36)	
2. Sealant	**Parts by weight**
PVC powder (K-value 73)	15.8
Sulphonic acid ester-plasticiser (Figure II-2.37)	21.7
Titanium dioxide	3.5
Xylene	4.7
Disperse at 50 °C for 1 h	
Prepolymer (see above)	28.5
Diurethane (Figure II-2.36)	25.8
Total	100

Table II-2.16: Formulation of a hot-applied, moisture-curing, 1-component polyurethane adhesive-sealant

1. Isocyanate-terminted prepolymer

Item	**Raw material**	**Parts by weight**
1	Polyether triol (molecular weight: 4700)	31.7
2	Polyether diol (molecular weight: 2000)	36.8
3	Toluylene diisocyanate (TDI)	8.0
4	Diisooctyl phthalate (DOP)	23.5
5	Diazabicyclooctane (catalyst)	0.03
Total		approx. 100

Stir at 80 °C until a constant isocyanate content of about 1.1 % is reached

2. Adhesive sealant

Item	**Raw material**	**Parts by weight**
1	Prepolymer (see above)	72.5
2	Carbon black	13.7
3	Kaolin (filler)	11.8
4	Silane adhesion promoter	0.6
5	DBTL (catalyst)	0.08
6	MDA-NaCl complex, 50 % in DOP ("Caytur 21")	1.3
Total		approx. 100

Up to item 5, this is a standard moisture-curing, 1-component polyurethane adhesive-sealant

Table II-2.17: Formulation of a moisture-curing, 1-component adhesive-sealant for windscreen bonding [69]

Item	Raw material	Parts by weight
1	Polyetherdiol	11.2
2	Polyethertriol	27.7
3	Diphenylmethane-4,4'-diisocyanate (MDI, solid)	7.1
4	Plasticiser (e.g. octyldecyl phthalate)	32.8
5	Carbon black (reinforcing)	21
6	Desiccant (e.g. tosyl isocyanate)	0.1
7	Catalyst (e.g. organobismuth compound)	0.1
Total		100

especially suitable (e.g. DBTL or bismuth octoate), because they ensure that the reactive prepolymers have good shelf stability. The rheology can be modified with carbon black, gellable PVC pastes, low-molecular urea derivatives and or hydrophobic, fumed silica. Additives containing water are unsuitable. Commercial prepolymers are easy to modify. Table II-2.14 shows a starting formulation for a 1-component PUR adhesive [32], Table II-2.15 for a 1-component construction sealant and Table II-2.16 and Table II-2.17 for a 1-component adhesive-sealants for windscreen bonding.

The filler is dispersed in the liquid prepolymer. The rheology is adjusted

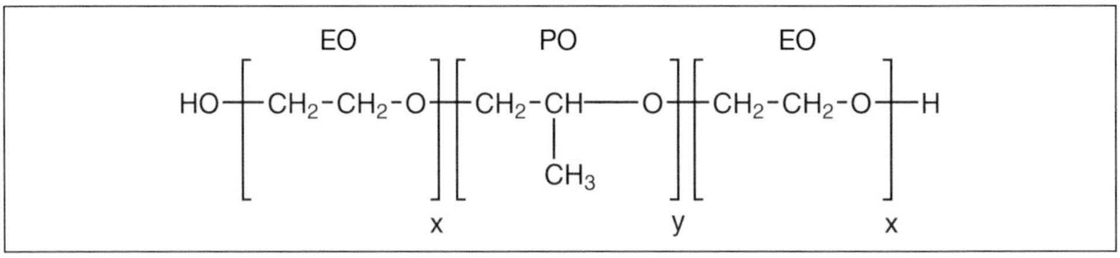

Figure II-2.35: EO-PO-EO block copolymer (y » x)

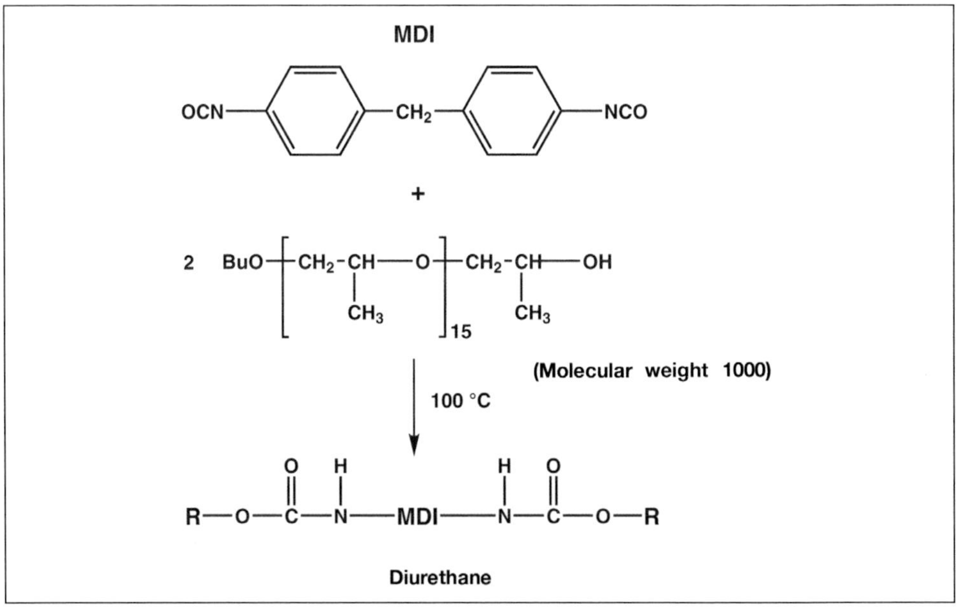

Figure II-2.36: Diurethane (simplified diagram)

by admixing a small amount of diamine dissolved in a little bit of polyol or solvent. Urea forms quickly, conferring the desired rheological properties.

A special method for making and using **1-component PUR** adhesives consists in dissolving isocyanate-terminated prepolymers in a solvent. Compared with traditional solvent-based adhesives, this approach facilitates much higher solids concentrations due to the lower molecular weight of the prepolymers, while offering the advantages of a reactive system. It also ensures that even those polyols which are solid at room temperature, such as crystalline polyesters, can be used. Application can take the form of either wet bonding or, after solvent evaporation, heat activation within a certain time window. Even though the use of solvents is a drawback, the properties of the reactive system are beneficial.

Table II-2.18: Water-containing formulation for a curing agent

Item	Raw material	Parts by weight
1	Cellulose powder	15
2	Plasticiser	65
3	Carbon black	10
4	Water	10
Total		100

Additionally, the curing agent may contain a non-ionic wetting agent and/or a polyol [67]

Accelerating pastes for 1-component polyurethanes

One-component polyurethanes can be accelerated with water-based pastes and processed like two-component polyurethanes.

Component A is essentially similar to the 1-component polyurethane adhesive-sealant described Table II-2.17; it additionally contains 0.4 wt% ethylenediamine adsorbed on 3.8 wt% of a 4 A° molecular sieve.

The curing agent is a water-containing formulation, with the water bound to a carrier, such as cellulose powder or starch flour, to increase pot life. These two carriers are able to reversibly bind water up to a ratio of 1 : 1. A formulation is presented in Table II-2.18. [67]

Mixing of component A and curing agent in the volume ratio 100 : 14 is carried out with the help of a static "Kenics" mixer fitted with 12 elements. [68] The processing time (maximum dwell time in the static mixer) is 15 min; gelling occurs after 30 min. Without cellulose powder, the mixture gels after 2.5 min. The tensile shear strength is 3.9 MPa after just 4 hours, and 4.6 MPa after 24 hours.

It should be noted that the higher the number of mixing elements, the more energy is required for mixing; i.e. the fewer mixing elements there are, the easier it is to mix – at the expense of mixing effectiveness.

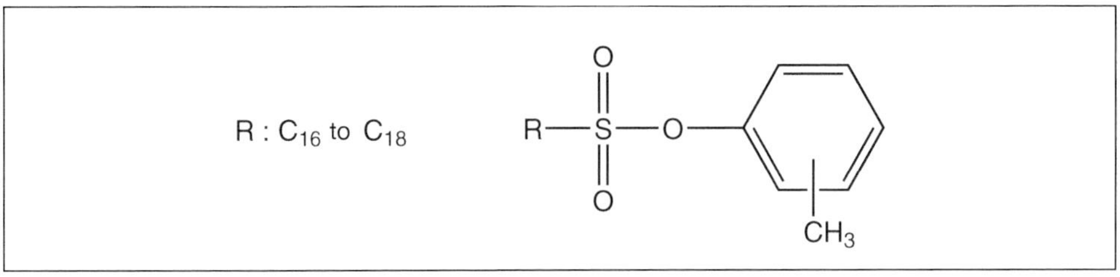

Figure II-2.37: Plasticiser based on sulphonic acids

Curing reaction and CO_2 bubbling

The reaction between isocyanate and water has already been described in detail but needs to be considered again at this point. In order for a reaction to occur, water must reach the reactive isocyanate groups. This restricts applicability to substrates which exhibit a certain degree of water permeability. Panel-like metal/metal laminates with thin adhesive gaps are therefore ruled out. The curing reaction starts from the surface and slowly proceeds by diffusion into the depths of the adhesive. The surface reaction which changes the paste adhesive into a cured polyurethane can be regulated over a wide range by means of catalysts (tack-free time). The reaction rate within the adhesive layer itself also depends on the rate at which water diffuses to the site of the reaction with the isocyanate groups. At deeper layers, the reaction rate is therefore a diffusion-controlled process. Diffusion of water into the polymer has to proceed through the already cured adhesive. Diffusion rates are determined by the polarity, the glass transition temperature and the proportion of filler. The polarity of the system is mainly determined by the type of polyol. Products based on hydroxyl-functionalised polybutadienes therefore cure much slower than those made with polyethers or polyesters.

The glass transition temperature is a function of the type of polyol and crosslink density. Diffusion rates in cured adhesives are quite unfavourable, because, unlike the case for sealants, adhesives require high cohesion. Additionally, the curing reaction itself leads to an increase in glass transition temperature, and so diffusion of water into deeper layers of the adhesive deteriorates even more. Diffusion rates decline proportionally to the proportion of filler, because the filler particles themselves have no permeability. During the chemical conversion, two isocyanate groups react with water to form a urea linkage and CO_2, which has to be transported out of the adhesive layer again by diffusion. Otherwise, when the solubility is exceeded, bubbles are generated. The diffusion characteristics of CO_2 in polyurethanes are worse than in the more polar water, and the gas cannot readily reach the surface from the point of generation, as it has to diffuse through the already cured polyurethane. Consequently, the CO_2 formed locally will exceed the solubility limit and microbubbles will be formed. Due to the resultant lower surface energy, these microbubbles create larger bubbles and the adhesive starts foaming. This foaming leads to a loss of cohesion and changes the dimensions of the bond to an undesirable extent for many applications. Consequently, bubbles can be avoided by thin adhesive layers, a low NCO content and slow curing.

The literature is full of measures for avoiding bubbling, yet a one-hundred-percent solution for this problem is yet to be found. There is limited scope for minimising the NCO content. The low-molecular polyols essential for high cohesion require a certain excess of isocyanate. Although fillers and plasticisers thin the isocyanates to a certain extent, the quantities which can be added are limited due to the resultant weakening of cohesion and lowering of the glass transition temperature. For most manufacturers of prepolymers, thin-layer vacuum evaporation of isocyanates is not technically feasible. The diffusion and solubility properties of CO_2 cannot be influenced to any great extent.

However, bubbling can be reduced by adding systems which react much faster with water than with the isocyanates themselves, and which liberate polyfunctional products that can subsequently react with isocyanates (see Figure II-2.38).

Suitable functional groups for such latent hardeners include polyfunctional oxazolidines, ketimines, aldimines and enamines [1, 35, 36]. All these products are derivatives of aldehydes or ketones and yield amines in the presence of water. The amines in turn quickly react with isocyanates to form ureas. Figure II-2.39 schematically shows a bisoxazolidine. For steric reasons, the nitrogen in the ring is of low basicity and not nucleophilic. Consequently, adequate storage stability is guaranteed, especially in the presence of aliphatic isocyanates. Reaction with water causes ring

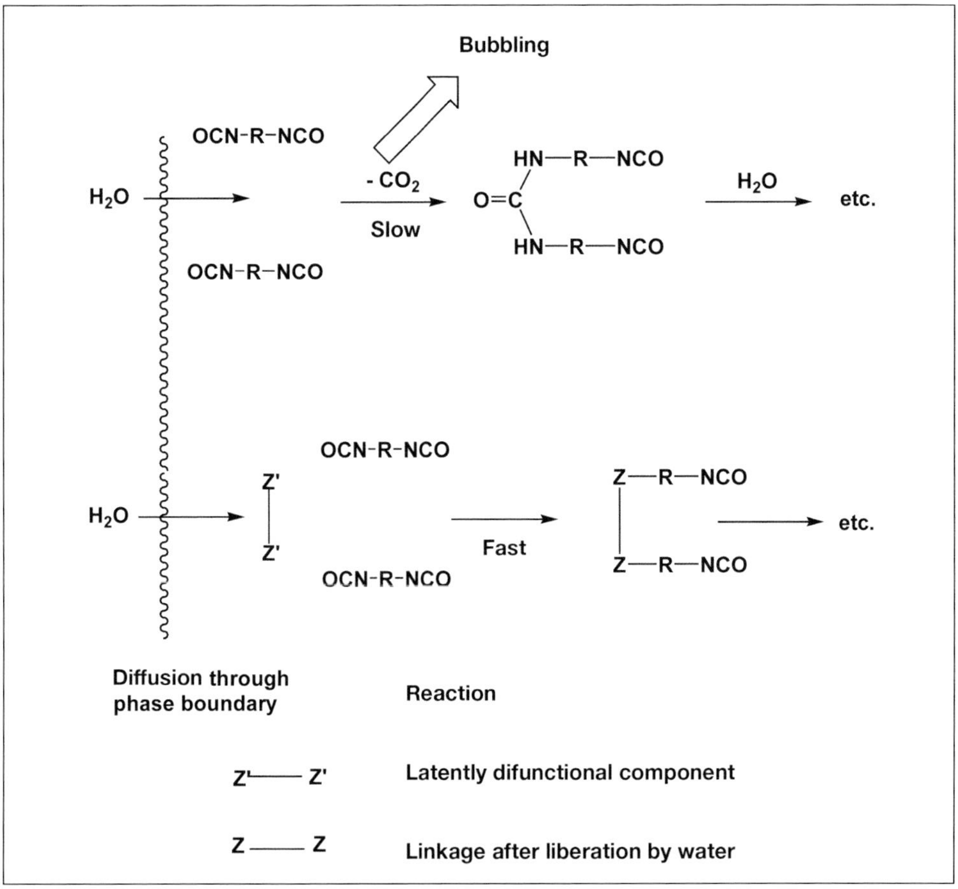

Figure II-2.38: Avoiding bubbling in 1-component PUR adhesives, schematic

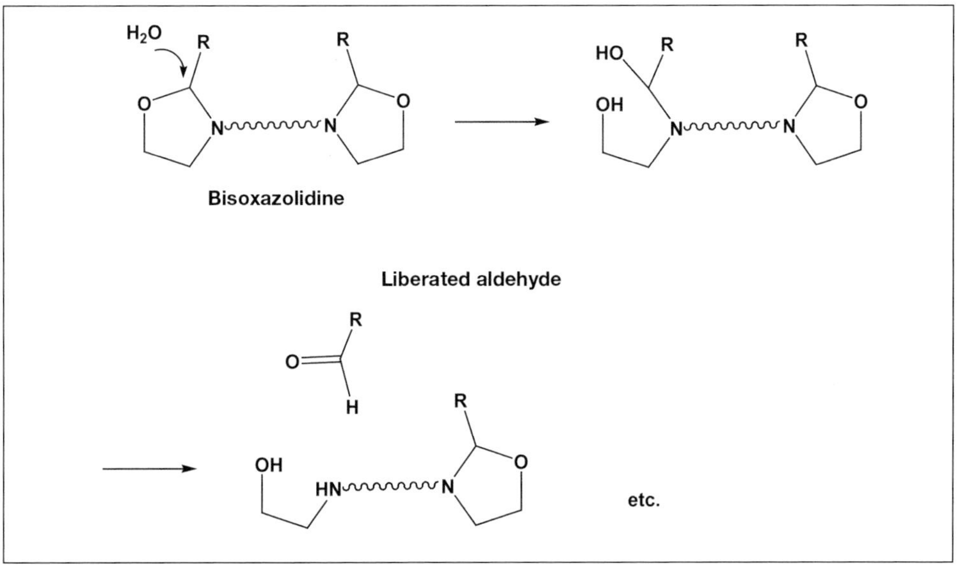

Figure II-2.39: Reaction of a bisoxazolidine with water

opening and the formation of a polyfunctional β-hydroxylamine, which reacts quickly with the isocyanate. At the same time, a low-molecular aldehyde (e.g. butanal) is liberated. However, this aldehyde is sometimes deemed by users to have an unpleasant odour.

A totally different approach is applied in a technology which exploits the reversible adsorption of amines on molecular sieves (see literature [31, 32, 35, 36] cited there). A suitable molecular sieve powder (dispersed in a plasticiser) is loaded with an amine. As this molecular sieve is present as a heterogeneous system in the adhesive formulation, it cannot react with the isocyanates. Where contact is made with water from the environment, the weakly adsorbed amine is liberated from the molecular sieve by the more strongly adsorbed water and an immediate reaction with the isocyanate groups of the prepolymer occurs. While ketimines, aldimines and enamines have failed to establish themselves on the market, systems based on oxazolidines, and amine-loaded molecular sieves have been successful for a number of years. The commercially available oxazolidines include, for example, "Hardener OZ" (Covestro).

Chapter II-2.2.1.11 (Table II-2.17 and Table II-2.18) describes a formulation which uses a modified form of amine-loaded molecular sieves. The corresponding 2-component PUR adhesive-sealant is based on isocyanate-terminated prepolymers. Therefore, like a 1-component system, it is capable of curing with ambient moisture.

Replacing isocyanate by alkoxysilane groups is another way to formulate moisture-curing systems (silane-terminated polyurethane prepolymers), in which the methanol liberated during the curing reaction does not cause bubbling (see Chapter II-2.4.3). The underlying polymer can be a polyurethane of similar structure, and so a similar product property profile will result.

Evaluation

Moisture-curing 1-component PUR adhesives offer broad formulating latitude for specific applications. However, they are unsuitable for bonding moisture-impermeable substrates. Where high bond strengths on metals are required, other bonding technologies should be used. The bubbling problem merits close attention, but the foaming process can sometimes be quite useful in terms of gap-filling properties and mechanical anchoring.

Advantages

+ 1-component system
+ good adhesion
+ good heat resistance
+ good hydrolytic stability
+ good low-temperature properties

Disadvantages

- slow curing
- moisture permeability limits choice of substrate
- no green strength
- foaming during curing
- more demanding manufacturing process
- less formulation latitude compared to 2-component systems

Fields of application

1-component PUR adhesives find applications in the **automotive industry**, where they are used in a flexibilised form as adhesive/sealants, e.g. for bonding windscreens. They serve in the manufacture of sandwich elements for caravans, wherever insulating materials based on expanded polystyrene have to be laminated with covering materials, such as SMC, aluminium and wood. 1-component PUR adhesives are generally very beneficial for bonding wood and similar materials in the **woodworking and furniture industry**, because the substrate has a high water content, good water permeability and functional groups. In the **construction industry**, facade elements, plasterboard walls, clinker and marble are bonded to expanded polystyrene in prefabricated houses [37].

2.2.1.12 Heat-curing, 1-component PUR adhesives

Curing

Technically there are various ways to formulate and produce heat-curing PUR adhesives, but significant commercial success has yet to be achieved. The major reason is that the scope for using heat-curing system is limited, because the processes are expensive and time consuming. In addition, heat curing only lends itself to heat-resistant substrates. Consequently, heat-curing processes are used in adhesives applications wherever the thermal energy is freely available and where there are no technical alternatives. For instance, 1-component epoxies are used in automotive body-in-white shops, where the heat is supplied by the paint-drying process and metal is bonded to metal. Heat curing is also employed to bond brake liners with phenolic resins (see Chapter II-2.4.1), because other adhesives are unable to deliver the required property profile. In the field of polyurethanes, heat-curing systems can technically be made by applying the principle of latent reactive systems, i.e. curing agents which react under the influence of heat. Some preconditions need to be met here. Every chemical reaction is accelerated by a rise in temperature, in accordance with the laws of Arrhenius (typically 2 to 4-fold increase for every 10 °C temperature increase). The degree of acceleration depends on the activation energy, and other factors. Systems which have a high activation energy exhibit high temperature dependence concerning the reaction rate. Typical temperatures for heat-curing systems range from 100 to 180 °C. A traditional Arrhenius system, which would cure in half an hour at 150 °C for example, would cure within half a year at 20 °C, by this estimation. This would be unacceptable because most industrial applications require a shelf life of half a year.

A further limitation is the commercial availability of the raw materials. The adhesives industry is a small industry and heat-curing technology constitutes just a small fraction thereof. Many products which are praised in the literature and which the major raw materials suppliers might even provide for trials, ultimately are not commercially available.

Formulation

The formulation of heat-curing PUR adhesives follows the same principles already described for 1- and 2-component polyurethanes. Besides the polyisocyanate/polyol system, which is blocked in the corresponding way, fillers, plasticisers and additives may be used. In many cases, there is a tendency to use reactive resins, such as epoxies, in addition to the isocyanate. The chemical processes by which such systems cure are naturally quite complex, but this allows the product's property profile to be modified over a wide range.

Blocked isocyanates

The reaction between isocyanates and nucleophiles is reversible. A number of isocyanate adducts can be cleaved thermally under quite mild conditions. If this is performed in the presence of polyfunctional nucleophiles which are capable of reacting with isocyanates to yield stable molecules, this can serve as a curing reaction. Examples of such blocked isocyanates are shown in

Table II-2.19: Deblocking of isocyanates

Isocyanate	Blocking agent	Deblocking temperature [°C]
Aromatic isocyanate	phenol	120 to 180
	butanone oxime	125 to 140
Aliphatic isocyanate	butanone oxime	140 to 160
	caprolactam	160 to 180
	diethyl malonate	100 to 120
	3,5-dimethyl pyrazole	110 to 120

Figure II-2.40, e.g. adducts with oximes, phenols, ε-caprolactam, diethyl malonate and 3,5-dimethyl pyrazole [38, 39].

The deblocking temperatures given in the literature can serve merely as a guide. Besides the temperature, they depend on the time available, the presence of catalysts (they catalyse cleavage in the same manner as adduct formation) and the presence of nucleophiles. Table II-2.19 shows guide temperatures for deblocking [2, 38, 39].

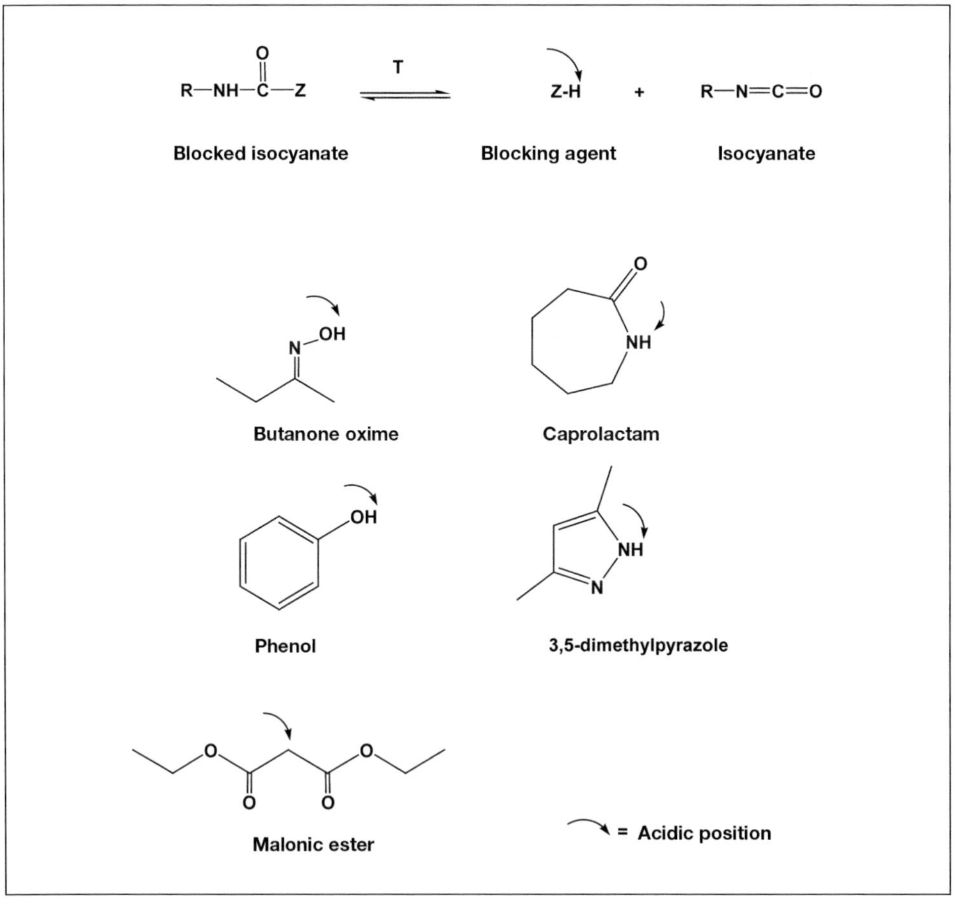

Figure II-2.40: Blocking agents for isocyanates

Table II-2.20: NX-2026 compared to other blocking reagents [64]

Blocking agent	Isocyanate type	Appearance	Viscosity at RT [cps]	Deblocking Temperature [°C]*
NX-2026	TDI	yellowish liquid	344	115
NX-2026	MDI	Yellowish waxy solid		145
NX-2026	HDI	amber liquid	115	155
NX-2026	IPDI	yellowish liquid	2895	130
NX-2026	p-MDI	yellowish liquid	151	170
Phenol	TDI	white solid		164
Phenol	MDI	white solid		204
Phenol	IPDI	white solid		160
Nonyl phenol	TDI	thick colourless liquid		185
MEKO	TDI	yellowish waxy solid		70
Ɛ-caprolactam	TDI	white solid		150

*Deblocking temperatures have been determined by DSC analysis (10°C/min)

Adducts based on aromatic isocyanates are generally less stable than those based on aliphatic types. Blocked isocyanates that are commercially available for the paint industry can basically also be used to formulate adhesives and sealants. Examples are the "Crelan" series from Covestro [40] (aliphatic isocyanates, lactam-blocked; curing conditions for powder coatings: 15 min at 160 °C) and the "Desmocap" series from Covestro.

Blocked isocyanates with a high proportion of biobased raw materials can be obtained by reaction with cashnew nut oil. One well-known product is NX-2026 from Cardolite, whose properties are summarized in Figure II-2.41 and Table II-2.20. The cashew nut oil separated after deblocking remains in the cured polyurethane and does not pose a hazard due to the low vapor pressure.

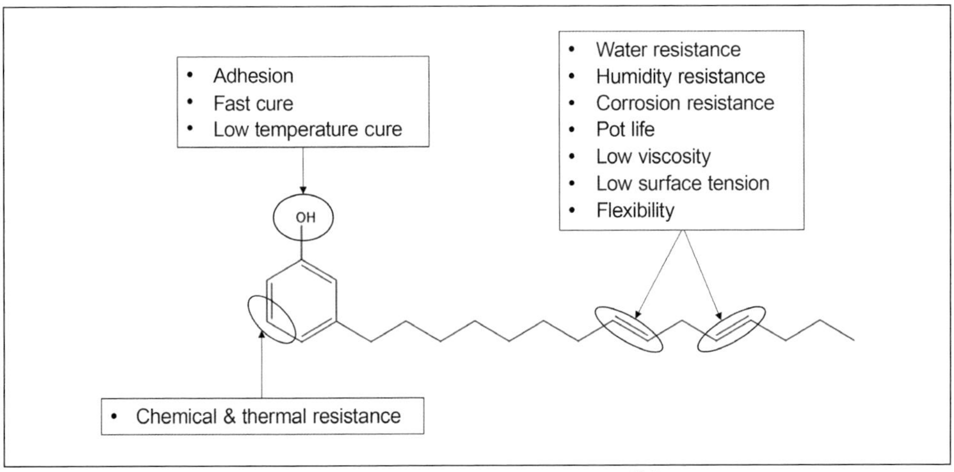

Figure II-2.41: Cachew nut oil as a blocking reagent for isocyanates [64].

Table II-2.21: Heat-curing adhesive based on uretdione

Item	Raw material	Parts by weight
1	Polyol 1, liquid	42.7
2	Polyol 2, solid	42.7
3	Catalyst	0.02
4	Uretdione	14.5

Item 1: Polypropylene oxide
M = 3000 g/mol, f = 3 val/mol, hydroxyl number = 56 g KOH/kg, OH equivalent weight = 1000 g/mol
Item 2: Polyester, ethanediol/adipate
M = 2000 g/mol, f = 2 val/mol, hydroxyl number = 56 g KOH/kg, OH equivalent weight = 1000 g/mol
Item 3: Borchi Kat 315, Borchers
Item 4: TDI dimer, "Thanecure T 9", TSE Industrie
M = 348 g/mol, f = 2 (+2 blocked) val/mol, isocyanate equivalent weight not blocked = 174 g/mol ≙ 174 g/val, isocyanate equivalent weight total = 87 g/mol ≙ 87 g/val

Uretdiones

Uretdiones are thermally unstable and undergo thermoreversible cleavage to yield isocyanates (the dimer of TDI cleaves at about 160 °C) [1–3, 38, 41]. If polyols or polyamines are present, this affords a simple way of effecting a thermal cure. Figure II-2.42 shows the uretdione of TDI and its thermal cleavage.

Uretdiones based on aliphatic and aromatic isocyanates are commercially available, e.g. "Crelan" series from Covestro (aliphatic isocyanate dimers for powder coatings, which cure in about 15 min at 180 °C [40]) and "Thanecure T9" from TSE Industries (aromatic isocyanate, dimer of TDI).

Guide formulation

Table II-2.21 shows the starting formulation for a heat-curing PUR adhesive based on uretdiones [42]. The corresponding polyols are incompatible, and the polyester is admixed as a powder. Curing is performed at 120 to 130 °C for 2 to 3 h.

Blocked amines

Amines are electron donors and so, in combination with certain metal ions, can serve as complex ligands. Such complexes are often thermally unstable, a fact what can be exploited to heat-cure PUR adhesives [43]. A corresponding complex made from 4,4'-diaminodiphenylmethane (MDA) and NaCl cleaves thermally at 100 to 150 °C. The liberated diamine cures the isocyanate-terminated PUR prepolymer (see also Chapter II-2.2.1.11).

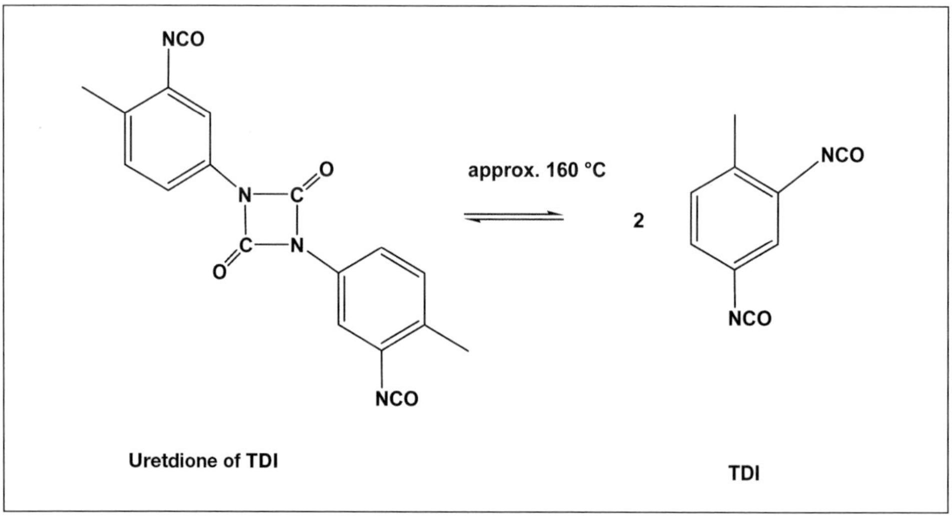

Figure II-2.42: Cleavage of a uretdione

Blocked polyols

Solid polyols, such as pentaerythritol and mannitol, can be formulated with isocyanate-terminated prepolymers. Curing is effected by heating at 110 to 140 °C [44].

Evaluation

When suitably formulated, heat-curing PUR adhesives exhibit a much more versatile product profile as regards polyurethane chemistry than heat-curing epoxies. However, their practical uses are limited due to process and substrate constraints. By comparison, epoxies offer better heat resistance.

Fields of application

The **automotive industry** offers some applications for heat-curing PUR adhesives. If they are applied to the cataphoretic base paint ahead of the filler paint, the subsequent paint-cure cycles (approx. 30 min at 160 °C, 30 min at 140 °C) can be used to cure the adhesive. Unlike traditional PVC plastisols, these adhesives have nearly constant mechanical properties over a wide range of temperatures because they are chemically crosslinked. They are thus genuine structural adhesives. Heat-curing PUR adhesives in some types of cars have slashed the number of spot welds, thereby increasing corrosion inhibition and conferring better crash behaviour.

2.2.1.13 Curing hot-melts

Concepts

For a variety of reasons (see also Chapter II-1.3), hot-melts have captured a sizable share of the market. Because they set rapidly, they lend themselves to high-speed, high-volume industrial manufacturing. Hot-melts do not require any application auxiliaries, such as solvents or water. Drawbacks include their sometimes moderate adhesion to many substrates and, especially, their limited heat resistance, which stems from their thermoplastic properties.

The first reactive hot-melts were developed and commercialised in the early 1980s. These polyurethane-based products were traditional hot-melts in terms of processing and setting characteristics. Apart from possessing the typical physical setting mechanism of hot-melts, they can enter into the chemical curing reaction that typifies reactive adhesives. Thus, they ideally combine the advantages of both types of adhesives [45].

There are all sorts of concepts for formulating reactive hot-melts. Basically, the different principles behind hot-melts and reactive adhesive systems can be combined almost arbitrarily, although not all combinations are suitable for practical requirements. Many thermoplastic polymers have hot-melt properties (see Chapter II-1.3). Classification is either by chemical composition (polyolefins, polyesters, polyamides, etc.) or setting mechanism, with a distinction being drawn between polymers with a glass transition temperature (Class A) or a crystalline melting point (Class B) above room temperature and polymers of high molecular weight but having a low glass transition temperature (Class C). If these polymers can be linked to the typical functional groups of reactive adhesives, the result is functionalised polymers which can combine both properties. Such functional groups are those already encountered in the field of reactive 1-component adhesives and sealants. They include isocyanates (see Chapter II-2.2.1.14), silanes (Chapter II-2.4.3), epoxies (Chapter II-2.2.2) and even double-bond systems, such as acrylics (Chapter II-2.3.1). Curing usually takes place after application, by a mechanism which varies with the functional group, i.e. with ambient moisture or oxygen, temperature increase or high-energy radiation. An overview is provided in Figure II-2.43.

By far the most important class of reactive hot-melts are moisture-curing isocyanate-functionalised polyurethanes based on polyesters or polyethers on account of their structural versatility. They will therefore be described in more detail.

2.2.1.14 Moisture-curing PUR hot-melts

Basic principles

Development of strength

PUR hot-melts are applied hot in the manner of traditional hot-melts. Due to rapid cooling and physical setting after mating, high green strength is quickly achieved. Consequently, bonded parts can be handled within seconds or minutes, without the need for mechanical fixturing. Once the adhesive has been applied, moisture from the environment or from the substrates triggers a chemical reaction with the isocyanate groups to afford a high-molecular, crosslinked polymer within hours or days, which is associated with a further increase in strength. Figure II-2.44 shows the change in bond strength over time [46].

Hot-melt type	Polymer (selection)
amorphous, T_g (Class A)	polyester polyamide polyacrylate polystyrene
crystalline, T_m (Class B)	polyester polyamide polyethylene
amorphous, high mn, low T_g (Class C)	polyisobutylene polyacrylate

Functional group	Reaction with or by
isocyanate R—N=C=O	moisture
silane 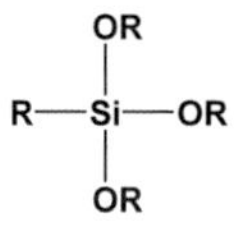	moisture
epoxy 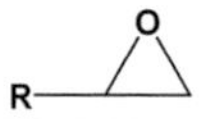	heat
double bond 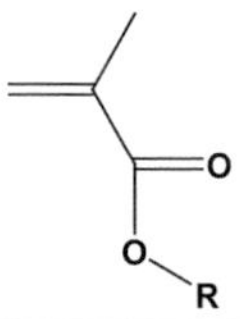	oxygen heat radiation

Figure II-2.43: Concepts for reactive hot-melts

Raw materials

PUR hot-melts mainly consist of isocyanate-terminated prepolymers, which can be obtained by making polyols react with excess isocyanate (see Chapter II-2.2.1.11). The chosen polyol mainly determines the setting characteristics while the system of isocyanate and catalyst is responsible for subsequent curing with ambient moisture. The desired hot-melt properties can be realised in various ways. The prepolymers can be synthesised from crystalline (e.g. hexanediol adipates) or amorphous polyesters (mixed polyesters based on aromatic dicarboxylic acids) as well as polyethers (polypropylene oxide or poly-THF). These polyols are treated with excess isocyanate to yield functionalised reactive prepolymers. Aromatic isocyanates are preferred, e.g. MDI or its derivatives. These are cheaper than the aliphatic types and have a relatively low vapour pressure which minimises exposure of the user to harmful isocyanate monomer vapours. Additionally, aromatic isocyanates are much more reactive with moisture than are aliphatics, a fact which can be important in the subsequent curing reaction. Chemical curing of the isocyanate groups occurs in the same way as described for 1-component PUR adhesives (see Chapter II-2.2.1.11), via the formation of urea linkages and the release of carbon dioxide. Side-reactions may lead to the formation of allophanate or biuret linkages, which contribute additional crosslinking.

PUR hot-melts can be optimised from various aspects. For users, the most important properties (besides price) include the processing conditions (temperature, viscosity, stability, setting characteristics, curing characteristics, handling safety and toxicology) and the ultimate bonding properties in a given application (adhesion to substrates, transfer of static and dynamic loads, resistance to heat, moisture and other influences).

Synthesis and formulation

Over the years, various technologies for synthesising reactive hot-melts emerged to accompany the various generations.

1st generation

High-molecular thermoplastic polymers, e.g. polyester-polyurethanes, already used in the synthesis of traditional hot-melts, can be mixed with isocyanate-terminated polyurethane prepolymers and other resins. The polymers mainly dictate the processing conditions and the setting characteristics of the hot-melt, while the isocyanate groups in the prepolymer are responsible for curing

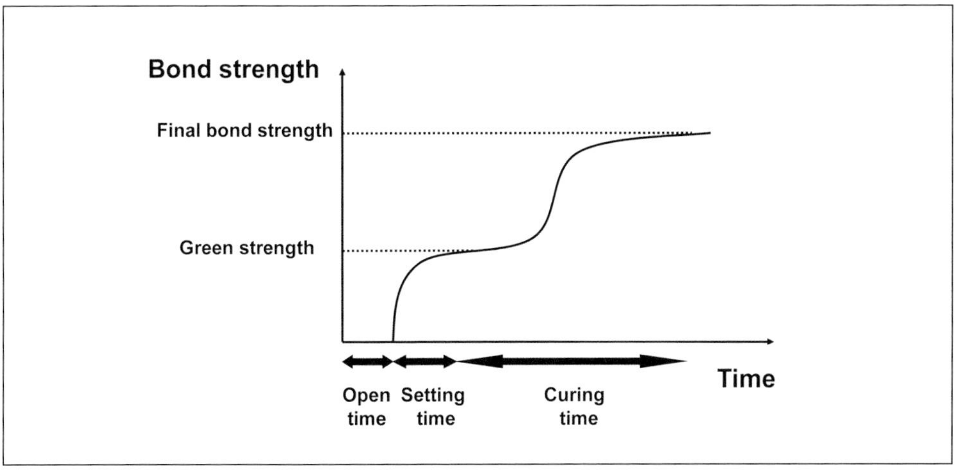

Figure II-2.44: Development of bond strength

Table II-2.22: PUR hot-melt, 1st generation

Adhesive components (reactive prepolymer, item 7)

Item	Raw material	Parts by weight
1	Diisocyanate	41
2	Polyesterpolyol	51.6
3	Chain extender	7.4

Polyurethane polymer (item 8)

Item	Raw material	Parts by weight
4	Diisocyanate	20.6
5	Polyesterpolyol	75.7
6	Chain extender	3.7

Adhesive formulation

item	Raw material	Parts by weight
7	Reactive prepolymer	50 to 65
8	Polyurethane-polymer	5 to 30
9	Resin	15 to 30

Item 1: "Desmodur PF" (MDI-derivate), NCO = 23 % ≙ 5476 mval/kg, Covestro
Item 2 and 5: Polyesterpolyol from butanediol, ethyleneglycol and adipic acid, M = 2000 g/mol, f = 2 val/mol ≙ 1000 mval/kg
Item 3: Tripropyleneglycol, M = 192 g/mol, f = 2 val/mol ≙ 10416 mval/kg
Item 4: "Desmodur CQ 44 M Liquid MS" (MDI), M = 250 g/mol, f = 2 val/mol ≙ 8000 mval/kg, Covestro
Item 6: 1,4-Butanediol, M = 90 g/mol, f = 2 val/mol ≙ 22220 mval/kg
Item 9: Ketone-aldehyd-condensation product

with ambient moisture. Due to the use of high molecular weight polymers, products of this generation exhibit very fast setting, and bonded adherends quickly develop excellent green strength, a fact which facilitates further handling in industrial manufacturing processes. Drawbacks include the high application temperatures (160 to 180 °C) typical of hot-melts. This limits the stability of the adhesive under processing conditions and in the worst case can cause emissions of monomeric isocyanates and even damage to heat-sensitive substrates. Products of this first generation failed to establish themselves on the market.

Example

Table II-2.22 shows a corresponding starting formulation. The polyurethane consisting of crystalline polyesters, dictates the hot-melt properties, while the isocyanate groups of the prepolymer are responsible for the curing properties after application. The resin and the prepolymer serve to lower the viscosity and improve tack. The application temperature of about 180 °C ensures that the high polymer is sufficiently liquid [47].

The NCO/OH ratio = Σ equivalents NCO/Σ equivalents for the reactive prepolymer is calculated as follows (see Chapter II-2.2.1.5):

41 g "Desmodur PF" corresponds to $41 \cdot \frac{5476}{1000} = 225$ mval NCO
51.6 g polyester polyol corresponds to $51.6 \cdot \frac{1000}{1000} = 52$ mval OH
7.4 g tripropylene glycol corresponds to $7.4 \cdot \frac{10416}{1000} = 77$ mval
$\rightarrow NCO/OH = \frac{225}{(52+77)} = 1.7.$

Similarly, the figures for the polyurethane polymer are:

20.6 g "Desmodur 44 MS" corresponds to $20.6 \cdot \frac{8000}{1000} = 165$ mval NCO
75.7 g polyester corresponds to $75.7 \cdot \frac{1000}{1000} = 76$ mval OH
3.7 g 1,4-butanediol corresponds to $3.7 \cdot \frac{22220}{1000} = 82$ mval OH
$\rightarrow NCO/OH = \frac{165}{(76+82)} = 1.03;$

in other words, the reaction is performed close to the equivalence point in order that a high molecular polymer may be obtained.

2nd generation

Low-molecular (M = 1000 to 6000 g/mol), crystalline polyesters (see Chapter II-2.2.1.2), e.g. hexanediol adipate s or butanediol adipates, are allowed to react with excess isocyanate (NCO/OH = 1.5 bis 2.5) to yield prepolymers. These crystalline prepolymers possess hot-melt properties because they are solids at room temperature and can be liquefied to low-viscosity products at temperatures of around 100 °C. Hexanediol adipates (m.p. 50 to 60 °C) are more resistant to hydrolysis than are butanediol adipates (m.p. 40 to 50 °C). Higher molecular polyesters (M = 3000 bis 6000 g/mol) crystallise faster than lower molecular polyesters (M = 1000 to 2000 g/mol). The viscosity of the hot-melt is dictated by the molecular weight of the polyester used and the NCO/OH ratio. The viscosity falls with increase in NCO/OH ratio. Addition of plasticisers is not recommended, as that might impair crystallisation and therefore the setting process. Mixed polyesters can also be used. These can reduce the degree of crystallinity in the prepolymers and so enable the open time and the setting characteristics of the hot-melt to be modified. Adhesion to certain substrates may be additionally improved due to reduced crystallinity and therefore reduced shrinkage upon solidification. The addition of higher-functionalised isocyanates yields a higher crosslink density and therefore the products feature greater heat resistance and improved hydrolytic stability. Overly excessive amounts of isocyanate have to be avoided, because the high quantities of monomeric isocyanate would remain in the product, requiring them to be classified as dangerous goods and creating the problem of environmental pollution when processed under heat. The cure rate can be varied within limits via the catalyst system. Suitable catalysts are those which are suitable for 1-component polyurethanes, such as DBTL. Provided that excessive quantities of catalysts are not used (< 200 to 1000 ppm) and that moisture is excluded, the corresponding melts have excellent stability, even at the application temperature. Cooling as part of the bonding process leads to solidification by crystallisation. Due to the low molecular weight of the prepolymers, products of this second generation exhibit quite low viscosities under mild application conditions (80 to 120 °C). This means that thermally sensitive substrates can also be bonded. According to the current state of knowledge, a combination of low application temperatures and a suitable formulation does not cause the release of relevant quantities of monomeric isocyanates [13]. A disadvantage is the poor reproducibility of crystallisation which is part of the setting process. Consequently, certain leeway as regards fixturing times is needed in industrial manufacturing processes. Not only that, but the cooled prepolymers are somewhat brittle (waxy), because they do not have a high molecular weight yet and therefore the bonds in this stage are sensitive to peel and impact. Crystalline products always show a certain tendency to shrink during crystallisation. This can affect adhesion to certain substrates. Thick-layer application and application to moisture-impermeable substrates are not recommended [48, 53].

Example

A typical formulation (according to [53, Ex. 3]) is shown in Table II-2.23. The polyol is a crystalline polyester based on hexanediol adipate.
The NCO/OH ratio $=\frac{\Sigma \text{ equivalents NCO}}{\Sigma \text{ equivalents OH}}$, according to Chapter II-2.2.1.5, is calculated as follows:

12 g "Desmodur 44 MS" corresponds to 96 mval NCO,
87.68 g "Dynacoll 7360" corresponds to 47 mval OH
$\rightarrow$ NCO/OH = $\frac{96}{47}$ = 2.0.

This leads to a low molecular weight for the prepolymers and therefore to a low application viscosity. The content of free monomeric MDI, which is of relevance for declaration as dangerous

Table II-2.23: PUR-Hot-melt, 2nd generation

Item	Raw material	Parts by weight
1	Diisocyanate	12
2	Crystallising polyester	87.68
3	Stabilizer	0.23
4	Catalyst	0.09

Manufacturing:
Item 1 to 3 are recated at approx. 80 °C for 1 h under stirring and exclusion of moisture. Item 4 is added, and the reaction is finished over another h at appr. 100 °C.
Optionally before filling vacuum is applied in order to minimize gass bubbles.
Item 1: "Desmodur CQ 44 M Liquid MS", M = 250 g/mol, f = 2 val/mol ≙ 125 g/val, NCO = 33.6 % ≙ 8000 mval/kg, Covestro
Item 2: "Dynacoll 7360", M = 3500 g/mol, f = 2 val/mol (manufacturer information) ≙ ca.1750 g/val, OH = 30 g KOH/kg ≙ 0.91 % ≙ 536 mval/kg ≙ 1865 g/val, Evonik
Item 3: "Additiv TI", Borchers Americas Inc.
Item 4: DBTL, Borchers Americas Inc.

goods and possible vapour exposure during manufacture and application, can be calculated in accordance with equation (4) in Chapter II-2.2.1.5, as follows:

$$m_{res}\,[\%] = [(NCO/OH - 1)/(NCO/OH)]^2 \cdot m_{orig}\,[\%] = [(2-1)/2]^2 \cdot 12 = 3\,[\%]$$

3rd generation

When low-molecular amorphous polyesters (M = 2000 to 6000 g/mol) with a glass transition temperature above room temperature react with excess isocyanate to yield prepolymers, the products have a glass transition temperature above room temperature [49, 50, 54]. These products are brittle, of course. So, a second prepolymer based on a polyester or polyether with a low glass transition temperature is added. Where possible, the two prepolymers should not be wholly compatible. The viscosities are adjusted via the NCO/OH ratio (typically 1.5 to 3). In addition, limited amounts of (mainly solid) phthalate plasticisers may be added. The cure reaction is again regulated via the quantity of catalyst (DBTL, 200 to 1000 ppm). The resultant adhesive can be readily applied at 100 to 150 °C. Solidification of amorphous systems on cooling occurs spontaneously in the mated substrates as soon as the temperature within the adhesive joint falls beneath the glass transition temperature of the higher-melting prepolymer. Besides fast setting, third-generation products possess high flexibility, even in the uncured state [55]. Compared with crystalline products, hot-melt adhesives offer improved adhesion to many substrates, because they do not shrink due to crystallisation as they cool. However, they have the disadvantage of tending to creep under load before curing. Using mixtures with prepolymers of crystalline polyesters reduces the creep tendency. Thick layers of amorphous PUR hot-melts can undergo undesirable bubbling during curing.

Example

Table II-2.24 presents a typical formulation, according to [49]. The polyols are two amorphous polyesters, one of which has a glass transition temperature above room temperature and is responsible for setting, the other having a low glass transition temperature and being responsible for flexibility.

The adhesive is synthesised in two steps. First, a prepolymer is made with the high T_g polyester. For this prepolymer, the NCO/OH ratio can be calculated as follows:

9.1 g "Desmodur 44 MS" corresponds to 72.8 mval NCO.
65.5 g polyester item 1 corresponds to 24.6 mval.
$\rightarrow NCO/OH = \frac{72.8}{24.6} = 3.0.$

The NCO/OH ratio for the second prepolymer, which is made not separately but rather within the first prepolymer and is a flexible, low-T_g prepolymer, is calculated as follows:

3.2 g "Desmodur 44 MS" corresponds to 25.6 mval NCO
21.8 g polyester item 4 corresponds to 11.7 mval OH

$$\rightarrow NCO/OH = \frac{25.6}{11.7} = 2.2.$$

For the overall adhesive formulation, the NCO/OH ratio is:

$$\frac{(72.8 + 25.6)}{(24.6 + 11.7)} = 2.7.$$

A high isocyanate excess makes for good processability and adequate post-crosslinking through the formation of allophanates and biurets. Bubbling during the curing reaction is a disadvantage.

Table II-2.24: PUR hot-melt, 3rd generation

Item	Raw material	Parts by weight
1	Diisocyanate	12.3
2	Polyester 1	65.5
3	Stabilizer	0.2
4	Polyester 2	21.8
5	Catalyst	0.05
6	Plasticizer	0.15

Manufacturing:
9.1 parts by weight of item 1 (≙ 73 %) are reacted with item 2 and 3 for 30 min at 120°C, the the rest of item 1 (3,2 weight parts (≙ 27 %)) and pos. 4 are added and reacted for another 30 min at 120 °C. Finally, item 5 and 6 are added and reacted for 30 min at 120 °C. Optionally before filling vacuum is applied in order to minimize bubbles.
Item 1: "Desmodur CQ 44 M Liquid MS", M = 250 g/mol, f = 2 val/mol ≙ 125 g/val, NCO = 33.6 % ≙ 8000 mval/kg, Covestro
Item 2: "Dynacoll 7140", T_g = 30 °C, M = 5500 g/mol, f = 2 val/mol, Evonik (manufacturer information) ca. 2250 g/val, OH = 21 g KOH/kg ≙ 0.64 % ≙ 375 mval/kg ≙ 2667 g/val
Item 3: "Additiv TI", Borchers Americas Inc.
Item 4: "Dynacoll 7231", T_g = -30 °C, M = 3500 g/mol, f = 2 val/mol, (manufacturer information) ≙ ca. 1750 g/val, OH = 30 g KOH/kg ≙ 0.91 % ≙ 536 mval/kg ≙ 1865 g/val, Evonik
Item 5: DBTL, Borchers Americas Inc.
Item 6.: "Mesamoll", Lanxess

The content of free MDI can be estimated with the help of equation (4) from Chapter II-2.2.1.5 if it is assumed that the reaction has been performed in just one step, in accordance with the stoichiometry of the formulation:

$$\text{Content of free MDI in } [\%] = \left[\frac{(2.7-1)}{2.7}\right]^2 \cdot 12.3 = 4.85\ [\%]$$

4th generation

When polypropylene oxides with molecular weights of about 100 to 8000 g/mol react with excess isocyanate to afford prepolymers, the result is the familiar prepolymers whose properties, such as the glass transition temperature, are very similar to those of the polyethers used (for polypropylene oxide: about -60 °C). However, when low-molecular polyethers react with just a moderate excess of isocyanate (NCO/OH ratio close to 1), isocyanate-functionalised prepolymers are obtained. These prepolymers have a much higher glass transition temperature than the polyols employed. The glass transition temperature of the prepolymers can thus be readily adjusted via the urethane concentration (as a result of the molecular weight of the polyethers) [51, 52].

Figure II-2.45 shows the influence of the molecular weight of polyethers on the glass transition temperature of the resulting prepolymer for an NCO/OH ratio of 1.2. Only polyethers of sufficiently low molecular weight yield prepolymers whose concentration of urethane groups is such that the glass transition temperature is increased to the room temperature range. Polyethers with a molecular weight of about 400 g/mol are ideal [52].

The NCO/OH ratio exerts a significant influence on the glass transition temperature. Only if it is quite close to 1 is there a sufficient quantity of urethane groups in the prepolymer. These urethane groups are responsible for the higher glass transition temperatures. Consequently, an NCO/OH ratio

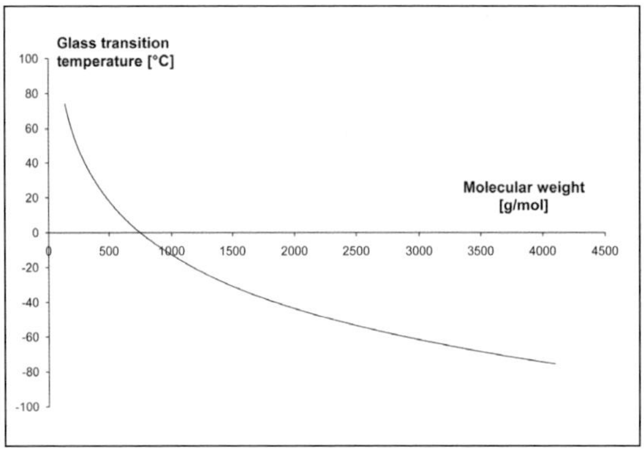

Figure II-2.45: Glass transition temperature of prepolymers as a function of the polyether molecular weight for an NCO/OH ratio of 1.2

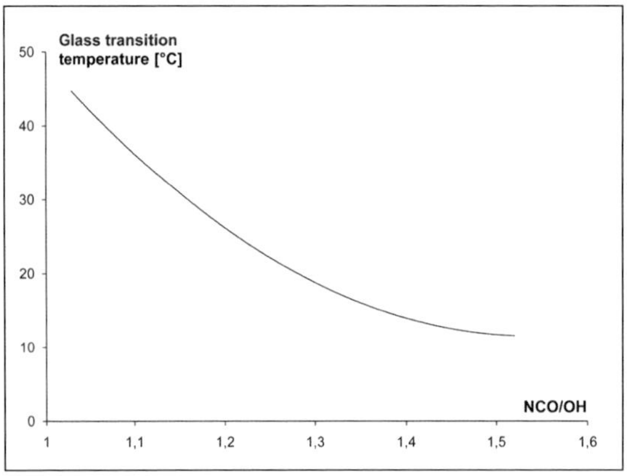

Figure II-2.46: Glass transition temperature of prepolymers as a function of the NCO/OH ratio for polyethers (PPO with 400 g/mol)

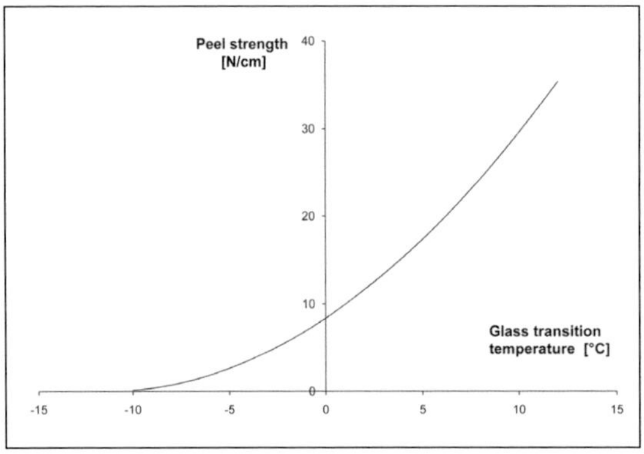

Figure II-2.47: Green strength as a function of glass transition temperature

of about 1.2 provides the necessary urethane concentration in the prepolymer, but there are also still enough reactive isocyanate groups for the subsequent curing reaction. Figure II-2.46 shows the influence of the NCO/OH ratio on the glass transition temperature for a polyether of molecular weight around 400 g/mol [52].

Adding a second prepolymer consisting of higher-molecular polypropylene oxide (M = 2000 to 4000 g/mol, diol or triol) or other less compatible polyols, such as hydroxyl-functionalised polybutadienes (see Chapter II-2.2.1.2) further flexibilises and toughens the adhesive formulation. To improve heat resistance, a small amount of triol or triisocyanate may be used, because it affords a way of increasing the crosslink density in the cured polymer.

Stabilisation of the prepolymers should be effected in any case during synthesis by adding the standard stabilisers (Chapter II-2.2.1.11) in a proportion of about 0.1 %. The application viscosity can be controlled via the NCO/OH ratio and the quantity of higher-molecular polyether. The cure rate can be adjusted via the catalyst concentration (DBTL, 200 to 500 ppm). Reactive hot-melts made in this way are relatively cheap, because they mainly consist of polypropylene oxide. They set rapidly and have high initial flexibility. Figure II-2.47 shows the green strength of such reactive hot-melts as a function of glass transition temperature. The green strength rises significantly with increase in glass transition temperature [52].

The disadvantages of polyether-based reactive hot-melts are their frequently moderate adhesion to important plastics, such as PVC, and, if applied in thick layers, a tendency to foam during curing.

Table II-2.25: PUR hot-melt, 4th generation

Item	Raw material	Parts by weight
Adhesive components: Prepolymer 1 (item 9)		
1	Diisocyanate	41
2	Polyether, low molecular mass	58.4
3	Stabilizer	0.5
4	Catalyst	0.1
Item 1 to 4 are reacted for 1 h at 120 °C		
Prepolymer 2 (item 10)		
5	Diisocyanate	11.7
6	Polyether, high molecular mass	87.6
7	Stabilizer	0.5
8	Catalyst	0.2
Item 5 to 8 are reacted for 15 min at 110 °C		
Formulation of the adhesive:		
9	Prepolymer 1	90.1
10	Prepolymer 2	9.9

Item 9 and 10 are mixed under vacuum for 1 h at 120 °C
Item 1 and 5: "Desmodur CQ 44 M Liquid MS", M = 250 g/mol, f = 2 val/mol ≙ 125 g/val, NCO = 33.6 % ≙ 8000 mval/kg, Covestro
Item 2: "Arcol PPG 425", M = 426 g/mol, f = 2 val/mol (manufacturer information), OH = 263 g KOH/kg ≙ 7,98 % ≙ 4696 mval/kg ≙ 213 g/val, Covestro
Item 3 and 7: "Additiv TI", Borchers Americas Inc.
Item 4 and 8: DBTL, Borchers Americas Inc.
Item 5: "Arcol PPG 4000", M = 4000 g/mol, f = 2 val/mol, OH = 28 gKOH/kg ≙ 0,85 % ≙ 500 mval/kg ≙ 2000 g/val, Covestro

Examples

Table II-2.25 presents a typical formulation, according to [51, ex. 1].

The adhesive is formulated from two separately synthesised prepolymers. For prepolymer 1, the hot-melt properties are adjusted by allowing a low polyether to react at an NCO/OH ratio close to 1:

41 g "Desmodur 44 MS" corresponds to 328 mval NCO,
58.5 g "Arcol PPG 425" corresponds to 274 mval OH
$\rightarrow$ NCO/OH = $\frac{328}{274}$ = 1.2.

For prepolymer 2, a higher polyether is used. The NCO/OH ratio is adjusted close to 2 in order that a sufficiently low viscosity may be obtained. It would even be possible to use polyols of higher functionality without having the problem of gelling.

11.7 g "Desmodur 44 MS" corresponds to 94 mval NCO,
87.6 g "Arcol PPG 4000" corresponds to 44 mval OH.
$\rightarrow$ NCO/OH = $\frac{94}{44}$ = 2.1.

The prepolymers are mixed to form an adhesive composed of approx. 10 % of the flexible prepolymer and 90 % of the solid prepolymer.

Table II-2.26: PUR hot-melt for edge bonding

Item	Raw material	Parts by weight
1	Amorphous copolyester	35
2	Liquid polyester	30
3	Highly crystallizing polyester	25
4	Slightly crystallizing polyester	10
5	4,4'-MDI	OH/NCO ratio: 1/2.2

Item 1: "Dynacoll 7150", solid, saturated, amorphous copolyester with high glass transition temperature, hydroxyl number 38 - 46 mg KOH/g, molecular weight 2600 - from hydroxyl number, Covestro
Item 2: "Dynacoll 7231", liquid, saturated copolyester, hydroxyl number 27 - 34 mg KOH/g, molecular weight 3500 - from hydroxyl number, Covestro
Item 3: "Dynacoll 7380", solid, highly crystalline, saturated copolyester, hydroxyl number 27 - 34 mg KOH/g, molecular weight 3500 - from hydroxyl number, Covestro
Item 4: "Dynacoll S 1402", saturated, linear copolyester, slightly crystalline, hydroxyl number 4 - 10 mg KOH/g, Covestro
Item 5: 4,4'- MDI
The product of the reaction between Dynacoll 7000 polyesters and an excess of diisocyanates is a reactive hot-melt (RHM). The reaction occurs in the melt. For characterization of the RHM data in this brochure the preparation was carried out under the following constant laboratory conditions: The polyester melt was evacuated in a flask, regardless of the actual (normally low) water content of the DYNACOLL 7000 products, in a vacuum of less than 10 mbar for 45 minutes at 130 °C. The polyesters are then reacted in an inert gas atmosphere (dried nitrogen or carbon dioxide) with the calculated amount of diisocyanate at 130°C. The reaction is complete when the theoretical free isocyanate content is obtained.
After 45 minutes the melt was degassed until it was free of bubbles. The reactive hot-melt was then filled into containers and stored under exclusion of moisture and light. Under production conditions, the reaction times should be individually adapted to the adhesive formulations. Drying is determined by the water content of all the components of the formulation and is generally recommended in order to prevent side reactions. [62]

Table II-2.27: PUR hot-melt for profile wrappings

Item	Raw material	Parts by weight
1	Solid polyester	44.5
2	Liquid polyester	17.0
3	Crystallizing polyester	22.0
4	Highly crystallizing polyester	16.5
5	4,4'-MDI	OH/NCO ratio: 1/2.0

Item 1: "Dynacoll 7131", solid, amorphous, saturated copolyester, hydroxyl number 31 - 39 mg KOH/g, molecular weight 3000 - from hydroxyl number, Covestro
Item 2: "Dynacoll 7231", liquid, saturated copolyester, hydroxyl number 27 - 34 mg KOH/g, molecular weight 3500 - from hydroxyl number, Covestro
Item 3: "Dynacoll 7360", solid, partially crystalline, saturated copolyester, hydroxyl number 27 - 34 mg KOH/g, molecular weight 3500 - from hydroxyl number, Covestro
Item 4: "Dynacoll 7380", solid, highly crystalline, saturated copolyester, hydroxyl number 27 - 34 mg KOH/g, molecular weight 3500 - from hydroxyl number, Covestro
Item 5: 4,4'- MDI
The product of the reaction between Dynacoll 7000 polyesters and an excess of diisocyanates is a reactive hot-melt (RHM). The reaction occurs in the melt. For characterization of the RHM data in this brochure the preparation was carried out under the following constant laboratory conditions: The polyester melt was evacuated in a flask, regardless of the actual (normally low) water content of the Dynacoll 7000 products, in a vacuum of less than 10 mbar for 45 minutes at 130 °C. The polyesters are then reacted in an inert gas atmosphere (dried nitrogen or carbon dioxide) with the calculated amount of diisocyanate at 130 °C. The reaction is complete when the theoretical free isocyanate content is obtained.
After 45 minutes the melt was degassed until it was free of bubbles. The reactive hot-melt was then filled into containers and stored under exclusion of moisture and light. Under production conditions, the reaction times should be individually adapted to the adhesive formulations. Drying is determined by the water content of all the components of the formulation and is generally recommended in order to prevent side reactions. [62]

Further examples

Further examples of sample formulations for reactive PUR hot-melts for typical applications can be found in the Table II-2.26 and Table II-2.27.

Hybrid types and modifications

The flexibility of the end products, along with their setting and adhesion characteristics, can be adjusted by blending polyethers and polyesters [56–59]. Compatibility then becomes a major consideration. If the polymers are fully compatible, the glass transition temperature of the mixture will be a function of the individual weight fractions (see Chapter II-1.1.4.3, Fox's equation). Where they are partially incompatible, the reaction must be suitably controlled to ensure that no macroscopic phase separation occurs during manufacture or application. If, for example, polyethers are co-condensed during the manufacture of polyesters, no macroscopic phase separation can occur.

Table II-2.28: Pseudoplastic PUR hot-melt

Item	Raw material	Parts by weight
1	Diisocyanate	11.4
2	Crystallizing polyester	84.5
3	Rheological additive	4
4	Catalyst	0.1

Item 1, 2 and 4 are mixed and reacted for 1 h at 100 °C. Then item 3 is added, and vacuum is applied still in the heat.
Item 1: "Desmodur CQ 44 M Liquid MS", M = 250 g/mol, f = 2 val/mol ≙ 125 g/val, NCO = 33.6 % ≙ 8000 mval/kg, Covestro
Item 2: "Dynacoll 7360", M = 3500 g/mol, f = 2 val/mol, (manufacturer information) ≙ ca.1750 g/val, OH = 30 g KOH/kg ≙ 0,91 % ≙ 536 mval/kg ≙ 1865 g/val, Evonik
Item 3: Fumed silica, "Aerosil R 202", Evonik
Item 4: DBTL, Borchers Americas Inc.

Crystalline polyols (like many polyesters or poly-THF) reduce the tendency of the uncured adhesive to creep along with the tendency to bubble during curing. However, they help to increase the brittleness of the prepolymer if they are used in high concentrations. Particularly-rapid-crystallising systems can be made with polyesters based on higher aliphatic polycarboxylic acids, such as dodecane dicarboxylates [60].

The rheological properties can be modified by formulation with 3 to 5 % hydrophobic fumed silica. The resultant products show sag resistance, i.e. will not flow down vertical surfaces [61].

Example

Table II-2.28 presents a PUR hot-melt of pronounced pseudoplasticity.

The formulation concept in this case is that of a 2nd generation adhesive. However, the principle of rheological modification can also be applied to other generations.

11.4 g "Desmodur 44 MS" corresponds to 91 mval NCO,
84.5 g "Dynacoll 7360" corresponds to 45.3 mval OH.
$\rightarrow$ NCO/OH = $\frac{91}{45.3}$ = 2.0.

Trends and Outlook

Recent developments are aimed at lowering the content of free monomer. Isocyanates bearing various reactive functional groups can be used to further reduce the NCO/OH ratio for prepolymer synthesis and so to further lower the monomer content (see also Chapter II-2.2.1.12). Reputable producers of isocyanates also supply prepolymers whose content of monomer has already been minimised by thin-layer evaporation.

It is also possible to utilise the large difference in reactivity between the two isocyanate groups of 2,4-MDI to synthesise prepolymers with a particularly low residual monomer content [63].

The use of silanes to replace isocyanate groups as the reactive function for moisture-curing has failed to catch on. As with MS polymers or silicones (see Chapter II-2.4.3), the technology for using silanol groups instead of isocyanate groups in prepolymers exists. However, numerous disadvantages (price, setting process, adhesion) have so far hampered large-scale market introduction. The newly introduced POR hot-melts (see Chapter II-2.4.2) employ the curing technology of silicones, but not the polymer structure of polyurethanes.

Delivery form and application

Moisture-curing reactive hot-melts have to be stored away from moisture after manufacture. Packaging varies with the intended application, but includes 300 ml cartridges, 1.5 l tins, 20 l pails and 200 l drums. For cost reasons, packaging is gradually evolving in the direction of laminated aluminium foil bags. They are applied by machines similar to those employed for traditional hot-melts. For small quantities, cartridge pistols are available, while large quantities are applied from pails or drums with the aid of drum melters. In any event, the applicators must be designed to minimise access of moisture to the product before application (e.g. drum punch pumps). Adhesive beads (optionally in foam form) [53] are applied through round nozzles. For two-dimensional application, slot dies nozzles, roll coating applicators or spin-sprayers are used. For laminar, two-dimensional applications, as is the case for traditional hot-melts, thermal reactivation is possible within a certain time window before the curing reaction has proceeded too far.

Products

Products currently offered on the market differ mostly in the following parameters:
- application temperature (100 to 180 °C)
- application viscosity (5000 to 100,000 mPa.s)
- reactivity/curing time (hours to days)
- stability under application conditions (varies with supplier)
- open time (seconds to minutes)
- setting time (seconds to minutes)
- adhesion properties on certain substrates (varies with supplier)
- content of free monomeric isocyanate (varies with supplier)
- price (polyester or polyether polyols).

One of the most underestimated risk factors in industrial applications is the products' reactivity with moisture. Even tiny quantities of moisture are sufficient to cause rapid skinning at the elevated application temperatures. Such skins can prove very troublesome for processing through nozzles. And yet, rapid curing to reach the end properties is desirable after application. Safe and straight-forward application are key to the further success of reactive hot-melt technology.

The free-MDI content of most commercial products is not critical, according to current knowledge, provided that application temperatures are not too high and sufficient ventilation is guaranteed. In any event, the best-possible safety measures should be adopted, because long-term exposure to even tiny amounts of isocyanates causes sensitisation. For a few years now, MDI has been suspected of being carcinogenic. The worst-case scenario in spin-spray application is that reactive aerosols may be formed. Accordingly, safety measures are imperative.

Evaluation

Moisture-curing reactive PUR hot-melts have a very interesting product profile. They combine many of the advantages of hot-melts with those of reactive adhesives. They boast the rapid setting of a hot-melt, a fact which is very important for industrial applications, but also exhibit many of the advantages of reactive adhesive systems.

Advantages

+ 100 % system
+ good storage stability
+ can be used for automated processes
+ mild application conditions prevent damage to substrates
+ rapid setting
+ good adhesion
+ good heat resistance
+ good low-temperature properties
+ good chemical and hydrolytic stability

Disadvantages

- price
- special packaging required
- need for an applicator
- moderate green strength compared to traditional hot-melts
- curing reaction may vary with weather conditions and substrates
- isocyanate is a harmful substance

Fields of application

The highly attractive product properties and scope for automated application are ideal for applications in the **automotive industry**, especially where non-metallic surfaces have to be joined. However, the success of applications usually hinges not so much on the adhesive properties as on process stability, which has not always been adequate in the years following the introduction of the adhesives. Successful applications include the bonding of headliners to the coated roof, lamination of door panels, dashboards and rear parcel shelves, lamination of car seats, and attachment of ABS hooks to interior door panels.

Utility vehicles boast various bonded injection-moulded parts. For reactive hot-melts, wood and timber-derived materials usually make good substrates because they already contain a great deal of humidity and are also moisture-permeable. Thus, applications are to be found in the **woodworking and furniture industry** for demanding edge-veneer bondings that must be resistant to heat and humidity. The combination of hot-melt with good adhesion properties and improved heat resistance have paved the way for reactive hot-melts in **book-binding** applications. They are the products of choice when papers which are difficult to bond must satisfy high quality requirements. Use on **construction sites** is usually too problematic due to the need for an applicator. However, vendor parts, such as flooring elements for computer rooms (metal-laminated particle board) and special roofing parts have been manufactured successfully with the aid of reactive hot-melts. The attachment of shoe soles to shafts, a very demanding adhesive process, which is mainly still done with solvent-based polyurethanes, has been successfully executed in various line trials, but the **shoe industry**, is fairly conservative and has been unwilling so far to make the necessary changes in products and processes. Even now, it prefers solvent-based products, with all their toxicological and ecological consequences (see Chapter II-1.1). In the **textiles industry** reactive hot-melts are being used increasingly to laminate high-quality fabrics. They make ideal products for the manufacture of durable laminates because of their low temperatures, and rapid, automated processing.

2.2.1.15 References

[1] Becker, G. W., D. Braun, Kunststoffhandbuch 7, Polyurethane, 3rd edition. Hanser (1993)
[2] Szycher, M., Szycher's Handbook of Polyurethanes, CRC Press (1999)
[3] Saunders, H. J., K. C. Frisch, Polyurethanes, Chemistry and Technology, Vol. 1 and 2, Interscience (1962)
[4] Uhlig, K., Polyurethan Taschenbuch, Hanser (1998)
[5] Somarathna, H. M. C. C., et al. "The use of polyurethane for structural and infrastructural engineering applications: A state-of-the-art review." Construction and Building Materials 190 (2018): 995–1014.
[6] Golling, Florian E., et al. "Polyurethanes for coatings and adhesives–chemistry and applications." Polymer International 68.5 (2019): 848–855.
[7] Sahoo, Swarnalata, Smita Mohanty, and Sanjay Kumar Nayak. "Biobased polyurethane adhesive over petroleum based adhesive: Use of renewable resource." Journal of Macromolecular Science, Part A 55.1 (2018): 36–48.
[8] Segura, D. M., et al. "Chemistry of polyurethane adhesives and sealants." Handbook of adhesives and sealants. Vol. 1. Elsevier Science Ltd, 2005. 101–162.
[9] Gadhave, Ravindra V., Prakash A. Mahanwar, and Pradeep T. Gadekar. "Bio-renewable sources for synthesis of eco-friendly polyurethane adhesives." Open Journal of Polymer Chemistry 7.04 (2017): 57–75.
[10] Schindler, M., Coating, 2003, 1, 22–24
[11] Fisseler-Eckhoff, Annette, et al. "Environmental isocyanate-induced asthma: morphologic and pathogenetic aspects of an increasing occupational disease." International journal of environmental research and public health 8.9 (2011): 3672–3687.
[12] safeusediisocyanates.eu
[13] Popp, Matthias, Katharina Teczyk, and Andreas Hartwig. "Emissionen von PUR-Hotmelts bei unterschiedlichen Applikationstechniken." Chemie Ingenieur Technik 77.12 (2005): 1978–1983.
[14] Bornholdt, Nick, Ralph Lehnen, and Elke Fliedner. "Cyclische Carbonate als Vernetzer für Bioklebstoffe." *adhäsion Kleben & Dichten* 64.7 (2020): 45–47.
[15] Niesiobędzka, Joanna, and Janusz Datta. "Challenges and recent advances in bio-based isocyanate production." Green Chemistry 25.7 (2023): 2482–2504.
[16] Jeska, Simone, and Khaled Saleh Pascha. Emergent timber technologies: materials, structures, engineering, projects. Birkhäuser, 2014.
[17] Brandmair, Anton, et al. "Ultraschnelle Montagetechnik: Mit Klebstoff punkten." Adhäsion Kleben & Dichten 55.12 (2011): 22–25.
[18] Flory, P. J., Principles of Polymer Chemistry, Cornell University Press (1953)
[19] Fogiel, A. W., Macromolecules, Vol. 2, No. 6, Nov – Dec (1969), 581 ff
[20] Bonart, R. et al., Colloid and Polymer Sci. 260 (1982), 518 ff
[21] Penati, Amabile, Carlo Maffezzoni, and Enrico Moretti. "Practical and theoretical molecular weights in polyaddition reactions of ethylene and propylene oxide." Journal of Applied Polymer Science 26.4 (1981): 1059–1071.
[22] Schmidheiny, in prochem.ch/html/forum/PURchem.pdf
[23] Hartshorn, S. R, Structural Adhesives, Plenum Press (1986)
[24] Skeist, I., Handbook of Adhesives, 3rd edition, Chapman & Hall (1989)
[25] G. Habenicht, Kleben, 3rd edition, Springer (1997)
[26] Sardon, Haritz, et al. "From lab to market: current strategies for the production of biobased polyols." ACS Sustainable Chemistry & Engineering 9.32 (2021): 10664–10677.
[27] Zimmer, Bettina, et al. "Chemistry, polymer dynamics and mechanical properties of a two-part polyurethane elastomer during and after crosslinking. Part I: dry conditions." Polymer 115 (2017): 77–95.
[28] Zimmer, Bettina, et al. "Chemistry, polymer dynamics and mechanical properties of a two-part polyurethane elastomer during and after crosslinking. Part II: moist conditions." Polymer 149 (2018): 238–252.
[29] Bayer, Baycoll, Brochure, dated 11/99
[30] Morrison, R. T., R. N. Boyd, Lehrbuch der Organischen Chemie, 3rd edition, Verlag Chemie (1973)
[31] March, J., Advanced Organic Chemistry, 4th edition, J. Wiley&Sons (1992)
[32] Bayer Material Science, One Pack PUR adhesive, MMX 0336, edition 2004-02-04
[33] EPA 0 289 901 A2
[34] Krobb, J., E. Wistuba, Adhäsion (1992) No. 5, p. 14–20
[35] Carter, N. G., in chemsoc.org/pdf/gcn/Industrial.pdf
[36] Kozakiewicz, J., Adhesion, 13 (1989), 114 ff
[37] Jowat AG, Power PUR, Techn. Information JN D044 06/2003, D008 05/2003
[38] Wicks, Z. W., Progress in Org. Coatings, 3 (1975), 73–99

[39] Baxenden Chemicals Limited, Blocked Isocyanates, info brochure
[40] Covestro, Crelan-Produkte, solutions.covestro.com/en/brands/crelan
[41] Müller, P., et al., Angew. Makromol. Chem., 65 (1977), 23–39
[42] DE 3 940 273
[43] DuPont, Adiprene, Caytur 21, information brochure E-37761
[44] EP 0 311 852
[45] Voithenberg, H. v., Europ. Adh.and Sealants, 1 (1984), 4, 28–32
[46] BOSTIK Supergrip 2000, Emhart Chemical Group, Bostik Technical Centre, company brochure (1985)
[47] DOS 3 236 313
[48] DOS 2 609 266
[49] EP 0340 906
[50] Huber, H. F., H. Müller, Adhesives Age, 11, 1987, 32–35
[51] EP 0 369 607
[52] Gilch, H., W. Rath, UTECH 90, Conference, The Hague (1990)
[53] EP 0 405 721
[54] company brochure, Evonik, Degussa, Dynacoll 7000, 2003
[55] BOSTIK Tucker GmbH, technical data sheet Supergrip 2000-9810 (1989)
[56] DOS 199 61 941
[57] DE 4 418 177
[58] DOS 195 04 007
[59] DOS 44 19 449
[60] EP 0 421 154
[61] EP 0 386 897
[62] products.evonik.com/assets/37/68/DYNACOLL_Product_Range_EN_Asset_1203768.pdf
[63] Krebs, Michael. "Neue monomerreduzierte Polyurethan-Hotmelts: Nicht als Gefahrstoff kennzeichnungspflichtig." adhäsion KLEBEN & DICHTEN 48 (2004): 15–19.
[64] cardolite.com/products/nco-blocking/
[65] pu-additives.com/en/brands/kosmos
[66] solutions.covestro.com/de/products/desmodur/desmodur-cq-n-7300_84603813-19833122?Selected-Country=US
[67] DE-OS 40 28 704 A1
[68] Kirk-Othmer, Encyclopedia of Chemical Technology, 3rd edition, Vol. 15, Mixing and Blending, p. 623, Wiley, New York (1981)
[69] DE-OS 40 38 815 A1
[70] Dollhausen, M. "Polyurethane Adhesives Based on Baycoll, Desmocoll and Desmodur." (1988).
[71] hobum.de/en/solutions/products/merginol/

2.2.2 Epoxies

Epoxy adhesives have been introduced commercially in 1946. Due to their exceptional and versatile product portfolio, especially the high achievable bond strength, they are used in various application areas such as transportation including aerospace, electrical and electronics, renewable energy and general industry, and their consumption continues to increase. Today the main driver is the automotive industry with the increasing demand for light weight construction allowing low-carbon emitting vehicles. Furthermore, the growing demand for electric vehicles is also having a positive effect. In 2020 the epoxy adhesive market was valued at USD 9.2 billion and is projected to reach USD 11.2 billion by 2027 [1].

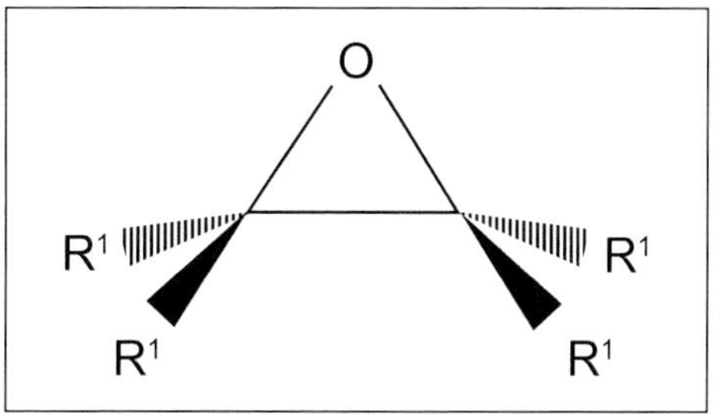

Figure II-2.48: Strained ring structure of the oxirane group

2.2.2.1 Reaction principle

It is the ability of the oxirane oxirane- (or the epoxide-)ring (Figure II-2.48), especially its reactions with nucleophiles and electrophiles which forms the chemical basis of reactive epoxy adhesives. The equilateral triangle of the oxirane ring is a highly strained structure, because the sp^3-hybridised carbon has to adopt an angle of about 60° instead of the ideal 109.5° for a tetrahedron.

The curing reaction of epoxies to yield the polymer always occurs under ring opening of the strained equilateral triangle. As shown in Figure II-2.49, ring opening can be initiated by an S_N2 attack of (polyfunctional) nucleophiles at the sterically less hindered carbon, such as primary and secondary amines.

2.2.2.2 Raw materials and reactions

The properties of epoxy adhesives can be adapted to the specific application within certain limits by selecting

- the appropriate epoxy resin or combination of resins
- the curing agent or a combination of curing agents
- the filler composition
- and additives

The properties of the various raw materials and their respective effect on the properties of the resulting adhesives are discussed in the following sections.

Epoxy resins

Epoxy adhesives are made from epoxy resins, i.e. monomeric or oligomeric reactants which have more than one epoxy group in the molecule, and which are capable of reacting with curing agents to form polymers. The most important epoxy resins for formulating adhesives are adducts of epichlorohydrin and bisphenol A [2, 3, 4].

Liquid diglycidyl ether of bisphenol A

For the synthesis of epichlorohydrin (Figure II-2.50) propylene is converted by a free-radical substitution into allyl chloride, which is converted into epichlorohydrin in the presence of a base and chlorine.

Bisphenol A is synthesized in an acid-catalysed reaction from phenol and acetone which both are obtained by the Hock synthesis of phenol (Figure II-2.51).

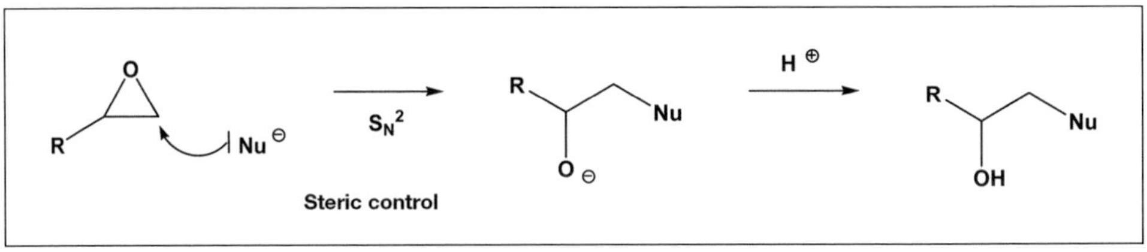

Figure II-2.49: Nucleophilic ring opening

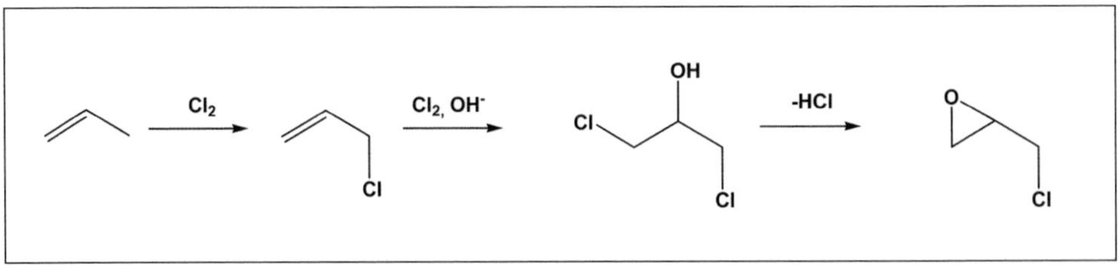

Figure II-2.50: Synthesis of epichlorohydrin

Thanks to their well-balanced properties profile epoxy resins formed from the reaction of bisphenol A and epichlorohydrin are most commonly used for the formulation of epoxy adhesives. They are known as diglycidyl ether of bisphenol A, also known by its IUPAC name 2,2'-[2,2-propanediylbis(4,1-phenylen-oxymethylene)] dioxirane or more commonly by the abbreviations DGEBA and BADGE, respectively. The synthesis of bisphenol A diglycidyl ether described in Figure II-2.52 is carried out by reacting bisphenol A with an excess of epichlorohydrin under basic conditions leading to bifunctional epoxy resins. Bifunctional oligomers are obtained by adjusting the mol-ratio of the reactants.

In accordance with their molecular weight indicated by the value of n, bisphenol A based epoxy resins are either liquids (n = 0 – 0.5), semi-solids (n ≈ 1) or solids (n = 2 – 18). An adduct with n = 0.15 (a standard liquid epoxy resin commonly used in the formulation of adhesives), for example, is a colourless liquid with a viscosity of approx. 10,000 mPas. A corresponding adduct with n = 2.15 is a solid resin which melts at around 60 °C, while an oligomer with n =5.15 melts at 90 °C [6]. A pure product with n = 0 is crystalline (m.p. 42 °C), which is not desirable for adhesives applications due to processing problems.

During curing, crosslinking is primarily achieved via the epoxy groups. Provided that an appropriate curing agent is present, additional crosslinking can be achieved using the OH groups. On the other hand, the polar OH group contributes to the good adhesion properties of epoxy adhesives. The aromatic bisphenol A group contributes to the relatively good heat and chemical resistance.

If formaldehyde is reacted with phenol instead of acetone, epoxy resins based on bisphenol F (Figure II-2.53) are obtained in a similar way. For steric reasons, these bisphenol F adducts are not only p,p'-disubstituted but they have also a certain o,p- and o,o'-content. Commercial mixtures of bisphenol A and F types are also available.

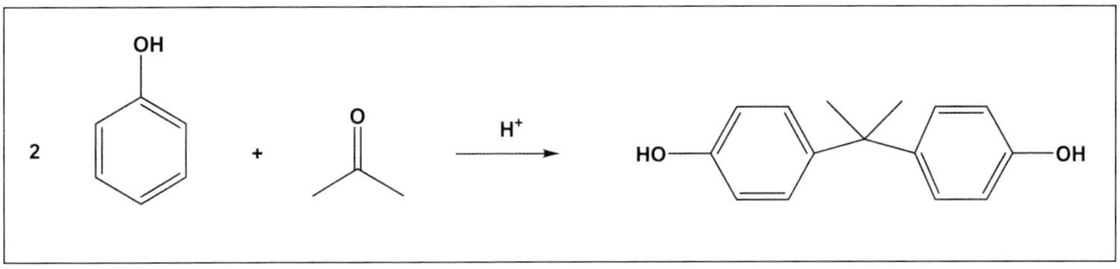

Figure II-2.51: Synthesis of bisphenol A

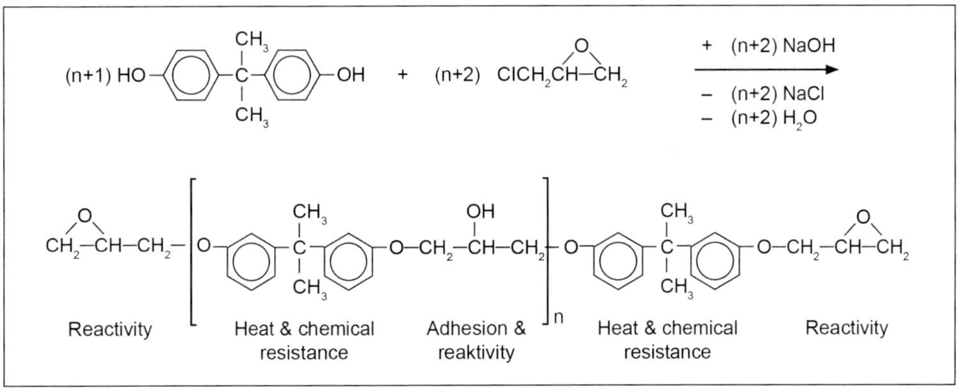

Figure II-2.52: Synthesis of bisphenol A diglycidyl ether

The logical consequence of the classification of bisphenol A as a CMR (**C**arcinogenic, **M**utagenic or toxic for **R**eproduction) substance, the residual content contained in bisphenol A-based epoxy resins is increasingly being viewed critically and resins with a significantly reduced residual bisphenol A content have been developed. For use in the manufacture of electronic devices, the residual content of chlorine, especially that of hydrolysable chlorine, is often of high importance due to its potential corrosive effect. Here, too, resins with significantly less than 200 ppm hydrolysable chlorine have been developed in recent years.

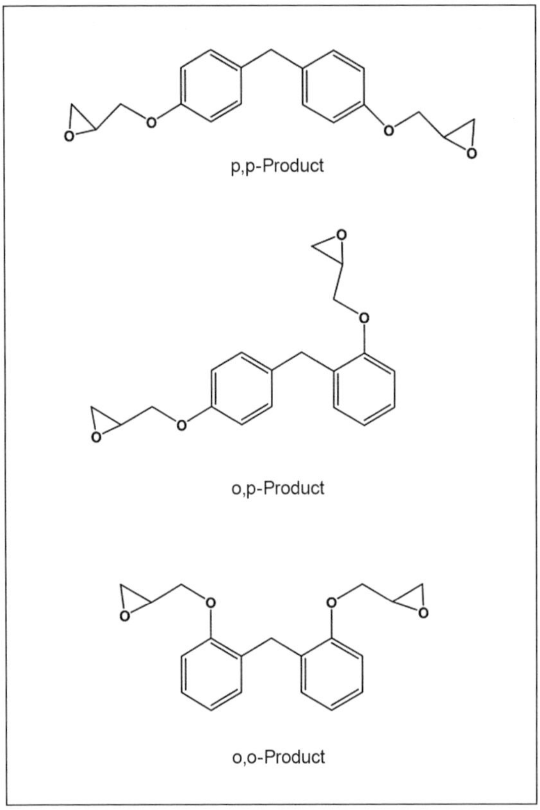

Figure II-2.53: Bisphenol F diglycidyl ether

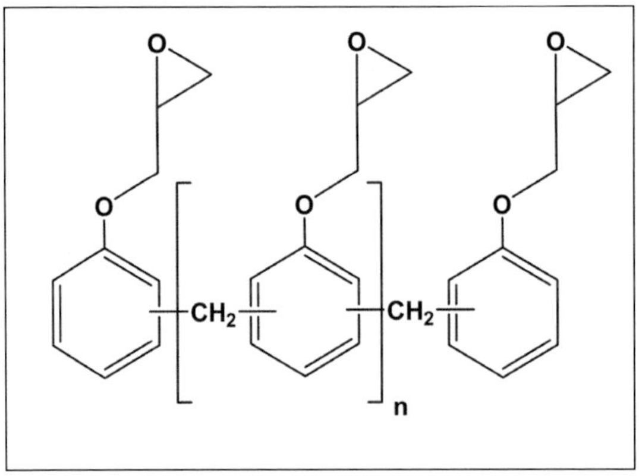

Figure II-2.54: Epoxy novolak resin

Novolac epoxy resin

Higher-functional aromatic epoxy resins (f = 2 – 6 val/mol) can be obtained, for example, by the analogous reaction between novolacs and epichlorohydrin. The reaction of phenol novolac with epichlorohydrin yields liquid resins, whereas solid resins are obtained with cresol novolac, so that the latter are of rather minor importance for the use in adhesives. The structure of phenol novolac resins is shown in Figure II-2.54.

Due to their higher functionality and good heat resistance, they are used in combination with standard resins to increase the crosslinking density or to improve the temperature resistance of the adhesives, and less frequently as base resins.

Epoxy resin based on renewable biomaterial

Relatively new on the market are epoxy resins based on cashew nutshell liquid (CNSL), a natural, non-food chain, and annually renewable biomaterial. CNSL is extracted from the shells of cashew nuts and is considered a byproduct of the cashew nut industry. It consists essentially of cardanol built up from an aromatic ring with an OH group and a long aliphatic side chain (Figure II-2.55). The aromatic ring provides a strong chemical resistant backbone while the OH-group can be epoxidized. The aliphatic side chain provides excellent water resistance, good flexibility and low viscosity. Based on cardanol, the bifunctional, relatively low-viscosity epoxy resin also shown in Figure II-2.55 and the novolac resin are available for the formulation of adhesives.

The chain of 8 carbon atoms separating aromatic groups in the difunctional cardanol-based resin allows its use in conjunction with conventional epoxy resins to increase flexibility and elongation, bond strength, as well as water and chemical resistance while minimizing the impact on mechanical strength. In addition, this resin provides improved hydrolytic stability.

The cardanol-derived polyglycidyl ether epoxy novolac resin can also be used in combination with standard epoxy resins. It improves flexibility, also at low temperatures and extends pot life without adversely affecting chemical and water resistance.[5]

Aliphatic and cycloaliphatic epoxy resin

Aliphatic epoxy resins are synthesized either from substances with carbon-carbon double bonds by oxidation, e.g. means of peracids, or hydrogen peroxide or by reacting aliphatic alcohols, polyols or carboxylic acids with epichlorohydrin. The resins of the first group are characterized by a low functionality and are typically used in combination with other epoxy resins to improve cure rate, flexibility, and heat deflection temperature of the adhesive. Examples are shown in Figure II-2.56. The epoxidation of aliphatic alcohols with epichlorohydrin allows the synthesis of mono-, di-, or higher functional aliphatic resins typically characterized by a low viscosity. They are used in so called "modified epoxy resins" as reactive diluent or to improve adhesion. Except of resins obtained from long chain polyols which are used to improve mechanical properties like impact resistance they show a lack of toughness and a certain water sensitivity. Resins based on glycerol and pentaerythritol are water soluble. Examples are also shown in Figure II-2.56.

Hybid resins

Various polymers such as nitrile, phenolic, nylon, polysulphide, polyester, polyol and polyurethane resins can be used to improve the properties of cured epoxy resins. These so-called hybrid resins are obtained by mixing the respective polymer in the epoxy resin (modified resin) or by a chemical reaction between reactive groups of the elastomer with those of the epoxy resin (epoxy prepolymer). The latter is shown in Figure II-2.57 using the example of the reaction of liquid, carboxyl-terminated rubber leading to a prepolymer widely used in combination with conventional epoxy

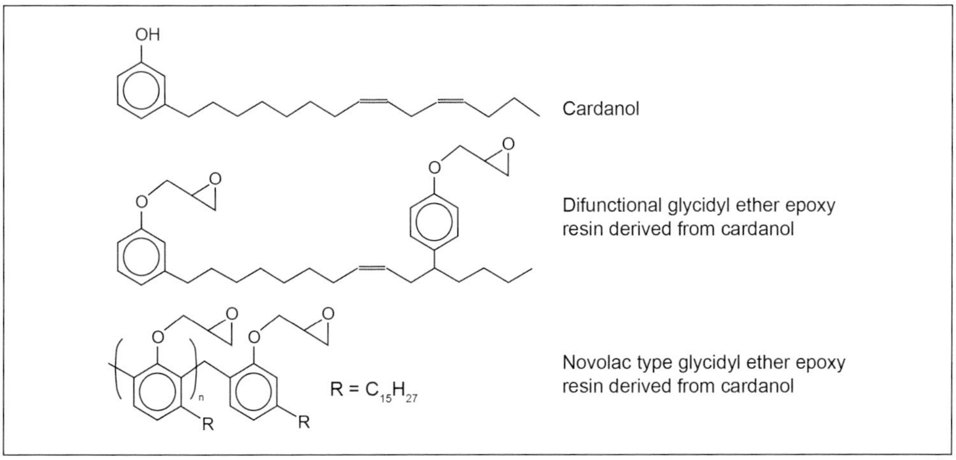

Figure II-2.55: Chemical structure of cardanol and glycidyl ether epoxy resins derived from cardanol [5]

resins to reduce brittleness and to improve impact resistance, low temperature properties, lap shear strength flexibility and thermostability without loss of other important properties. In this manner, prepolymers containing polyether, silicone or polyurethane segments can also be obtained. Table II-2.29 provides an overview of the properties that can be achieved by using hybrid resins in the formulation of epoxy adhesives.

Compared to the group of reactive diluents discussed later, hybrid resins are characterized by a higher molecular weight and thus a sometimes quite high viscosity. They are therefore sometimes regarded as critical when adhesives with good self-levelling properties are required. They also counteract a high filler content, used to lower the raw material price, but generally have less detrimental effects on the cured properties of the epoxy resin.

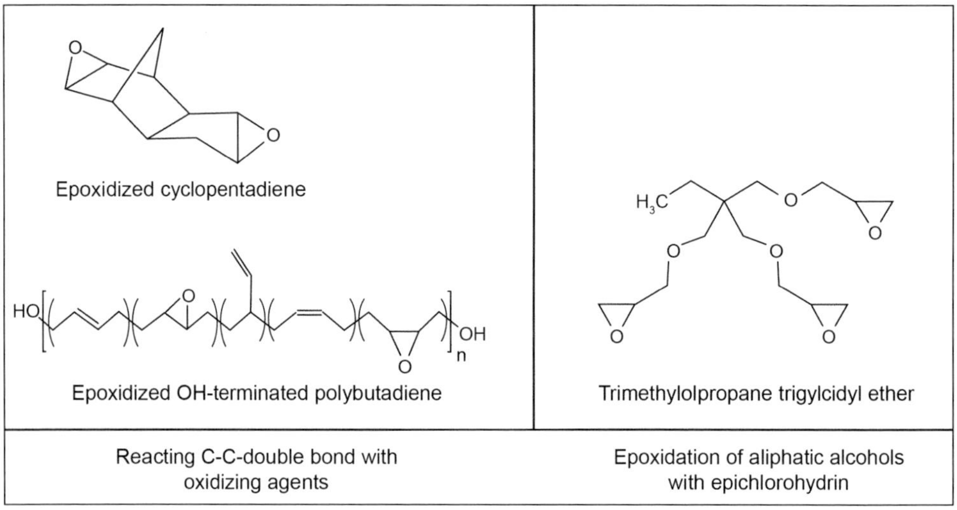

Figure II-2.56: Examples of aliphatic epoxy resins

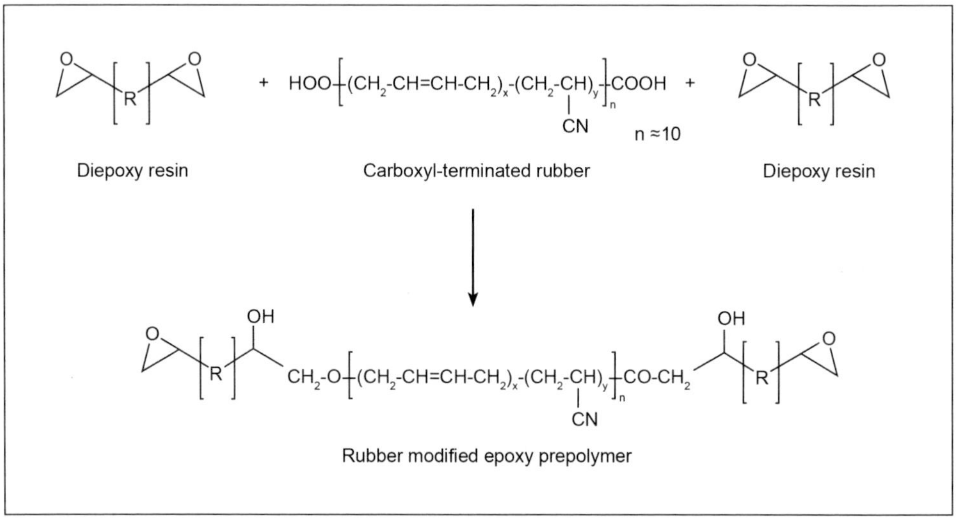

Figure II-2.57: Synthesis of rubber-modified epoxy prepolymer

Table II-2.29: Effect of the addition of hybrid-epoxy resins on the properties of epoxy adhesives

Nitrile rubber modified epoxy (DGEBA or DGEBF) prepolymer	Epoxy resin modified with carboxyl-terminated butadiene nitrile (CTBN) copolymer provides toughness, impact resistance and high peel strength without compromising heat and chemical resistance. On the other hand, the high viscosity of CTBN-modified resins has a detrimental effect on the viscosity of the adhesive formulation. Applications for these adhesives include the crash resistant bonding of automotive body structures
Polyurethane modified epoxy resin (DGEBA) Polyol modified epoxy prepolymer (DGEBA)	High level of toughness, high elongation, very good peel and impact strength improving low temperature properties at a still relatively high tensile strength. Similar to CTBN-modified resins they are used for bonding crash-resistant automotive body structures
Epoxy-functional polyester	Improving mechanical properties such as peel strength and impact resistance
Epoxy-vinyl	Vinyl components increase toughness and peel strength. The adhesives offer very good oxidation stability with moderate strength at temperatures of up to 150 °C. Applications for these adhesives include bonding safety glass, aerospace engine structures and structural panels.
Epoxy "Nylon"	High tensile shear strength and high peel strength. The adhesives exhibit improved low-temperature properties, but only moderate properties at high temperatures.
Epoxy-polythiol (mercaptan)	High elongation but a reduced tensile strength compared to unmodified epoxies. Relatively low viscosity. They are mainly used as sealants or coatings in the construction industry.
Epoxy-phenolic	High shear strength at temperatures above 150 °C, but the adhesives are relatively brittle and have low peel and impact strength. Typically, resole-type phenolics are used and the epoxy is only a minor constituent of the hybrid resin.
Epoxy-acrylate resins	Allows UV-cure via the acrylic function in a radical polymerization reaction (see Chapter II-2.3.1.3)

Reactive diluents

The relatively high viscosity of epoxy resins sometimes makes it difficult to blend the ingrediants during the production of the adhesives and to dispense the finished adhesive. This applies not only to applications requiring low-viscosity self-levelling adhesives, but also to those requiring adhesives with high yield points (sag resistance), which do not sag even when applied to vertical surfaces. The raw materials usually used to achieve high yield points, such as carbon black or fumed silica, also result in an undesirable increase in viscosity with regard to application and wetting properties. Likewise, the possibility of achieving certain properties or reducing raw material costs by adding appropriate raw materials, especially fillers, is severely limited.

Even if the use of solvents is in principle possible and in part also practiced, this not only leads to the undesirable labelling of the adhesive as flammable but also requires solvent evaporation during the bonding process.

Non-reactive diluents based on phthalates, nonylphenol or N-methylpyrrolidone (NMP), which have been used for a long time, are due to the high environmental and health hazards now hardly used anymore. Cardanol (cashew nutshell liquid (CNSL)), a natural, non-food chain, and annually renewable biomaterial) and Cardanol-based diluents (Figure II-2.58) can be considered as an environmentally friendly alternative. Additionally, they act similar as nonylphenol or NMP as accelerators between epoxy resins and amines. The long hydrophobic aliphatic side chain of the cardanol gives these products a very low viscosity, provides excellent water resistance and corrosion protection as well as a certain flexibility of the cured adhesive. Last but not least the aromatic ring contributes to an improvement in temperature and chemical resistance. [5].

The addition of typical non-reactive, especially phthalate-based plasticizers has the disadvantage that they

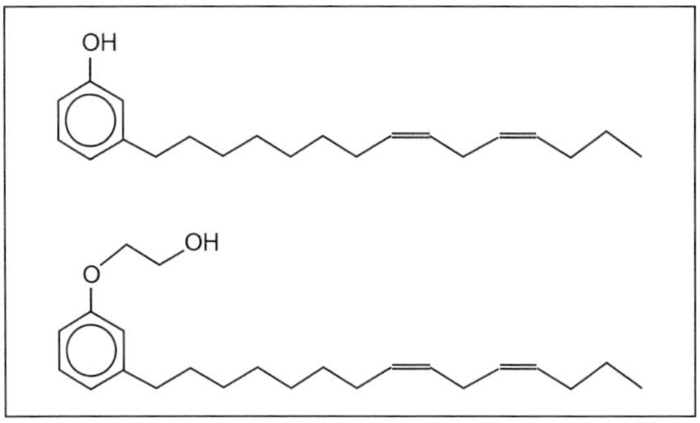

Figure II-2.58: Non-reactive cardanol based diluents [5]

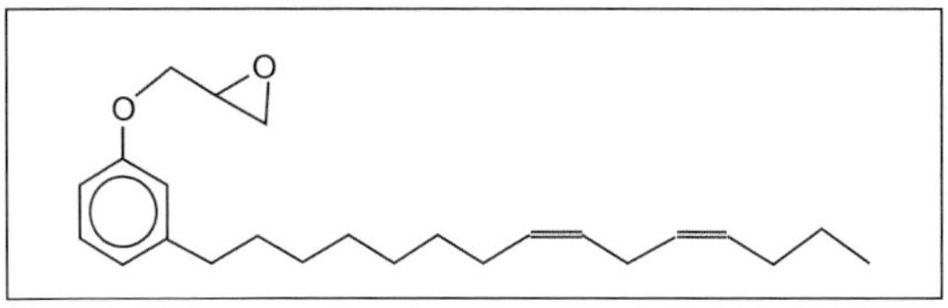

Figure II-2.59: Epoxidized cardanol as reactive diluent [5]

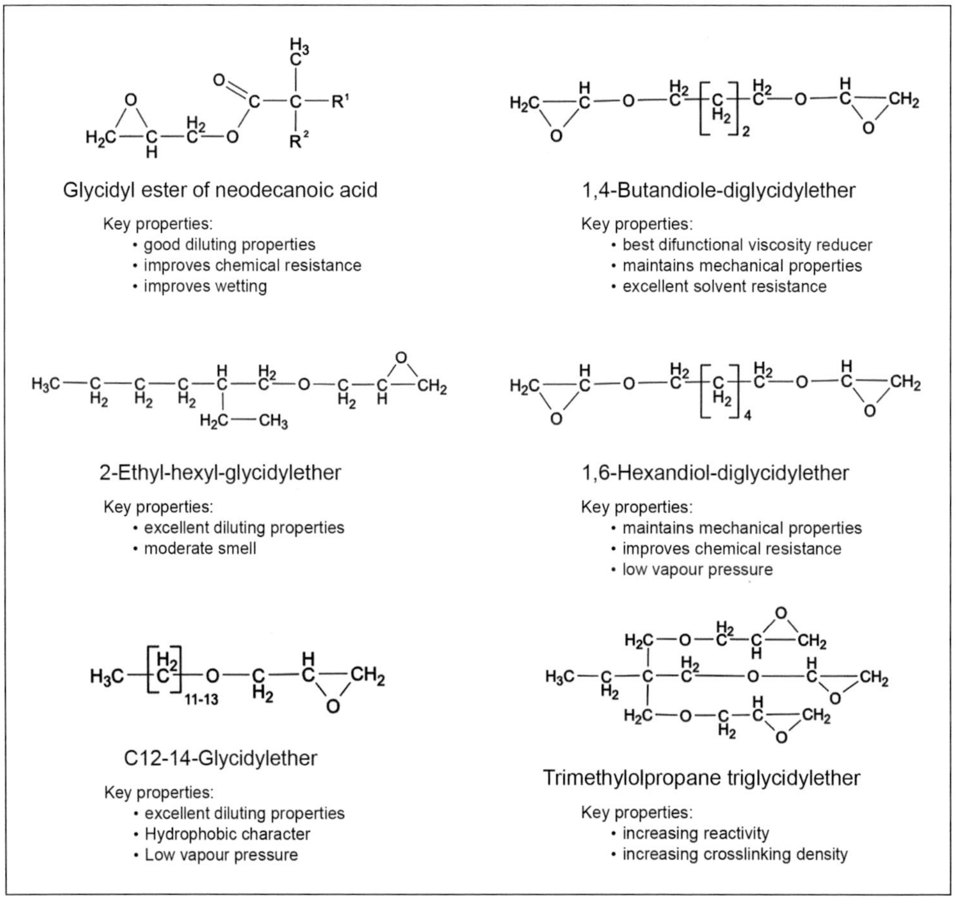

Figure II-2.60: Examples of reactive diluents [7, 8]

are not incorporated into the polymer network of the cured adhesive and can show a certain tendency to migrate, they can diffuse out of the adhesive over time, thus changing the mechanical properties of the adhesive.

Therefore, reactive diluents rather than non-reactive diluents or even solvents are used in modern epoxy adhesives. Reactive diluents are usually mono- or polyglycidyl ether resins produced by reacting epichlorohydrin with an alcohol, a phenol or a polyol. If epoxidation occurs on long-chain polyether polyols, such as polypropylene glycol, diepoxy resins with an internal polyether chain are formed, which both reduce viscosity and improve the flexibility of the adhesive. Likewise, glycidyl esters are formed if mono- or polyfunctional organic acids are used instead of alcohols. Cardanol-based reactive diluents (Figure II-2.59) are available as well. Again, the good hydrophobicity, chemical and thermal resistance, and flexibility inherited from the cardanol backbone ensures good performance of the cured epoxy system.

By their nature, monofunctional epoxy diluents are essentially chain stoppers, so in addition to reducing viscosity depending on the amount added, they have a certain effect on the mechanical properties of the resulting adhesive. They also increase thermal elongation of the cured adhesive, which, depending on the thermal elongation characteristics of the substrate material, can reduce or increase the internal stresses occurring in the adhesive joint during temperature fluctuations. In addition, an improvement in impact and thermal shock resistance and a slight reduction in heat resistance are frequently observed.

In contrast, di- and polyfunctional reactive diluents do not act as chain stoppers and therefore do not have a major effect on the properties of the cured resin they can be used in larger dosages. Many of the difunctional epoxy diluents are used in epoxy adhesive formulations to reduce viscosity and to improve flexibility. Tri- or higher-functional reactive diluents actually result in an additional crosslinking leading to an increase in crosslinking density and thus cohesion, and also often improving chemical resistance.

Figure II-2.60 shows some examples of common reactive diluents and their influence on the resulting adhesive properties.

Crosslinking agents

In addition to the already mentioned higher-functional epoxy resins (novolac), the likewise higher-functional reactive diluents and curing agents discussed in the following section, special products containing epoxy groups have been developed to increase the crosslinking density. Examples of these substances, referred to as crosslinking agents, are shown in Figure II-2.61. Naturally, these are typically used in relatively small dosages to improve resistance to elevated temperature and chemicals.

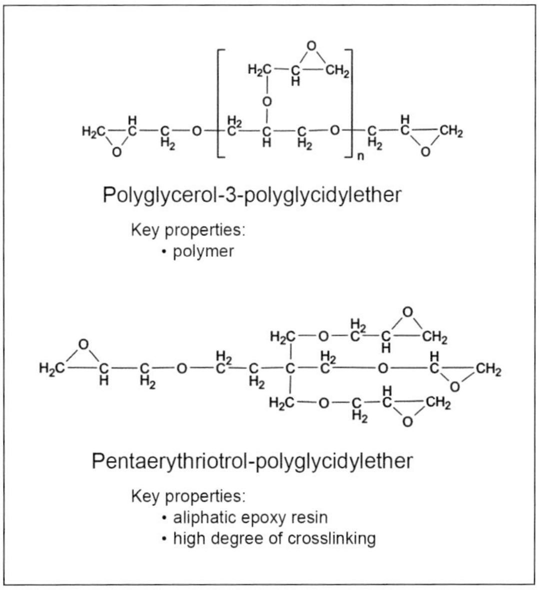

Figure II-2.61: Examples of crosslinking agents [8]

Table II-2.30: Types of curing agents and their application areas depending on the curing conditions

Curing agent family	Cure condition: Ambient cure	Elevated temp.	Heat cure	Number of components 1-C/2-C
Polyamides	X	X	–	2-C
Amidoamines	X	X	–	2-C
Aliphatic amines	X	(X)	(X) (mod. aliphatic amines)	2-C
Aycloaliphatic amines	X	X	–	2-C
Phenalkamine/ -alkamide	X	X	–	2-C
Aromatic amines	(X)	X	X	2C
Polysulfides/polythiols/ mercaptanes	X	–	–	1-C/2-C
Catalytic curing agents	accelerator	accelerator	X	1-C/2-C
Anhydride	–	–	X	1-C
Dicyandiamide (dicy)	–	–	X	1-C
Substituted urea	–	–	accelerator for "dicy"	1-C

Curing agents (curative, hardener)

The selection of the curing agent or the mixture of curing agents affects both the cure properties, such as pot life and cure speed, and the properties of the cured adhesive, such as tensile strength, glass transition temperature and hardness. Therefore, selecting the right type and amount of curing agent is very important to achieve optimum epoxy resin performance in adhesive formulation.

A large number of substances are capable of reacting with epoxide groups under ring opening and are suitable as curing agents for epoxy adhesives. A distinction is made between polyfunctional curing agents, which act in a stoichiometric reaction, and catalytic curing agents, which first open the epoxide ring and subsequently cause homopolymerization of the resin.

Table II-2.30 provides an overview of the different curing agent families and their areas of application with regard to curing conditions.

By nature, the use of curing agents that are already effective at room temperature is limited to the formulation of 2-component epoxy adhesives, i.e. the resin and curing agent components are prepared and stored separately and homogeneously mixed together in a defined mixing ratio immediately before use. Curing takes place at different rates depending on the curing agent used, with acceleration by heat being possible but usually not essential.

In contrast, curing agents that are inactive at room temperature, i.e. do not react with the epoxy resin, are frequently used for 1-component systems. Epoxy resin and curing agents are mixed in the required ratio already during the adhesive production. When heated after joining, the resin and curing agent react and the adhesive cures. To ensure sufficient storage stability, the curing temperature must be well above the maximum allowable storage and transport temperature.

Table II-2.31: Comparison of advantages and disadvantages of various families of curing agents used in epoxy adhesives [9, 11 – 16]

Curing agent	Advantages	Disadvantages
Aliphatic amines	– fast cure at room or elevated temperature cure – low viscosity – moderate to good chemical resistance – low formulation cost	– critical mix ratio – short working life, exothermic – reduced bond strength above 80 °C – low peel and impact strength, rigid to brittle – relatively water sensitive – high vapor pressure – strong skin irritant
Cycloaliphatic amines	– room and low temperature cure – good adhesion – good chemical resistance – high T_g if heat cured	
Aromatic amines	– moderate heat resistance – moderate chemical resistance	– slow cure at elevated temperature – solid at room temperature – poor peel and impact strength, rigid to brittle – EH&S issues, toxicity
Polyamides	– room temperature cure, long pot life – good bond strength, good water/ corrosion resistance – good flexibility, toughness, moderate peel and impact strength – low toxicity	– high viscosity – long cure time at room temperature – low heat and chemical resistance – high formulation cost
Amidoamines	– low viscosity – convenient mix ratio – good adhesion – low vapor pressure	– lack of elevated temperature performance – incompatible with some epoxy resins
Polysulfides, polythioles and mercaptans	– fast room temperature cure – good moisture resistance – high flexibility	– lack of elevated temperature performance – relatively low tensile strength – unpleasant odor
Dicyandiamide (DICY)	– 1-component heat curing adhesives with relatively long shelf life – good elevated temperature performance – good chemical resistance	– long cure time at high cure temperature – not soluble in epoxy resin
Imidazoles	– can be used as accelerator for other curing agents and as sole curing agent – suitable for 1C heat curing adhesives, rapid cure – high T_g cured resin, – high thermal and chemical resistance	– relatively short shelf life
Anhydride	– good heat resistance – good chemical resistance	– slow cure at elevated temperature – rigid to brittle
Catalytic curing agents	– long pot life – high heat resistance	– slow cure at elevated temperature – lack of moisture resistance – rigid to brittle

Storage stability can of course be improved by temperature controlled or even refrigerated transport and storage.

1-component epoxy adhesives can also be prepared by using photointiators that cause cationic ring-opening polymerization under UV radiation.

Due to the large number of curing agents available on the market, estimated there are well over a hundred different products available, it is not possible to discuss the properties of the individual ones. For this purpose, reference must be made to the product literature especially the data sheets provided by the different manufacturers. [5, 6, 8, 11 – 16, 41]

Nevertheless, some typical examples from the different families of curing agents will be described in the following sections and their typical properties are summarized in Table II-2.31

Primary, secondary and cycloaliphatic amines

Primary or secondary aliphatic or cycloaliphatic amines like ethylenediamine (EDA), diethylenetriamine (DETA), triethylenetetramine (TETA), tetraethylenepentamine, diaminoethylcyclohexane, piperazine, isophoronediamine (IPDA), N-aminoethylpiperazine, and diaminomethylbenzenes (Figure II-2.62) can serve as curing agents for the reaction with epoxy resins. Each NH-function of an amine can react with an epoxy-group. Higher-functional amines are used to create crosslinking sites in the polymer. Figure II-2.63 shows examples of flexibilizing, amine-based curing agents, such as amino-terminated polyethers ("Jeffamine "[21], see also Chapter II-1.3.2.3) and amino-terminated, acrylonitrile-butadiene copolymer ("ATBN ") [21, 22]

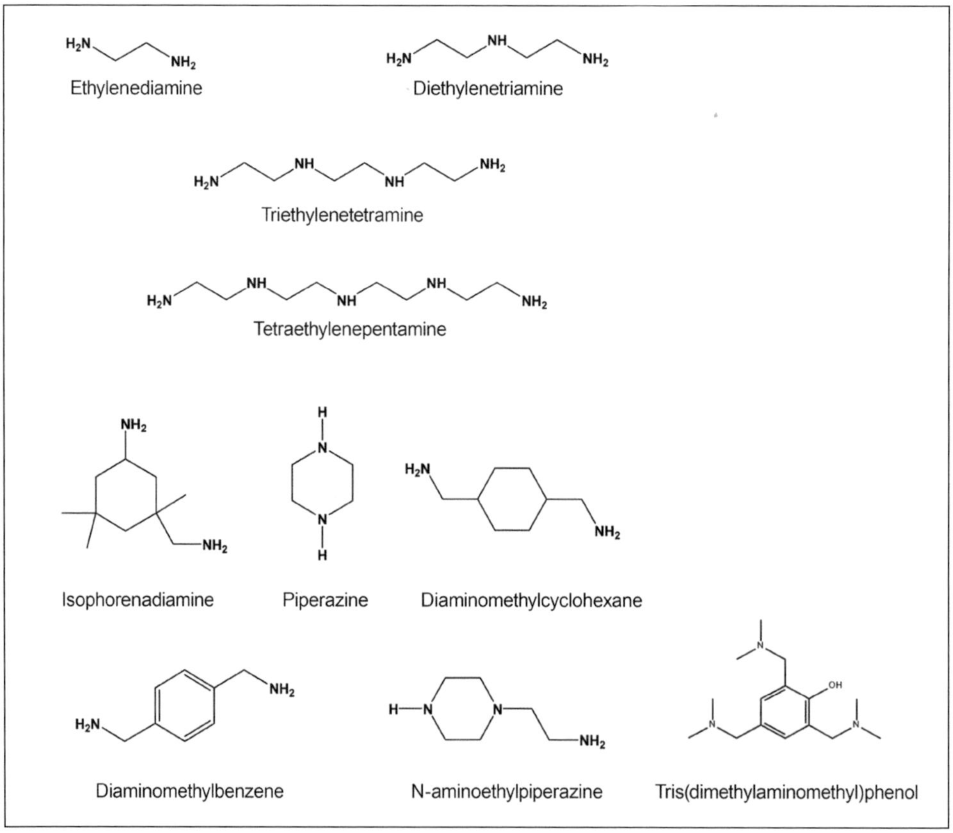

Figure II-2.62: Examples of aliphatic amine curing agents

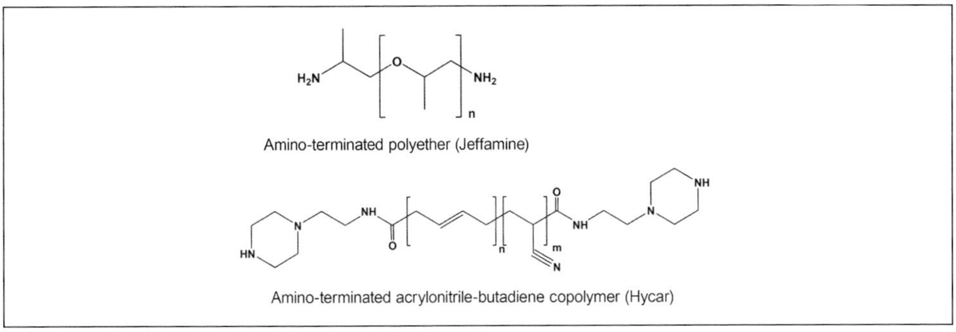

Figure II-2.63: Examples of flexibilizing curing agents

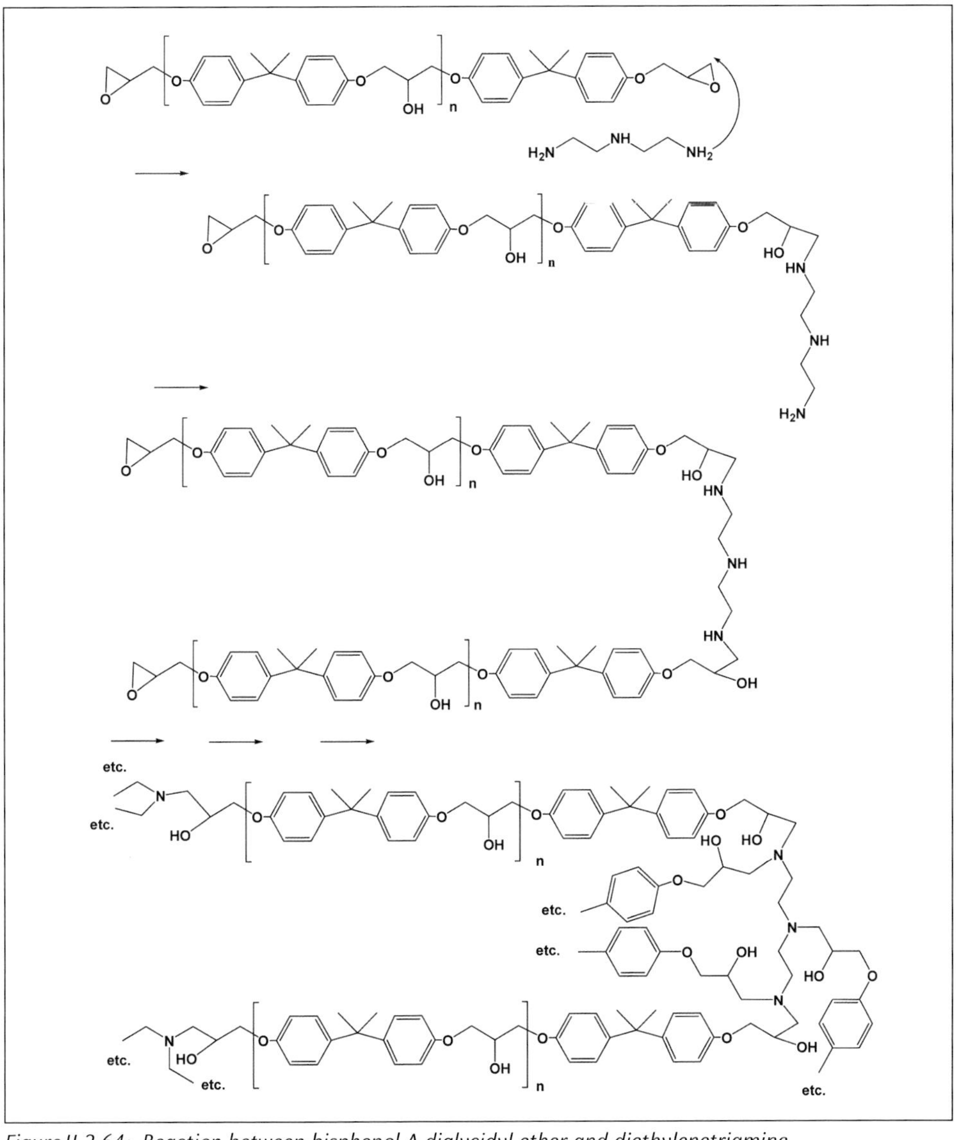

Figure II-2.64: Reaction between bisphenol A diglycidyl ether and diethylenetriamine

The reactions occurring during curing are shown in Figure II-2.64, on the example of the curing of a bisphenol A epoxy resin with diethylenetriamine.

When epoxy resins react with excess amines in the right stoichiometric ratio, amino-terminated prepolymers are obtained, which can also serve as curing agents.

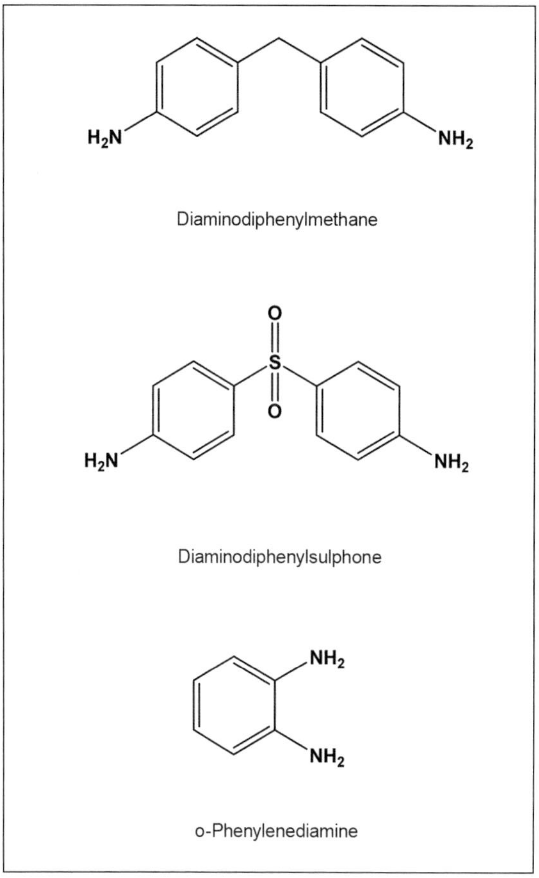

Figure II-2.65: Aromatic amines

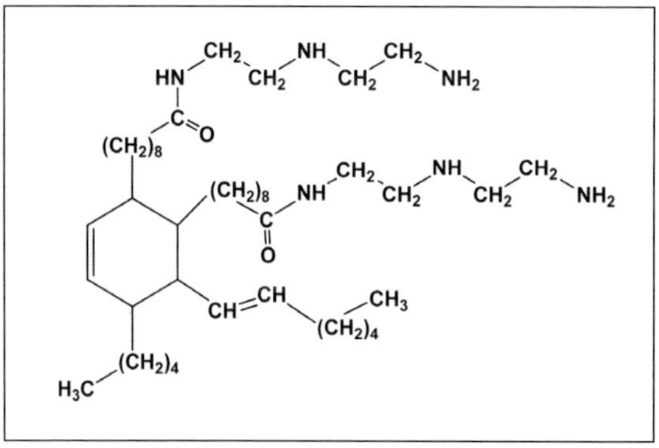

Figure II-2.66: Polyamide made from a dimer fatty acid and diethylenetriamine

Aromatic amines

Aromatic amine curing agents, such as those shown in Figure II-2.65, are hardly used anymore because of their high health hazards (toxic, carcinogenic). They are usually solid and can be blended with other amines when a long pot live is required.

Polyamides

Condensation of dicarboxylic acids with excess diamines leads to amino-terminated oligomeric polyamides. Figure II-2.66 shows typical products of the reaction between dimer fatty acid and diethylenetriamine (see also Chapter II-1.3).

This introduces a bulky, oil-compatible carbon group between the amine sites, due to which the cured epoxy resin gains increased flexibility. By varying the molar ratio of amine to acid, different viscosities can be achieved.

The curing reactions between the terminal primary and internal secondary amine groups along the polyamide backbone and epoxies are identical with those of primary and secondary amines. The amide groups are practically non-reactive.

Polyamide-cured DGEBA epoxies offer better flexibility, moisture resistance and adhesion than aliphatic amines alone, but generally, show lower shear strength and heat resistance due to the lower crosslink density. Because of their lower crosslinking density, they rapidly lose structural strength with increasing temperatures and are generally limited to applications below 65 °C. Another disadvantage of polyamide curing agents in certain applica-

Table II-2.32: Mix ratio effect on a DGEBA epoxy resin cured with a polyamidoamine from tall-oil fatty acid and triethylenetetramine [9]

	Polyamidoamine [pph]			
Property	**50**	**75**	**100**	**150**
Pot life [min]	45	63	57	60
Exotherm [°C]	209	190	169	133
Tensile strength [mPa]	58	41	12	1.4
Elongation at break [%]	9	12	70	93
Shore D hardness (after 24 h)	80	74	56	10
Ultimate Shore D hardness/elapsed time [days]	80/2	76/2	60/2	15/2

tions is that they are much darker in colour than polyamines. Another advantage is that they are less critical as regards to toxicology and the mixing ratio is less critical. By increasing the curing agent content, adhesion and flexibility of the cured resin can be further improved, but at the expense of heat deflection temperature and chemical resistance.

Amidoamines and Polyamidoamines

Amidoamines are prepared by the reaction of monobasic fatty carboxylic acids and aliphatic polyamines. Figure II-2.67 shows as an example the amidoamine from ricinoleic acid ((9Z,12R)-12-Hydroxyoctadec-9-enoic acid) and diethylenetriamine. The amidoamine or polyamidoamine curing agents exhibit similar reactivity to DGEBA epoxy resins as the polyamides and offer the advantage of lower viscosity and colour. Since they have only one amide group per molecule, the amine functionality is lower as well. Similar to the polyamides, the properties of the cured epoxy resin can be varied considerably by changing the mixing ratio.

Using the example of a DGEBA epoxy resin cured with a poly-amidoamine of tall oil fatty acid and triethylenetetramine, Table II-2.32 shows the quite significant influence of the different curing agent concentrations on the properties of the cured resin.

Both the polyamide and amidoamine curing agents can be accelerated to a certain extend by the addition of a tertiary amine such as tris(dimethylamino-methyl)phenol.

Mannich bases

For decades, the industry standard for fast curing agents were the so-called "Mannich bases" (β-aminocarbonyl compounds), which are accessible by an aminoalkylation of phenol or an alkyphenol by a primary or secondary amine in the presence of formaldehyde (Mannich reaction).

Mannich-based curing agents are highly reactive curing agents for epoxy systems with excellent performance even at low temperatures. Due to the content of phenolic ingredients, which are considered critical from an HSE (health, safety and environmental) point of view, their use is subject to significant restrictions. In addition, epoxy systems with these curing agents often show a strong tendency to yellowing.

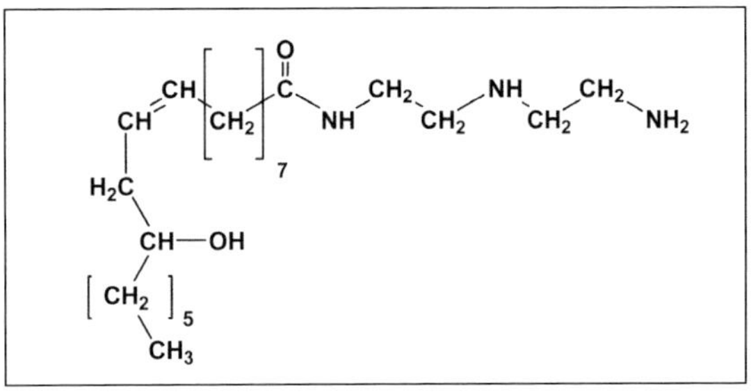

Figure II-2.67: Polyamide made from ricinoleic acid and diethylenetriamine

Phenalkamine and Phenalkamide curing agents

In analogy to epoxy resins and reactive diluents, cashew nutshell liquid can also be used for the production of cardanol-based amine curing agents (Figure II-2.68). Both phenalkamines and phenalkamides are available.

The long cardanol chain in the backbone of phenalkamine epoxy curing agents ensures low viscosity and thus enables formulation of adhesives with good flow properties or vice versa higher filler contents. They provide fast development of bond strength to various substrates even under demanding conditions such as low temperatures and humid environments (sometimes even under water). In addition, phenalkamines make cured adhesives more tolerant to non-optimally prepared surfaces, which reduces the risk of application errors.

Phenalkamides combine the advantages of phenalkamines and polyamides in one technology. In addition to the fast curing and high corrosion protection of phenalkamines, even at relatively low temperatures, they offer an improved flexibility and UV resistance of polyamides.

Polythiols (Polymercaptans)

Polythiols, or formerly also called polymercaptans due to their ability to bind mercury, are aliphatic oligomeric substances containing sulfohydro-groups (-SH). In the presence of tertiary amines these groups are able to react with epoxy groups at room temperature. Figure II-2.69 shows the thiol/epoxy reaction mechanism. In a first step (1), the thiol is ionized by a tertiary amine catalyst. The ionized thiol is strongly nucleophilic which makes it very reactive towards the epoxide group causing opening of the epoxy ring forming first a thiolate anion (2) which subsequently forms the base catalyst in a proton exchange reaction (3).

Thiol curing agents offer, especially in combination with tertiary amine catalysts like tris-(dimethylaminomethyl) phenol very fast cure ("5-minute epoxies") or allow low temperature cure and provide flexibility and a relatively high strength to the cured adhesive. Due to residual impurities thiol curing agents often exhibit a characteristic smell, which is often considered as unpleasant and sometimes leads to concerns about possible health hazards. The extent to which these impurities are removed during production has a considerable influence on the intensity of the smell.

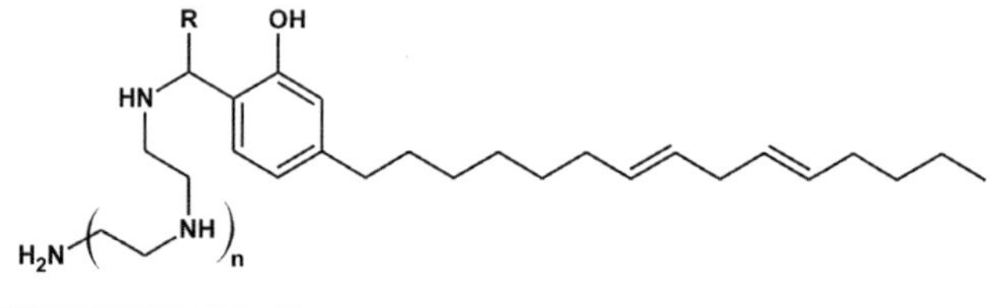

Figure II-2.68: Phenalkamine curing agent [5]

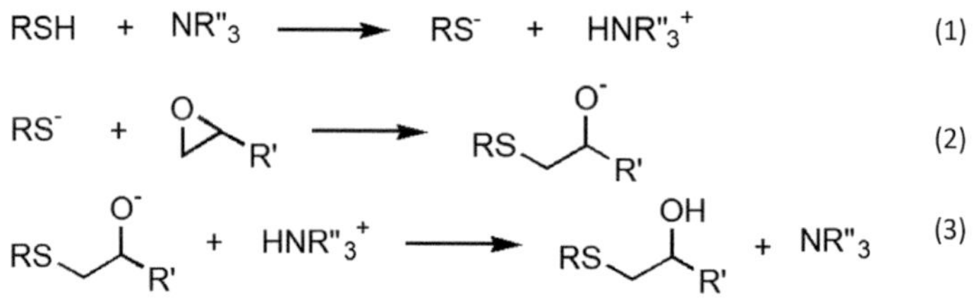

Figure II-2.69: Epoxy-thiol reaction mechanism [10

Both primary and secondar thiol curing agents are in use. Although the reactivity of secondary thiols is lower than that of primary thiols, it is always significantly higher than that of typical amine curing agents. Figure II-2.70 shows examples of thiol epoxy curing agents.

Dicyandiamide (DICY)

By far the most important curing agent for 1-component heat-curing epoxy resins is dicyandiamide (DICY), a solid (with a melting point of 208 – 211 °C) insoluble in epoxy resins at room temperature and thus non-reactive. When heated to temperatures above 100 °C, DICY slowly starts to dissolve in the epoxy resin and thus becomes reactive. However, on being exposed to temperatures above 140 °C, the dicyandiamide becomes soluble in the epoxy resin and cure progresses rapidly.

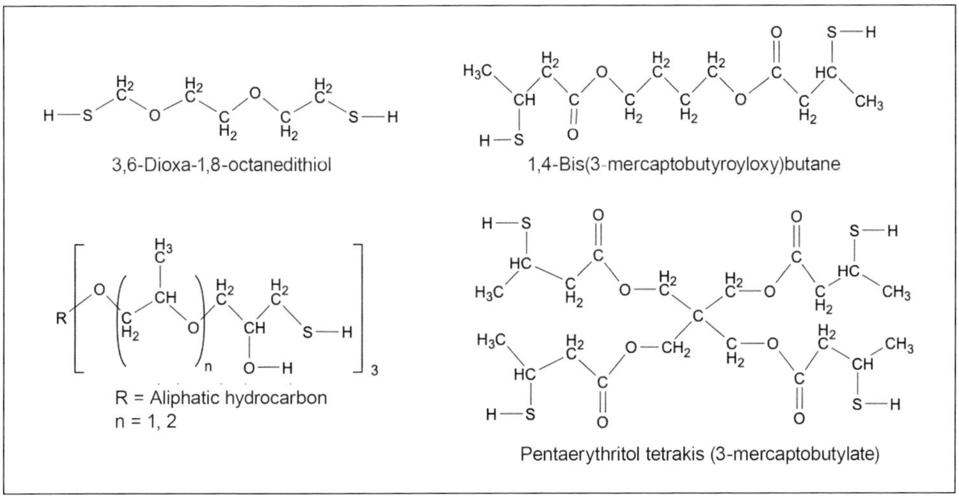

Figure II-2.70: Primary and secondary thiol curing agents

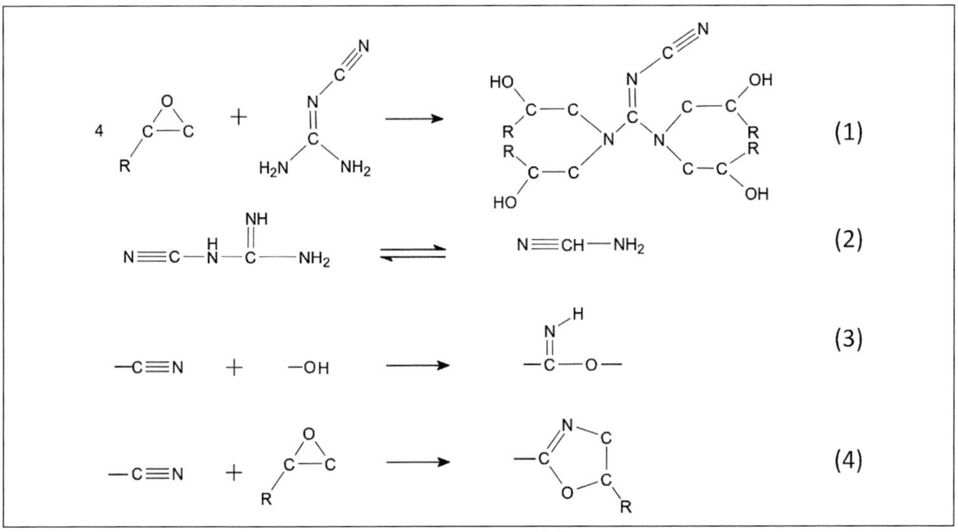

Figure II-2.71: Basic reactions of DICY with epoxy groups [16]

Figure II-2.71 shows the basic reactions of DICY with epoxy groups. DICY reacts with epoxide groups via the primary amine groups and the secondary amine functions formed in the process under ring opening (1). At temperatures above 100 °C, DICY gradually decomposes into amide-like fragments such as cyanamide (2). The nitrile-groups can react with the OH groups formed in reaction (1) or directly with an epoxide group (3) and (4).

Adhesives formulated with DICY as the curing agent alone require a curing temperature of at least 180 °C. To lower the curing temperature while maintaining or improving the physical properties of the adhesive, DICY is often used in conjunction with an accelerator. These include modified aliphatic amines, imidazole derivates and, in particular, substituted urea such as Monuron and Diuron. During heat cure, the urea catalyst decomposes into an isocyanate and dimethylamine. (Figure II-2.72)[18].

As shown in Figure II-2.73 the isocyanate can react with epoxy groups to yield an oxazolidone (1). The dimethylamine reacts to yield a tertiary amino alcohol (2). The latter exhibits basic properties which are relevant to the subsequent curing reaction. This base deprotonates DICY in an equilibrium reaction and a new N-nucleophile, which adds to the epoxy ring, is generated (3). After proton rearrangement, a new N-nucleophile, which adds to a second epoxy ring, is formed (4). This adduct undergoes intramolecular ring formation to yield an oxazolidine (5). The oxazolidine liberates cyanamide (6) and, after deprotonation, reacts with another epoxy group (7). The liberated cyanamide again reacts to yield DICY (8) and can in this way participate in the curing reaction once more. However, recent studies have suggested that a curing reaction should also be possible without the thermal decomposition of the urea catalyst [19]. In total, one mole of cyanamide (half of a DICY) is consumed in the reaction with three epoxy groups. This agrees with experimental results, namely that 1 mole of DICY is needed to cure 6 to 7 moles of epoxy groups [20].

DICY comes in the form of a white crystalline powder. The grades offered by manufacturers differ mainly in particle size, but also in the amount of flow control agent present. When manufacturing the adhesive, care must be taken to ensure that the DICY powder is homogeneously dispersed into the adhesive. This is essential to achieve a uniform curing and, ultimately, a homogeneous adhesive film in terms of its properties. If the necessary dispensing technology is not available, it is advisable to use products already pre-dispersed in epoxy resin.

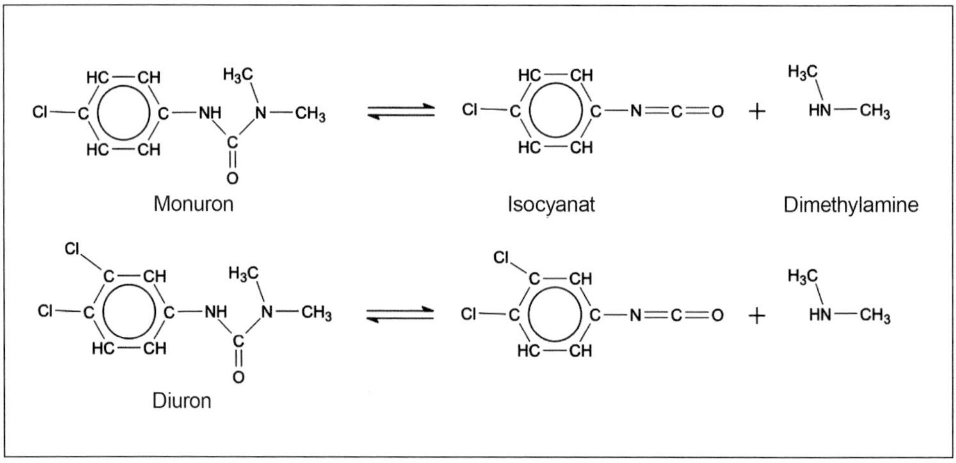

Figure II-2.72: Thermal decomposition of Monuron and Diuron

Anhydride curing agents

Due to the long pot life and low exotherm, the resulting relatively high heat distortion temperature typically in the range of 280 – 290 °C and the good physical properties of the cured resin, anhydride curing agents are widely used for curing epoxy casting and moulding resins as well as for the production of composites.

Since anhydride curing agents, unlike amine curing agents, require a heat curing process with high curing temperatures in excess of 150 °C in the absence of an accelerator and long curing times, it is not surprising that the use of anhydride curing agents in adhesive formulations is generally limited to applications where the resulting high crosslinking density and thermal resistance or the long pot life is required. In addition, the potential differences between the coefficients of thermal expansion of the epoxy adhesive and the adherend can lead to considerable stresses in the bonded joint on cooling, which in turn can have a detrimental effect on the load bearing capacity and durability of the bonded joint. In principle both acid and base catalysts are suitable to act as accelerator, but base catalysts are more common with tertiary amines or imidazole derivates are most common.

The curing reactions are quite complex. Figure II-2.74 shows the reaction process in the absence of an accelerator. In a first step, the OH groups of an oligomeric bisphenol A based epoxy resin react by opening the anhydride ring forming a hemi-ester (1). The resulting carboxylic acid groups react with further epoxy rings and form an ester and a new alcohol (2), which in turn can react with an anhydride under ring opening.

As every time a carboxylic acid group reacts with an epoxy group, another aliphatic hydroxyl is created, which is then available to react with another anhydride group, it is not really necessary to observe strict stoichiometry in the formulation of an adhesive. Nevertheless, empirical optimisation is recommended [9, 33].

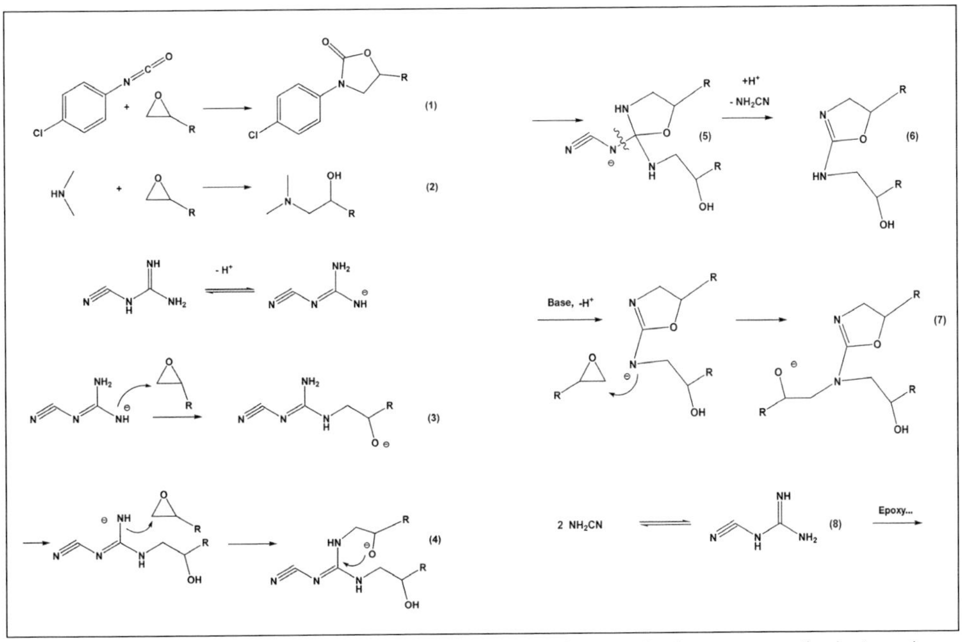

Figure II-2.73: Reaction pathways in the curing of epoxides by DICY in the presence of substituted urea

Figure II-2.75 shows the idealized structure for a bisphenol A epoxy resin cured with an anhydride, in this case hexahydrophthallic anhydride (HHPA) [26].

In the presence of an accelerator, the first step is the opening of the anhydride ring by the base catalyst, e.g. an imidazole, followed by epoxy ring opening, ether and ester formation (Figure II-2.76).[24]

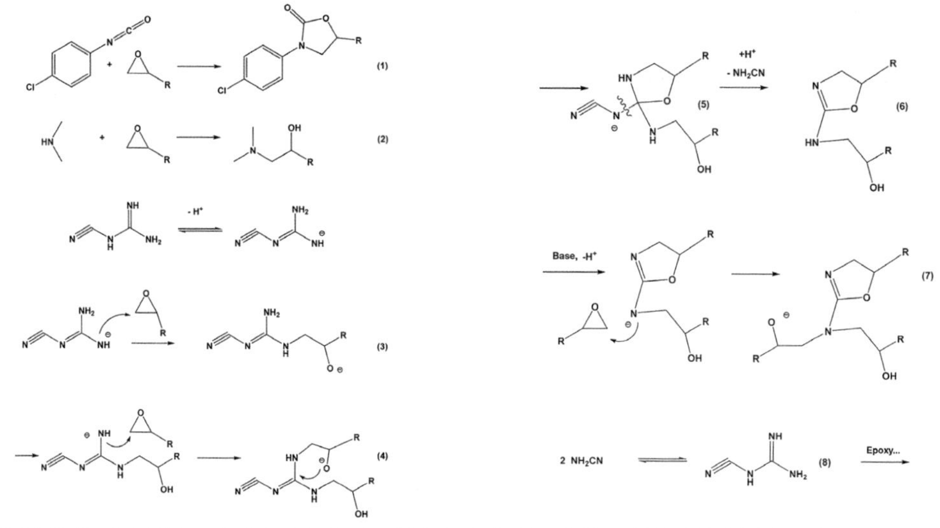

Figure II-2.74: Reaction pathways in the curing of epoxides with anhydrides

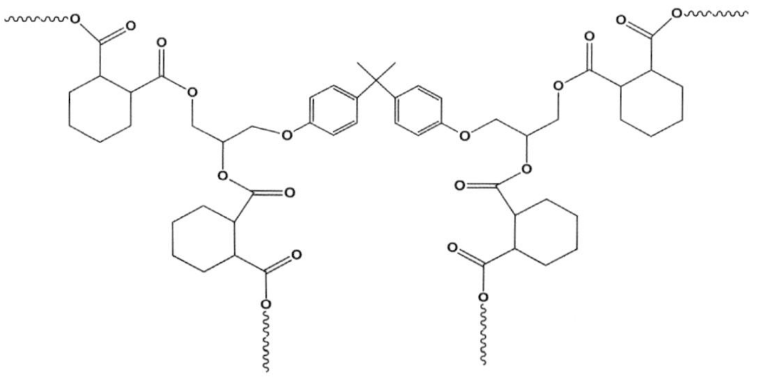

Figure II-2.75: Idealized structure for a bisphenol A epoxy resin cured with hexahydrophthallic anhydride (HHPA) [26

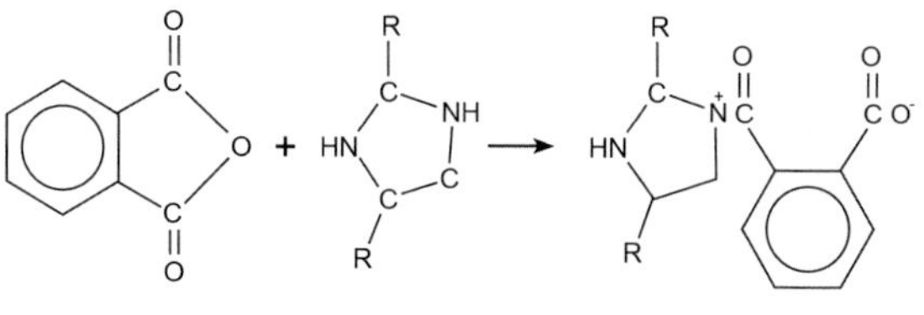

Figure II-2.76: Ring opening reaction of an anhydride curing agent by imidazole [24]

Figure II-2.77 shows the chemical structures of some common anhydride curing agents. According to [9], the most important anhydride in epoxy adhesive formulations is **pyromellitic dianhydride (PMDA)**. Due to the two anhydride groups symmetrically attached to a benzene ring, very high crosslinking densities and thus high heat and chemical resistance can be achieved. Epoxy adhesives cured with PMDA are characterized by a heat resistance in the range of 280 – 290 °C. PMDA is at room temperature insoluble in DGEBA resins but becomes soluble at elevated temperature. In order to prevent a premature reaction with gel formation when mixing at elevated temperatures, a suspension of PMDA in DGBEA prepared at room temperature can be used; the curing reaction then only starts during the curing process when the PMDA begins to dissolve in the resin. Another option is the use of a PMDA-glycol adduct that is already soluble in the resin at room temperature. As already mentioned, it is not necessary to maintain the exact stoichiometric resin/curing agent ratio due to the complexity of the curing reaction of the anhydrides. In fact, PMDA is usually used sub-stoichiometrically, so that the pot life can be adjusted within certain limits via the amount of PMDA.[9]

Nadic methyl anhydride (NMA) (Figure II-2.77) is probably the most versatile of all anhydrides. It is not only readily soluble in epoxy resins at room temperature but also has a certain thinning effect due to its low viscosity of < 300 cPs. A content of 60 – 90 pph has proven to be effective for curing liquid DGEBA epoxy resins. At a content of 60 pph and without catalyst, the pot life is around 2 months; if 0.5 pph of Lewis base. e.g. 2,4,6-tris-(dimethylaminomethyl) phenol (DMP-30) is added as an accelerator, the pot life is reduced to 4 – 5 days.

Not only the pot life but also the properties of the crosslinked resin can be varied over a wide range via the resin-curing agent ratio, the type and concentration of the accelerator and the curing conditions. The highest degree of cross-linking and the greatest hardness is achieved by using stoichiometric mixtures of NMA and long curing times at high temperatures (approx. 2 hours at 220 – 260 °C). A reduction in the anhydride content and curing temperature leads to an improvement in toughness, but also to a reduction in heat resistance. Curing time and temperature can be reduced by adding an imidazole accelerator. For example, with an NMA content of 90 pph and the addition of an imidazole accelerator, practical cure is achieved by curing for 2 hours at 80 – 100 °C plus 4 hours at 140 – 150 °C. The excellent electrical properties and high temperature performance make NMA-cured epoxies outstanding materials for encapsulating electronic components and circuits [25, 26].

Hexahydrophthalic anhydride (HHPA) (Figure II-2.77) is a solid with a low melting point of 36 °C. It can be liquefied at temperatures of 50 to 60 °C and is easy to mix with hot epoxy resins. It can be easily mixed with hot epoxy resins at temperatures of 50 to 60 °C. The mixed resins have a low viscosity (in the case of a DGEBA resin about 200 cP at 40 °C), long pot life and low exotherm. Due to its low reactivity, HHPA is generally used with an accelerator, usually benzyldimethylamine (BDMA) or 2,4,6-tris-(dimethylaminomethyl) phenol (DMP-30).

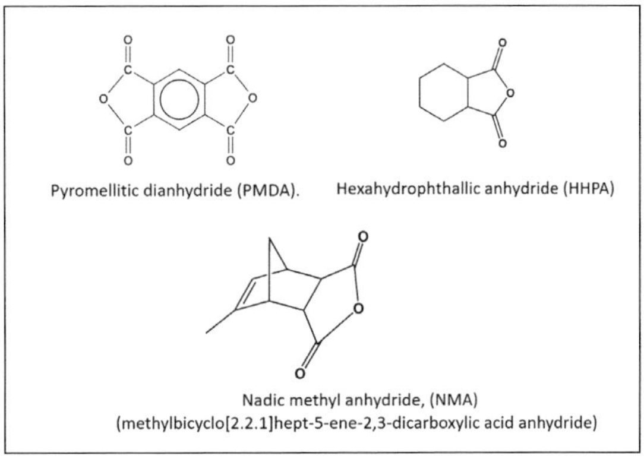

Figure II-2.77: Chemical structure of some common anhydride curing agents

HHPA is generally used at a concentration between 55 and 80 pph, depending on the type of epoxy resin.

A typical cure schedule for a system lightly catalyzed with BDMA (0.5 to 2 %) or 2,4,6-tris-(dimethylaminomethyl) phenol is 2 hours at 80 °C plus 1 hour at 200 °C. As is typical for all anhydride curing agents, the cured epoxy resin exhibits high heat deflection temperature and excellent chemical resistance. [26]

Imidazoles

Imidazole and its derivates (Figure II-2.78) are another group of amine curing agents used for 1-component heat curing epoxy adhesives showing a good stability at room temperature together with a rapid cure beyond activation temperature. The choice of the groups marked as R^1, R^2 and R^3 which creates steric and/or solubility effects, allows the formulator to control degree of latency as well as reaction activation temperature. The use of imidazoline adducts instead of simple imidazolines offers the possibility of a further improvement in shelf life.

The cured adhesives are characterized by a high thermal and chemical resistance. Heat deflection temperatures (HDT) of > 200 °C are possible.

Investigations into the curing mechanism of epoxy curing with imidazolines revealed that curing initially takes place at both N atoms, but that propagation probably occurs preferentially from only one centre. The mechanism is more complicated than shown in Figure II-2.79, with various reactions occurring in parallel in some cases, refer to [42] for more details.

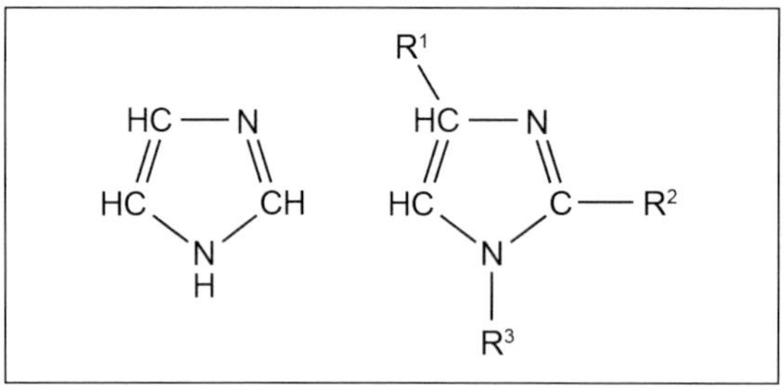

Figure II-2.78: Chemical structure of imidazoline (left) and imidazoline derivates (right)

Cationic polymerization by Lewis-acids

Even though the Lewis acid-initiated ring opening (Figure II-2.80) does not follow the mechanism of a polyaddi-

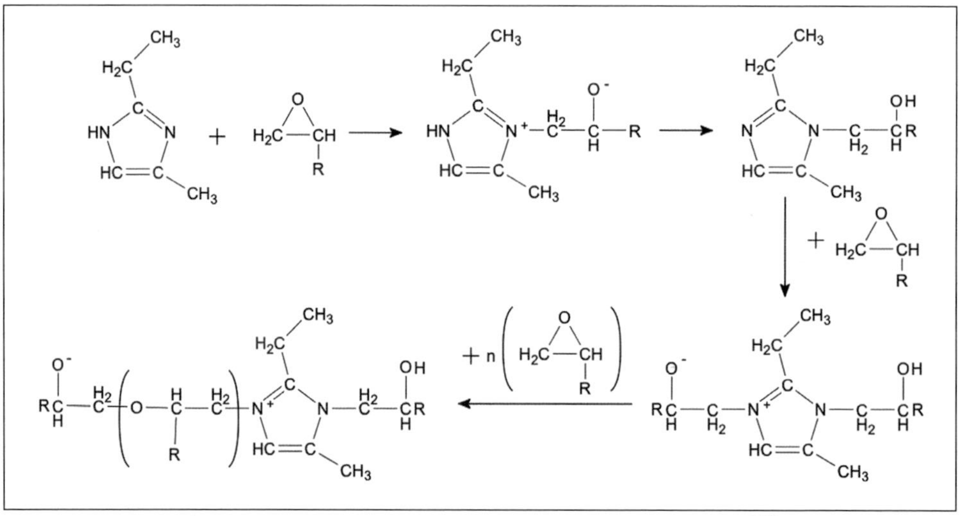

Figure II-2.79: Simplified reaction mechanism of epoxy curing with imidazolines [42]

tion but rather that of a polymerization, it appears appropriate to cover this group of adhesives in this chapter.

Lewis acids, such as boron trifluoride-amine complexes are strong electrophiles that act as curing agents for epoxy resins. As shown in Figure II-2.80 the first step is the attack of a Lewis acid (electrophile) on the free electron pairs of the oxygen of the epoxide group, which facilitates the subsequent cleavage of the C-O bond. The second step is the ring opening of the epoxide group and the charge transfer to the next epoxide group. During chain growth, the third step, charge transfer again takes place to a new epoxide monomer or to an already formed chain. The reaction comes to an end when all monomers are consumed, or the reaction is stopped.

Unlike free radical polymerization where a free radical is transferred from monomer to monomer during chain growth, cationic polymerization transfers a charge from monomer to monomer.

Boron trifluoride monoethylamine (BF_3-MEA) is the by far most important Lewis acid used commercially to cure epoxy resin systems by polymerization. The complex, which can be obtained from boron trifluoride and monoethylamine, is a solid that melts at a temperature close to its dissociation temperature (80 – 85 °C). Contact with moist air should be avoided as it is hydroscopic and its suitability as a curing agent is lost when it hydrolyses.

For its use, the BF_3-MEA complex must be dissolved in the epoxy resin. For this purpose, it can either be stirred into the resin pre-heated to 85 °C or a masterbatch is first prepared by mixing about 3 parts by weight of the BF_3-MEA complex into about 5 parts by weight of the resin pre-heated to 50 °C. The resulting paste is then added to the rest of the resin pre-heated to 85 °C.

The intention by using the BF_3-MEA complex instead of just BF_3 is to reduce the high reactivity of BF_3 to such an extent that BF_3-MEA can be considered as a latent catalyst. Concentrations of 2 – 4 pph have proved effective for curing DGEBA epoxy resins. Single component adhesives with a shelf life of up to 12 months at room temperature are feasible. Due to the intentionally reduced reactivity, curing temperatures of at least 100 – 125 °C are required. As curing at 120 °C is very fast and is accompanied by a high exotherm, it is advisable to perform curing in two stages, e.g. for 2 hours at 105 °C, followed by post-curing at 150 – 200 °C for 4 hours. [9]

UV-curing

Using a suitable photoinitiator the cationic polymerization can be started by UV-light. Sulfonium or iodonium salts (Figure II-2.81) are usually used as photointiators for UV curing of epoxy adhesives in quantities of less than 1 up to 20 % by weight. They consist of a cationic and an anionic part. The cationic part is responsible for the absorption of UV radiation (depending on the specif-

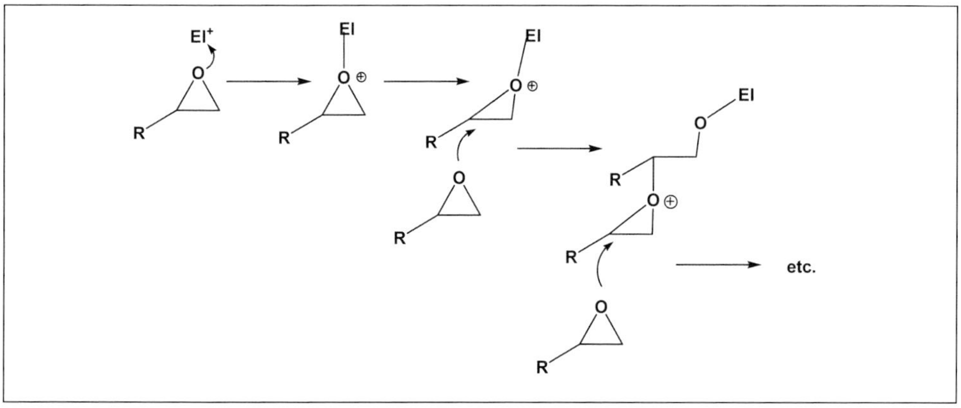

Figure II-2.80: Cationic ring-opening polymerization of epoxy resins

ic product they absorb UV radiation in the wavelength band of λ = 230 – 280 nm) determined by its molecular structure surrounding the sulphur or iodine atom. The anionic part of the molecule, on the other hand, becomes a strong acid through UV absorption.

In a first step the energy of the UV radiation (hγ) absorbed by the cationic photo initiator is used to generate an excited electron state. Subsequently the molecule undergoes both a homolytic cleavage, which produces several free radicals, and a heterolytic cleavage, which produces cationic species. The products of the heterolytic cleavage stabilize themselves by reacting with monomers, producing the strong acid that subsequently initiates cationic polymerization. The stronger the acid the higher the reactivity. The hexafluoroantimonate ion has a very high pK_a-value of approximately -30 (for comparison, hydrochloric acid pK_a = - 6) and is therefore also referred to as superacid. As the real concentration in UV-curing epoxy resins is only in the range of a few hundredths of a percent by weight, this is nothing to be worried about.

Cycloaliphatic epoxy resins like 3,4.epoxycyclohexylmethyl-3',4'-epoxycyclohexane carboxylate (ECC) or Diglycidyl-1,2-cyclohexanedicarboxylate (Figure II-2.82) are preferably used in the formulation of UV-curing adhesives.

Figure II-2.83 shows that UV crosslinking of ECC molecules (Figure II-2.81) alone produces a tight, extremely brittle network with a large number of polar groups. The brittleness can be counteracted by either adding flexibilized epoxy resins or reactive components with a flexibilizing effect, e.g. from the group of oxetanes that also act as diluent and significantly increase the cure speed (Figure II-2.84), polyols that act as chain extenders (e.g. di-, try- or tetrafunctional ethoxylated polyols, polycaprolactone oligomers) or the relatively new aliphatic dendtritic polyester polyols (Figure II-2.85).[31, 32, 33]

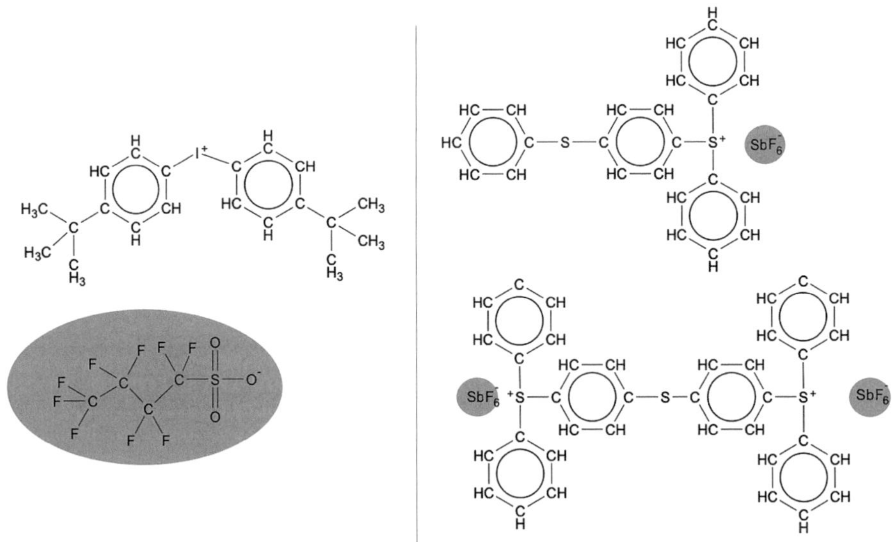

Bis(4-tert-butylphenyl)iodonium perfluoro-1-butanesulfonate - λ_{max}: 237 nm

Bis(4-(diphenylsulfonio)phenyl)sulfide bis(hexafluoroantimonate) - λ_{max}: 271 nm

Figure II-2.81: Examples of common cationic photo initiator: iodonium and sulfonium salts. (The cationic parts are marked by grey ellipses, the parts not marked represent the anionic parts relevant for UV adsorption.) [30, 34]

In contrast to the free radicals formed during radical polymerization and triggering the polymerization reaction the acids generated by light from cationic photo initiators have a significantly longer active lifetime and can trigger polymerization over a period of up to several days, whereas the active lifetime of free radicals is only seconds or fractions of a second.

This long active lifetime means that once reactive sites are created by UV irradiation, they continue to migrate through the monomers and polymerize them until all monomers are consumed without the need for continuous UV irradiation. Unlike radical polymerization, chain growth does not stop when the UV source is switched off. It is therefore also possible to bond

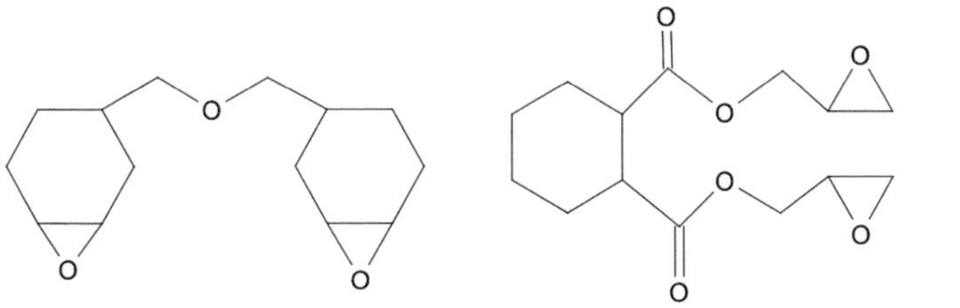

3,4-Epoxycyclohexylmethyl-3',4'-epoxycyclohexane carboxylate (ECC)

Diglycidyl-1,2-cyclohexanedicarboxylate

Figure II-2.82: Examples of cycloaliphatic epoxy resins used in the formulation of UV-curing epoxy adhesives

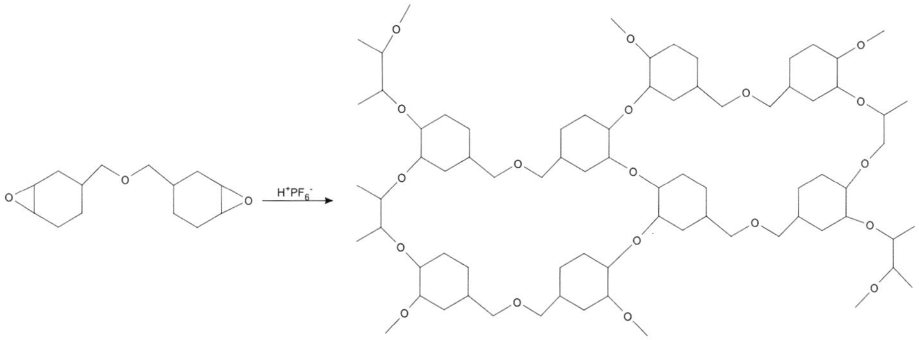

Figure II-2.83: Principle Structure of an UV-cured homopolymer from ECC [31]

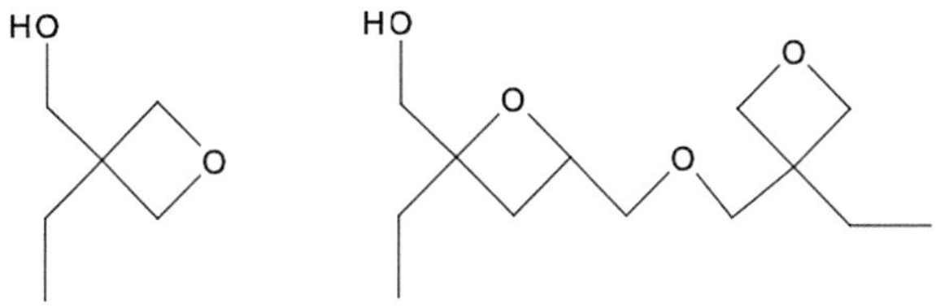

3-Ethyl-3-hydroxymethyl oxetane

3,4-Epoxy cyclohexyl methyl-3,4 epoxy cyclohexane carboxylate

Figure II-2.84: Examples of thinning, accelerating and flexibilizing oxetanes [32, 33]

substrates that are not UV-permeable with cationic-curing UV adhesives. After the adhesive is applied, it is initially exposed to UV light and the curing reaction starts; the adhesives have been activated. Similar to the pot life of 2-component adhesives, there is a time window, known as the open time, within which the reacting adhesive still has sufficient wettability for the second adherend, i.e. the bonding process must be completed. Similarly, this photo-acid migration effect, also known as the dark reaction, enables the complete curing of complex geometries, in shadow areas and in areas where parts of the component must be protected from direct UV radiation.

A further advantage over radical polymerization is that the presence of oxygen does not impair the cationic polymerization process, so there is no risk of the UV-cured adhesive not being completely cross-linked on the surface, which can lead to reduced strength especially in thin-adhesive layer bonding or, for example, to an unwanted and disadvantageous surface tack when encapsulating electronic parts.

In addition, the monomers link together during curing, reducing the distance between neighbouring monomers (shrinkage), but on the other hand the length of the molecule increases due to the opening of the cyclic epoxy group. This reduces or even compensates the shrinkage effect resulting in a by far less overall shrinkage compared to what is often observed in radical polymerization.

It should not be underestimated that to a certain degree the moisture sensitivity of cationic polymerizing systems can be seen as a disadvantage. Although curing is not prevented, it can be delayed. This is not a major problem in controlled production environments where the ambient humidity is controlled and kept below 70 – 80 %. In environments where the humidity varies

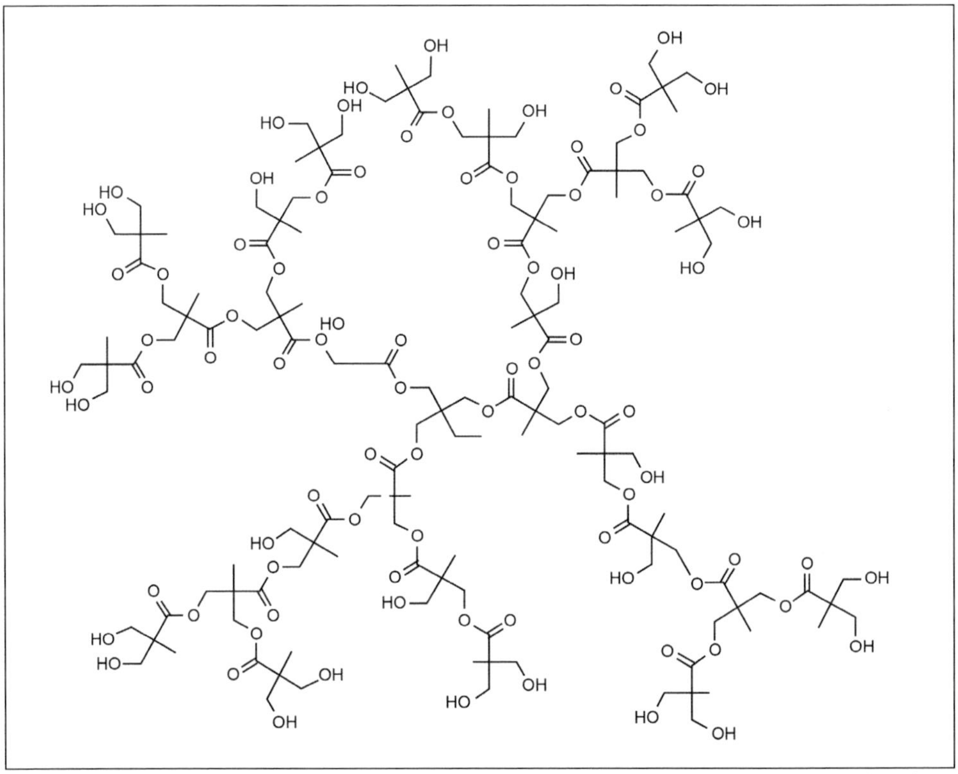

Figure II-2.85: Example of an aliphatic dendritic polyester polyol ("Boltorn H2004", Perstorp) [32, 33]

greatly, the effects of humidity must be taken into account, e.g. by extending the curing time. It is also possible to remove moisture from the resin by local air conditioning immediately before and during curing, e.g. using a dry air blanket or heating elements. [9, 31]

Catalysts/accelerators

Catalysts are used in non-stoichiometric amounts. They catalyse ring-opening polymerization of the epoxies but are not consumed during the reaction. Catalytic curing agents are often used in combination with curing agents that enter into nucleophilic addition and subsequently give rise to synergistic effects.

Tertiary amines – anionic ring opening

Examples of **tertiary amines**, such as dimethylaminophenol, benzyldimethylamine, and 2,4,6-tris(dimethylaminomethyl)phenol (Mannich adduct) are shown in Figure II-2.86.

Mild conditions are sufficient for tertiary amines to catalyse the non-stoichiometric ring-opening polymerization of epoxies shown in Figure II-2.87 [2, 18]. Tertiary amines are frequently used together with other curing agents, as they polymerize excess epoxy resin and thus ensure complete curing.

As a sole catalyst, dimethylaminophenol is used only in applications where a short pot life can be tolerated and where maximum physical or chemical properties are not required. It is typically used at concentrations of 4 – 10 pph with liquid DGEBA epoxy resins showing a fairly fast cure overnight, even at room temperatures [9].

When benzyldimethylamine (BDMA) is used as the sole curing agent for DGEBA epoxy resins, adhesives with a pot life of 1 – 4 hours are obtained at a curing agent content of 6 – 10 pph. The final strength is achieved after about 6 days at room temperature and the cured adhesives show a significantly higher temperature resistance. However, BDMA is mostly used as an accelerator for anhydride- and dicyandiamide-cured epoxy resins.

Besides ring-opening polymerization, tertiary amines can also catalyse the nucleophilic addition of acidic nucleophiles, such as thiols (Figure II-2.69).

Imidazoles

Imidazolines are not only used as the sole curing agent as described above but are also often used in combination with other curing agents.

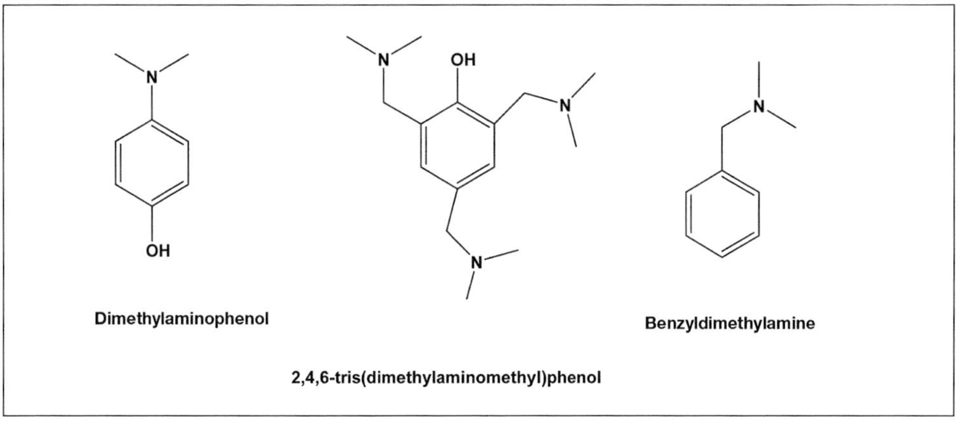

Figure II-2.86: Tertiary amine catalysts

Urea derivates

As mentioned above in connection with the heat curing of epoxy resins by means of dicyandiamide, heat-activated urea derivatives such as Monuron (m.p. 174 °C), Diuron (m.p. 158 °C), Fenuron (m.p. 131 °C) and the adduct of dimethylamine at TDI (m.p. 182 °C) (Figure II-2.88) are used to lower the required cure temperature or reduce the cure time, but unfortunately at the expense of shelf life [2, 20, 37, 46].

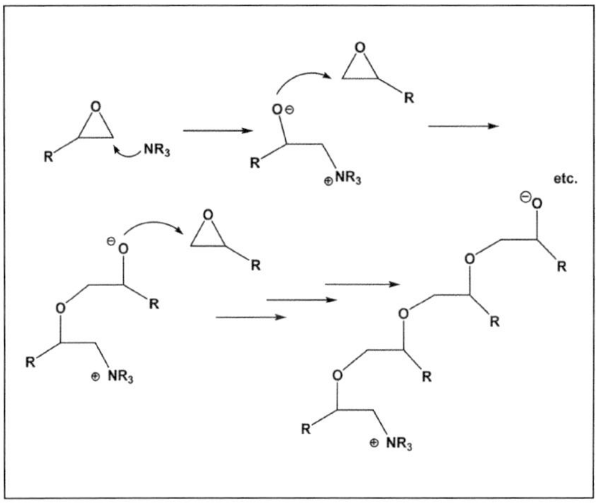

Figure II-2.87: Epoxy curing with tertiary amines by anionic ring opening polymerization

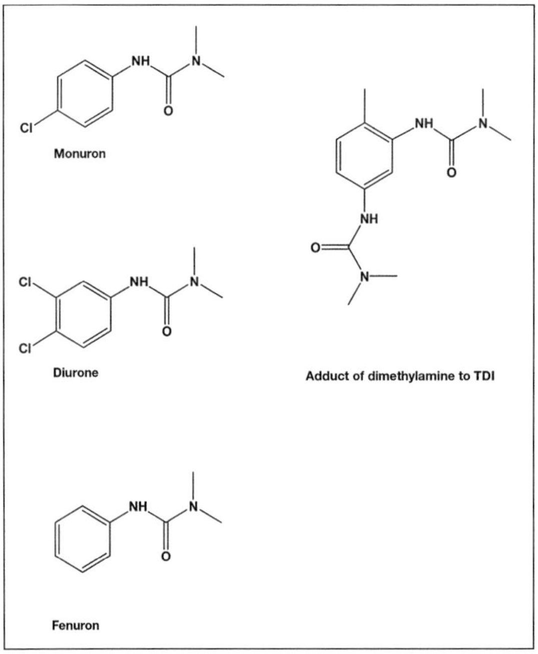

Figure II-2.88: Accelerator for dicyandiamide curing of epoxy resins

Toughening agents

Cured epoxy adhesives are amorphous and highly cross-linked materials, which is why they have many useful properties such as a high modulus of elasticity, high tensile strength, low creep behaviour and good chemical as well as temperature resistance. Unfortunately, this microstructure also results in the usually undesirable property of high brittleness. This is reflected in a low resistance to crack formation and growth, which leads to inadequate impact and peel resistance for many applications.

These problems can be overcome by skilful formulation of the adhesives. There are different ways to do this. Firstly, the cured polymer can be made more flexible by an intelligent choice of epoxy resin and curing system and by the addition of flexibilizing diluents. The increase in flexibility is mainly achieved by a reduction in the cross-linking density or incorporation of flexible domains into the polymer backbone resulting essentially in an improved elongation of the cured adhesive. Possibilities for this have already been described in the description of the respective raw material groups used in the formulation of epoxy adhesives. As a result, forces acting on the bond under mechanical load are distributed over a larger area and flexibilized epoxy resins are better able to compensate for differences in the thermal expansion or modulus of elasticity of the adherends and typically show improved peel resistance.

However, the resulting incorporation of flexible groups into the polymer backbone not only leads to the intended flexibilization of the cured adhesive, but unfortunately often also to a loss of valuable properties such as a reduction in modulus and tensile strength, a certain loss of resistance to media (moisture, chemicals) and temperature resistance as well as a reduced glass transition temperature.

The second way to overcome the problems associated with the brittleness of cured epoxy adhesive is to improve toughness, resulting in improved fracture-, impact-, fatigue- and stress-resistance with minimal change in cross-linking density while modulus and tensile strength, resistance to media and temperature resistance as well as glass transition temperature remain in principle unaffected.

Tougheners improve the properties through small elastic particles (diameter in the order of 1 μm) enclosed in the epoxy matrix. If stress is applied to the cured system, it can be dissipated uniformly in all directions when encountering such an elastomer domain. If a crack has already occurred, the elastomer domains stretch perpendicular to the crack and preventing crack propagation as they are chemically bonded into the epoxy matrix (Figure II-2.89).

The different technologies commonly used for toughening of epoxy adhesives are described in the following sections.

Reactive liquid rubber

The most used technology is based on prepolymers produced by reacting end-capped reactive elastomers, like carboxyl-terminated butadiene acrylonitrile (CTBN), with an excess of epoxides (Figure II-2.57). The resulting terminated prepolymers are compatible with epoxy resins and the other common formulation components. Because of the high viscosity of CTBN itself it is difficult to handle and especially difficult to mix with epoxy resins. Therefore, various manufacturers offer "ready-to-use" prepolymers based on standard resins, usually DGEBA or DGBEF with a significantly lower and easier-to-handle viscosity.

During curing, compatibility progressively decreases with the extent of curing to the cross-linked polymer. This results in microscopic phase separation with the formation of rubber domains in the adhesive matrix. The rubber domains (whose diameter is in the micrometer range) are well dispersed as a separate phase but are incorporated into the high-strength epoxy matrix and are connected to each other by covalent bonds. Forces acting on the adhesive are dissipated by the rubber particles in all directions by forming the so-called shear-bands (Figure II-2.89). Stress-induced microcracks in the epoxy matrix end at the elastomeric rubber phase, the effect is known as crack-bridging and illustrated in Figure II-2.89 as well. The formation of a fully dispersed phase depends on a delicate balance between the miscibility of the elastomer or its adduct with the resin and the resin-curing agent mixture and suitable precipitation during the crosslinking reaction.

The advantage of this toughening technology is that the high modulus and T_g of the epoxy resin is maintained. The cured adhesive shows two glass transition temperatures (T_g). One is typically at temperatures well above room temperature, depending on the epoxy resin and curing agent used, and is therefore an indicator of the epoxy

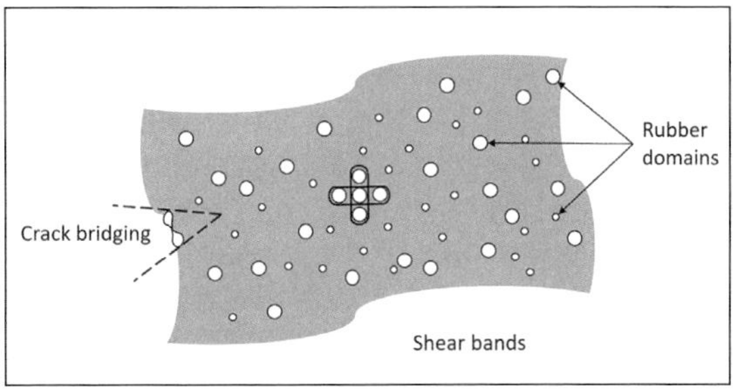

Figure II-2.89: Principle mode of action of elastomer tougheners

resin used as a matrix. The second is in the significantly negative temperature range of around - 40 to - 50 °C and is attributable to the rubber domains.

Compared to a formulation formulated without CTBN adduct (control), the toughened formulation has a significantly higher peel strength, a moderately higher lap shear strength and a significantly improved impact resistance [9]. Although it can also be used to formulate 2-component systems, CTBN is used in particular to formulate 1-component systems that cure with dicyandiamide at elevated temperatures. These adhesives are used, for example, in automotive engineering for bonding crash-relevant car body structures.

Similar effects are achieved with liquid amine-terminated butadiene-acrylonitrile (ATBN). ATBN tougheners which are synthesized by reacting a CTBN polymer with an amine. Since the terminal secondary amine groups undergo the typical reactions for amine groups, i.e. they also react with epoxy groups, ATBN tougheners are usually not incorporated directly into the epoxy component of a 2-component adhesive. This would lead to crosslinking and a significantly shorter shelf life. Instead, they are usually added to the curative containing component of 2-component systems and are rarely used for the production of 1-component systems. The use of ATBN tougheners in 2-component systems with conventional room temperature curing agents such as fatty polyamides, amidoamines and amines leads to an increased tensile shear and peel strength as well as improved environmental resistance of the cured adhesive. However, it should be noted that the improvement in toughness measured by peel tests is less significant when using faster curing agents. [9]

Core-shell elastomer-modifiers

Apart from CTBN and ATBN, so-called core-shell particles can also be used, which utilize the same toughening mechanism. These particles essentially consist of small elastomer beads (e.g. silicone rubber, butyl acrylate rubber) which are surrounded by an outer shell that enables good anchoring with the resin matrix, either via covalent bonds or intramolecular interactions (Figure II-2.90). In order to achieve good properties, it is important that the particles not only have the "correct" particle size distribution but are also well dispersed in the matrix.

The use of epoxy resin-compatible nanoscale surface-modified SiO_2 particles also leads to a certain increase in fracture toughness and impact strength. In addition, the modulus of elasticity is increased, and the shrinkage upopn cure is reduced while the thermal properties are maintained. Due to the small particle size of approx. 20 nm, high filling levels can be achieved with only a marginal increase in viscosity [35. 36].

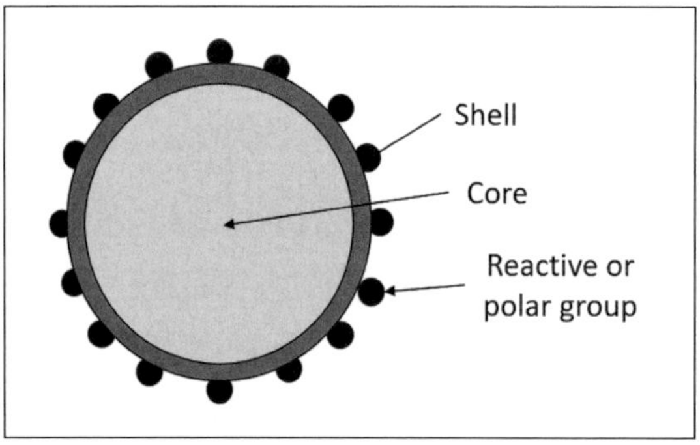

Figure II-2.90: Schematic representation of a core-shell toughening agent

Interpenetrating polymer network (IPN)

The properties of the cured adhesive can also be modified by forming so-called interpenetrating networks during cure. Many properties of the adhesive can be adjusted without affecting the thermal stability. In addition to the epoxy resin and the respective curing agent, these adhesives contain reactive substances (monomers, oligomers or pre-polymers) that can form their own network via a different chemical mechanism without or only reacting with the epoxy monomer to a certain extend. During the cure, two independent networks are formed that penetrate each other but are not or just slightly covalently bonded to each other.

One example from this group are the polysiloxane-epoxy block copolymers. This newly rediscovered class of raw materials is made up of long, flexible polysiloxane segments that alternate with short epoxy, hard segments. Terminally the molecules are epoxy functional. Due to their good compatibility with epoxy resins phase separation does not occur so that they can be mixed with epoxy resins in practically any ratio. During cure, which can be carried out using common epoxy curing agents, both an interpenetrating network as well as a mixed network is formed.

The properties of the adhesives containing such polysiloxane-epoxide block copolymers, in particular the degree of flexibilization, can be varied over a wide range both by adjusting the ratio of epoxy resin to polysiloxane block copolymer and through the molecular weight of the polysiloxane segments. The combination of the good electrical, thermal and chemical properties of the polysiloxane with the good mechanical properties of the epoxides, together with their typically good adhesion properties to copper and other materials, makes the adhesives ideal for electronic applications. [35, 36]

Inorganic particles/filler

Various fillers such as glass beads, fly ash, aluminium oxide trihydrate and silicon dioxide have also been considered as toughness improvers for epoxy materials and were used early on. It has been shown that the fracture energy of filled epoxies reaches a maximum at a certain concentration determined by the respective filler. Due to the partially high hardness (silicon dioxide, fly ash) and the resulting abrasive effect on the components used for both the production and application of the adhesives, their use has not become established. In the case of fly ash, the composition of the fly ash, which is highly dependent on the composition of the respective fuel, containing with CaO corrosive material and in particular its potentially toxic contents, such as heavy metals and dioxins, which must be removed before use in adhesives, are further complicating factors. [9]

Filler

Many epoxy resins contain fillers for both commercial and technical reasons (adjustment of certain rheological properties, reduction of exothermic reactions, reduction of curing shrinkage, increase of thermal or electrical conductivity). Fillers, their properties and applications are discussed in Chapter IV-1.4.

2.2.2.3 Formulation of reactive epoxy adhesives

Structure-property-relationships

When starting into the development of an epoxy adhesive the requirement profile should be prepared. Important questions that arise are about the bonding process parameters and, in particular, the curing conditions (curing at room temperature, curing at elevated temperature, heat curing). Linked to this is the question of whether a 2-component or 1-component adhesive should be developed, with the latter necessarily requiring elevated temperature to hot cure conditions. Of course, the rheological requirements resulting from the application process must also be considered.

In addition to the requirements resulting essentially from the bonding process, the adhesive properties resulting from the requirements to be met by the adhesively to be bonded part for the respective application must of course also be fulfilled, taking into account the expected service life. These are usually the mechanical properties such as tensile and peel strength, flexibility,

Table II-2.33: Epoxy curing agents for 2-part adhesives – comparison of key properties

Property		**Order of gradation**	
Viscosity	low	CYL < AA < AM < PS - PA	high
Colour	light	CYL ~ AA ~ PS < PA < AM	dark
Speed of cure	fast	PS < AA < CYL < PA < AM	slow
Adhesion	fair	AA ~ CYL < AM < PA	excellent
Tensile strength	low	PS < AA <AM < PA	high
Film flexibility	brittle	AA < CYL < AM < PS - PA	good
Heat resistance	poor	PA ~ PS ~ AM < CYL ~ AA	good
Acid resistance.	poor	PA < AM < AA < CYL	excellent
Solvent resistance	poor	AM < PA < CYL < AA	excellent

AM: aliphatic
AM: aminoamide amine
CYL: cycloaliphatic amine
PS: polysulfide/mercaptane/polythiol
PA: polyamide

toughness and impact resistance, the thermal properties such as the glass transition temperature, the required chemical resistance, the required adhesion properties, and probably various additional application-specific requirements.

The strengths and weaknesses of the various raw materials used in the formulation of epoxy adhesives have already been discussed in the previous sections. The following section provides a brief summary about strengths and weaknesses of 1-component and 2-component epoxy adhesives.

Epoxy resins based on bisphenol A diglycidyl ether and short-chain polyamines cure to form amorphous, brittle polymers that have a glass transition temperature of up to approx. 130 °C. They exhibit only a very slight tendency to creep under load and they adhere very well to a large number of substrates. Due to their aromatic ring in the polymer chain, cured epoxies have very little chain mobility, and the use of short-chain curing agents leads to highly crosslinked thermosets. The co-use of epoxy novolacs of higher functionality can further increase the network density and thus improve thermal resistance. The polymer chains mainly contain ether linkages, which chemically are very inert. This explains the excellent chemical and thermal resistance of epoxies.

Room-temperature curing 2-component epoxy adhesives

Room-temperature-curing 2-component epoxy adhesives usually offer a relatively high strength level but do not achieve the strength level of heat-curing 1-component systems. The reason is, that during the cure the glass transition temperature increases continuously, and the system progressively freezes, i.e. the chain mobility decreases. Consequently, not all reactive groups are able to reach a partner by means of chain diffusion. If aliphatic amines serve as curing agent, primary amines exhibit high reactivity due to their greater nucleophilicity. The cured polymers possess good chemical resistance. Cycloaliphatic amines cure more slowly, but the resulting polymers have better temperature resistance (permanent exposure up to 100 °C). Aromatic amines (often used in combination with other curing agents) lead to long pot lives. Polyamides, usually made from dimer fatty acid and excess polyamine, react more slowly than small aliphatic amines, but the resulting polymers are more flexible. The heat resistance of just around 70 °C can be a disadvantage sometimes. In the presence of tertiary amines, thiols react very rapidly. They cure to form flexible polymers which have only moderate heat resistance. Table II-2.33 provides a comparison of the adhesive properties that can be achieved with the different curing agents.

Heat-curing 1-component epoxy adhesives

Heat-curing 1-component adhesives can achieve very high bond strength. If only bisphenol A diglycidyl ether is used as the epoxy resin, they are impact-sensitive and brittle due to their high glass transition temperature and high crosslinking density. Due to the limited selection of available curing agents compared to 2-component adhesives (dicyandiamide is the most widely used heat curing agent for epoxy adhesives) 1-component heat curing epoxy adhesive are containing flexibilized resins or reactive diluents with a flexibilizing effect are often added. Structural adhesives must withstand dynamic forces, including impact loads at low temperatures, too. For this reason, the toughening agents described above are often used in the formulation of epoxy adhesives, especially of heart curing 1- component adhesives.

Stoichiometric considerations

As already explained in Chapter II-2.1.3.1 of this book, maintaining the stoichiometric resin-curing agent ratio as precisely as possible is of crucial importance for the quality of the resulting bond in the case of adhesives curing by a polyaddition reaction.

Since the majority of the epoxy adhesives described cure via the polyaddition mechanism, the following sections describe the calculation of the respective stoichiometric resin-curing agent ratio.

The calculation is usually based on the epoxy equivalent weight (EEW) provided by the respective manufacturers for the epoxy resins and the amine equivalent weight (AWE) provided for the amine curing agents.

Epoxy resins

The epoxy equivalent weight (EEW) is the quantity of an epoxy resin in [g] that corresponds to one equivalent [val] of epoxy functions. It can be calculated by dividing the molecular weight M in [g/mol] by the functionality f in [val/mol].

$$\text{EEW [g/val]} = \text{M [g/mol]/f [val/mol]}$$

Alternatively, concentrations can be specified in [mval/kg] (sometimes misleadingly in [mmol/kg]) or in [%].

An epoxy group, i.e. 1 val (-C_2H_3O), corresponds to 43 g. That means 1000 mval/kg epoxy corresponds to 43 g epoxy/kg or 4.3 %.

$$4.3\ \%\ \text{epoxy} \triangleq 1000\ \text{mval/kg}$$

Examples

Bisphenol A diglycidyl ether: (Figure II-2.52 with n=0)

$$C_{21}H_{24}O_4,\ M = 340\ \text{g/mol},\ f = 2\ \text{val/mol}$$

$$\text{EEW} = \frac{M}{f} = \frac{340}{2} = 170\ \text{g/val},$$

where 1000 mval ≙ 70 g.

→ $\frac{1000 \cdot 1000}{170} = 5880$ mval ≙ 1 kg

where 1000 mval/kg ≙ 4.3 %

→ 5880 mval/kg ≙ 25.3 %

Since for the liquid epoxy resins normally used for the formulation of adhesives n is in the range of 0 – 0.5, the EEW is correspondingly higher. The standard resins offered by various manufacturers, for example, have an EEW of approximately 190 g/val.

Given a functionality of f = 2 val/mol, this corresponds to a molecular weight of 380 g/mol or a polymerization degree of n = 0.15 (see Figure II-2.52). This yields the following characteristic data:

where 1000 mval ≙ 190 g
→ $\frac{1000 \cdot 1000}{190}$ = 5260 mval ≙ 1 kg
where 1000 mval/kg ≙ 4.3 %
→ 5260 mval/kg ≙ 22.6 %

Bisphenol F diglycidyl ether: (Figure II-2.53)

$C_{19}H_{20}O_4$, M = 312 g/mol, f = 2 val/mol
EEW = $\frac{312}{2}$ = 156 g/val

where 1000 mval ≙ 156 g.
→ $\frac{1000 \cdot 1000}{156}$ = 6410 mval ≙ 1 kg
where 1000 mval/kg ≙ 4.3 %
→ 6410 mval/kg ≙ 27.6 %

Epoxy novolaks: (Figure II-2.54 with n =1)

$C_{29}H_{30}O_6$, M = 474 g/mol, f = 3 val/mol
EEW = $\frac{474}{3}$ = 158 g/val

where 1000 mval ≙ 158 g
→ $\frac{1000 \cdot 1000}{158}$ = 6330 mval ≙ 1 kg
where 1000 mval/kg ≙ 4.3 %
→ 6330 mval/kg ≙ 27.2 %

Commercial products deviate slightly from this. The epoxy equivalent weights range from 170 – 240 g/val for functionalities of f = 2.5 to 8 val/mol.

Amine and amine derivatives

The amine equivalent weight (AEW) is the quantity of amine in [g] that contains one equivalent [val] of reactive NH groups. It can be calculated by dividing the molecular weight M in [g/mol] by the functionality f in [val/mol].

$$\text{AEW [g/val]} = \frac{M\left[\frac{g}{mol}\right]}{f\left[\frac{val}{mol}\right]}$$

Alternatively, concentrations can be specified in [mval/kg] (sometimes misleadingly in [mmol/kg]).

To characterize the basic properties of amines, the amine number is often used as an equivalent to KOH.

1 val = 1000 mval KOH ≙ 56 g (see Chapter II-2.2.1.5).
1000 mval basic amine/kg corresponds to an amine number of 56 g KOH/kg.

Primary, secondary, and tertiary amines are equivalent in terms of amine number and are all just monofunctional in terms of an acid/base reaction.

A different picture emerges regarding the reaction of amines and their derivatives with epoxides. Primary amines react bifunctionally with epoxides, while secondary amines only react monofunctionally. A tertiary amine has no functionality for an addition reaction to epoxies, but only has a catalytic effect.

Examples

The two ways of defining concentrations referring to different reactions, which will be illustrated by the following examples. (units for f in [val/mol]).

Hexylamine

$C_6H_{15}N$, M = 101 g/mol $f^{epoxy} = 2$ $f^{acid/base} = 1$

→ $AEW^{epoxy} = 50.5$ g/val $AEW^{H+} = 101$ g/val

→ 19,800 mval/kg 9900 mval/kg

Dipropylamine

$C_6H_{15}N$, M = 101 g/mol $f^{epoxy} = 1$ $f^{acid/base} = 1$

→ $AEW^{epoxy} = 101$ g/val $AEW^{H+} = 101$ g/val

→ 9900 mval/kg 9900 mval/kg

Triethylamine

$C_6H_{15}N$, M = 101 g/mol f^{epoxy} 0. $f^{acid/base} = 1$

→ $AEW^{H+} = 101$ g/val

→ 9900 mval/kg

All three isomeric amines have the same amine number, namely $9900 - \frac{56}{1000} = 554$ g KOH/kg. Characteristic data for selected amine-curing agents are shown in Table II-2.34.

Dicyandiamide (DICY)

Due to the complex reaction process of epoxy curing with dicyandiamide (Figure II-2.71), a functionality of 6 or 7 is usually assumed. With a molecular weight of 84 g/mol, the AEW is therefore 12 or 14 g/val, corresponding to 83.33 or 71.43 mval/kg.

Anhydride curing agents

According to the reaction mechanism for epoxy curing with anhydrides (Figure II-2.74), one anhydride group is required per epoxy group. The anhydride equivalent weight (AhEW) required analogous to the amine equivalent for the calculation of the resin-curing agent ratio is thus calculated by dividing the molecular weight of the anhydride curing agent by its functionality that means the number of anhydride groups.

Polysulfide (mercaptan)/polythiol curing agents

In analogy to the amines, the thiol equivalent weight is calculated by dividing the molecular weight by the functionality, i.e. the number of SH groups.

Table II-2.34: Characteristic data for amine-curing agents

Amine	Molecular formula	M [g/mol]	f [val/mol]	NH AEW [g/val]	NH [mval/kg]	Amine number [g KOH/kg]
Ethylenediamine	$C_2H_8N_2$	60	4	15	66,660	1865
Diethylenetriamine	$C_4H_{13}N_3$	103	5	21	48,540	1630
Triethylenetetramine	$C_6H_{18}N_4$	146	6	24	41,100	1533
Tetraethylenepentamine	$C_8H_{23}N_5$	189	7	27	37,040	1481
Diaminomethyl cyclohexane	$C_8H_{18}N_2$	142	4	35.5	28,170	789
Piperazine	$C_4H_{10}N_2$	86	2	43	23,260	1303
Isophorone diamine	$C_{10}H_{22}N_2$	170	4	42.5	23,530	660
N-aminoethylpiperazine	$C_6H_{15}N_3$	129	3	43	23,260	1303
Diaminomethylbenzene	$C_8H_{12}N_2$	136	4	34	29,400	823
Diaminodiphenyl methane	$C_{13}H_{14}N_2$	198	4	49.5	20,200	566
Diaminodiphenylsulphone	$C_{12}H_{12}N_2O_2S$	248	4	62	16,130	452
Phenylendiamine	$C_6H_8N_2$	108	4	27	37,040	1037
Amidoamine 1 (amide from ricinoleic acid and ethylenediamine)	$C_{20}H_{40}N_2O_2$	340	2	170	5880	165
Amidoamine 2 (amide from ricinoleic acid and diethylenetri-amine)	$C_{22}H_{45}N_3O_2$	383	3	127	7830	290
Amidoamine 3 amide from ricinoleic acid and triethylenetetramine	$C_{24}H_{50}N_4O_2$	426	4	106.5	9390	395
Amidoamine 4 amide from ricinoleic acid and tetraethylene-pentamine	$C_{26}H_{55}N_5O_2$	469	5	94	10,660	478
Polyamide 1 amide from dimer fatty acid and 2 mol ethylene-diamine	$C_{40}H_{76}N_4O_2$	644	4	161	6210	174
Polyamide 2 amide from dimer fatty acid and 2 mol diethylen-etriamine	$C_{44}H_{86}N_6O_2$	730	6	122	8220	307
Polyamide 3 amide from dimer fatty acid and 2 mol triethyl-enetetramine	$C_{48}H_{96}N_8O_2$	816	8	102	9804	412
Polyamide 4 amide from dimer fatty acid and 2 mol tetraethyl-enepentamine	$C_{52}H_{108}N_{10}O_2$	904	10	90.4	11,060	495

Calculation of the resin to curing agent ratio

The calculation of the amount of curing agent required for curing an epoxy resin is explained below using the example of amine curing agents. If anhydrides or thiols are used as curing agents, the calculation is carried out analogously using the anhydride or thiol equivalent weight.

The quantity of amine in [g] needed to cure a given quantity of a given epoxy resin in [g] can be calculated as follows:

Equation II-2.2.1

$$m_{amine}\,[g] = \frac{AEW\,[g/val] \cdot m_{epoxy}\,[g]}{EEW\,[g/val]}$$

m_{amine} : Quantity of amine curing agent
m_{epoxy} : Quantity of epoxy resin
AEW : Amine equivalent weight
EEW : Epoxy equivalent weight

If mixtures of several epoxy resins with different epoxy equivalent weights are used, the equivalent weight of the mixture is calculated as follows:

Equation II-2.2.2

$$EEW_{mixture}\,[g/val] = \frac{m_{mixture}\,[g]}{\frac{m_1\,[g]}{EEW_1\,[g/val]} + \frac{m_2\,[g]}{EEW_2\,[g/val]} + \frac{m_3\,[g]}{EEW_3\,[g/val]} + \ldots + \frac{m_i\,[g]}{EEW_i\,[g/val]}}$$

Analogously, the amine equivalent weight of a mixture of different amine curing agents is calculated according as follows:

Equation II-2.2.3

$$AEW_{mixture}\,[g/val] = \frac{m_{mixture}\,[g]}{\frac{m_1\,[g]}{AEW_1\,[g/val]} + \frac{m_2\,[g]}{AEW_2\,[g/val]} + \frac{m_3\,[g]}{AEW_3\,[g/val]} + \ldots + \frac{m_i\,[g]}{AEW_i\,[g/val]}}$$

Examples

Calculation of the stoichiometric quantity of the curing agent

15 g of a liquid bisphenol A epoxy resin (EEW = 190) is to be cured with triethylenetetramine (AEW = 25 g/val). How much curing agent is needed?

Equation II-2.2.4

$$m_{amine}\,[g] = \frac{25\,[g/val] \cdot 15\,[g]}{190\,[g/val]} = 1.97\,[g]$$

Calculation of the EEW of a mixture of an epoxy resin and a reactive diluent

100 g (m_1) of a liquid epoxy resin (EEW_1 = 190 g/val) is mixed with 20 g (m_2) of a reactive diluent (EEW_2 = 130 g/val). The epoxy equivalent for the mixture is calculated as follows:

Equation II-2.2.5

$$EEW_{mixture}\,[g/val] = \frac{120\,[g]}{\frac{100\,[g]}{190\,[g/val]} + \frac{20\,[g]}{130\,[g/val]}} = 176.5\,[g]$$

Calculation of the amount of curing agent required for curing a liquid bisphenol A resin with dicyandiamide

As mentioned above a a functionality of 6 or 7 is usually assumed for dicyandiamide. With the molecular weight of 84 g/mol, the AEW is therefore 12 or 14 g/val, corresponding to 83.33 or 71.43 mval/kg.

To cure 100 g of a standard liquid epoxy resin (EEW = 190 g/val), the amount of DICY, where the AEW is assumed to be 12, is calculated as follows:

Equation II-2.2.6
$$m_{DICY}\,[g] = \frac{12\,[g/val] \cdot 100\,[g]}{190\,[g/val]} = 6.32\,[g]$$

However, experience shows that the amount of hardener should not exceed 5 – 6 % in relation to the epoxy content.

Guide formulations

The development of an adhesive formulation usually starts with the compilation of the requirements profile, whereby health and safety and environmental protection must be considered in addition to the requirements placed on the adhesive.

The next step is the selection of the raw materials, which essentially involves the epoxy resin, the curing agent and any fillers used. The typical properties of the various resin and curing agent technologies described above provide valuable initial information and can be used, together with the exemplary guide formulations given below and the product-specific information provided by the manufacturers in the respective technical data sheets, to select the raw materials used for initial experimental formulations. After technical application testing of the experimental formulations, the properties obtained will be compared with those specified in the requirement profile in order to specifically eliminate the existing deficits in subsequent optimization steps. For curing agents in particular, some manufacturers offer valuable diagrams which, based on a standard curing agent, show alternative curing agents with which certain properties, such as the viscosity of the adhesive, the pot life or curing speed, the temperature resistance, or the flexibility of the adhesive in the cured state can be optimized [37].

2-component epoxy adhesives

Basics principles

Important parameters to be considered in the development of a 2-component epoxy adhesive essentially resulting from the intended application process have already been described in Chapter II-2.1. These include the mixing of the components, the mixing ratio, the viscosities of the individual components, the application and mixing technique, the quality of the mixing process and, in particular, the mixing ratio tolerance. Other parameters relate to the time course of the curing reaction, i.e. the pot life, the time until a certain handling strength has been reached and the time to reach full cure. Typically, 2-component epoxy adhesives undergo a polyaddition reaction with the curing agents. This is a stoichiometric reaction, which is why the mixing ratio should be observed as precise as possible in order to achieve a reproducible result. Using too much curing agent ultimately leads to a polymer that remains soft because a certain proportion of monomers remain either unreacted and can act as plasticizers or polymers with a reduced chain length are formed. For processing, a mixing ratio of 1 : 1 (by volume) and viscosities that are as equal as possible should be aimed for. A long pot life in combination with fast curing would be ideal, but this is a fundamental contradiction that always requires a compromise. Moderate heating of the adhesive, ideally after joining the components, can be helpful to achieve the required handling strength and final strength faster without shortening the pot life. Figure II-2.91 shows the typical strength building process over time for 2-component epoxy adhesives, one at 20 °C and one at 30 °C. From the inflection point of the curing curve, also known as the gel point, the curing speed slows down and ultimately comes to a halt. Not all reactive groups are able to react as the mobility of the polymer chains becomes severely restricted due to the increasing chain length and cross-linking density. If the adhesive is exposed to elevated temperature at a later point in time, the resulting increase in the mobility of the polymer chains can lead to a post-curing process, resulting in an increase in glass transition temperature, tensile strength but also brittleness. This may have a detrimental effect on the bond, so it is advisable to take into account the temperature loadings that occur during the lifecycle of the adhesively bonded component when determining the adhesive properties.

Advantages and disadvantages

The basic advantages and disadvantages of 2-component adhesives are provided with Table II-2.35.

Fields of application

2-component epoxy adhesives are used for many different applications in industrial production and craftsmanship, especially when warm or hot curing is not desired or not possible. High-strength metal-, plastic- or even hybrid bonds with good long-term resistance can be produced.

Without claiming to be exhaustive, only a few exemplary applications can be briefly described here. In the manufacture of rotor blades for wind turbines, for example, the two moulded shells are usually bonded together using 2-component epoxy adhesives. In the transportation industry, they are used in bonding of fibre reinforced composite parts e.g. from SMC, in the manufacture of rear lids or front hoods, but also for bonding CFRP body components, whether

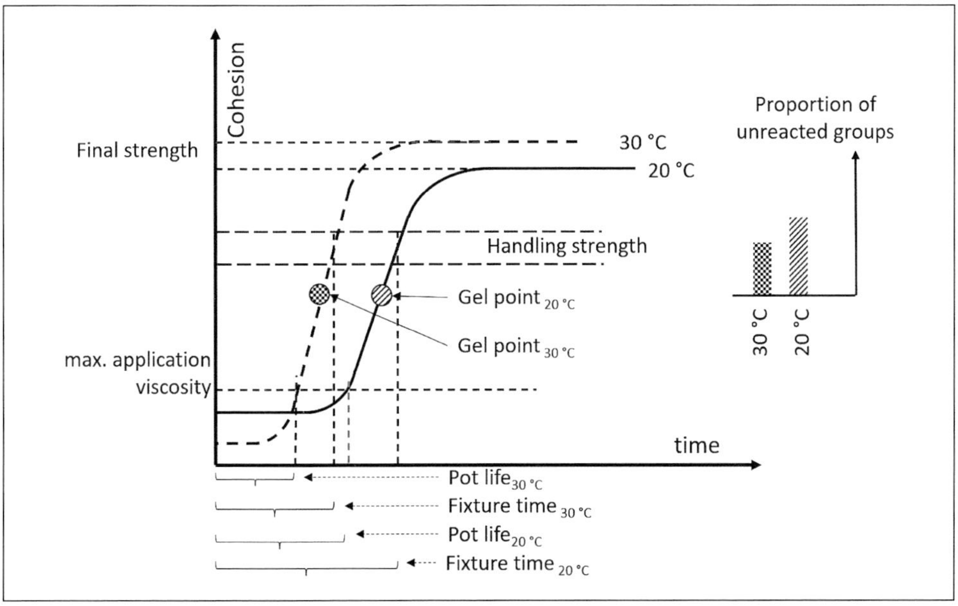

Figure II-2.91: Time course of curing of 2-component epoxy adhesives (cure profile)

Table II-2.35: Advantages and disadvantages of 2-component epoxy adhesives

Advantages	**Disadvantages**
Very variable formulation options due to a wide range of raw materials	2-component process requires dosing and mixing of the components
Polyaddition curing, no potential adverse effect on the adherends by released cleavage products	Polyaddition curing, strict requirements are placed on the dosing quantities.
Room or slightly elevated temperature cure	Pot life must be observed
Good adhesion to many materials	Fixation of adherends required during cure
Medium to high tensile strength	When cured relatively brittle and low impact resistance, limited improvement possible requires special raw materials
Good thermal properties	Amine curing agents and epoxy resins are hazardous substances, the resulting occupational health and safety and environmental protection regulations must be applied
Low creep	
Good resistance against hydrolytic, chemical exposure and UV-light	
Moderate shrinkage on curing	

to other composite parts or to metal body parts. Aerospace and, more recently, boat and shipbuilding are also important areas of application. Another increasingly important field of application is without question the electrical and electronics industry with a wide range of very diverse applications. They are used for assembling electronic circuit boards, for bonding SMD (surface-mounted device) components, for protecting components with so-called glob tops, in the dam & fill process or purely for the mechanical stabilization of flip chips. As potting compounds, they reliably protect electronic components from moisture, contamination or the influence of chemicals. In the construction industry, 2-component epoxies are used, for example for concrete repair and for chemical anchoring (securing fastening elements in masonry or the floor, e.g. for securing machines).

Guide formulations

General remarks on the raw materials

The resin component, often referred to as the A component, essentially contains the epoxy resin, usually a liquid bisphenol A resin (DGEBA-resin) or a mixture of different epoxy resins and possibly reactive or non-reactive diluents. The curing agent component, often referred to as the B component, essentially contains the curing agent or a mixture of different curing agents. In addition to the various amine curing agents, polyamides and amido amides, these also include polythiols (mercaptans) for very fast-curing adhesives are common.

Even though fillers and their use are described in detail in Chapter IV-1.4, a brief overview of the fillers commonly used in 2-component epoxy adhesives shall be provided here.

Fillers such as wollastonite, China Clay, chalk, talk, quartz powder, carbon black, aluminium oxide and aluminium powder as well as pigments, can be added to both components. They are not solely used to reduce the raw material cost of the formulation but help to reduce the exotherm of the cure reaction and shrinkage on curing. Other important reasons for adding fillers are to achieve certain rheological properties adapted to the respective application. To achieve a high yield stress which is characterising a sag resistant adhesive typically hydrophobic fumed silica is used. Thermal conductivity required for applications as thermal interface materials in electronics can be achieved, for example, by aluminium oxide, aluminium powder or boron nitride, electrical conductivity for example by the addition of silver powder or, if only low conductivity is required, graphite, for example. As the fillers usually do not react with either the resin or the curing agent, the mixing ratio can be influenced by intelligently splitting the filler quantity between the two components. This is particularly important for adhesives that are intended for processing from 2C-cartridges, as empty cartridges are essentially limited to volumetric mixing ratios of 1 : 1, 2 : 1 and 10 : 1. The use of lightweight fillers, e.g. hollow glass beads, expands the possibilities for influencing the volumetric mixing ratio through the specific gravity of the components.

Table II-2.36: Straight forward formulation with flexibilization via modified epoxy resin [3]

Component A

Item	Raw material	Parts by weight
1	Standard liquid epoxy resin	46.7
2	Longer chain polyglycol di-epoxide liquid resin, toughening	20
3	Filler, aluminium oxide	33.3
Total		100.0
Component B		
4	Aliphatic amine	8.6

Mixing ratio: A/B = 100/8.6

Item 1: "D.E.R. 331", Olin Corporation, EEW = 190 g/val

Item 2: "D.E.R. 732P", Olin Corporation, EEW = 320 g/val

Item 3: "Alumina T-60", Almatis

Item 4: "Ancamine TETA", curing agent, Evonik, AEW = 27 g/val

It needs to be considered that, except of special reinforcing fillers the addition of fillers does more or less reduce cohesion and therefore tensile strength. The effect may give the misleading impression of better adhesion to the adherend.

Fillers can be mixed into the components relatively easily using mixers adjusted to the resulting viscosity. To avoid air inclusions, it is recommended to either mix under vacuum or to degas the components under vacuum after mixing. Even if high shear forces are generally used to achieve a good, uniform filler distribution, fumed silica should always be mixed at low shear rates in order not to destroy its structure which is the reason why it is typically added in the last step. This also applies, albeit to a lesser extent, to hollow glass beads.

Pigments are mainly used to give the adhesive a certain colour. To avoid mixing up the components and to allow visual control of the mixing quality, the A and B components are often given different colours.

Straightforward 2-component epoxy adhesive formulation flexibilized by a modified epoxy resin (Table II-2.36)

The A component is a combination of a liquid standard epoxy resin and a flexibilized polyether-based epoxy resin. Triethylenetetramine is used as the only curing agent (B component). An ultra-fine ground calcium carbonate-based mineral filler (e.g. "Omyacarb GU", OMYA International AG) can be used as a less abrasive alternative to aluminium oxide, whereby the quantity can be adjusted if necessary to set the desired viscosity. The resulting mixing ratio and the different viscosities of the filled A component and unfilled B component are unfavourable in regard to processing.

The epoxy/amine ratio is calculated as follows

46.7 g epoxy resin (Item 1) corresponds to $46.7 - \frac{1000}{190} = 246$ mval

20 g epoxy resin (Item 2) corresponds to $20 - \frac{1000}{320} = 62.5$ mval

Table II-2.37: 2-component epoxy adhesive flexibilized by a polyamide curing agent [38]

Item	Raw material	Parts by weight
Component A		
1	Epoxy resin, standard liquid resin	50
2	Filler, aluminium	50
Total		100
Component B		
3	Polyamide	28

Mixing ratio: A/B = 100/28
Item 1: "Epon 828", Westlake Epoxy, EEW = 188 g/val
Item 2: "120 atomized powder", Eckardt GmbH
Item 3: "Epicure 3125", polyamide based on dimer fatty acid, Westlake Epoxy, AEW = 103 g/val

This makes a total of 308.5 mval epoxy.

8.66 g of the curing agent corresponds to 8.66 – $\frac{1000}{27}$ 1000/27 = 320 mval

The resulting epoxy /amine ratio is with $\frac{308}{320}$ = 0.96 very close to the aimed for optimum of 1.

Simple 2C-epoxy adhesive formulation flexibilized by a polyamide curing agent (Table II-2.37)

The resin component of the formulation provided with Table II-2.37 just consists of a standard liquid epoxy resin, with toughening provided by a general purpose, medium viscosity reactive polyamide curing agent, based on dimerized fatty acid and polyamines. Here too, as an alternative to the relatively abrasive aluminium oxide, an ultra-fine ground calcium carbonate-based mineral filler (e.g. "Omyacarb GU", OMYA Internati-

Table II-2.38: 2-component epoxy adhesive, flexibilized, 1 : 1 mix ratio [28]

Item	Raw material	Parts by weight	Specific gravity [kg/m³]
Component A			
1	Epoxy resin, liquid standard resin	50.5	1.16
2	Reactive diluent	7	0.89
3	Filler, talc	40	2.6
4	Rheological additive	2.5	2.2
Total		100.0	1.75
Component B			
5	Polyamide curing agent 1	38	0.99
6	Polyamide curing agent 2	15	0.94
7	Filler, aluminium	20	2.7
8	Filler, talc	25	2.6
9	Rheological additive	2	2.2
Total		100.0	1.74

Mixing ratio 1:1 (by weight and by volume)
Item 1: "Epon 828", Westlake Epoxy, EEW = 188 g/val
Item 2: "Epodil 748", Alkyl($C_{12\text{-}14}$ glycidyl ether), Evonik, EEW = 290 g/val
Item 3: "Microtalc Talc", talc Specialty Minerals.
Item 4: "Cab-O-Sil TS-720", fumed silica, Cabot Corporation
Item 5: "Ancamide 910", versatile, flexible polyamide curing agent, Evonik, AEW = 230 g/val
Item 6: "Ancamide 2482", modified polyamide curing agent, Evonik, AEW = 125 g/val
Item 7: "Toyal 101", Toyal America
Item 9: "Cab-O-Sil TS-720", fumed silica, Cabot Corporation

omnal AG), or talc (e.g. "Talkum HB-M15B", Prechjel GmbH) can be used. The amount may have to be adjusted to achieve the desired viscosity.

2-component epoxy adhesive formulation flexibilized by a mixture of two polyamide curing agent adjusted to 1:1 mix ratio (Table II-2.38)

2-component adhesives are easiest to apply when the mixing ratio, both by volume and weight, is 1 : 1. Table II-2.38 shows how this in principle can be achieved through the use of higher-molecular (and thus more flexible) curing agents and a suitable combination of fillers.

The epoxy/amine ratio is calculated as follows: 50.5 g epoxy resin (Item 1) corresponds to 266 mval, 7 g epoxy resin (Item 2) corresponds to 24.1 mval, yielding a total of 290 mval epoxy. 38 g curing agent (Item 5) corresponds to 165 mval, 15 g curing agent (Item 6) corresponds to 120 mval, yielding a total of 285 mval and an epoxy/amine ratio close to 1.0.

The specific gravity of the two components calculated based on the specific gravity data of the individual raw materials and their weight proportions are almost identical, so that the mixing ratio for this formulation is 1:1, both by weight and by volume.

2-component car-body repair adhesive (Table II-2.39)

Table II-2.39 shows the formulation for a 2-component car-body aftermarket repair adhesive that is equally suitable for bonding metals like steel, e-coated steel and composite materials such as SMC. The toughening required for automotive applications was achieved by combining a core-shell additive pre-dispersed in a standard liquid epoxy resin, a low-viscosity ATBN-based curing agent and the CNSL-based reactive diluent. It must be pointed out at this point that the toughening properties of the formulation are only conditionally sufficient for the repair of not crash-relevant components, as for the repair of crash parts the mechanical properties have to fully match those of the 1-component heat-curing adhesive used for this part during vehicle production. The adhesive has a pot life of approx. 90 minutes and achieves lap shear values of around 10 MPa with predominantly cohesive failure.

Fast curing 2-component epoxy adhesive (Table II-2.40)

The formulation provided with Table II-2.40 is a fast-curing 2-component adhesive. To achieve the fast curing of the standard bisphenol A resin, a polymercaptan curing agent was used in combination with a Lewis base as a catalyst. The curing speed can be adjusted via the Lewis base content. With the amount of catalyst shown inTable II-2.40, the pot life is 6 to 7 minutes. Due to the filler composition, the adhesive shows a certain sag resistance. To make it easier to visually distinguish between the two components and to control the mix-quality, the B-component was coloured black using carbon black.

Heat-curing, 1-component epoxy adhesives

Basics principles

The advantage of a 1-component adhesive for the user is that no dosing and mixing is required during application. The disadvantage is that heat is required for curing. For manufacturing processes, such as automotive construction, where appropriate heating ovens are required anyway, e.g. for a painting process and can be utilized for adhesive curing, the use of hot curing adhesives are adapted to the existing oven conditions is an obvious choice. Due to the intentional high viscosity of the adhesives the use of planetary type or another suitable high shear mixer is recommended for mixing. To avoid entrapped air vacuum capable mixers are preferred.

Table II-2.39: 2-component epoxy adhesive for car body repair, 2 : 1 mix ratio

Item	Raw material	Parts by weight	Specific gravity [kg/m³]
Component A			
1	Liquid standard epoxy resin	35.0	1.16
2	CNSL based reactive diluent	7.0	1.06
3	Core-shell impact modifier	1.6	1.15
4	Epoxy-functional silane adhesion promoter (beta-(3,4-epoxycyclohexyl)ethyltrimethoxysilane)	2.0	1.07
5	Silane coated quartz fine flour	12.0	2.65
6	Light weight filler, hollow microspheres	26.9	0.42
7	Rheological additive	2.0	2.2
8	Solid glass microspheres, spacer	1.5	2.54
Total		100.0	1.17
Component B			
9	Polyamide curing agent	33.0	0.99
10	Activated aliphatic amine curing agent,	27.5	0.97
11	ATBN-based toughening agent	16.5	0.94
12	Tris-(dimethylaminomethyl) phenol Lewis base catalyst	7.2	0.97
13	Silane coated quartz fine flour	11.0	2.65
14	Rheological additive	3.58	2.2
15	Solid Glass Microspheres, spacer	1.0	2.54
Total		100.0	1.22

Mixing ratio 2:1 by volume, 1,3:1 by weight
Item 1: "Kukdo YD-128" Kukdo Chemical Co.,Ltd, EEW = 184 g/val
Item 2: "Cardolite NC 513" Cardolite Corporation, EEW = 500 g/val
Item 3: "Kane Ace MX 257" Kaneka, EEW = 294 g/val
Item 4: "Silquest A-186", Momentive Performance Materials GmbH
Item 5: „Silbond W12EST", Quarzwerke GmbH
Item 6: "Q-Cell 6042S", Potters Europe
Item 7: "Aerosil R208", Evonik
Item 8: "SpheriGlass 2024", Potters Europe
Item 9: "Ancamide 910" Evonik, AEW = 230 g/val
Item 10: "Ancamine 1768" Evonik, AEW = 95 g/val
Item 11: "HYPRO 1300x42", CVC Thermoset Specialties, AEW = 450
Item 12: "Kukdo KH-3001 Kukdo Chemical Co. Ltd
Item 13: "Silbond W12EST", Quarzwerke GmbH
Item 14: "Aerosil R208", Evonik
Item 15: "SpheriGlass 2024", Potters Europe

Advantages and Disadvantages

The basic advantages and disadvantages of 1-component heat curing epoxy adhesives are provided with Table II-2.41.

Table II-2.40: Fast curing 2-component epoxy adhesive, 1 : 1 mix ratio

Item	Raw material	Parts by weight	Specific gravity [kg/m³]
Component A			
1	epoxy resin, liquid standard resin	77.0	1.16
2	filler, calcium carbonate	12.4	2.7
3	filler, silane coated quartz fine flour	9.0	2.65
4	rheological additive	1.6	2.2
Total		100.0	1.50
Component B			
5	Modified polymercaptan curing agent	78.0	1.14
6	Tris-(dimethylaminomethyl) phenol Lewis base catalyst	0.5	0.94
7	Filler, calcium carbonate	14.0	2.7
8	Filler, silane coated quartz fine flour	5.5	2.65
9	Rheological additive	1.0	2.2
10	Carbon black pigment	1	1.8
Total		100.0	1.46

Mixing ratio 1:1 (by weight and by volume)
Item 1: "Epon 828", Westlake Epoxy, EEW = 188 g/val
Item 2: "Millicarb BG", Omya
Item 3: "Silbond W12EST", Quarzwerke GmbH.
Item 4: "Cab-O-Sil TS-720", fumed silica, Cabot Corporation
Item 5: "Epikure 02511", AEW = 185 g/val
Item 6: "Kukdo KH-3001" Kukdo Chemical Co.,Ltd
Item 7: "Millicarb BG", Omya
Item 9: "Cab-O-Sil TS-720", fumed silica, Cabot Corporation
Item 10. "Printex 60", Orion Engineered Carbons

Table II-2.41: Advantages and disadvantages of 1-component heat curing epoxy adhesives

Advantages	Disadvantages
+ 1-component, no dosing and mixing required during application	- curing requires high temperatures
+ Good wetting and good adhesion properties bonding of oiled sheet metal possible	- limitation to heat-resistant substrates
+ Good adhesion to many materials, especially metals	- sometimes long curing times required
+ Moderate shrinkage on curing	- fixation of adherends required during cure
+ Good thermal properties	- curing agents and epoxy resins are hazardous substances, the resulting occupational health and safety and environmental protection regulations must be observed
+ Low creep	
+ Good resistance against hydrolytic, chemical exposure, good UV-stability	

Fields of application

In the automotive industry, epoxy adhesives have been used successfully for a long time, initially limited to bonding the hem flange but now used much more widely often in combination with spot welding as fixation till the adhesive has been heat cured. After applying the adhesive in the area of the future bondline, the outer panel is flanged over the inner part, which serves as a stiffening profile, e.g. of a front hood, a door or a rear lid, whereby the adhesive is crimped and almost completely fills the gap (see Figure II-1.58). As a result, and in most cases due to the subsequently applied hem flanged seam seal, the penetration of moisture into the gap and thus corrosion is prevented. Based on the positive experience gained, the range of applications was extended to include additional bonding on the car body. See also the excursus "Adhesives and sealants in car body construction" in Chapter II-1.4.

In the aerospace industry, 1-component epoxy adhesives are often used in the form of reactive films, usually in conjunction with riveting for joining aluminium components, e.g. the production of so-called dobblers for local reinforcement of the fuselage or wings. With the increasing use of carbon fibre composites, bonding has gained in importance as the sole joining process, as a combination with riveting is not acceptable due to the damage to the fibre structure caused by rivet holes. In addition to film adhesives, paste adhesives are also increasingly being used but usually as 2-component systems.

The electrical and electronics industry is also an important area of application. For example, heat-curing epoxy adhesives are used for bonding strain gauges or for bonding magnets in the manufacture of electric motors. They are also used for moisture sensitive components by potting.

Guide formulations

General remarks on raw materials

Heat-curing 1-component epoxy adhesives contain a stoichiometric amount of a latent curing agent, typically a powdery solid that is insoluble in epoxy resins at room temperature. Latent curing agents show a solubility gradient in the epoxy resin that increases strongly as a function of temperature. As the solubility at room temperature is very low, a chemical reaction can only take place to a small extent at the phase boundary between resin and curing agent so that the adhesives show an acceptable shelf life. As the temperature rises, solubility increases, and curing can take place quickly. Some latent curing agents become active at higher temperatures by releasing a component that carries a functional group that can react with epoxy resins. During formulation, care must be taken to ensure that the adhesives do not cure ahead of time when exposed to elevated temperatures, e.g. under transport or storage conditions. It is also important to ensure that the mixing of the raw materials results in a homogeneous dispersion of the finest possible particles of the curing agent and that agglomerates of curing agent particles are not maintained as otherwise inhomogeneous curing would occur during heat cure. The use of commercially available masterbatches containing the curing agent pre-dispersed in a defined epoxy standard resin simplifies the manufacturing process.

Since some of these adhesives, especially those used in automotive body construction, must have a high sag resistance, they are sometimes characterized by a very high viscosity, which requires slight heating both during their production and application. It goes without saying that this must not lead to an early cure.

This adhesive group also primarily uses epoxy resins based on bisphenol A diglycidyl ether, possibly mixed with bisphenol F diglycidyl ether or other resins to achieve special properties.

As with 2-component epoxies, common fillers (up to approx. 50 %) such as aluminium oxide, kaolin, chalk, talc and aluminium powder can be added, although their abrasiveness should be taken into account.

Dicyandiamide (DICY) (Figure II-2.71) is the most commonly used latent curing agent. The speed of curing of DICY systems can be accelerated by substituted urea derivatives such as diuron and monuron (Figure II-2.72), but this usually also reduces the storage stability.

In addition to dicyandiamide, imidazoles, such as 2-ethyl-4-methyl-imidazole (EMI), are also used as latent curing agents. These formulations based on bisphenol A digycidyl ethers can be stored for more than 6 months at < 40 °C, but gel quickly at temperatures of 120 to 170 °C. After curing for 30 minutes at 120 °C, the adhesive exhibits a high degree of thermal and chemical resistance.

In principle, the boron trifluoride-monoethyleneamine complex (BF_3-MEA) can also be used as a latent curing agent. However, due to its disadvantages already described before in the curing agent section, it is rarely used as the sole curing agent.

Table II-2.42: 1-component heat curing epoxy metal bonding adhesive for car body construction

Item	Raw material	Parts by weight
1	Standard liquid epoxy resin	45.3
2	Low viscosity CTBN modified epoxy prepolymer based on DGEBA	13.4
3	Dicyandiamide, 50 % pre- dispersed in a standard liquid epoxy resin	13.0
4	Filler, ultrafine surface treated precipitated calcium carbonate	16.5
5	Filler, Muscovite-based	8.3
6	Carbon black	2.2
5	Hydrophobic fumed silica	0.5
Total		100.0

Item 1: "Epon 828", Westlake Epoxy, EEW = 188 g/val
Item 2: "Struktol Polydis 3605" Schilkl+Seilacher Struktol GmbH, EEW = 300 g/val
Item 3: "Dyhard D50EP", AlzChem GmbH, EEW of the contained resin = 188 g/val
Item 4: "Socal U1S2-G", Imerys Performance Additives
Item 5: "Mica-MU 101", Imerys Performance Additives
Item 6: "Printex 60", Orion Engineered Carbons

Flexibilization or toughening components (approx. 15 % of the polymer) are used for compounds that have to withstand dynamic forces. These include the carboxy-terminated polyacrylonitrile-butadiene prepolymers (CTBN) and the types of toughening agents already described above in the section about toughening agents.

To increase the yield point and thus improve sag resistance, hydrophobized fumed silica is generally used, sometimes in combination with carbon black.

1-component heat curing epoxy metal bonding adhesive for car body construction

Table II-2.42 shows the formulations for a moderately toughened 1-component heat-curing epoxy adhesive for use in car body construction. The use of a masterbatch of dicyandiamide already dispersed in epoxy resin as a curing agent simplifies the homogeneous mixing of the curing agent, which is important for a uniform curing. Due to the typical curing conditions in automotive construction (approx. 180 °C, 30 minutes), acceleration via monuron or diuron can be avoided in favour of shelf life. By using a filler mixture of calcium carbonate, muscovite, carbon black and hydrophobized fumed silica, a good ratio between viscosity and sag resistance and the common black pigmentation is achieved. The lap shear strength on untreated slightly (< 3 g/mm^2) oily cold rolled steel is about 12 MPa (layer thickness 0.2 mm, curing 30 minutes at 180 °C in a convection oven).

When calculating the amount of dicyandiamide required for curing, in this case in the form of a masterbatch, the EEW of the epoxy resin contained in the masterbatch, in this case 188 g/val, must of course also be taken into account.

Table II-2.43: Advantages and disadvantages of 1-component cationic UV-curing epoxy adhesives

Advantages	Disadvantages
+ 1-component, no dosing and mixing required during application	- curing requires corresponding equipment
+ Fast cure	- relatively high raw material costs
+ No limitation to UV-permeable adherends	- often brittle in the cured state, requires addition of tougheners
+ Low shrinkage on curing	- epoxy resins are hazardous substances, the resulting occupational health and safety and environmental protection regulations must be applied
+ Good adhesion to many materials, especially metals and many plastics	
+ Good transparency, yellowing resistant	
+ Thermal post cure possible	
+ No impairment of radical polymerization by atmospheric oxygen, formation of tack-free surfaces	

UV curing, 1-component epoxy adhesives

Basic principles

The ability of the epoxy function to react in a cationic induced ring-opening polymerization (Figure II-2.80) has paved the way to UV curing, 1-component epoxy adhesives. UV curing is primarily used when rapid curing adhesives are required (typically 20 – 60 s). It is often followed by a thermal post-cure.

Advantages and Disadvantages

UV-curing epoxy adhesives have a favourable property profile outlined in Table II-2.43 f

Fields of application

Cationic UV-curing epoxies are widely used in the electronics industry, e.g. in the manufacture of hard disks for fastening and securing electronic components, for sealing housings, including pin sealing and in the production of memory chips using the dam & fill process. They are also used in optoelectronics, e.g. in the production of LED flash modules. [39]

Guide formulations

General remarks on raw materials

A typical cationic UV adhesive formulation contains an epoxy resin (typically > 70 %), often an aliphatic cure accelerating resin, a reactive or non-reactive diluent and the photoinitiator. In some cases, polyols (low-molecular polyols, polyethers, polyesters or castor oil) and up to 15 % copolymerisable vinyl ethers (e.g. triethylene glycol divinyl ether) or high-boiling solvents (e.g. propylene carbonate) are added to keep raw material costs low or to improve flexibility.

UV curable cationic epoxy laminating adhesive

Table II-2.44 and Table II-2.45 showing two guide formulation for laminating adhesives. The adhesive of Table II-2.44 offers good adhesion to various plastics like polyethylene, polyester, and polycarbonate but also to aluminium and steel. The laminating process can start immediately after UV exposure (dosage 33 – 75 mJ/cm^2. Full cure is achieved after 1 hour.

For structural bonds high molecular weight thermoplastics can be added to increase fracture toughness with no reduction in glass transition temperatures, which often comes as an unwanted side effect with the use of reactive flexibilizers and rubbers. Besides the polyester-type thermoplastic polyurethane used in the formulation provided with Table II-2.45 products such as "Udel P-1700" from Syensqo, or "Ultem 1000" from Sabic can be used.

Table II-2.44: UV curable cationic epoxy laminating adhesive [43]

Item	Raw material	Parts by weight
1	Modified cycloalipohatic epox EEW app. 134 g/eq	87.3
2	Tripropylene glycol	9.7
3	Photoinitiator	3.0
Total		100.0

Item 1: "Uvacure 1534", Solvay/Syensqo
Item 2: "Tripropylenw glycol" Dow Chemical
Item 3: "Uvacure 1590", Solvay

Table II-2.45: UV curable cationic epoxy laminating adhesive

Item	Raw material	Parts by weight
1	Cycloapliphatic epoxy resin, EEW app. 126 g/eq	80
2	Polyester-type thermoplastic polyurethane	10
3	Oxetane monoalcohol cationic reactive diluent	8
4	Iodonium hexafluoroantimonate type cationic photoinitiator, absorption maximum @ 230 nm	2
Total		100

Item 1: "UviCure S105", Arkema
Item 2: "Pearlkstick 5701" Lubrizol
Item 3: "Curalite Ox", Perstorp
Item 4: "SpeedCure 937", Arkema

2.2.2.4 References

[1] Markets and Markets Report "Epoxy Adhesvies 2020 – 2027 https://www.marketsandmarkets.com/Market-Reports/epoxy-adhesive-market-142980020.html?gclid=Cj0KCQjwiIOmBhDjARIsAP6YhSXz-rHlaiRQPt8QIsOdI0ukH3Jcu7kyQ6yN_i5r4J387_B9R2mnz9VIaAlCLEALw_wcB, accessed 2023-07-26
[2] Hartshorn, S. R., Structural Adhesives, Plenum Press (1986)
[3] Skeist, I., Handbook of Adhesives, 3rd edition, Chapman & Hall (1989)
[4] Habenicht, G., Kleben, 3rd edition, Springer (1997)
[5] Cardolite Corporation, Adhesives Product overview, https://www.cardolite.com/, accessed 2025-01-11
[6] Bakelite, Binders for the Adhesives Industry, company brochure, 2004
[7] Hexion, technical datasheet Cardura E-10-P https://www.hexion.com/CustomServices/PDFDownloader.aspx?type=tds&pid=9bce8a2c-5814-6fe3-ae8a-ff0300fcd525 accessed 2023-12-31
[8] Ipox chemicals GmbH, product range https://ipox-chemicals.de/produktuebersicht/, accessed 2023-08-05
[9] Petrie, Edward M., Epoxy Adhesives Formulations, McGraw-Hill, New York, 2006
[10] Chem. Rev, 56, 571 (1956)
[11] Westlake Epoxy, Epoxy curing agents https://www.westlakeepoxy.com/chemistry/epoxy-resins-curing-agents-modifiers/curing-agents, accessed 2023-08-05
[12] Kukdo, Company brochure Chemical Hardeners
[13] TriiSO, https://www.tri-iso.com/documents/Kukdo_Epoxy_Brochure.pdf, accessed 2023-08-05
[15] Evonik https://crosslinkers.evonik.com/en/products/ancamide, accessed 2023-08-05
[16] Evonik https://crosslinkers.evonik.com/en/products/ancamine, accessed 2023-08-05
[17] Odagiri, Nobuyuki, et. al., Cure Path Dependency in Crosslinked Structure of DGEBA/DICY/DCMU Thermosetting Resin, Materials System, Vol.38, page 21ff https://www.jstage.jst.go.jp/article/materialssystem/38/0/38_21/_pdf/-char/ja, accessed 2023-08-06

[18] Barwich, J., et al., Adhäsion 1989, 33 (5), 27-30
[19] Groß, A., et al., Adhäsion 1988, 11, 16–24
[20] Degussa Fine Chemicals, Dyhard, company brochure, March 2003
[21] Huntsman brochure: Jeffamine® Polyetheramines https://www.huntsman.com/docs/Documents/JEFFAMINE®%20Polyetheramines%20Formulation%20Guide.pdf, accessed 2023-08-06
[21] Huntsman brochure: HYPRO® 1300x16 ATBN, https://www.huntsman.com/docs/Documents/Hypro1300X16ATBN_US_e.pdf accesses 2023-08-06
[22] Dodiuk, H., et al., J. Adhesion, Vol 22 (1987), 227-251
[23] Dow, Liquid Epoxy Resins, company brochure. 2004
[24] Hartshorn, S. R., (editor), Structural Adhesives, Chapter 3, Plenum Press, New York, 1986
[25] Dixie Chemical, technical data sheet Nadic Methyl Anhydride (NMA), https://www.tri-iso.com/documents/Anhydride_Curative_NMA_TDS.pdf accessed 2023-12-08
[26] Dixie Chemical, Technical Bulletin, Formulating anhydride-cured epoxy systems, https://www.tri-iso.com/documents/Formulation_Guide_Anhydride_Curatives_for_Epoxy_Systems.pdf
[27] Frihart et al., Cognis (Henkel) in www.adhesivesmag.com
[28] Air Products,Epoxy Additives for Adhesive Applicartions, company brochure, 2001
[29] DuPont, Techn. Information (3/02), W-400030
[30] Lamone, L., UV Curing Part Five. Cationic Photopolymerization, 2016-02-08 https://polymerinnovationblog.com/uv-curing-part-five-cationic-photopolymerization/ accessed 2023-12-21
[31] Lamone, L., UV Curing Part Six. Cationic Photopolymerization, 2016-02-15 https://polymerinnovationblog.com/uv-curing-part-six-cationic-resin-curing-party-not-stop-lights-go/ accessed 2023-12-21
[32] Perstorp Holding AB, Presentation "Perstorp Catinioc offerings – General Introduction to Radiation curing", www.perstorp.com (Perstorp cationic technology) accessed 2023-12-22
[33] James, D., Cationic UV Cure on Polyolefins, Perstorp White Paper. Downloaded from https://www.perstorp.com/en/products/resins_and_coatings/cationic_curing 2023-12-22
[34] Double Bond Chemical Ind. Co., Ltd, Taiwan, Company Brochure: UV Curable Materials 2022.12
[35] Sprenger, S., et. al., With polysiloxane-epoxy copolymers towards flexible electronics, adhesion 4| 2022, page 18
[36] Sprenger, S., et. al. Adhäsion kleben + dichten 3 (2003), 24-31
[37] Evonik Operations GmbH, Product selection charts: Epoxy curing agents for adhesive applications, https://crosslinkers.evonik.com/en/down-loads/brochures e.g. ECA for 2K Adhesive Application pdf accessed 2023-12-28
[38] Resolution Products, Starting formulations under www.resins.com/resins/am/product_html (website no longer available)
[39] DELO Industrial Adhesives, Bond it – Bonding Technology and its industrial application, 6th Edition, 2020
[40] UCB Chemicals, website no longer available
[41] https://www.westlakeepoxy.com/applications/ accessed 2025-01-17
[42] Wright, C. D., Muggee, J. M., Epoxy Structural Adhesives in Hartshorn, S. R., Structural Adhesives, Plenum Press New York, 1986
[43] https://adhesives.specialchem.com/formulation/uv-curable-cationic-epoxy-laminating-adhesive, acessed 2025-01-018

2.2.3 Polysulphides

As far back as 1840, elastic "Thioplaste" were synthesised in Germany, but they fell into obscurity due to a lack of suitable applications. In the 1920s, solid polysulphides were marketed in the USA under the name "Thiokol" [1]. Nowadays, solid polysulphides no longer play an important role, as liquid polysulphides can be produced by reductive cleavage of disulphide bridges (Figure II-2.92).

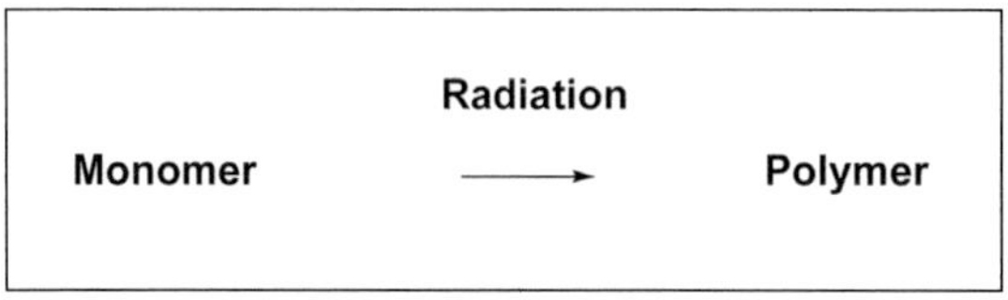

Figure II-2.92: Synthesis of liquid polysulphides by reductive cleavage of disulphide bridges

The chains can be branched (crosslinked) by adding trichloropropane. As shown in Figure II-2.92, liquid polymers terminated with thiol groups are obtained in a controlled molecular weight ranging from 1,000 to 8,000 g/mol (the figure varies with the concentrations of NaSH and Na_2SO_3).

A comparatively up-to-date overview of the chemistry and properties of polysulphides as sealants can be found in [6, 7, 9].

Advantages

- + excellent resistance to weathering and ageing (including ozone)
- + excellent resistance to fuels, oils, chlorinated hydrocarbons and numerous chemicals
- + permanent elasticity from -50 to 90 °C
- + relatively tolerant of errors during mixing of the 2-component system [4].

Table II-2.46: Formulation for component A of an insulating glass secondary seal based on polysulphide

1	Liquid polysulfide	100
2	Plasticizer	33.5
3	Calcium carbonate (fine)	82.5
4	Calcium carbonate (coarse)	82.5
5	Additives*	1.5

Item 1: "Thioplast G112", molecular weight: 3.900-4.400 g/mol, Nouryon
Item 2: "Santicizer 278", Valtris Specialty Chemicals
* Adhesion promoter (silane), antioxidant
Recommended use in combination with managanese(IV)oxide curing agent (5.7-6.3 g MnO_2/100 g).

Applications

One of the first major applications was and remains the sealing of fuel tanks in aircraft wings, as well as other applications in aircraft construction [4, 8]; shipbuilding; construction sealants; insulating glass sealants.

2.2.3.1 2-component insulating glass secondary seal

The liquid polymers can be converted into sealants by adding fillers, plasticisers and additives. Crosslinking occurs by admixing a curing agent (2-component system). The active agent in the curing component is always an oxidant (e.g. manganese dioxide) which oxidises the thiol groups to disulphide bridges. This oxidative coupling, shown in Figure II-2.93, is essentially a reversal of the reductive chain cleavage presented in Figure II-2.92.

As a result of this heterogeneous oxidative cure, major deviations from the value of the mixing ratio will still yield sufficient strength in many instances [4]; this 2-component system is therefore tolerant of mixing errors. Tetramethylthiuram disulphide (TMTDS) can be added as an accelerator (Figure II-2.94).

Guide formulation

A formulation for an insulating glass secondary seal based on polysulphide is presented in Table II-2.46 [1].

For industrial applications in insulating glass manufacture, two-component polysulphide sealants are usually formulated so as to yield a processing time or pot life of 30 min; after 1.5 to 2 hours, the insulating glass units have cured enough to be processed further [4].

The curing agent (component B) consists of manganese dioxide in plasticiser and added vulcanisation accelerator. The mixing ratio of component B and curing agent is 10 : 1.

The binder : plasticiser : filler ratio computes to 100 : 55* : 165

(* includes estimated quantity of plasticiser from the curing agent).

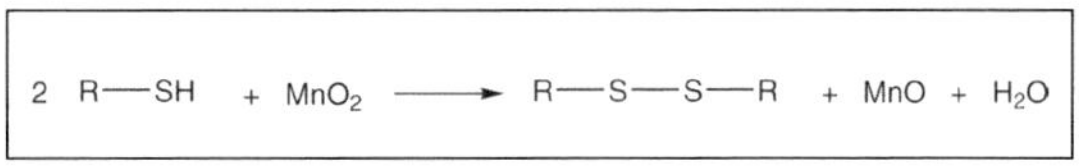

Figure II-2.93: Oxidative coupling of thiols

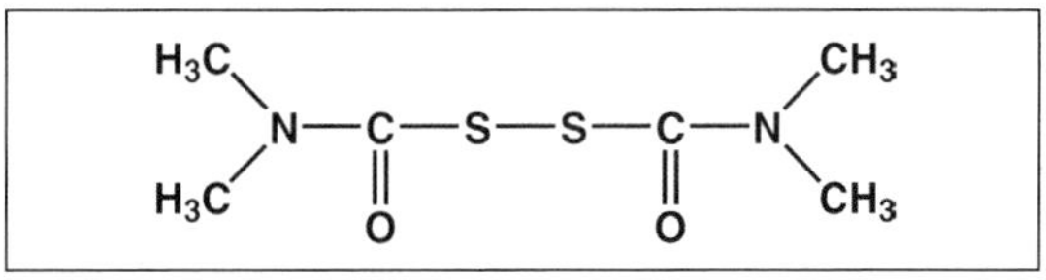

Figure II-2.94: Tetramethylthiuram disulphide (TMTDS)

2.2.3.2 2-component construction sealants

2-component polysulphide systems have been commercially available as high-performance sealants for construction applications since 1955 [5].

Construction sealants based on polysulphide have the following advantages

Advantages

+ very good resistance to water and chemicals
+ good resistance to weathering and ageing
+ 2-component systems are less sensitive to mixing errors (as regards quantity of curing agent) than 2-component polyurethanes
+ good mechanical properties (25 % permissible total deformation)
+ lower water-vapour permeability than 1-component polyurethanes

Disadvantage

- slow curing rate of the 1-component systems

Figure II-2.95 shows a polysulphide expansion joint on the outside of a building of washed concrete slabs that was built around 1970. Because of the long weathering period, cracks are visible (see [6]). When the sealant is cut with a knife, it is found that the embrittlement is only in the surface layer – the interior is still elasto-plastic. Figure II-2.96 shows two scanning electron micrographs of this sealant surface. The plan view shows filler particles in the surface laid open by photooxidative polymer degradation; construction sealants are highly loaded with fillers (see

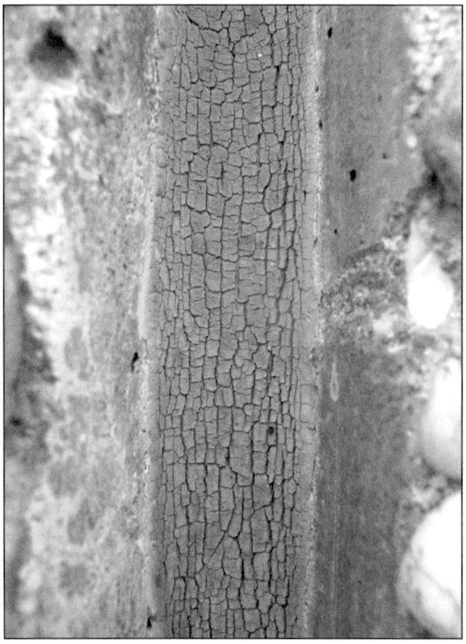

Figure II-2.95: Expansion joint based on polysulphide on the outside of a building (width = 2 cm)

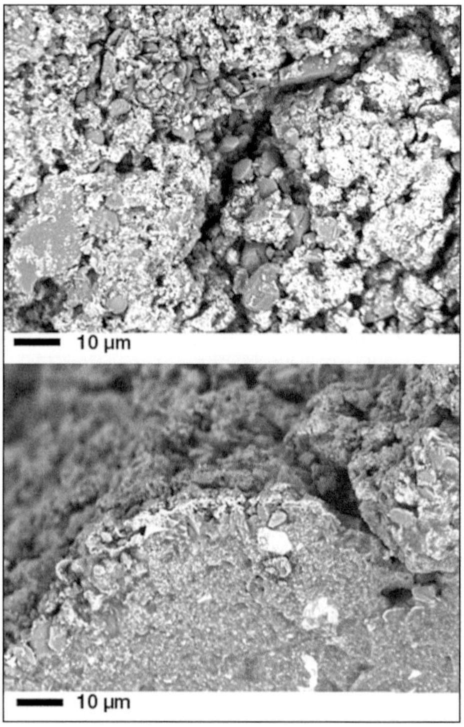

Figure II-2.96: Scanning electron micrographs of the sealant surface (from Figure II-2.95), plan view (top) and edge of a freeze fracture (bottom).

Table II-2.46). The edge of a freeze fracture in Figure II-2.96 shows that the damage caused by weathering is only on the surface and that the sealant in the interior is still intact.

Guide formulation

The chemistry of 2-component polysulphide sealants is described in detail in Chapter II-2.2.3.1 (insulating glass applications). We can therefore proceed directly to a guide formulation (Table II-2.47).

The binder : plasticiser : pigments (+ fillers) ratio is roughly 100 : 70 : 80.

Table II-2.47: Formulation for a 2-component polysulphide sealant for building construction [7]

Item	Raw material	Parts by weight
Component A		
1	Liquid polysulfide	27.3
2	Plasticizer	9.0
3	Chalk	53.6
4	Adhesion promoter	1.0
Component B (curing agent)		
5	Plasticizer	3.6
6	Manganese dioxide	2.7
7	Accelerator	0,3
8	Chalk	2,5
	Total	100

Item 1: "Thioplast G21" pre-polymer, end-capped by SH-functional end groups, Nouryon
Item 2 and 5: "Santicizer 278", Valtris Specialty Chemicals
Item 4: 3-Glycidyloxypropyl)trimethoxysilane
Item 7: Tetramethylthiuramdisulfid
* Benzylbutyl phthalate is now classified as toxic; if possible, an alternative plasticiser should be used

2.2.3.3 1-component polysulphides

With polysulphides, 1-component systems are less important than 2-component types. 1-component polysulphide sealants cure by an oxidation reaction similar to that of 2-component systems (see Figure II-2.93), albeit with peroxides activated by water (humidity). The curing agents are frequently combinations of peroxides and barium oxide. Accelerating amines are sometimes added to 1-component systems; they are adsorbed on molecular sieves and are released by humidity.

2.2.3.4 References

[1] Kittel, H., Lehrbuch der Lacke und Beschichtungen (in German), Vol. I/1 (1971) p. 188 f
[2] Usmani, A. M. "Chemistry and technology of polysulfide sealants." Polymer-Plastics Technology and Engineering 19.2 (1982): 165–199.
[3] Ghatge, N. D., S. P. Vernekar, and S. V. Lonikar. "Polysulfide sealants." Rubber Chemistry and Technology 54.2 (1981): 197–210.
[4] DE-OS 42 05 358 A1
[5] Irfan, M. H., and M. H. Irfan. "Polysulphides in the construction industry." Chemistry and Technology of Thermosetting Polymers in Construction Applications (1998): 203–226.
[6] Müller, B., Understanding Additivee, Vincentz Network (2019), Chapter 8
[7] nouryon.com/product/thioplast-g21/
[8] Song, Bo, et al. "Formulation and processing of conductive polysulfide sealants for automotive and aerospace applications." 2019 IEEE 69th Electronic Components and Technology Conference (ECTC). IEEE, 2019.
[9] Pirayesh, Amin, et al. "Polysulfide polymers: synthesis, blending, nanocomposites, and applications." Polymer Reviews 59.1 (2019): 124–148.

2.3 Polymerization adhesives

2.3.1 Reactive acrylic adhesives

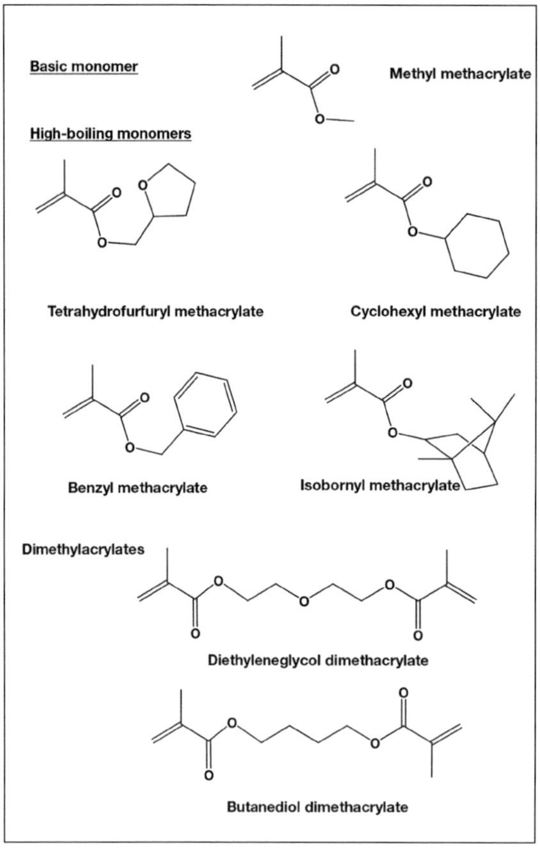

Figure II-2.97: Monomers for acrylic adhesives

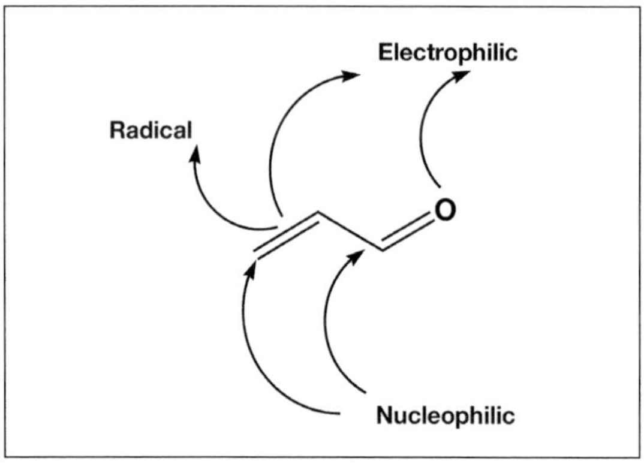

Figure II-2.98: Reactions of α,β-unsaturated carbonyl compounds

2.3.1.1 Basic principles

Acrylic adhesives include besides the classic 2-component adhesives the 1-component anaerobic curing adhesives, radiation-curing acrylic adhesives and cyanoacrylate adhesives, which differ in their properties and processing conditions, but despite these differences have some similarities. Except as for cyanoacrylate adhesives curing by anoinic polymerization which are discussed in Chapter. II-2.3.5 curing takes place by means of a radical polymerization reaction, which, in contrast to adhesives that crosslink in a polyaddition reaction (Chapter II-2.2) or polycondensation reaction (Chapter II-2.4), places lower demands on the stoichiometric accuracy of the resin-hardener ratio.

2.3.1.2 Raw materials and functional group

Reactive monomers

Reactive acrylic adhesives are based on monomeric or oligomeric mixtures of the esters of acrylic or methacrylic acid. Figure II-2.97 shows examples of the most important monomers. For the synthesis of methacrylic acid, see also Chapter II-1.1.4.4.

The monomers are usually low-viscosity, colourless liquids with a characteristic to pungent odour, which is often perceived as unpleasant and occasionally annoying when processing the adhesives. Esters of higher alcohols and esters of diols do not have an „noticeable odour. As shown inTable II-2.48, the boiling point of the esters increases with increase in the number of carbons in the ester alcohol [4, 5].

Chemically, acrylates are α,β-unsaturated carbonyl compounds. As shown in Figure II-2.98, they can react with nucleophiles (at the β-carbon or directly at the carbonyl function) and with electrophiles (at the C=C double bond or at the non-bonding electrons on the oxygen).

The C=C double bond (conjugated with the acid or ester function) is the most commonly used function for both the synthesis of polymers and the curing of acrylic adhesives. As shown in Figure II-2.99, the monomers can be polymerized via a radical attack on the C=C double bond.

As shown in Figure II-2.100, stoichiometric vinylogous nucleophilic addition at the β-position of the carbonyl group can serve as an alternative curing reaction. This addition reaction is also used for curing epoxy acrylates.

Table II-2.48: Boiling points of monomeric acrylates

Monomer	**bp [°C]**	**T_g polymer [°C]**
Methylacrylate	80.5	22
Ethylacrylate	99	-8
Methyl methacrylate	100	105
Ethyl methacrylate	119	67
Acrylic acid	141	130
Butylacrylate	145	-43
Isobutyl methacrylate	155	64
Methacrylic acid	163	162
Butyl methacrylate	164	32
Tetrahydrofuryl methacrylate	178	75
Cyclohexyl methacrylate	210	104
Benzyl methacrylate	>200	54
Isobornyl methacrylate	>200	110
Ethyleneglycol dimethacrylate	>200	
Diethyleneglycol dimethacrylat	>200	
Butanediol methacrylate	>200	

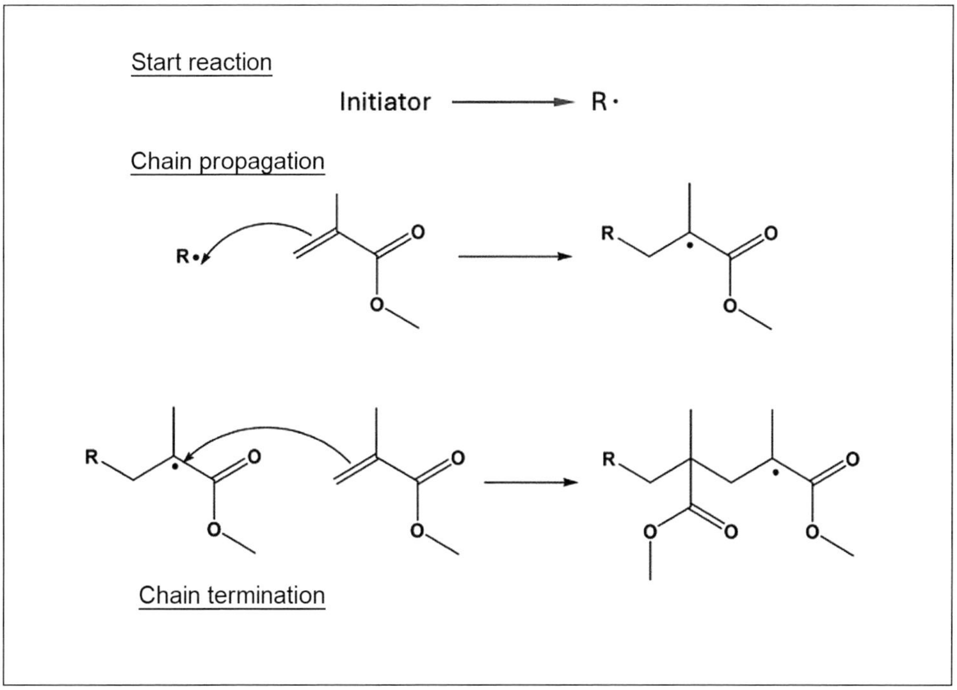

Figure II-2.99: Polymerization of acrylates

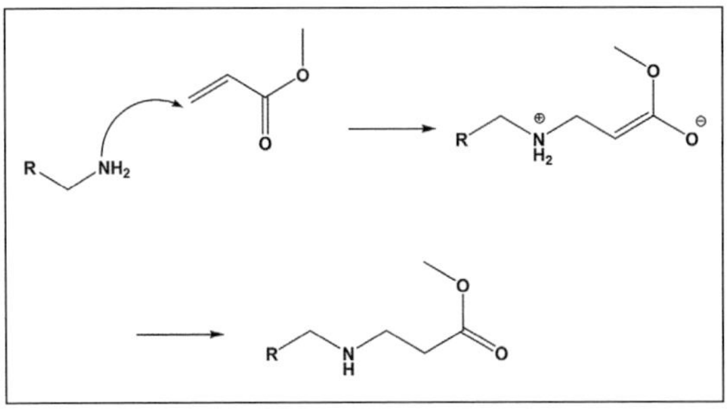

Figure II-2.100: Vinylogous nucleophilic addition to acrylates

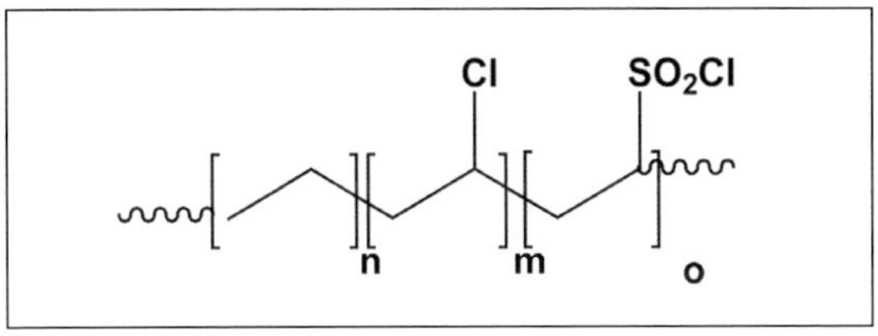

Figure II-2.101: Chlorosulphonated polyethylene

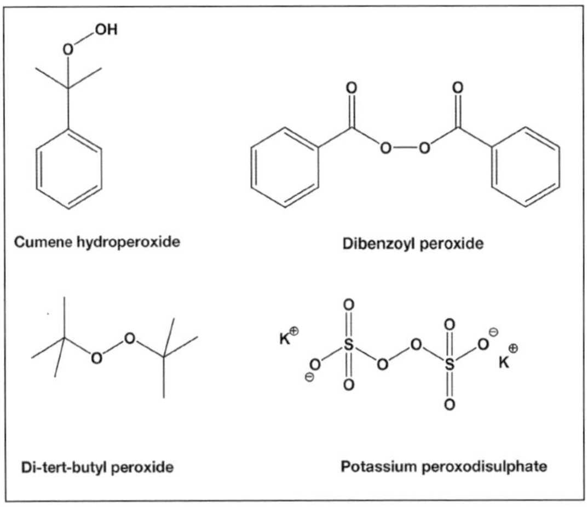

Figure II-2.102: Examples of suitable peroxide curing agents

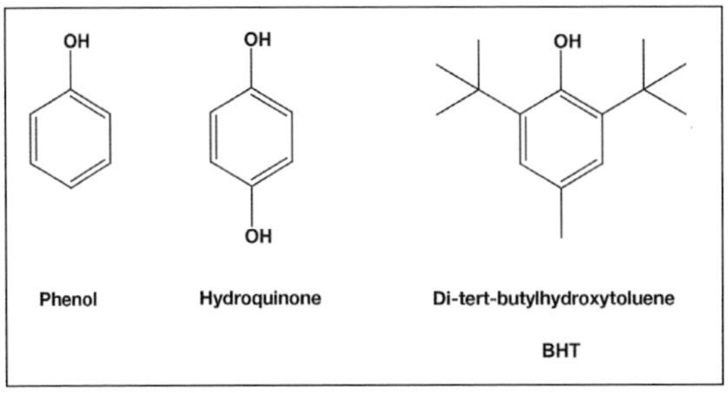

Figure II-2.103: Examples for suitable stabilisers

Crosslinker

Acrylate esters of diols (diacrylates, dimethacrylates; see Figure II-2.97) are polyfunctional in respect of a free-radical reaction. They are added in judicious amounts to regulate the crosslink density in the cured adhesive.

Toughening agent

Adhesives expected to have good impact resistance and good low-temperature properties require tougheners. These are primarily modified polyolefins (e.g. chlorosulphonated polyethylene [7, 9, 11, 21], Figure II-2.101) which have limited compatibility with the cured polymer and lead to phase separation during polymerization with the formation of domains acting as crack stoppers (Figure II-2.89).

Other potential toughening agents are besides others polychloroprene, polybutadiene, polybutadiene/acrylonitrile adducts, urethane elastomers, acrylic elastomers, polybutadiene/isocyanate adducts, vinyl-terminated butadiene/acrylonitrile adducts and polyurethane methacrylates [21]

Curing agents (initiators, starters)

The curing agents are typically substances based on organic peroxides, which are able to form free-radicals. The free-radicals for initiating the polymerization can be generated by heat, amine catalysis or UV radiation. Important peroxide-curing agents are shown in Figure II-2.102.

Stabilisers

Small amounts of antioxidants, such as hindered phenols, hydroquinone and quinones (see Figure II-2.103), are added to the reactive resin to prevents premature initiation of the free-radical chain reaction and to improve storage stability [14].

Accelerators and activators

Many amines catalyse the decomposition of peroxides to free radicals. Especially tertiary amines, such as N,N-dimethyl-p-toluidine or capped amines like the imine formed from butanal and aniline are used to facilitate acrylic adhesives to cure at room temperature. Additionally, salts of transition metals, such as copper octanoate (many transition metals can readily change oxidation number and so are able to catalyse peroxide decomposition), which are soluble in organic milieu, make effective accelerators. Increasing use has recently been made of dihydropyridine derivatives [6, 7], because they are reputed to be less health hazardous. (Figure II-2.104)

2.3.1.3 Technologies of curing

The low-molecular weight reactive monomers in the adhesive joint can be converted into a polymer (or copolymer) at a particular time in different ways [1 – 3, 8, 10, 12].

2-component acrylic adhesives

As is usual with 2-component adhesives, these adhesives are supplied in the form of two separately packaged components, which must be mixed immediately before use. One component contains the monomer mixture and the accelerator, the second contains the curing agent and, if the curing agent is only active in the presence of the accelerator, also monomers. In order to achieve certain properties, fillers and functional additives can be added to the components as additional formulation ingredients.

Within this group, a distinction is made between adhesives in which the components are mixed before application to an adherend and those in which the two components are applied unmixed to one adherend each and are only brought into contact with each other during joining.

Anaerobic acrylic adhesives

The anaerobic 1-component adhesives are also essentially based on the typical acrylate or methacrylate monomers shown in Figure II-2.97, possibly with the addition of dimeric or modified acrylate esters. Due to the inhibiting effect of oxygen, which prevents curing, curing in a polymerization reaction only occurs in the adhesive joint in the absence of atmospheric oxygen and simultaneous contact with metal ions.

Radiation-curing acrylic adhesives

In addition to the reactive mixture of acrylate monomers and oligomers, this 1-component system contains so-called photoinitiators, which decompose upon irradiation with light of a certain wavelength to form reactive products that initiate polymerization.

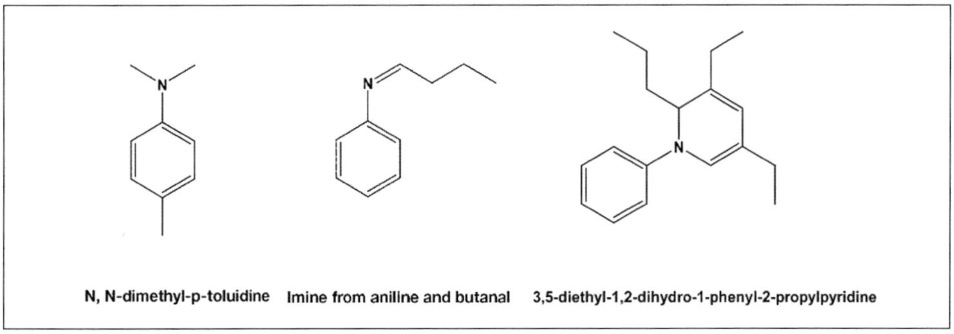

Figure II-2.104: Activator for room temperature curing acrylic adhesives

For the bonding process, the adhesive is applied, the adherends are joined and the adhesive layer is exposed to radiation. Therefore, at least one of the adherends has to be transparent to UV radiation. The thick layers typically employed for sealants cannot be cured in this way.

2.3.1.4 General structure/property relationships

Polymeric acrylates are amorphous substances for which the glass transition temperature can be controlled in a wide range via the composition of the monomers. According to equation (I),

$$T_{g\,(polymer)} = \left[\frac{1}{\left[\frac{c_{(1)}[\%]}{100 \cdot T_{g(1)}[K]} + \frac{c_{(2)}[\%]}{100 \cdot T_{g(2)}[K]} + \cdots \frac{c_{(n)}[\%]}{100 \cdot T_{g(n)}[K]} \right]} \right] - 273.16\,K$$

where

$T_{g\,(polymer)}$: Glass transition temperature of the Polymer

$c_{(n)}$: Concentration by weight of the monomers

$T_{g(1)n}$: Glass transition temperature of the monomers

the glass transition temperature of the polymer can be calculated from its quantitative monomer content. For adhesives in general, the glass transition temperature should be outside the temperature application range of the adhesive bond. As adhesives with a glass transition temperature well below room temperature show a very good elongation capacity but just a low strength level in the usual temperature application range, adhesives with a high glass transition temperature are required for high-strength bonds (see Table II-2.48 and Chapter II-1.1.4.4, Table II-1.13 and Table II-1.14).

The most important monomer for reactive acrylate adhesives is methyl methacrylate. The corresponding polymer has a glass transition temperature of 105 °C. Adhesives based on polymethyl methacrylate (PMMA) have the high modulus desired for high strength adhesives but are hard and brittle. If they are mainly linear polymers, they are thermoplastic, meltable (usually under decomposition) and soluble in polar solvents. The chemical and thermal resistance required by structural adhesives applications can be improved by controlled crosslinking with higher-functional acrylates (dimethacrylates; see Figure II-2.97). Due to thermal depolymerization, however, their heat resistance is limited typically to 170 –200 °C, whereby the exact temperature depends on the respective formulation.

Chemically, polymeric acrylates are polyolefins with pendant ester groups. Esters are relatively stable to chemical and hydrolytic influences but may be hydrolysed under vigorous conditions to the corresponding acid. As the ester group is located in the side-chain and is not part of the polymer chain, hydrolytic degradation, of the polymers is hardly ever encountered.

To improve the mechanical properties, it is common to add polymers to the adhesive formulation which are soluble in the low-molecular monomeric components, but which have only limited solubility in the corresponding polymeric acrylates. For this reason, phase separation occurs during cure. If the added polymers have a low glass transition temperature, they can improve toughness, low-temperature properties and peel strength, as already discussed for epoxy adhesives in Chapter II-2.2.2.

2.3.2 2-component acrylic adhesives (MMA adhesives)

These adhesives are often referred to as methyl methacrylate (MMA) adhesives, as methyl methacrylate makes up the majority of the reactive monomers besides other acrylate or methacrylate monomers.

2.3.2.1 Chemistry of cure

This group of reactive acrylic adhesives are supplied as already mentioned above in the form of two separately packaged components. One component contains the mixture of acrylate and methacrylate monomers and the activator/initiator. Depending on the desired properties, they may also contain tougheners and fillers. The main ingredient of the second component is the curing agent, typically a peroxide. In order to achieve an easy to handle mix ratio it is also possible to add besides certain fillers monomers together with certain stabilizers to improve shelf life.

Figure II-2.105 shows the curing reaction using the example of methyl methacrylate as the sole monomer, cumene hydroperoxide as curing agent and a dihydrpyridine derivative as initiator or accelerator. As soon as the two components are mixed the amine activator reacts with the peroxides to release reactive free radicals (1). These free radicals then interact with the C=C double

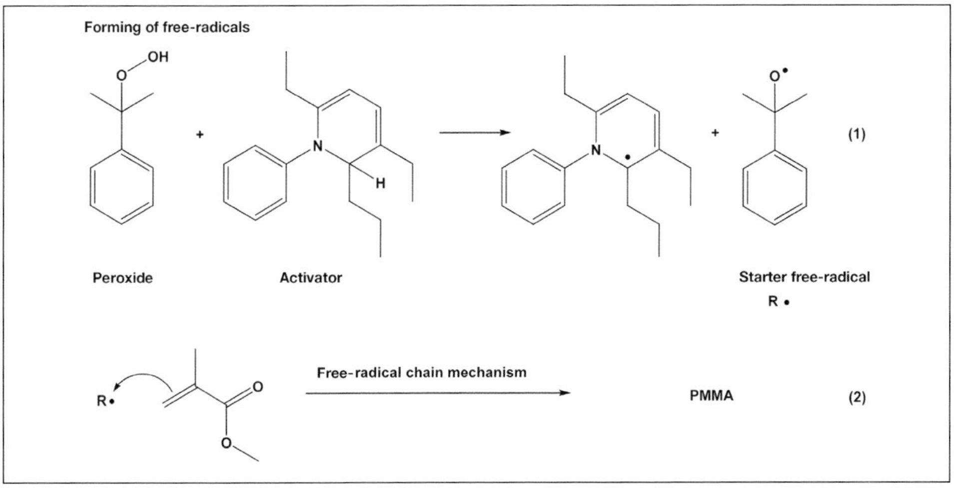

Figure II-2.105: Curing reaction between methyl methacrylate, cumene hydroperoxide and a dihydrpyridine derivative

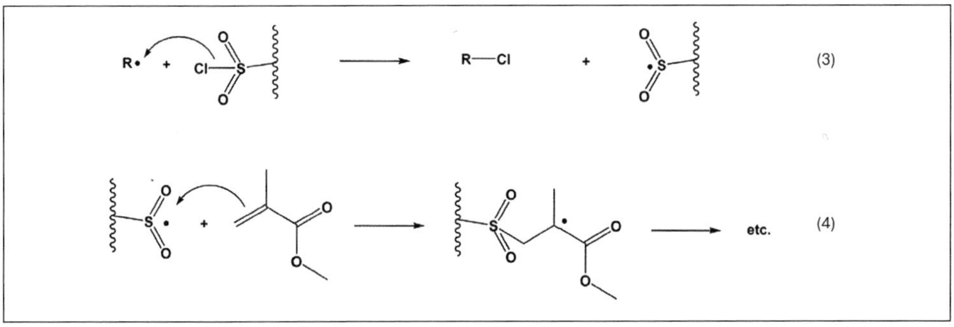

Figure II-2.106: Parallel graft copolymerization between chlorosulphonated polyethylene and methylmetharylate monomer

bonds in the (meth)-acrylate monomers, resulting in a rapid, exothermic polymerization reaction and the curing of the adhesive (2).

In the presence of chlorosulphonated polyethylene as toughener (Figure II-2.106) the starter radical can attack the chlorosulphonyl groups of the elastomer and create free-radicals at the surface (3), which serve as the starting point for graft polymerization of further monomeric methyl methacrylate (4). As a result, the acrylic phase and the elastomer phase are covalently linked to each other.

Since reactive acrylate adhesives are less sensitive to fluctuations in the mixing ratio than 2-component adhesives that cure in a polyaddition or polycondensation process, the cure profile can be adapted to a certain extent to the requirements of the adhesive during formulation. It is therefore possible to formulate the adhesive so that it initially has a slower reaction speed after mixing and thus offers a longer processing time or pot life. Nevertheless, these adhesives achieve the strength sufficient for further handling or processing of the bonded part faster as they show an approximately exponential strength build, which leads in the end compared to polyadditions or polycondensation adhesives to a more favourable ratio between processing and curing time (see Figure II-2.107).

2.3.2.2 Methods of application

Due to the basic difference between curing in a chain reaction by polymerization and curing in a polyaddition or polycondensation reaction, reactive acrylic adhesives are much more tolerant in terms of strict observance of the stoichiometric mixing ratio and quality of mixing of the components compared to polyaddition and polycondensation adhesives. Furthermore, under certain conditions they can also be used without prior mixing of the components, i.e. according to the so-called A/B or the no-mix method.

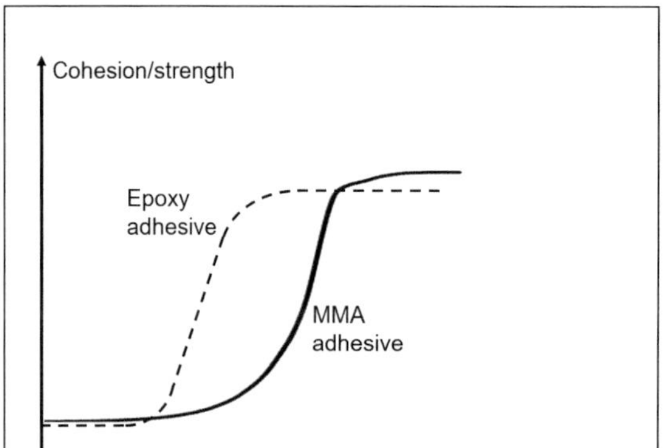

Figure II-2.107: Strength build comparison MMA-adhesive vs. epoxy adhesive

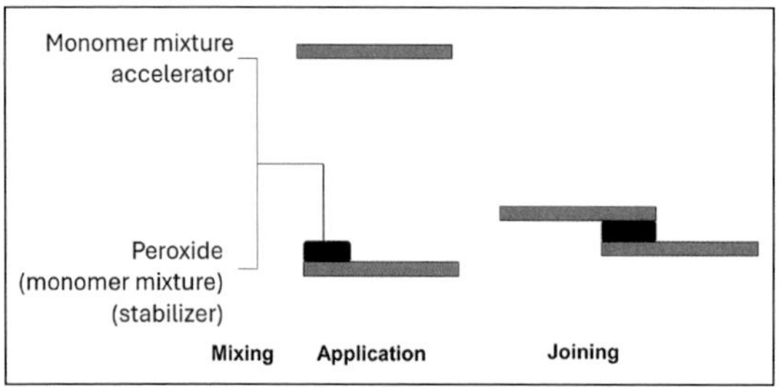

Figure II-2.108: Processing of 2C acrylic adhesives according the standard 2C-process

Application in a standard 2-component process

The two components which are typically in a liquid or paste form are dispensed considering the mixing ratio either from pails or drums using a meter-mix machine or from twin cartridges through a mixer, in most cases a static mixer to one of the adherends (Figure II-2.108). In rare cases or when products where the hardener is sometimes in powder form, the components are weighed manually and also mixed manually.

Early 2-component acrylate adhesives contained the monomer mixture together with the accelerator in one component and the peroxide in the second component. When the two components come together, the peroxide is activated according to the reaction path shown in Figure II-2.105 finally starting the chain reaction of the monomers. Such adhesives often had an unfortunate mixing ratio, i.e. only very small amounts of peroxide were required. Current systems usually have a volumetric mixing ratio of 10 : 1 or 1 : 1 which is achieved by either using more stable peroxides and/or adding stabilizers to the peroxide-component which allows to add monomers also to the peroxide component without reducing the shelf life.

Due to the limited processing time or pot life, the time available for applying the adhesive, joining and clamping the adherends is limited and needs to be observed. The mixed adhesive cures uniformly, even in thick layers, and gap-filling properties and sag resistance can be achieved by adding fillers.

Application in an A/B-process

For thin adhesive joints, typically below 0,7 mm, the use of 2-componet acrylate adhesives specially formulated for processing in the so-called A/B process (see Figure II-2.109) is possible. The advantage is that a pot life does not have to be considered and therefore handling strength can be reached very fast and thus short cycle times can be realized. Typically, these systems develop handling strength in about two minutes and full strength in 4 to 24 hours.

Here too, the two components are packaged separately. Mixing before applying the adhesive is not necessary; instead, one adhesive component is applied to the first adherend and the second component to the other one. During joining, the two components come into contact with each other, and the polymerization reaction starts because both components intermingle along the contact surfaces. Full cure will only be achieved with thin adhesive layers and when the adhesive components are sufficiently low in viscosity. Therefore fillers can only be added to a limited extent and the adhesives offer just a limited gap filling capability and sag resistance.

Even if acrylate adhesives are relatively tolerant with regard to maintaining the exact mixing ratio, the dosing of the adhesive components in the A/B-process presents a certain challenge. In order to avoid a difficult to achieve mix ratio it is highly recommended to, so to say dilute the activator by adding monomers. A significant excess of activator may prevent the adhesive from forming an intimate bond with the substrate.

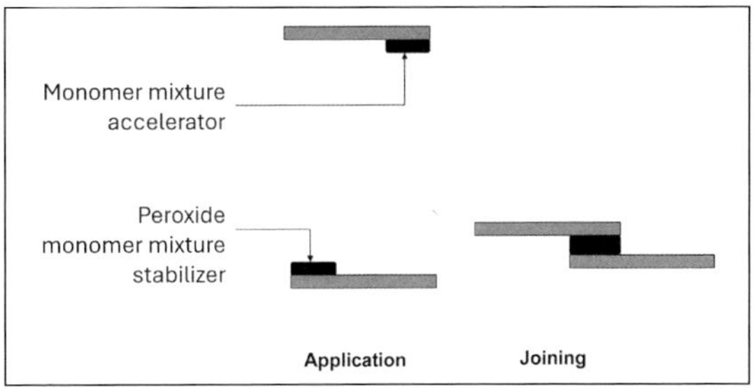

Figure II-2.109: Processing of 2-component acrylic adhesives in the A/B-process

Application according to the no-mix process

In this process, the accelerator, which is dissolved or dispersed in a solvent, is first applied in a thin layer to one of the adherends. Once the solvent has evaporated, the pre-coated adherends have a practically unlimited shelf life, but

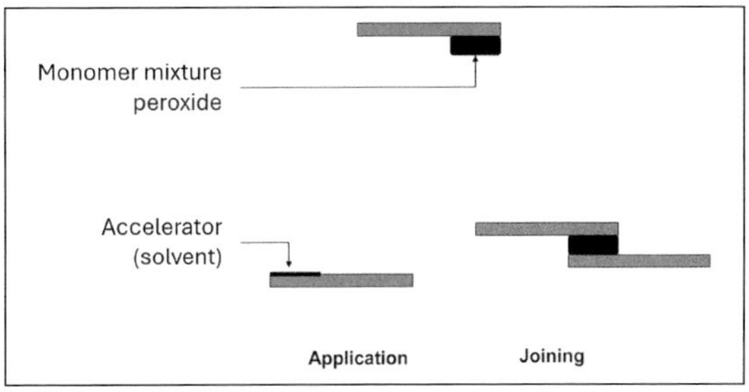

Figure II-2.110: Processing of 2-component acrylic adhesives according to the no-mix-process

are already prepared for subsequent bonding. In addition to the monomer mixture, the second component of the "no-mix adhesive" contains the peroxide required for polymerization. The required storage stability is achieved by using appropriate types of peroxide, e.g. cumene hydroperoxide, possibly in conjunction with suitable stabilizers. During the actual bonding process, this second component is applied to the second adherend and, after joining, contact between the peroxide and the amine released from the activator initially results in the formation of radicals, which ultimately initiate polymerization (Figure II-2.110). Consequently, during the bonding process, there is no need for mixing and metering, because at this point the adhesive can be applied like a 1-component system. As polymerization proceeds from the interface of the substrate treated with the activator, there are certain limitations regarding the thickness of the gap (typically maximum 0.5 mm).

Comparison of the different application methods

The most important advantage of the adhesives that can be processed using the A/B or no-mix method compared to standard 2-component adhesives is that they combine a very long open time, whereas the no-mix systems even have a practically infinite open time, with short to very short curing times. However, the field of application of both the A/B and no-mix systems is limited by their system-related restriction to thin adhesive joints, typically < 0.5 to 0.7 mm, and their low viscosity. Even if the requirement for the accuracy of the mixing ratio for polymerization adhesives is not as high as for polyaddition or polycondensation adhesives, a certain range which is not always easy to maintain is required for those used in the A/B-process.

As a result, and certainly also due to the advances in the application technology of 2-component adhesives the MMA adhesives used in the standard 2-component process have pushed the A/B- and no-mix-systems into the background, both for low-volume applications from cartridges as well as for high volume applications from drums, kegs or pails. The greater formulation options are also an important factor.

Fields of application

In addition to a broad adhesion spectrum and a certain tolerance to slight contamination of the surfaces of the adherends, 2-component acrylic adhesives offer a good combination of strength, toughness, good resistance to environmental factors and fatigue as well as an easy processing, so that they have found a wide range of applications in many industries, mainly in the field of structural bonding. They are capable to bond metals as well as both thermoset and most thermoplastics. For example, they are used in the manufacture of traffic signs, in boatbuilding for bonding the deck to the and hull as well as for bonding bulkheads, replacing time-consuming and material-intensive laminating. In automotive industry, 2-component MMA adhesives are used to bond exterior plastic parts like spoilers, hoods and lids.

The disadvantages of the early MMA adhesives, brittleness, high exotherm and high shrinkage on curing have largely been overcome by appropriate formulation. However, this only applies to a limited extent to the typical odour, which is often perceived as unpleasant. Even the so-called low-odour products have an odour that cannot be neglected.

2.3.2.3 Guide formulation

MMA adhesive to be applied by the common 2C-process or the A/B process

Table II-2.49 shows a formulation for a 2-component MMA adhesive suitable for use in the A/B process as well as for the standard 2-component process with mixing of the components before application to adherends. The mixing ratio is 1 : 1 the cured adhesive has a lap shear strength on Xenoy 1102 (PBT/PC-copolymer) of about 10 MPa, a tensile strength of about 40 MPa and elongation at break of about 10 %.

By substituting part of the MMA by other methacrylate, acrylate or monomers properties can be tailored. Using high boiling monomers is e.g. helpful to reduce the smell of the uncured adhesive. The addition of free radical polymerizable ethylenically unsaturated mono or poly carboxylic acids like acrylic or methacrylic acid often improvers adhesion. [20]

Suitable elastomers to reduce the brittleness of the cured adhesive are besides chlorosulonated polyethylene (Figure II-2.101) especially block copolymers of styrene and butadiene or isoprene but also other elastomers having a low glass transition temperature and are soluble in the monomer mix can be used.

The core-shell graft copolymers useful as impact modifier are characterized by a "rubbery" core and a "hard" shell. They swell but do not dissolve in the monomer compositions. The "core" or backbone polymer of the graft copolymers has a glass transition temperature substantially below ambient temperatures. The "shell' polymer which is grafted onto the backbone polymer has a glass transition temperature substantially above ambient temperature. Examples of useful core-shell graft copolymers are those where "hard' monomers, such as styrene, acrylonitrile or methylmethacrylate, are grafted on to a rubbery core made from polymers of "soft" or "elastomeric" monomers, such as butadiene or ethyl acrylate.

The curing agent are free radical generators. Typically peroxides, hydroperoxides, perester or peracids (Figure II-2.102) are used in the amount of 0,1 – 3 % by weight of the adhesive composition. In order to enhance the reactivity tertiary amines and imines (Figure II-2.104) are added as activator.

The addition of fillers such as chalk or talc and rheological additives such as fumed silica can improve the gap filling capacity and sag resistance, requiring processing in the standard process.

Table II-2.49: Reactive acrylic adhesive, standard 2-component-process or A/B-process [20]

Resin component

Item	**Raw material**	**Parts by weight**
1	Methyl methacrylate monomer	54.40
2	Toughening elastomer	23.30
3	Core-shell impact modifier (I)	12.05
4	Adhesion promoter	9.70
5	Curing agent	0.30
6	Stabilizer	0.25
Total		100

Activator

7	Activator	7.55
8	Methyl methacrylate monomer	74.65
9	More-shell impact modifier (II)	8.20
10	Resin	9.60
Total		100

Item 1: MMA, Evonik
Item 2: Chlorosulphonated polyethylene, „Hypalon 30", DuPont
Item 3: "Paraloid BTA-751U", Dow Chemical
Item 4: MAA methacrylic acid, Evonik
Item 5: Cumene hydroperioxide
Item 6: 2,6, Di-t-butyl-4-methylphenol
Item 7: Aniline butyraldehyde, "Vanax 808 HP", Vanderbilt Chemicals
Item 8: MMA, Evonik
Item 9: "Paraloid KM 330", Dow Chemical
Item 10: Acrylic-styrene-acrylonitrile (ASA) resin, "Geloy 1020", Sabic

No-mix acrylic adhesive

Table II-2.50 provides a starting point formulation for a reactive, no-mix acrylic adhesive [11]. Besides the usual ingredients, the formulation contains a small quantity of acrylic acid and a crosslinker for improving heat resistance. The relatively high amount of elastomer confers good impact resistance.

Table II-2.50: Reactive acrylic adhesive, no-mix process [11]

Resin component

Item	Raw material	Parts by weight
1	Methyl methacrylate	44
2	Methacrylic acid	7.5
3	Crosslinker, di-methacrylate	1
4	Curing agent	0.5
5	Toughening polymer	47
Activator		
6	Imine	
7	Solvent	

Item 1: MMA, Evonik
Item 2: MAA methacrylic acid, Evonik
Item 3: ethyleneglycol dimethacrylate, "SR 206", Sartomer
Item 4: cumenehydroperoxide, "Luperox CU 90", Atofina
Item 5: chlorosulfonated polyethylene, "Hypalon 20", DuPont
Item 6: imine from butanal and aniline, "Vanax 808", Vanderbilt Chemicals
Item 7: ethyl acetate

For production of MMA adhesives, the liquid raw materials are usually mixed in a first step and then the solid components are added step by step. Alternatively, individual masterbatches can first be made from the monomer mixture or one monomer and the soluble or swellable raw materials, which are then mixed accordingly before the remaining ingredients are added. The product is generally degassed after mixing and before packaging.

2.3.3 Anaerobic acrylic adhesives

2.3.3.1 Basic principles

The term anaerobic, derived from the ability of organisms to live without oxygen, has found its way into adhesives technology, too. Anaerobic adhesives ("anaerobics") are 1-component reactive adhesives which require the absence of oxygen (oxygen acts as an inhibitor) to cure and can therefore be stored in the presence of oxygen. When oxygen is excluded, i.e. when the (oxygen-impermeable) adherends are joined, the curing reaction is initiated by metal ions on the surface of the adherend (Figure II-2.111).

2.3.3.2 Formulation

Anaerobic adhesives are formulated by simply mixing the components. These consist of the reactive monomer system (mainly dimethyl acrylates, diacrylates and acrylates), the curing agent (peroxide), an accelerator (amines and amides) as well as stabilisers (complexing agents, phenols). Toughening agents, fillers, plasticisers and rheological additives are sometimes used as well.

The properties of the cured anaerobic adhesive are essentially triggered by the monomers used. Commonly highly reactive dimethacrylic substances, like

- methacrylate half esters obtained by the reaction of 2-hydroxyethyl methacrylate and acid anhydrides,
- polyurethane acrylates
- dimethacrylates of polyalkoxy polyols obtained by esterification of the corresponding acids or transesterification of standard methacrylates with diols such as tri- and tetraethylene glycol
- urethane dimethacrylates obtained by a reaction between hydroxymethacrylates and diisocyanates, such as TDI, as shown in Figure II-2.112

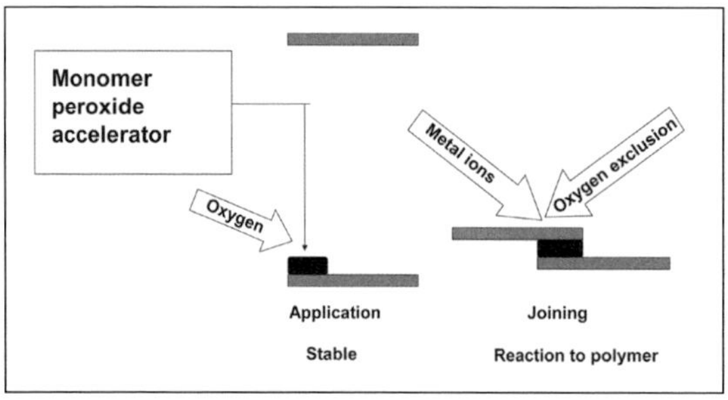

Figure II-2.111: Schematic illustration of the use of anaerobic adhesives

offering a certain elasticity of the cured adhesive depending on the length of the chain between the two reactive acrylic functions
- dicyclopentenyl methacrylate, characterised by a relatively low inherent odour
- triallylcyanurates, in combination with conventional methacrylic monomers which enable products with a high temperature and water resistance

are used. A comprehensive list of various monomers that have proven to be suitable for the formulation of anaerobic adhesives can be found, for example, in [21].

The purpose of the **curing agent or initiators** is to supply the free-radicals needed for polymerization. This are again peroxides, such as cumene hydroperoxide, di-tert-butyl peroxide and potassium peroxodisulphate (Figure II-2.102). The peroxide and activator together constitute the initiator system.

Activator catalyse the decomposition of the peroxides in a controlled manner. Important activator are tertiary amines, such as triethylamine and dialkylarylamines, such as N,N-dimethyl p-toluidine (see Figure II-2.104), often in combination with a co-activator based on amides and sulphonamides, such as saccharin (Figure II-2.113).

Since anaerobic adhesives require access to oxygen during storage to prevent premature reaction and cure of the adhesive, they are usually supplied in just partly filled thin-wall LDPE bottles. This permits oxygen (acting as an inhibitor) to gain access to the monomers. As the amount of dissolved oxygen is not necessarily sufficient to prevent premature curing, stabilizer are added. A distinction is made between stabilisers and inhibitors in terms of their mode of action. The term stabilizer refers to substances that block chemical substances that can initiate the release of free radicals. For example, some chelating agents, such as EDTA are able to mask traces of metal ions contained in many monomers through the formation of chelates, preventing premature decomposition of the peroxides and polymerization of the monomers. Inhibitors like oxygen are substances that stop the polymerization once it has started.

For stability reasons, anaerobic adhesives are usually formulated without **fillers**. In special instances, organic fillers may be used, such as polyethylene powder.

Plasticisers, such as polyethylene oxide octanoate, can be used to adjust viscosity.

Thickeners used are polymers, such as polyalkyl acrylates and polystyrene, which are dissolved in the monomer system. Rheological additives, such as fumed silica (up to approx. 5 %), can be used in the standard way to regulate flow properties. Monofunctional acrylate monomers can serve as reactive diluent to lower the viscosity.

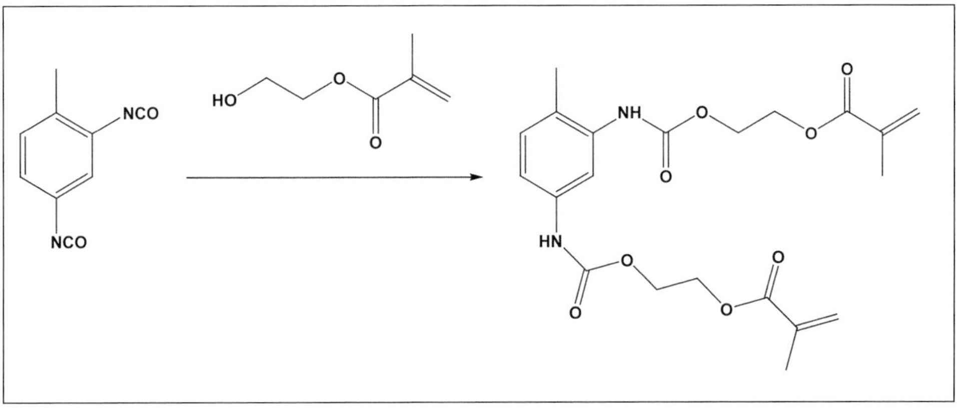

Figure II-2.112: Synthesis route for an urethane dimetharylate

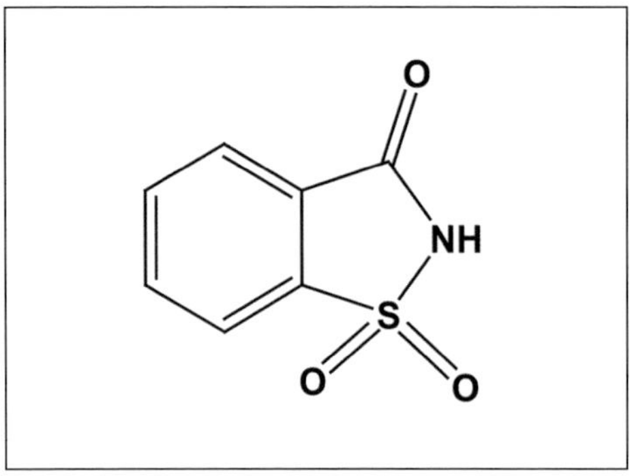

Figure II-2.113: Saccharin

To achieve a high sealing effect, temperature and chemical resistance, fluorinated organic substances such as PTFE can be added, as well as certain inorganic fillers with a lubricating effect, such as graphite or molybdenum disulphide. It should be pointed out here that perfluorinated and polyfluorinated alkyls (PFAS), which include PTFE, are very difficult or impossible to remove once they have been released into the environment. As some substances from this group have toxic properties, far-reaching restrictions on their use are currently under

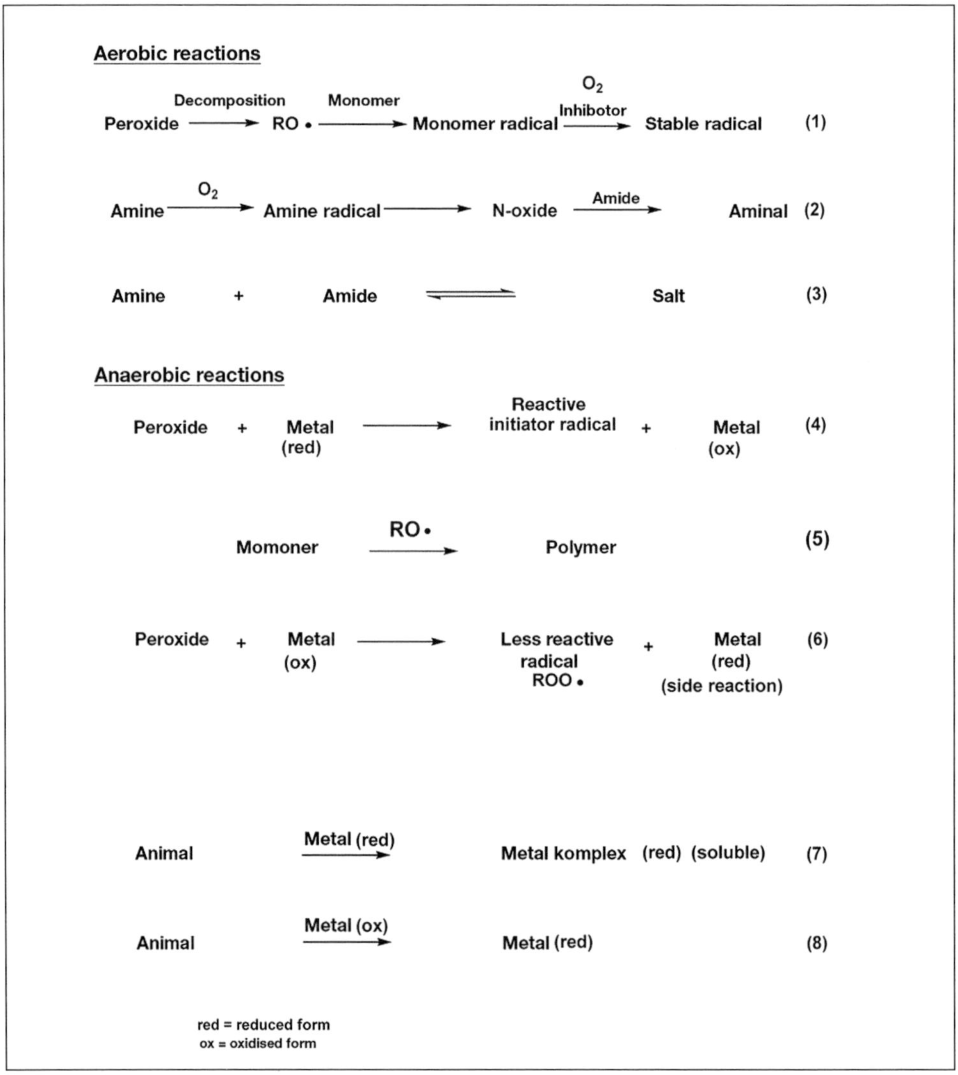

Figure II-2.114: Overview of the reaction steps involved in the curing of anaerobic adhesives

discussion. The hazard of PTFE in adhesives for humans and the environment is less to be seen during use, as it is bound in long polymer chains in the cured adhesive matrix. Problems are more likely to arise during subsequent thermal disposal of the adhesive and during disassembly or reassembly of the screw connection if PTFE is either thermally degraded in certain temperature ranges to form critical short-chain PFAS decomposition products or mechanically degraded at room temperature into relevant microparticles.

For special applications, like the use in the manufacture of self-locking screws and nuts or for the use as a flange sealant, the monomer system, the curing agent or the accelerator is microencapsulated so that the adhesive can be applied to the thread during screw manufacture or to the flat metal surface of a flange. The system only cures when the screw is screwed in or a load is applied by tightening the flange, which destroys the microcapsules initiating the formation of free radicals from the peroxide. To prevent premature curing, the microcapsules must not only have adequate mechanical stability, but also sufficient chemical stability against the various components of the formulation. Hard polymers or resins, especially those based on urea/formaldehyde resins, have proven their worth.

2.3.3.3 Chemistry of curing

The chemical processes that take place during curing are quite complex and differ depending on the initiator and redox system used. The N,N-dimethyl-p-toluidine/saccharin/cumene hydroperoxide system has been studied in detail [13] and will be used here for an overview of the individual reaction steps (Figure II-2.114).

The peroxide in the adhesive decomposes slowly, producing free radicals. However, under aerobic conditions, their concentration remains low due to the overriding influence of the inhibiting oxygen and the added inhibitor. The free radicals formed in (1) are relatively stable and not reactive enough to trigger radical chain polymerization (Figure II-2.115).

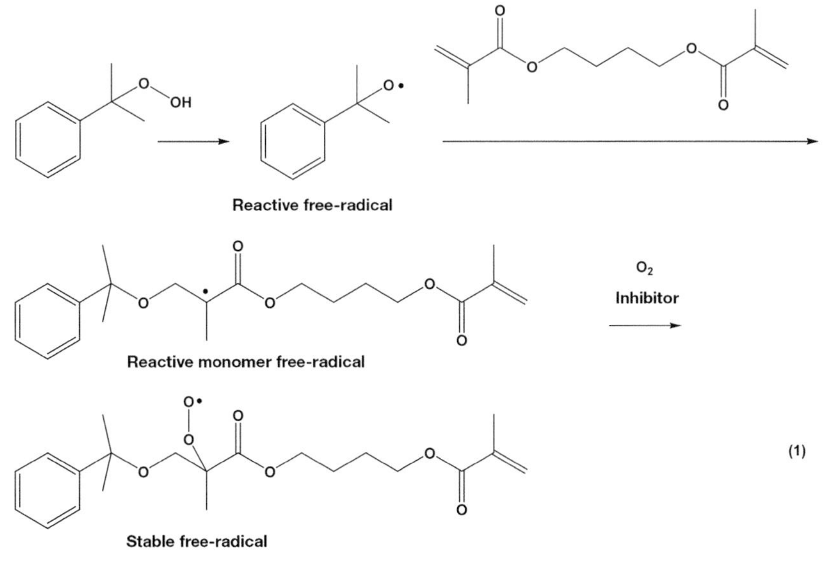

Figure II-2.115: Mode of action for oxygen as an inhibitor (step (1) in Figure II-2.114)

Under the influence of oxygen, the amine reacts with the saccharin (amide or sulphonamide) by a free-radical mechanism and via a nitrogen oxide to form an aminal (2), which dictates the further course of the reaction (Figure II-2.116).

Simultaneously, the concentration of the free amine is reduced by salt formation (3) with the acidic saccharin (pk_a saccharin = 2). Since amine-catalysed decomposition of the peroxide plays a secondary role as a result, the adhesive's storage stability improves (Figure II-2.117).

When the adhesive is applied during the bonding process to one of the adherends (e.g. copper) and the second adherend is mated, anaerobic conditions are generated. Simultaneously, traces of metal ions in the form of the complex salts are transported into the adhesive layer. The reduced form of the transition metal catalyses the decomposition of the peroxide (4). Reactive free-radicals are formed and, as there is no longer any inhibiting oxygen present, they can initiate free-radical polymerization (Figure II-2.118).

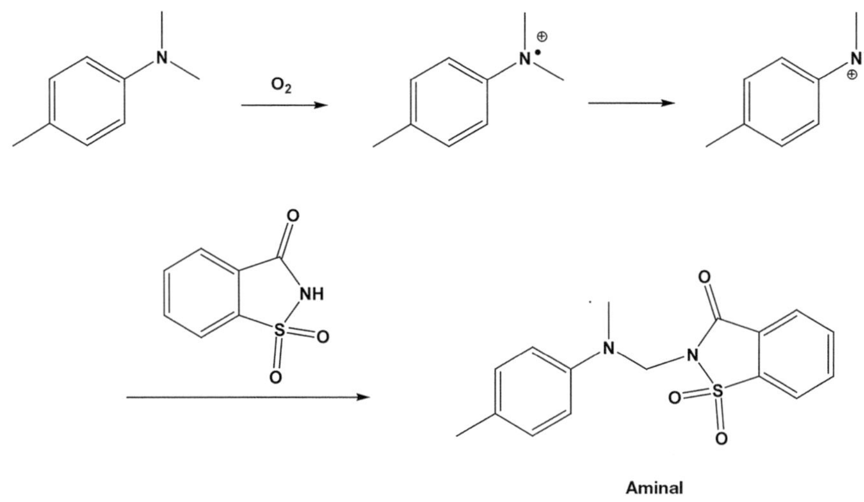

(2)

Figure II-2.116: Aminal formation (step (2) in Figure II-2.114)

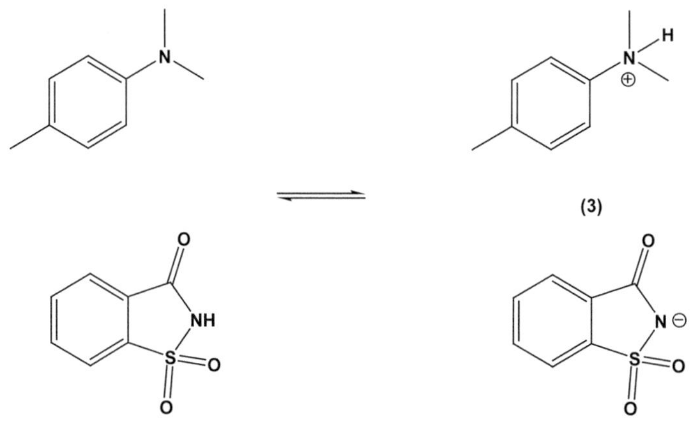

(3)

Figure II-2.117: Salt formation (step (3) in Figure II-2.114)

Many transition metals are particularly effective, because they can easily change their oxidation number. The oxidised form of the metal can initiate decomposition of the peroxide, too (6). However, the free-radicals formed in this way have low reactivity and so the reaction is quite slow and of secondary importance (Figure II-2.119).

The aminal formed under aerobic conditions has a decisive role in the further course of the reaction. First, it acts as a complexing agent (7), and can support dissolution of the catalytically active metal ions. Second, it is a reducing agent (8) and can transform the oxidised form of the metal ions back into the reduced form, which is important for the decomposition of the peroxides (Figure II-2.120).

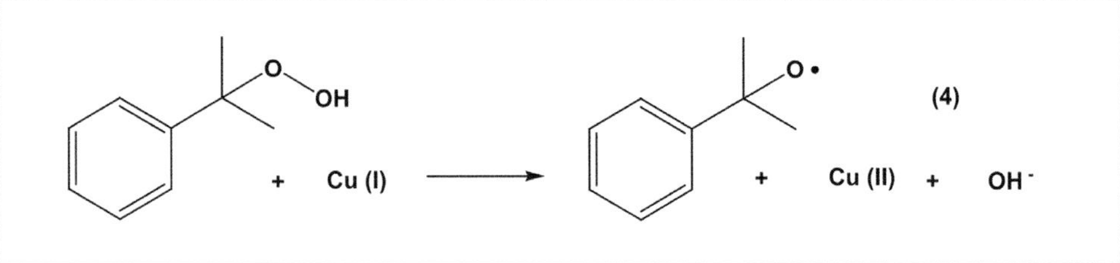

Figure II-2.118: Metal ion catalysed formation of free radicals from the initiator peroxide (step (4) in Figure II-2.114)

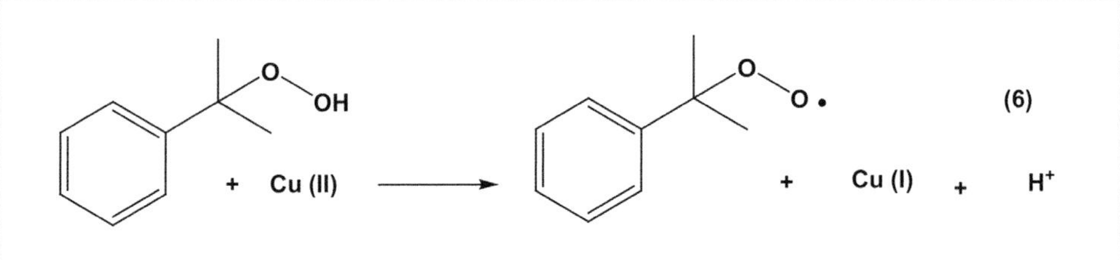

Figure II-2.119: Formation of less reactive free radicals induced by oxidized metal ions (step (6) in Figure II-2.114)

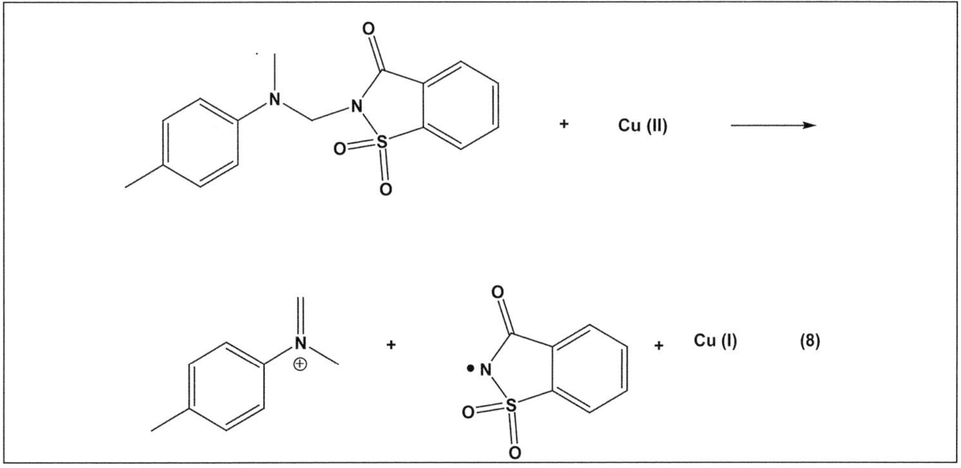

Figure II-2.120: Reduction of the metal ion by the aminal formed under aerobic conditions (step (8) in Figure II-2.114)

Table II-2.51: Advantages and disadvantages of anaerobic adhesives

Advantages	Disadvantages
– 1-component system – Rapid cure (5 to 10 min to reach initial strength) – High final bond strength – Good hydrolytic stability – Good chemical resistance – Good heat resistance	– application limited to certain metal substrates – only suitable for thin gaps (< approx. 0.1 mm) – concerns over toxicology of acrylates

Table II-2.52: Anaerobic acrylic adhesive [22]

Item	Raw material	Parts by weight
1	Dimethacrylate monomer	97.6
2	Hardener	0.5
3	Accelerator	0.8
4	Co-accelerator	1.1

Item 1: diethyleneglycol dimethacrylate, "DEGDMA", Evonik, Carbone Scientific Co., Ltd
Item 2: cumenehydroperoxide, "Luperox CU 90", Arkema
Item 3 N,N-dimethyl-p-toluidine, Merck
Item 4: saccharine, Hommel GmbH

2.3.3.4 Evaluation

Anaerobic acrylic adhesives are easy to use, are rapid-curing and possess properties which can be tailored to the application. They yield durable bonds of high mechanical strength, and therefore are classified as structural adhesives.

As curing requires both the absence of oxygen and contact with metal ions, their application is limited to non-porous active metallic surfaces. Metals such as copper or iron lead to rapid curing whereas curing on less active metals such as zinc or tin is slow. Special primers containing metal salts soluble in solvents, e.g. copper naphthenate, can be used to accelerate the cure reaction on these materials or to enable the cure on plastics. The general pros and cons are summarised with Table II-2.51:

2.3.3.5 Fields of application

Anaerobic adhesives are mainly used for metal bonding [12]. The adherends typically provide the catalytic metal ions required for the cure reaction, and the absence of oxygen, another condition for the curing reaction, is guaranteed. Applications are mainly found in mechanical engineering. These include screw locking (bonding of screw connections to prevent unintentional loosening due to vibrations), sealing of flanges and creating shaft-hub connections.

2.3.3.6 Guide formulation

Table II-2.52 presents a starting formulation for an anaerobic acrylic adhesive [22]. Adding small amounts (approx. 200 ppm) of stabiliser like EDTA can further extend the shelf life [14].

Table II-2.53 shows the formulation for a toughened anaerobic adhesive.

The solution of a compound acting as a source for copper ions shown in Table II-2.54, can serve as an activator for anaerobic adhesives and thus enable their use even on less active metal surfaces or plastics.

2.3.4 Radiation-curing acrylic adhesives

2.3.4.1 Basic principles

The free-radical polymerization of acrylic or methacrylic substances can also be initiated by high-energy radiation. Radiation curing was initially used in the paint and varnish industry, but is now also state of the art in the bonding sector and has proven itself in many applications despite some limitations. The prerequisite is that at least one of the adherends must be sufficiently transparent for the radiation used and it goes without saying that the process is essentially limited to unfilled adhesives. Even with transparent adhesives, thick adhesive layers are difficult to realise. Both visible light (VIS) UV light (wavelengths from 200 to 500 nm) and electron beams can be used as radiation –sources, whereby UV light with a wavelength in the range of 315 – 380 nm has become the standard for many applications. Due to the insufficient permeability of some plastics for UV light, light from the visible range (400 – 550 nm) is used instead. Electron beam curing has not been able to establish itself due to the high equipment costs and the extensive occupational safety precautions required and is only used in exceptional cases. [2, 3, 24].

In contrast to the epoxides cured by UV light-induced cationic polymerization discussed in Chapter II-2.2.2.3 chain growth in radical polymerization stops as soon as the radiation source is removed.

Table II-2.53: Toughened anaerobic acrylic adhesive [23]

Item	Raw material	Parts by weight
1	Methacrylic monomer (I)	55.0
2	Methacrylic monomer (II)	10.0
3	Functionalised toughening agent	10
4	Toughening agent, non functional	20
5	Stabilizer, inhibitor of free radical generation (I)	0.2
6	Stabilizer, inhibitor of free radical generation (II)	0.8
7	Accelerator	1.0
8	Co-accelerator	1.0
9	Free radical inhibitor	2

Item 1: methyl methacrylate monomer (MMA), Evonik
Item 2: methactylic acid, Evonik
Item 3: methacryl functionalised polybutadiene, "Hypro 2000x168 LC", Huntsman Corporation
Item 4: styrene butadiene copolymer, "Kraton D1155E", Kraton Polymers Nederland B.V.
Item 5: naphtoquinone 5 % in polyethylene glycol methacrylate (PEGMA),
Item 6: metal salt of ethylene diamine tetra-acetic acid (EDTA), 3.5 % in PEG
Item 7: 1-acetyl-2-phenylhydrazine (APH), TCI Deutschland GmbH
Item 8: saccharin, Chemische Werke Hommel GmbH
Item 9: cumene hydroperoxide, "Luperox CU 90", Arkema

Table II-2.54: Activator solution for anaerobic acrylic adhesive

Item	Raw material	Parts by weight
1	Solvent	98
2	Metal ion source	2

Item 1: propan-2-ol
Item 2: copper(II)-neodecanoate, Leap Chem Co. Ltd, Hongkong, Hubei Xingyan New Material Technology Co., Ltd

2.3.4.2 Formulation

Radiation-curing acrylate adhesives are essentially formulated in the same way as the reactive acrylate adhesives already mentioned. Depending on their structure, the reactive oligomers, e.g. urethane acrylates, provide mechanical strength and elasticity, while the monomeric acrylates or acrylic acid improve the specific adhesive properties. Additives may be added to adjust the rheology or to improve adhesion to certain substrates. However, care must be taken to ensure that the transmittance of the radiation intended for curing is not reduced unduly.

The additional photoinitiators contained in UV-curing adhesives are essential for the curing reaction. They are able to absorb electromagnetic radiation of a suitable wavelength, whereby they are split into free radicals, which in turn initiate radical polymerization. Figure II-2.121 shows selected photoinitiators for UV-curing adhesives. Electron beam curing adhesives do not require the addition of photointiators, as the high-energy electrons themselves are capable of triggering a radical reaction.

The range of available photoinitiators allows to customise UV-curing adhesives for materials that are only permeable to visible light or only to UV light. The selection of the photoinitiator depends primarily on the UV source intended for curing and in particular on its emission spectrum, its power, the distance to the adhesive and the possible irradiation duration. The technology of the radiation source must also be taken into account. In addition to the LED lamps mainly used today, mercury arc lamps are still in use.

It is possible to combine UV curing with a second curing mechanism so that initial strength is achieved by UV irradiation and final curing, including in potential shadow areas, takes place either thermally, by humidity or anaerobically.

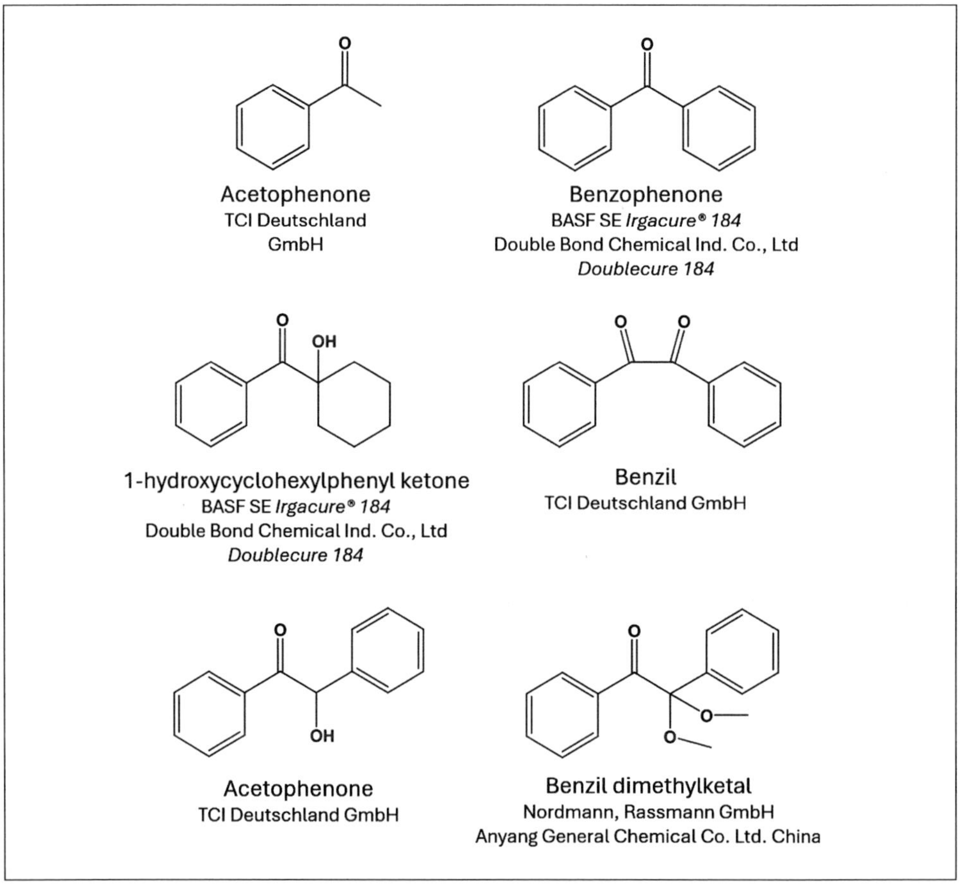

Figure II-2.121: Typical used photointiators for UV-curable acrylic adhesives
Note: The reference sources and trade names given are examples

2.3.4.3 Chemistry behind curing

During UV curing, free radicals are initially generated from the photointiators under the influence of UV radiation. A distinction is made between 2 reaction mechanisms, the Norish reactions type 1 and type 2. The type 1 Norrish reaction is characterized by a homolytic cleavage of the C-C bond of the photoinitiator adjacent to the carbonyl group into two radical fragments (Figure II-2.122). The free radicals then start the polymerization of the monomers or recombine to form the photoinitiator .

For a free-radical formation according to type 2 Norrish-reaction, a hydrogen donor is required in addition to the photoinitiator. In most cases, these are amines (amine synergists). By UV irradiation, the type II photoinitiator removes a hydrogen atom from the amine synergist and forms two radicals (Figure II-2.123). These radicals then initiate the polymerization. [25]

As typical for radical polymerization, oxygen can act as an inhibitor and thus disrupt the curing process on the surface.

2.3.4.4 Evaluation

The main advantage of UV-curing adhesives and sealants over other adhesive technologies is their extremely fast curing. Curing times play a key role in processes that require short cycle times or a high degree of precision due to higher quantities, such as applications in the medical, optical and electronics sectors. In addition, the adhesives are also suitable for all applications that require high stability while maintaining their flexibility and, due to their high transparency, they offer the possibility of producing invisible bonds, so to speak.

Due to the large number of available mono-, di- or multi-functional monomers, and various oligomers (epoxy, polyester and urethane oligomers), it is possible to adjust the mechanical properties of the cured adhesive and the adhesion spectrum to the respective application which makes UV-curing adhesives also suitable for many applications that require high stability while maintaining their flexibility.

The general pros and cons are summarised inTable II-2.55:

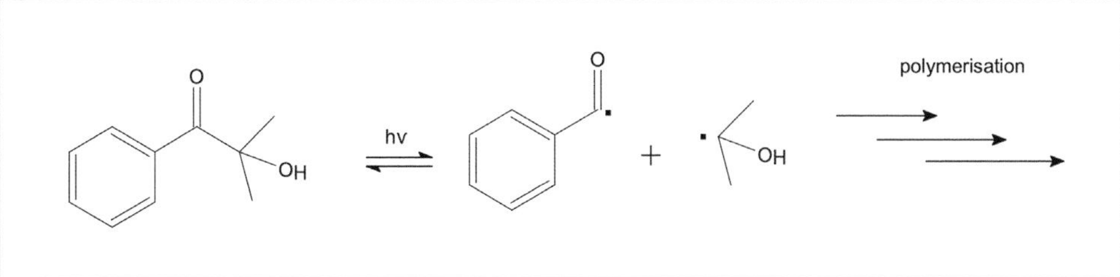

Figure II-2.122: Norrish type 1 free-radical formation [25]

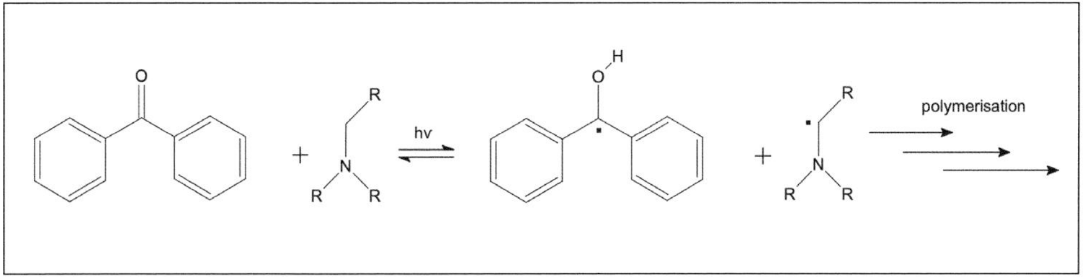

Figure II-2.123: Norrish type 2 free-radical formation [25]

Table II-2.55: General advantages and disadvantages of UV-curing acrylic adhesives

Advantages	Disadvantages
– 1-component system – Rapid cure – Mechanical properties adjustable over a wide range – Good adhesion to many substrates – Good hydrolytic stability – Good chemical resistance – Good heat resistance – Good transparency	– suitable UV light source required – at least one adherend must be sufficiently transparent for the UV light used for curing – limited bond line thickness – packaging and application technique must protect the adhesive from light – shrinkage during curing must be considered

Table II-2.56: General formulation for a UV curing adhesive [12]

Item	Raw material	Parts by weight
1	Urethane acrylate	33
2	Isobornyl acrylate	33
3	Trimethylolpentane triacrylate	25
4	Pigment, stabiliser	5
5	Photoinitiator	4

2.3.4.5 Fields of application

Due to the numerous advantages, it is therefore not surprising that UV-curing acrylic adhesive are used in various industries for a wide range of applications. Examples can be found in the **optical industry**, e.g. in the bonding of lenses, fibre-optic assembly; in **medical technology** in the manufacture of wearables, the bonding of cannulas into the syringe body of disposable syringes, respirators, anaesthesia bags, flexible tubular systems, permanent venous catheters; in **electronics and electrical engineering** with the encapsulation and thus the protection of sensitive components using the Dam & Fill process, the bonding of loudspeakers for mobile phones and the assembly of cover discs in the manufacture of displays and as a so-called cured in place gasket; in the **automotive industry** in the manufacture of camera modules or the bonding of the mirror base to the windscreen; in the **construction sector** for the production of invisible glass bonding, e.g. the bonding of hinges for shower enclosures [19].

2.3.4.6 Guide formulation

Table II-2.56 presents a general formulation for a UV curing adhesive.

2.3.5 Cyanoacrylate adhesives

2.3.5.1 Basic principles

Cyanoacrylate adhesives are a relatively young class of adhesives. Cyanoacrylate monomers were first synthesized in 1942 by the American chemist Harry Coover in connection with the development of optical prisms for weapons. He was experimenting with the transparent acrylate plastics widely used in aircraft construction and replaced a methyl group by a cyano group. The extreme stickiness of the substance, which interfered with processing, initially prevented its industrial use [26]. In the 1950s, he worked with Fred Joyner on the development of heat-resistant coatings for jet cockpits and remembered the cyanoacrylate monomers. It is said that during the routine characterization of a newly synthesized cyanoacrylate monomer when measuring the refractive index, the prisms of the refractometer were stuck together within a few seconds. Coover recognised the importance of cyanoacrylate monomers as the base for instant adhesives. Cyanoacrylate was patented in 1956 and with Eastman 910, the first adhesive on this basis, was introduced into the industrial adhesives market in 1958 [27]. Due to stability problems during production, packaging and transport on the one hand and the low interest of adhesive users on the other, it took until the 1970s for this type of adhesive to gain importance both for industrial use and for the DIY market. Not only was it possible to overcome the stability problems, but products from other manufacturers, in particular Loctite, also came onto the market.

Today, cyanoacrylate monomers form the basis for the very fast-curing adhesives that are often referred to colloquially as 'superglues' or 'instant adhesive' are well established in many industries as well in DIY. They are reactive 1-component systems that cure within seconds through a highly exothermic anionic polymerization mechanism triggered by the water film typically present on the surface of the adherends. Furthermore, detailed information on cyanoacrylate adhesives can be found in [1, 2, 15].

2.3.5.2 Raw materials

Monomers

From a chemical point of view, the cyanoacrylate monomers are esters of cyanoacrylic acid with alkyl chains of different lengths. Synthesis of the reactive monomers usually occur in a two-step Knoevenagel condensation reaction (Figure II-2.124). Although the reaction appears relatively simple, it is difficult to control on an industrial scale due to the high reactivity of the cyanoacrylic acid esters and their highly exothermic polymerization.

In a first step the corresponding cyanoacetic ester is made to react in solution (toluene or acetone) with formaldehyde (paraformaldehyde) in the presence of basic catalysts (e.g. piperidine, NaOH, CaO) to yield an oligomeric product. Once the polycondensation reaction is complete, the basic or nucleophilic substances such as catalysts are carefully removed from the polymer. After

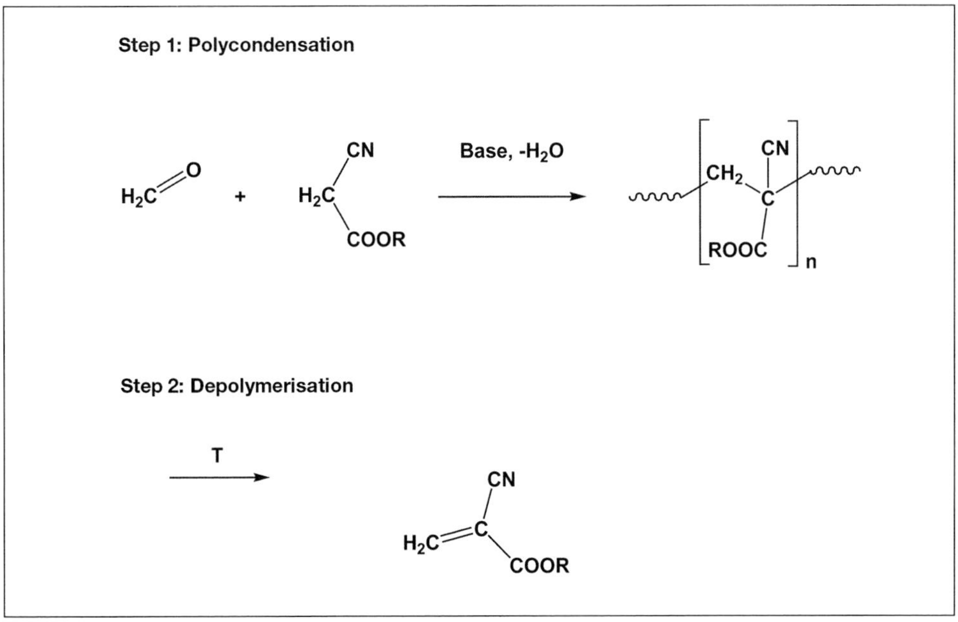

Figure II-2.124: Two-step reaction for the synthesis of cyanoacrylate monomers

drying, the oligomer is thermally depolymerised to the cyanoacrylic ester under high vacuum at 150 – 300 °C. High boiling transfer agents such as tricresyl phosphate are often added to reduce the viscosity of the polymer to improve the heat transfer to allow a reduction in the depolymerization temperature.

During the synthesis, care has to be taken in the second step to avoid contact with water and alkaline products (including the alkaline surface of glass) to avoid polymerization. Glass reactors therefore must be rinsed with acids in advance and a stream of, e.g., sulphur dioxide, passed through during the synthesis. For final cleaning of the depolymerised product, distillation in the presence of stabilisers is performed. The purity and therefore the quality of the monomer is primarily determined by the distillation process, which in turn has a significant influence on the quality of the adhesive and its shelf life.

In addition to alkyl cyanoacrylates (Figure II-2.125), which normally make up the majority of monomers used in cyanoacrylate adhesives, biscyanoacrylates and occasionally 2-vinyl-substituted 1-cyanoacrylates (Figure II-2.126) are also used. However, these cannot be synthesized using the same process as the alkyl cyanoacrylates. During the synthesis of biscyanoacrylates, the C-C double bond reactions are first blocked to prevent the formation of a highly cross-linked thermoset that can no longer be depolymerised. Viny substituted-1-cyanoacrylates are less reactive with regard to anionic polymerization and can therefore be synthesised from acrolein and a cyanoacetate ester under Lewis acid catalysis, e.g. zinc oxide. [1, 2].

The monomers are low-viscosity, colourless liquids that sometimes have a characteristic, pungent odour. Adhesives can be formulated with various esters. The monomers differ mainly as regards to volatility (boiling point), reactivity and influence on final adhesion properties. Table II-2.57 lists the important properties of cyanoacrylate monomers.

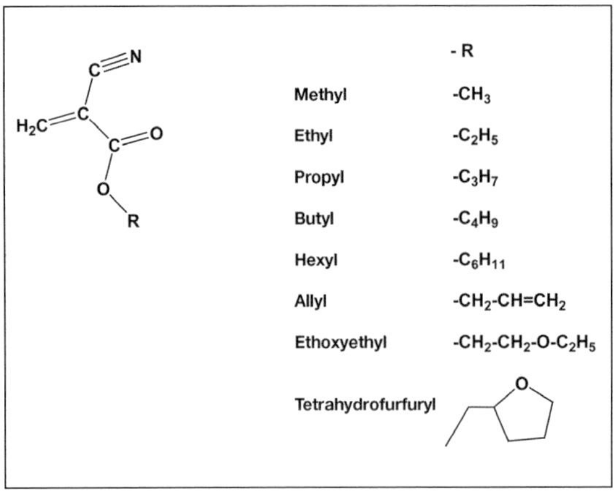

Figure II-2.125: Common cyanoacrylate monomers

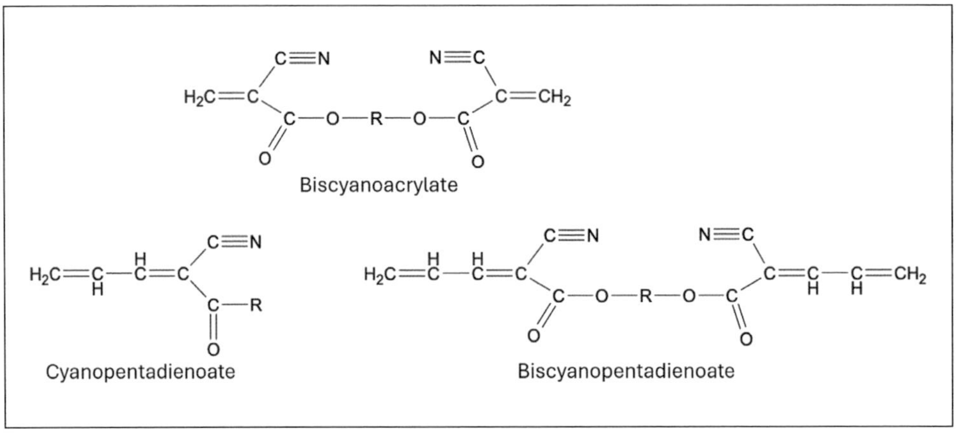

Figure II-2.126: Chemical structure of biscyanoacrylate and vinyl-substituted cyanoacrylates

Table II-2.57: General properties of cyanoacrylate monomers [1]

Monomer	bp [°C] at pressure in [hPa]	Smell	Reactivity
Methylcyano acrylate	48 (3.3)	strong	very high
Ethylcyano acrylate	54 (3.5)	strong	high
n-propylcyano acrylate	80 (8.0)	strong	high
Iso-propylcyano acrylate	53 (2.6)	strong	high
n-Butylcyano acrylate	83 (4.0)	strong	moderate
Iso-butylcyano acrylate	71 (2.5)	strong	moderate
Allylcyano acrylate	74 (5.62)	strong	moderate
Ethoxyethylcyano acrylate	100 (4.0)	slight	moderate
Tetrahydrofurfurylcyano acrylate	108 (2.0)	slight	moderate

2.3.5.3 Chemistry of curing

Cyanoacrylate esters can polymerise easily via various mechanisms. Anionic polymerization as shown in Figure II-2.127 is the most important and most common route for the use as adhesives. Although free-radical polymerization is possible, it is of no practical importance for curing the corresponding adhesives.

The high cure speed of cyanoacrylate adhesives is caused by the strong electrophilic effect of the cyano and ester groups. The resulting polarisation (Figure II-2.128) promotes the attack of nucleophilic substances like hydroxy groups found on the surface of the adherends and stabilises the carbanion formed in the start reaction. The polymerization proceeds via an anionic mechanism. The carbanion attacks another monomer generating a dimeric species, which reacts with more monomers until the available monomers are consumed or until a chain transfer or termination step intervenes.

Even water, a fairly weak nucleophile, is capable of starting this reaction. Most surfaces of substrates are covered by a thin film of water, and so the component for initiating the reaction is already present when the substrates are mated. Alkaline surfaces accelerate polymerization, while acidic surfaces, such as some types of wood, retard polymerization. In this case, the use of an al-

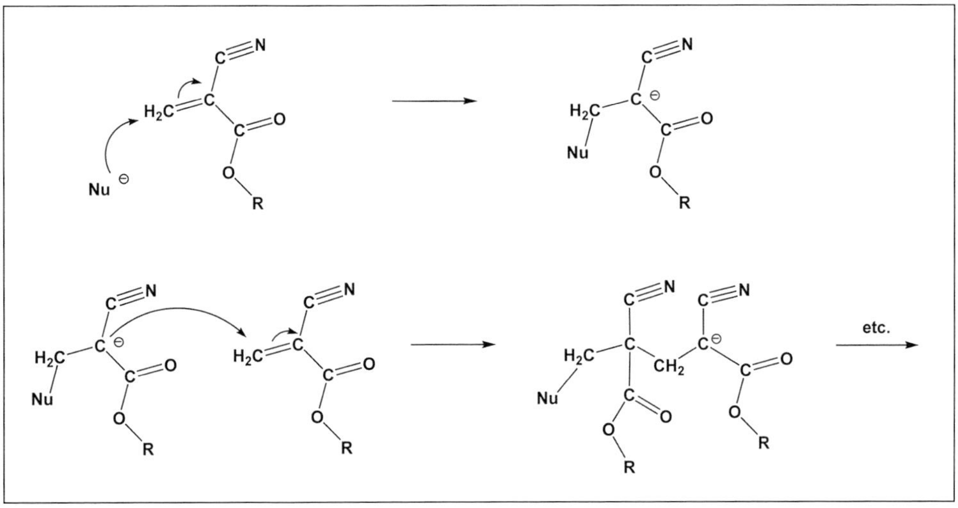

Figure II-2.127: Anionic polymerization of cyanacrylate

Table II-2.58: Glass transition temperatures of polycyanoacrylates

Polymer	**Glass transition temperature [°C]**
Polymethylcyano acrylate	165
Polyethylcyano acrylate	135
Polybutylcyano acrylate	85
Polyallylcyano acrylate	115

kaline primer (e.g. an amine) can support curing. The climatic conditions during bonding can also influence the curing behaviour; if the relative humidity is low, curing is slowed down. Conversely, if the humidity is too high, the amount of water on the surface can be that high that a so-called shock curing occurs. This can result in high stresses at the molecular level within the adhesive joint, making the cured adhesive susceptible to damage due to temperature-induced expansion or contraction of the adherends or to hydrolytic attack by moisture.

2.3.5.4 Structure/property relationships

Cured cyanoacrylate adhesives have a high molecular weight (M = 10^5 – 10^6 g/mol), are amorphous and usually thermoplastic. The glass transition temperature is mainly determined by the type of monomer. Table II-2.58 provides a list of glass transition temperatures of often used polycyanoacrylates.

As a consequence of the high glass transition temperatures, the corresponding bonds are characterized by high polymerization; however, they are sensitive to impact and peel forces. The strongest metal bonds are achieved with methyl cyanoacrylates while ethyl cyanoacrylates are typically recommended for bonding both plastics and metal. Due to the thermoplastic character the polymers are usually sensitive to polar solvents, such as acetone, and show a decrease in strength with elevated temperature. The operating temperature limit for standard cyanoacrylate adhesives is typically limited to about 80 °C. In order to improve heat resistance crosslinking agents like biscyanoacrylates (Figure II-2.126) can be used. An alternative are monomers from the group of alkenyl cyanoacrylate like allyl cyanoacrylate (Figure II-2.129). Small amounts of the alkenyl cyanoacrylates contained in the monomer mixture are incorporated into the polymer during curing by anionic polymerization. Even without the presence of a radical initiator, additional crosslinking takes place via the pendant alkenyl groups at temperatures above approx. 80 °C by means of radical polymerization, which improves the temperature resistance.

Even if this additional crosslinking of polycyanoacrylates does not prevent their basic sensitivity to hydrolytic attack, even very small amounts of hydroxy ions, especially under slightly alkaline conditions, are sufficient to initiate polymer degradation, the additional crosslinking results in an improved moisture resistance of the bond. Crosslinked polymers can simply suffer much more hydrolytic dam-

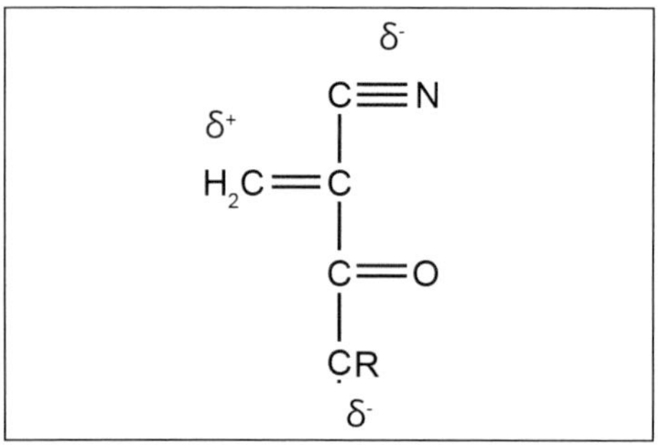

Figure II-2.128: Charge distribution in cyanoacrylate ester

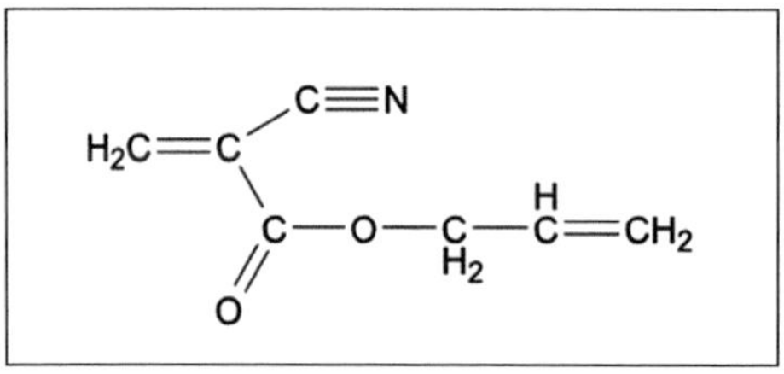

Figure II-2.129: Chemical structure of ally cyanoacrylate

age than just linear polymers before the molecular weight dependent properties, like tensile strength are affected.

2.3.5.5 Formulation

Cyanoacrylate adhesives mainly consist of one or more than one reactive **monomer**. Of the various esters of cyanoacrylic acid available, the **methyl ester** are preferred for very rapid-setting adhesives for bonding metals and thermoset materials. They show a good short term temperature resistance and a good resistance against chemicals. The most common monomer are the **ethyl esters** which are characterised by good adhesion to both metal and plastic including elastomers. Both types suffer from the characteristic, sometimes pungent odour.

Due to their reduced exotherm **butyl esters** are often used for skin adhesion medical applications. The **alkoxy-esters** especially methoxy and ethoxy-ester are preferred for low-odour adhesive applications; however, these esters have longer curing times and do not yield the same high level of bond strength. In contrast to methyl or ethyl ester they do not show the sometimes undesired so-called blooming, formation of a whitish deposit near the adhesive, which is caused by the precipitation of excess monomers.

According to [29] an improved water resistance can be achieved by the addition of bis- and monosilanes.

Although the anionic polymerization is the by far most important cure reaction for cyanacrylate adhesive, cyanoacrylate monomers can also polymerise by a free-radical mechanism affecting shelf life. Therefore, acidic **stabilizers**, such as SO_2 (a few ppm), aromatic sulphonic acids, phosphoric acid or phosphorus pentoxide and free-radical scavengers based on hydroquinone or hindered phenols (0.1 – 1 %), are often part of the formulation. For practical purposes, stabilisation with SO_2 is highly recommended, because it outgases spontaneously during application and so enables rapid polymerization during the bonding process.

The monomers typically used are characterized by a viscosity of less than 10 mPas. This can pose difficulties if the adhesive is to be applied to porous materials, vertical surfaces or if certain gap filling properties are required. In such cases, the viscosity is adjusted by incorporating polymers, such as polymethyl methacrylate and polyvinyl acetate in proportions of up to 25 %, into the reactive monomers to act as **thickeners**. The use of nitrile rubber, methyl methacrylate-butabiene-styrene (MBS) and acrylonitrile-butadiene-styrene copolymers has also been reported [1]. The use of hydrophobised fumed silica as a **rheological additive** leads to products that show a good sag resistance and can be used on porous materials as well as vertical surfaces without problems. Such cyanoacrylate gels also possess improved gap-filling properties. [15 - 17].

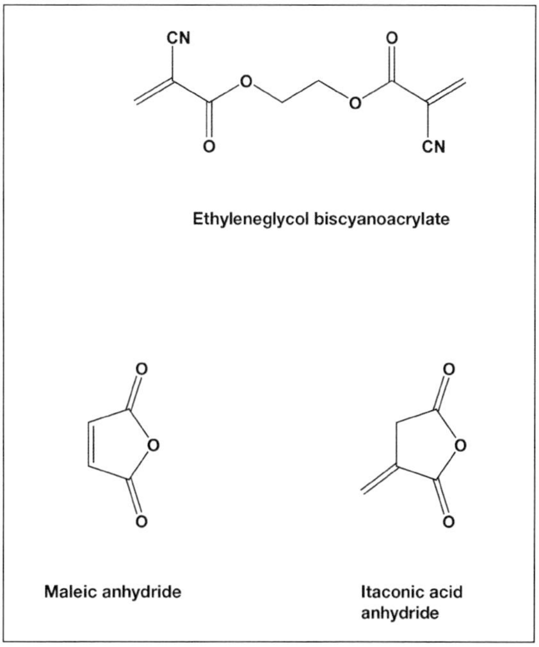

Figure II-2.130: Chemical structure of potential crosslinkers

Table II-2.59: General advantages and disadvantages of cyanoacrylate adhesives

Advantages	**Disadvantages**
– Easy to use – 1-component system – Solventless, 100 % system – Rapid cure, within seconds – High bond strength – Good adhesion to a wide range of materials, including EPDM – Resistant to non-polar solvents – Good biodegradability	– inherent smell of some monomers – only suitable for small bonds due to fast curing – low gap filling properties if not thickened – brittleness and poor impact restsinace if not toughened – limited heat resistance (app. 80 °C) – sensitive to solvents – sensitive to hydrolytic damage – can initiate stress corrosion cracking of sensitive plastics – risk of skin adhesion in the event of improper handling

Adding polyethylene oxides or crown ethers (0.5 – 1 %) accelerates curing even further and is therefore especially used on low-acid surfaces, without affecting the storage stability [1, 2].

Due to their high glass transition temperature, polymeric cyanoacrylates are quite brittle at ambient temperatures. Their impact resistance can be improved by formulation with elastomers (10 – 15 %), e.g. based on ABS (acrylonitrile-butadiene-styrene copolymers) [1, 18]. Polymers which have a low glass transition temperature and are not fully compatible with the polycyanoacrylate make ideal toughening agents. Flexibilization can also be conferred by adding typical plasticisers (carboxylates, phosphates, lactones, up to 20 %) [1].

Adding organic acids or anhydrides greatly improves metal adhesion [61]. Functionalised silanes have not proved successful for this.

In addition to the already mentioned use of bifunctional cyanoacrylates (Figure II-2.126) or allyl cyanoacrylates (Figure II-2.129) to increase the crosslinking density and thus the temperature resistance, formulations containing ethyleneglycol biscyanoacrylate or unsaturated anhydrides such as maleic or itaconic acids (Figure II-2.130) also show good effects, probably due to additional crosslinking through the double bond of the anhydrides. [1, 2, 15].

2.3.5.6 Evaluation

Cyanoacrylates are very fast-setting, general-purpose adhesives that exhibit high bond strength and good adhesion to a large number of substrates. As the chemical reaction is initiated directly by the adherend's surface, these 1-component systems are very user-friendly. The high bond strength meets the typical requirements of structural adhesives. However, polycyanoacrylates are brittle polymers, a fact which is unfavourable for dynamic loads. As a thermoplastic, the polymer is especially soluble in polar solvents. Depending on the formulation it may undergo degradation in harsh mainly alkaline hydrolytic conditions. Heat resistance is also limited due to the thermoplastic behaviour of the polymer. Even though basic formulations lose much of their strength already at temperatures above approx. 80 °C, there are more recent formulations that can withstand significantly higher temperatures. (Table II-2.59)

2.3.5.7 Fields of application

Cyanoacrylate adhesives are mostly used when high strength is required quickly and easily on a small bonding surface. The broad adhesion spectrum allows a wide range of materials to be bonded, including elastomers such as EPDM, engineered plastics including certain low surface energy plastics, metals, wood, leather and paper.

Cyanoacrylate adhesives are frequently used in many different applications in industrial production lines as well as in DIY, examples are:

- Electronics industry: bonding of printed circuit boards, fastening of components
- Automotive industry: repair of plastic parts, fastening of interior parts
- Medical technology: manufacture of medical devices, bonding prostheses, skin adhesive, wound sealing, treatment of bone fractures
- Furniture: attaching magnets to locking systems
- Packaging: visually appealing blister packs
- General industry: bonding of rubber gaskets, e.g. butt bonding to create of O-rings
- Model making: assembling model kits, repairing models
- Jewellery making: fixturing semi-precious stones on (noble) metal bases
- Cosmetics: attaching artificial fingernails, eyelash extensions
- Repair: repairing household items, assembling furniture

2.3.5.8 Guide formulations

Table II-2.60 presents a cyanoacrylate adhesive based on ethoxyethyl cyanoacrylate. This monomer has a low vapour pressure and therefore has only a slight inherent odour. The viscosity is adjusted with a polymethyl methacrylate, while stabilisation is achieved by hydroquinone and phosphorus pentoxide. Fumed silica serves as rheological additive [16].

Table II-2.61 shows a formulation based on for a cyanoacrylate adhesive with improved thermal, chemical and hydrolysis resistance.

Table II-2.60: Cyanoacrylate adhesive

Item	Raw material	Parts by weight
1	Cyanoacrylate monomer	87.95
2	Thickening agent	9
3	Rheological additive	3
4	Stabilizer 1	0.03
5	Stabilizer 2	0.02

Item 1: ethoxyethyl cyanoacrylate
Item 2: polymethyl methacrylate, "Plexidon M 449", Evonik, Röhm
Item 3: fumed silica, hydrophobic, "Aerosil R 972", Evonik
Item 4: phosphorous pentoxide
Item 5: hydrochinone

Table II-2.61: Cyanoacrylate adhesive with improved thermal, chemical and hydrolysis resistance accord based on [28]

Item	Raw material	Parts by weight
1	Cyanoacrylate monomer	84.4
2	Multicyanoacrylate monomer	0.5
3	Unsaturated carboxylic acid anhydride	0.9
4	Radical polymerization inhibitor	0.9
5	Acid stabilizing agent	0.01
6	Thickenining agent	10.3
7	Thixotropic agent	2.6
8	Accelerating agent	0.4

Item 1: ethoxyethyl cyanoacrylate
Item 2: bis-cyanoacrylate monomer
Item 3: maleic anhydride
Item 4: hydroquinone:
Item 5: sulphur dioxide
Item 6: polymethylmethacrylate, "Plexidon M 449", Evonik, Röhm
Item 7: fumed silica, hydrophobic, "Aerosil R 972", Evonik
Item 8: 18-crown-6-ether

2.3.6 References

[1] Millet, G. H. in S. R Hartshorn, Structural Adhesives, Plenum Press (1986)
[2] Skeist, I., Handbook of Adhesives, 3rd edition, Chapman & Hall (1989)
[3] Habenicht, G., Kleben, 3rd edition, Springer (1997)
[4] www.gifte.de
[5] https://gestis.dguv.de/search, accessed 2024-03-28
[6] Dunn, D., Reactive toughened Acrylic Adhesives Poised for Growth, http://www.bondch.com/bzzt/030521.html
[7] Reilly Industries, http://www.reillyind.com
[8] Wuich, W., GAK 7/1990, 43, 389-92
[9] Reactive Acrylic Bonding System based on Hypalon, Dupont, Technical Information, 2003
[10] Allen, K. W., Acrylates as reactive Adhesives, Int.J. Adhesion and Adhesives, Vol. 9, No.2, 1989, 103-5
[11] Hypalon, Technical Information, DuPont, 2003
[12] Endlich, W., Zeitgemäße Acrylatklebstoffe, Uta Groebel, Limeshain, 1985
[13] Wellmann, S., et al., Int. J. Adhesion, Vol. 14, No. 1, January 1994, 47 – 55
[14] US Patent 3.043.820
[15] Shanta, K. L., Developments and Applications of Cyanoacrylate Adhesives, J Adhes. Sci. Technol. 3 (1989), 4, 237 – 60
[16] EP 0 323 720
[17] US Patent 4 845 151
[18] Kotsev, D., Kabaivanov, V., Adhesion, 12 (1988), 82 – 105
[19] www.dymax.de
[20] US Patent 4,942,201 https://patents.google.com/patent/US4942201A/en accessed 2025-01-24
[21] Boeder, CH. W., in Hartshorn, S. R., (editor), Structural Adhesives, Chapter 5, Plenum Press, New York, 1986
[22] https://adhesives.specialchem.com/formulation/anaerobic-acrylic-adhesive-for-fastener-locking
[23] EP 3 436 537 B1
[24] DELO Bond It – Bonding Technology and It's Industrial Application; DELO Industrial Adhesvies, Windach 2020
[25] https://www.photoinitiators-platform.org/about-photoinitiators/, accessed 2024-06-09
[26] https://lemelson.mit.edu/resources/harry-coover, accessed 2024-06-11
[27] https://de.wikipedia.org/wiki/Cyanacrylate, accessed 2024-06-11
[28] EP 4 194 523 A1
[29] US Patent 9,096,779 B2

2.4 Polycondensation adhesives

2.4.1 Phenolic resins

2.4.1.1 Basic principles

Phenolic resins, the condensation products of phenol and aldehydes, form the basis for Bakelite one of the oldest synthetic polymers. For more than 80 years, phenolic resins have also been used in adhesives. They are easy to synthesise and have a very good property profile that lends itself particularly to adhesives. Detailed information on phenolic resins can be found in [1, 2, 3–5].

Raw materials

The most important base product for the synthesis of phenolic resins is phenol (made from benzene by the Hock process) and formaldehyde (made by oxidation/catalytic dehydrogenation of methanol). As shown in Figure II-2.131, substituted phenols, such as cresol, xylenols, alkylphenols and resorcinol, are used in addition to phenol [6].

Chemically, phenol is an electron-rich aromatic compound which undergoes electrophilic substitution at the o- and p-positions with formaldehyde to yield hydroxymethyl-substituted phenols. The further course of the reaction between phenol and formaldehyde varies with the type of catalysis and the stoichiometric conditions.

One bio-based phenol source is lignin, which has been described in academic papers for several years with regard to its usability in adhesives [10-11].

Reaction in highly acidic conditions (pH 1 to 4)

As shown in Figure II-2.132, formaldehyde reacts with phenol at low pH in the presence of excess phenol to yield a **novolak**.

Figure II-2.133 shows how formaldehyde is catalysed electrophilically by protons (1) and then attacks the phenol in an S_E reaction (2). In the presence of acid, the intermediate hydroxymethyl phenol is protonated to a benzyl alcohol, which splits off water to yield a carbenium ion. This ion again attacks another phenol in a subsequent S_E reaction (alkylation) (3).

Phenol has three reactive sites on the aromatic ring where formaldehyde can attack. At a low degree of oligomerisation, novolaks are mainly linear, thermoplastic molecules with a molecular weight ranging from 250 to 5000 g/mol. They are amorphous and have a glass transition temperature of 40 to 110 °C. The individual phenol molecules are linked to each other in the o- or p-position by methylene groups. The molecular weight is regulated by the stoichiometric excess of phenol, and therefore the oligomer has phenolic end groups. As these end groups are stable and unreactive, novolaks are stable in storage. To make novolaks cure chemically, reactive crosslinkers must be added. The pH can be regulated with acids such as hydrochloric, sulphuric, p-toluenesulphonic acid and oxalic acid. The last of these has the advantage of decomposing at about 180 °C.

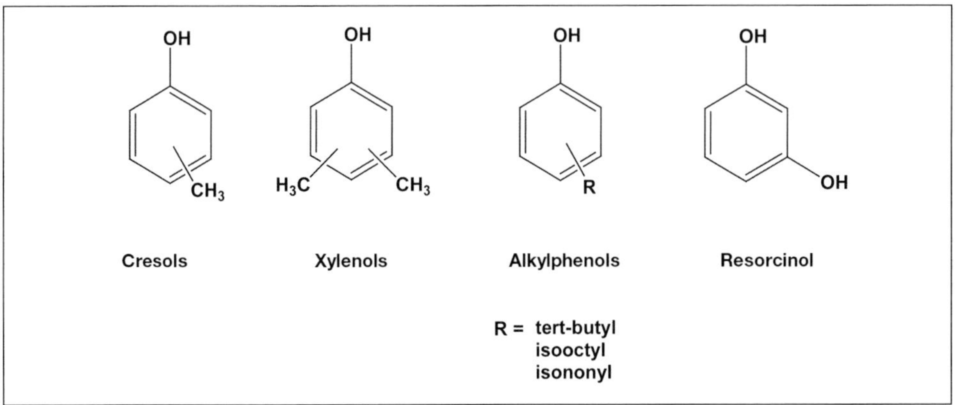

Figure II-2.131: Substituted phenols as raw materials

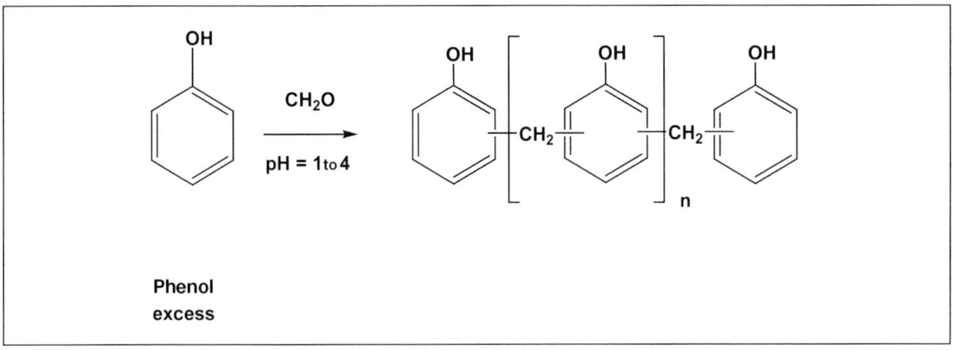

Figure II-2.132: Formation of novolak

Synthesis of a novolak [7, 8]

Some 1880 parts of phenol is melted in a reactor equipped with thermometer, stirrer and, optionally, a reflux or Liebig condenser. Then 50 parts of oxalic acid is added and the mixture heated to 100 °C. This is followed by addition of 1560 parts of a 30 % water-based formaldehyde solution in portions over a period of about 2 hours. The reaction is continued for another 5 hours at 98 °C. Finally, the volatile components are made to distil off under vacuum until, at a pressure of 2 kPa, the residue reaches a temperature of approx. 160 °C. The result is 1968 parts of a novolak with a melting point of 124 °C.

Reaction under weakly acidic conditions (pH 4 to 7)

The presence of certain metal carboxylate catalysts (multivalent ions, such as Zn, Mg, Al) affords novolaks with a distinctive o- linkage. This preferred ortho-reaction is believed to proceed via the coordinative influence of the metal ion as the formaldehyde is attacking the phenol. Compared to standard novolaks, ortho-novolaks exhibit greater reactivity (in the sterically more accessible p-position) with further formaldehyde in curing reactions. Figure II-2.134 shows the directing effect of the metal ions on the reaction between phenol and formaldehyde to yield novolaks.

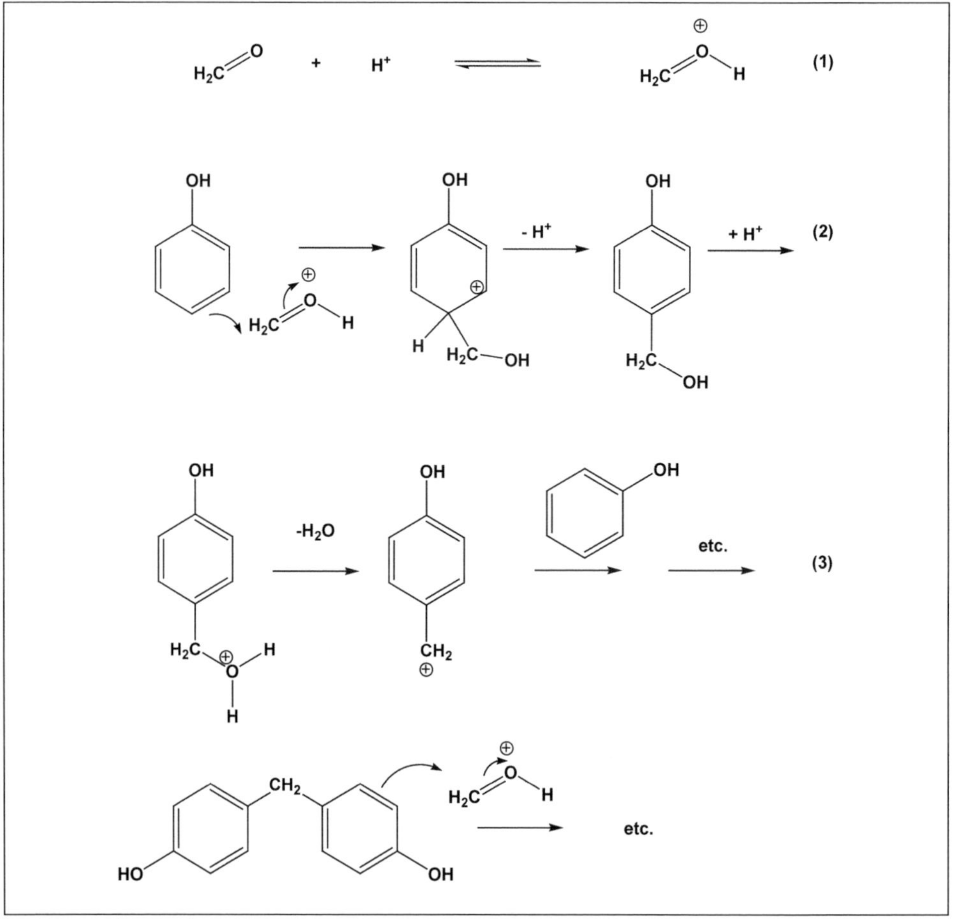

Figure II-2.133: Mechanism behind the formation of novolak

Reaction under alkaline conditions (pH 7 to 11)

When phenol is made to react with excess formaldehyde under basic conditions (dilute sodium hydroxide) at moderately elevated temperatures (see Figure II-2.135), and then gently distilled under vacuum to remove water and excess formaldehyde, the product is reactive hydroxymethyl-terminated oligomers (**resols**).

In alkaline conditions, phenol, in the form of the highly nucleophilic phenolate ion, reacts with formaldehyde to yield hydroxymethyl phenolates (Figure II-2.136).

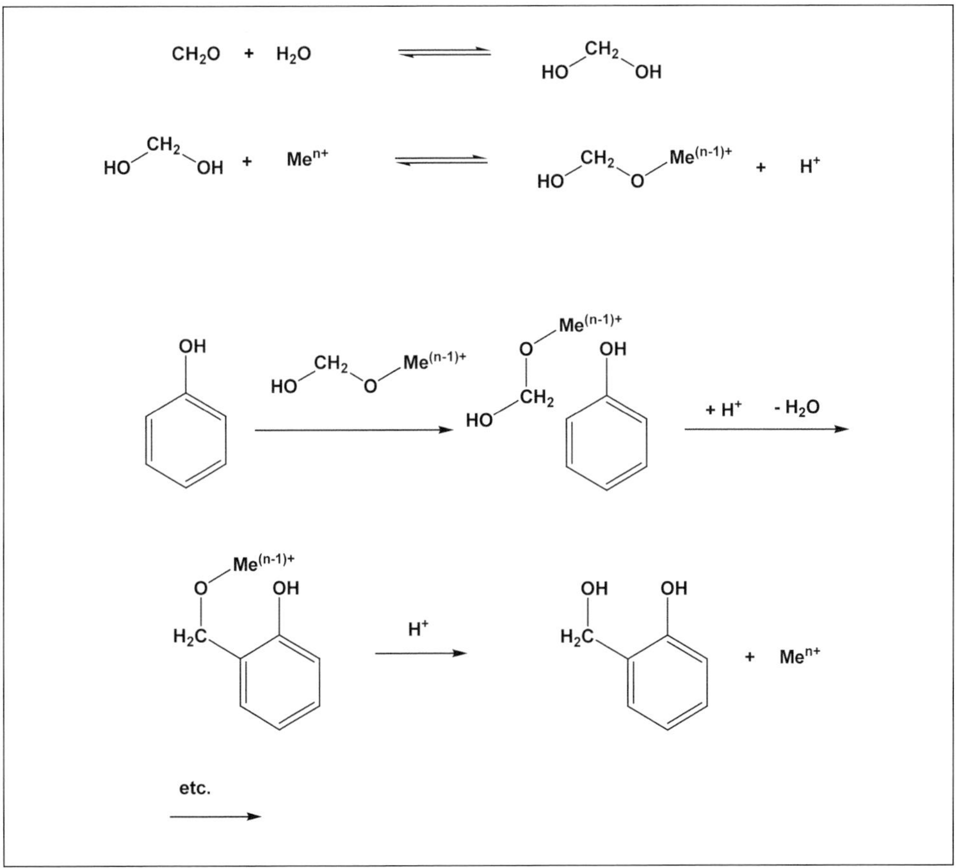

Figure II-2.134: Formation of ortho-novolaks

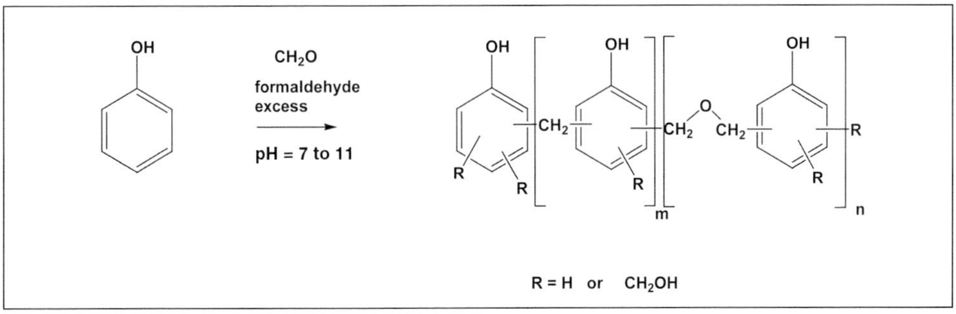

Figure II-2.135: Formation of resols

As may be seen in Figure II-2.137, these hydroxymethyl phenolates readily condense to oligomers having methylene or dimethylene ether bridges.

Resols can be either liquid or solid at molecular weights of 200 to 600 g/mol. They dissolve in water under alkaline conditions due to the formation of phenolate ions. Hydroxymethyl groups

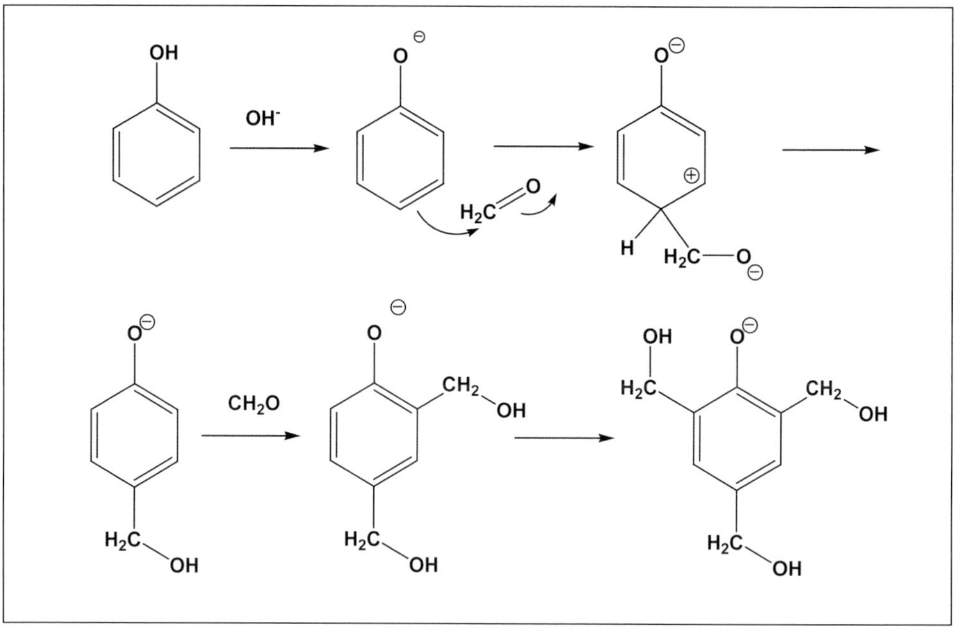

Figure II-2.136: Hydroxymethylation of phenol under alkaline conditions

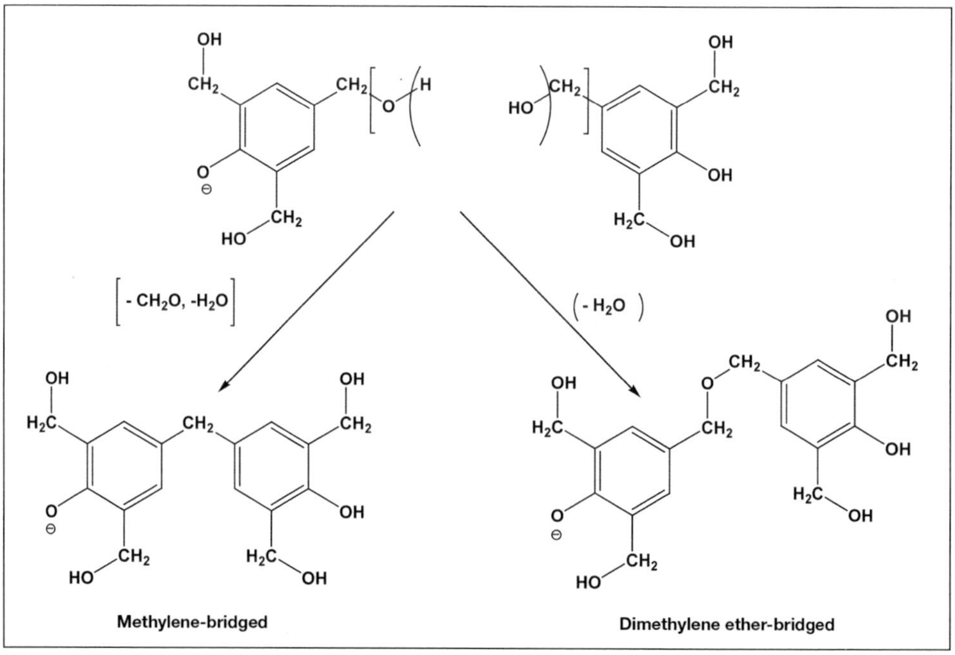

Figure II-2.137: Condensation of hydroxymethyl phenolates

readily undergo polycondensation. The corresponding adhesive cures rapidly, with curing readily initiated by adding acids; heating to above approx. 130 °C yields highly crosslinked polymers. Even at room temperature, a slow stepwise reaction occurs, which is why resols have a limited shelf life. Resols are soluble in aqueous alkaline milieu and in polar organic solvents. As with the synthesis of o-novolaks, resols with a high ortho content can be obtained in the presence of zinc hydroxides.

Synthesis of a resol [7, 8]

Some 846 g phenol and 1600 g formaldehyde (as a 30 % aqueous solution) are mixed at room temperature. Concentrated sodium hydroxide solution (approx. 110 ml, 38 %) is used to adjust the pH to 9.5 to 10. The mixture is heated for 4 hours at 60 °C. After cooling, the excess of phenol is gently made to distil under vacuum, so that a maximum inner temperature of 60 °C is reached at a pressure of 1.5 kPa. The residue (1360 g) is a resol.

Resorcinol

Some of the phenol can be replaced by resorcinol, which is more reactive. Resorcinol enters into a mainly bifunctional reaction (at positions 4 and 6) with formaldehyde at room temperature. The outcome is mainly linear oligomers. Resorcinol-phenol novolaks can be cured with formaldehyde under very mild conditions.

Alkylphenols

In alkylphenols, the functionality of phenols relative to formaldehyde is reduced to two (the two o-positions) due to blocking of the p-position. In addition, the resultant oligomers formed with formaldehyde are less polar and so are more soluble in organic solvents. Corresponding oligomeric novolaks or resols are thermoplastic. They serve as resins for increasing tack and heat resistance in solvent-based adhesives (see Chapter II-1.1, e.g. in polychloroprene and nitrile rubber adhesives).

Chemistry behind curing

The curing of phenolic resins is a polycondensation reaction which usually affords highly crosslinked polymers (resites). As small molecules, such as water and formaldehyde, have to be split off during the polycondensation, this fact must be taken into consideration for joining two

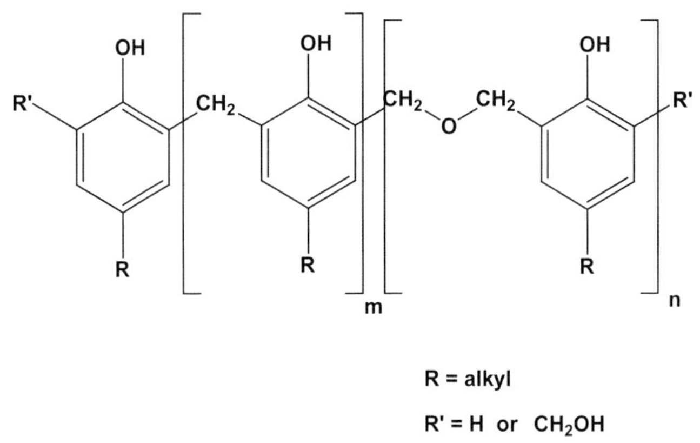

Figure II-2.138: Alkylphenol resin, resol type

substrates. The adhesives are typically cured at elevated temperatures, under application of pressure to the substrates.

Curing of novolaks

Novolaks possess few reactive phenolic groups. For curing to occur, therefore, a reactive crosslinker is required. Typically, hexamethylene tetramine (also called "hexa", "hexamine", "HMTA" or "urotropine") is added in proportions of 5 to 15 %, expressed in terms of the solids content. Urotropine readily decomposes in several steps under protic catalysis to yield formaldehyde derivatives, such as methyleneimine [9]. Finally, the aromatic rings are crosslinked above approx. 180 °C, with liberation of ammonia. Figure II-2.139 shows the acid-catalysed thermal decomposition of hexamethylene tetramine and the subsequent reaction with the novolak phenolic groups. Intermediates such as di- and tribenzyl amines are formed during the reaction [1]. At temperatures above 180 °C, the polyalkylamino bridges are converted into methylene bridges, with liberation of formimine [3].

Novolaks can also be cured by adding resols and then applying heat, because the resols bear excess methylol groups needed for the curing reaction.

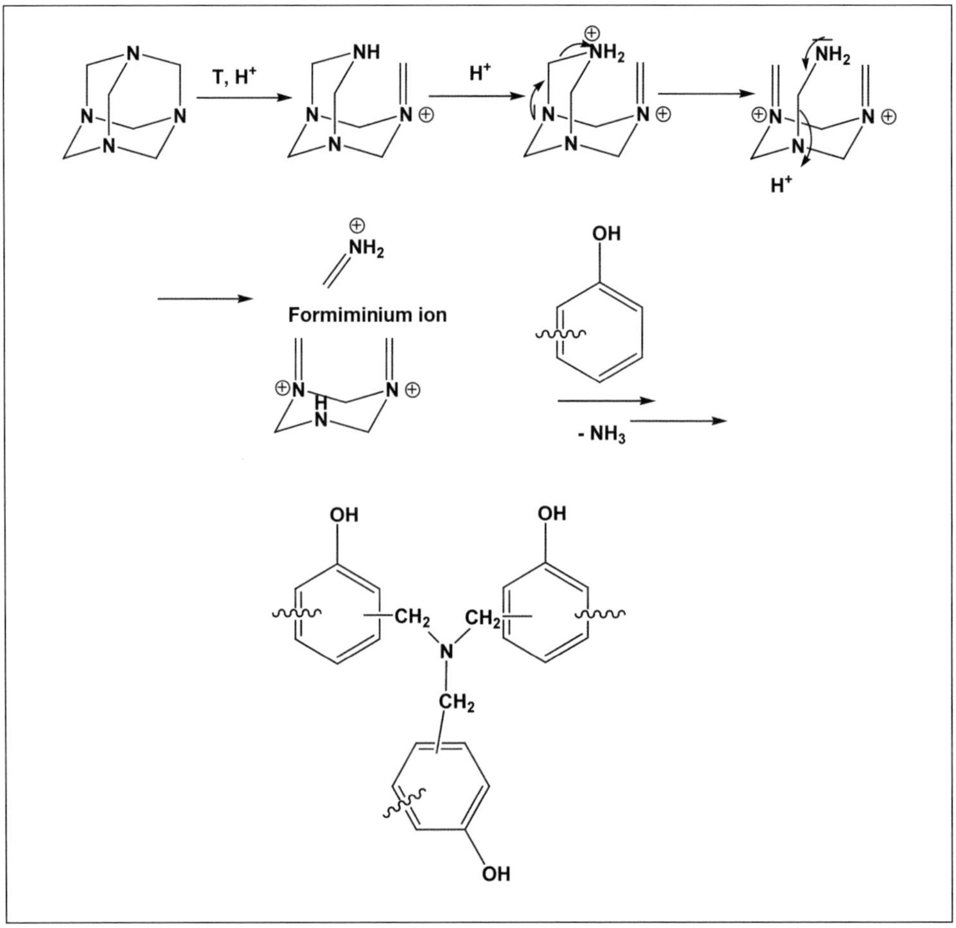

Figure II-2.139: Curing of novolaks

Curing of resols

Resols cure slowly at ambient temperatures and rapidly at 130 to 200 °C to produce highly crosslinked structures (resites). The starting reaction is believed to be a 1,4-elimination to yield reactive quinone methides (α,β-unsaturated carbonyl compound) [1].

In the presence of acid, curing proceeds via the formation of methylene or dimethylene ether bridges. At temperatures above approx. 180 °C, the latter are converted into methylene bridges, with liberation of formaldehyde. Figure II-2.141 schematically shows the polymeric network of a resite obtained when novolaks or resols are cured.

2.4.1.2 Structure/property relationships

Cured phenolic resins are polymers whose chain consists only of C-C bonds and aryl components. As shown in Figure II-2.141, they form highly crosslinked systems. This leads to high mechanical

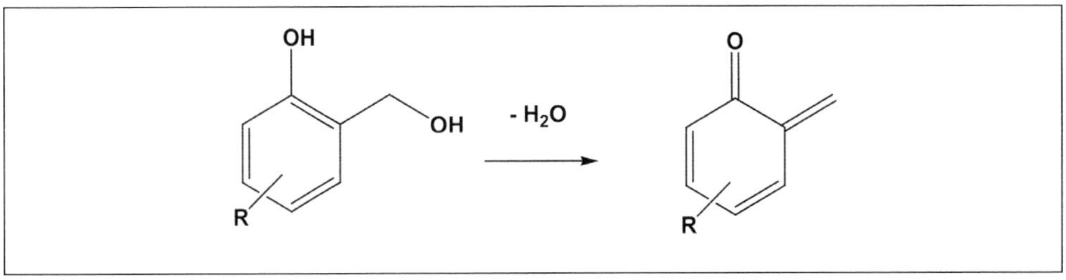

Figure II-2.140: Formation of quinone methides

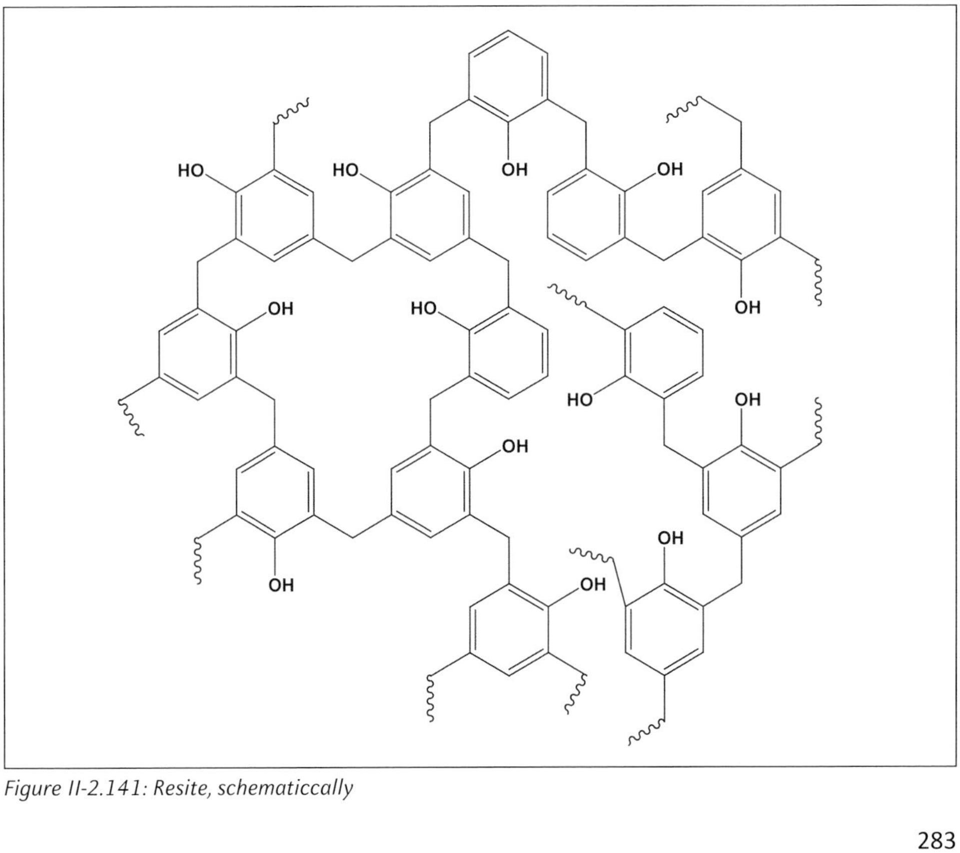

Figure II-2.141: Resite, schematiccally

strength and very good resistance to chemical and hydrolytic influences. The polymers are totally insoluble and practically incapable of swelling in any organic solvents. Their heat resistance is superior to that of other organic polymers and easily exceeds 300 °C, depending on time and other influences. Thermally initiated degradation starts with the liberation of some formaldehyde, water and phenol. Above 300 °C, stepwise degradation commences, yielding fragments, such as phenols, cresols, carbon monoxide, carbon dioxide and similar products. As may be seen in Figure II-2.142, oxidative degradation starts in the benzyl position with the formation of hydroperoxides [1].

The glass transition temperature of the polymer closely approximates the chosen curing temperature. In isolated cases, it can reach approx. 250 °C. During curing, the system progressively freezes at the curing temperature. Each phenol molecule can have up to three linkages, but not all can be utilised for crosslinking. Nonetheless, the resulting polymer is a typical thermoset with short network arc lengths.

Pure phenolic resins are very brittle after curing their mechanical properties can be improved by adding toughening polymers which are incompatible with the phenolic resin. Nitrile rubber (acrylonitrile butadiene copolymer) is frequently employed. During curing, this forms microscopic, incompatible phases which act as tougheners and improve the mechanical properties of the

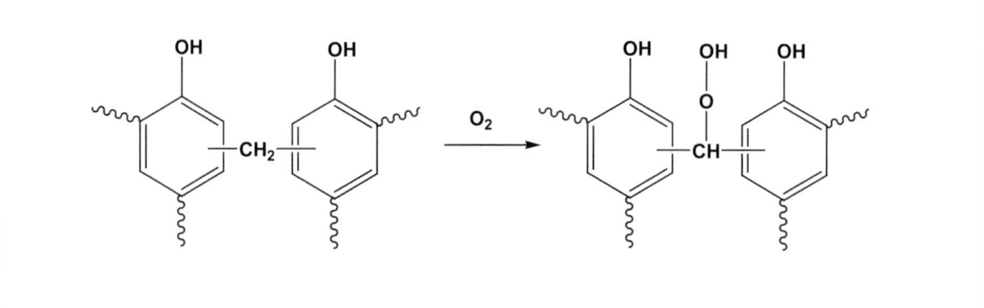

Figure II-2.142: Formation of hydroperoxides at the benzyl position

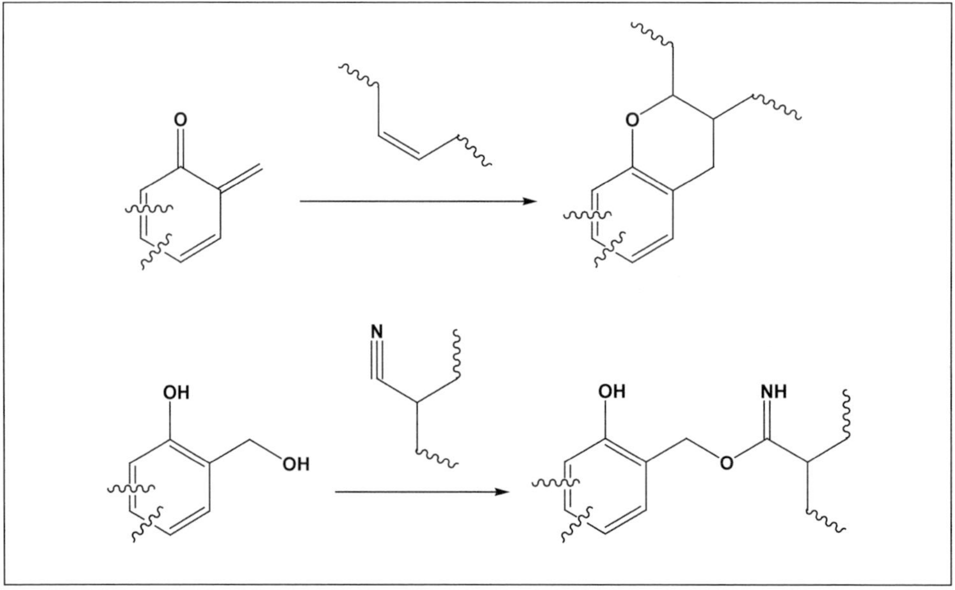

Figure II-2.143: Reactions between phenolic resins and nitrile rubber

polymer (see Chapter II-2.2.2). At the high curing temperatures, various functional groups form covalent bonds between the phenolic resins and the rubber (see Figure II-2.143). The quinone methides seem to have special importance here because, as dienes, they can enter into a Diels-Alder reaction with double bonds of the rubber phase. Nitrile groups, too, can react with hydroxy groups on phenolic resins to yield isomeric amides [1].

2.4.1.3 Formulation

The present book's approach to classifying adhesives by their setting mechanism or application method does not lend itself readily to phenolic resins. Instead, a more chemical approach is needed. Phenolic resins can assume various functions in adhesives formulations. Frequently, unreactive, oligomeric novolaks (often based on substituted alkylphenols) are added as **tackifying agents** for solvent-based adhesives. In some cases, a combination of phenolic resins and MgO or ZnO is added to improve the heat resistance of polychloroprene adhesives. They then constitute auxiliaries for improving special properties. This improvement is not achieved by a polycondensation reaction, but rather by complexation with the corresponding metal ions (see Figure II-1.13).

For this reason, in this book, a **reactive phenolic resin** adhesive will be taken to mean formulations in which the phenolic resin enters into a polycondensation reaction during the bonding process to yield a highly crosslinked thermoset. As these resites are very brittle in the pure form and unable to withstand dynamic forces, they would not be suitable for structural adhesives applications. They therefore require toughening with rubber polymers, such as nitrile rubber. In such cases, the thermally reactive phenolic resin system is formulated with the toughening polymers in solution or in water-based dispersion along with other additives. The resulting adhesives are therefore solvent-based or water-based. However, the relevant end property, e.g. high thermal resistance, is not achieved by evaporation of a solvent or water, but mainly by thermal polycondensation of the reactive resin as the adhesive cures. Consequently, these solvent-based phenolic resin/nitrile rubber adhesives are counted as reactive phenol resin adhesives. In some cases, a mixture of phenolic resin (novolak/urotropine) and nitrile rubber is formulated in solution and the solvent is already evaporated by the adhesive manufacturer, so that a solid, reactive adhesive film remains. For bonding, this film is simply placed between the two substrates and heat-cured. The adhesive in that case is a 100 % system; it differs from a solvent-based adhesive of comparable composition merely in its application form.

Heat-curing, reactive phenolic resin adhesives therefore consist of the **phenolic resin** (10 to 25 % novolak/urotropine or resol) and in most cases a **dispersion medium** (60 to 80 %), such as water, or an organic solvent, such as acetone, ethyl acetate, or butanone (MEK). The mechanical properties, especially the impact resistance, are modified with **organic polymers** (5 to 20 %), e.g. based on masticated nitrile rubber (see also Chapter II-1.1.2). **Fillers** (5 to 15 %), such as fumed silica, carbon black or simple chalk, may also be added. While the first two are active fillers, i.e. they improve the mechanical and thermal resistance, the latter merely keeps costs down. Fumed silica and carbon black also exert a strong influence on the rheology, which might be important for application by roller or doctor blade. Resorcinol-based phenolic resins can be cured with excess formaldehyde at room temperature in the form of a 2-component system due to the high reactivity of resorcinol.

Evaluation

Phenolic resin adhesives have found many applications. They exhibit high mechanical strength, excellent adhesion to many substrates and excellent thermal and hydrolytic stability. These are ideal preconditions for structural adhesives applications. Drawbacks include the need for relatively lengthy application of high temperatures and high pressure to the mated substrates. Additionally, curing releases toxic products, such as formaldehyde and phenols.

Advantages

+ low price
+ good adhesion, mainly to metals, such as steel and aluminium
+ high mechanical strength
+ high thermal resistance (briefly up to > 300 °C)
+ good hydrolytic stability
+ good chemical resistance

Disadvantages

- polycondensation needs time and the application of heat and pressure
- heat-sensitive substrates cannot be bonded
- liberated molecules can lead to bubbling in the adhesive layer
- liberated products, such as formaldehyde, are of toxicological concern
- brittleness of the bonds

Fields of application

Binders

Substantial quantities of phenolic resins (often in the form of alkaline water-based solutions of resols) serve as binders in adhesives-related areas. Curing occurs at 100 to 180 °C. In the manufacture of **particle board**, phenolic resins are used to bind the wood chips. Although this is a bonding process in the broader sense, these high-volume applications of reactive phenolic resins are not counted as adhesives applications. These products are called binders or binding agents, because an almost infinite number of small, undefined substrates are bonded instantaneously to a new raw material. The same is true of the manufacture of **moulded parts** based on timber-derived materials for automotive interior applications and the manufacture of furniture. Other applications of low-cost water-based phenolic resins (resols or novolak/urotropine) as binders are the bonding of sand in **casting moulds** for the manufacture of engine blocks etc. **Insulation materials**, such as glass and rock-wool, are sprayed with phenolic resins (often water-based resols) and thus made into insulating mats. Phenolic resins are the basis of **grinding materials**, such as sand-paper, and of **friction linings** for brakes and clutches in the automotive industry.

Adhesives

Phenolic resins can serve as 2-component adhesives for bonding wood [7]. Reactive phenolics made from resorcinol-based resins are stable in aqueous alkaline solution. As is typical of a 2-component adhesive, excess formaldehyde is added immediately before application. Even at low temperatures, crosslinked polymers with good bonding properties are obtained. Toughened reactive phenolic resin adhesives are the only types employed for reliably bonding friction linings, such as clutch and brake liners, to the metal base in cars. They exhibit excellent adhesion to both the metal surface and the friction material. They possess high mechanical strength and very good thermal resistance. For bonding brake liners, they act as "structural adhesives" in the truest sense of the phrase. In the **aircraft industry** phenolic resins are often used in the form of heat-reactive films to bond aluminium and steel. As production quantities in the aircraft industry are very small compared to those in the automotive industry, complex manufacturing procedures, such as the application of pressure and heat, are acceptable. Again, what is important is the ability to bond substrates reliably and durably.

2.4.1.4 Guide formulation

Phenolic resin adhesive modified with nitrile rubber

Table II-2.62 lists a reactive solvent-based, phenolic resin adhesive modified with nitrile rubber [1].

Table II-2.62: Phenolic resin adhesive modified with nitrile rubber

Item	Raw material	Parts by weight
1	Phenolic resin	15
2	Nitrile rubber	15
3	Solvent	69.7
4	Coupling agent	0.3

Item 1: novolak/urotropine-mixture
Item 3: acetone
Item 4: mercaptosilane, "Silquest A-189", Momentive

2.4.1.5 References

[1] Hartshorn, S. R., Structural Adhesives, Plenum Press (1986)
[2] Skeist, I., Handbook of Adhesives, 3rd edition, Chapman & Hall (1989)
[3] Bakelite AG, Bakelite Phenolharze, Leitfaden 2003
[4] Jordan, R., Hinterwaldner, R. Klebharze, R. Hinterwaldner Verlag, München
[5] Ullmann's Encyclopedia of Industrial Chemistry, release 2001, 6th edition
[6] Jordan, R. Hinterwaldner, Klebharze, R. Hinterwaldner Verlag, München
[7] Ullmann's Encyclopedia of Industrial Chemistry, release 2001, 6th edition
[8] Houben-Weyl, 4th ed., XIV/2, 272–291
[9] Rath, W., Dissertation Universität Stuttgart, 1984
[10] Gong, Xiaoyu, et al. "A review on lignin-based phenolic resin adhesive." Macromolecular Chemistry and Physics 223.4 (2022): 2100434.
[11] Sarika, P. R., et al. "Bio-based alternatives to phenol and formaldehyde for the production of resins." Polymers 12.10 (2020): 2237.

2.4.2 Moisture-curing POR hot-melts (polyolefin reactive hot-melts)

2.4.2.1 Basic principles and structure

Innovative reactive hot-melts based on silane-grafted amorphous poly-α-olefins have become commercially available in recent years. Amorphous α-olefins (copolymers of ethylene, propylene, 1-butene etc; see Chapter II-1.3) have good hot-melt properties. They are rapid-setting and for most applications have adequate heat resistance. Silanol groups can be grafted onto them to create reactive functional groups. After application, these silanol groups can be used for crosslinking reactions by condensation with moisture. POR hot-melts have an interesting property profile that sets them apart from the widely accepted PUR hot-melts. Figure II-2.144 shows their chemical structure and curing reaction [1–3, 7].

2.4.2.2 Formulation

Commercial silane grafted α-polyolefins (e.g. "Vestoplast 206", Evonik [4] can be formulated with resins, waxes and additives in the usual manner (see also Chapter II-1.3). Resins increase tack and lower melt viscosity, while waxes reduce cost and viscosity. Furthermore, non-functionalised α-polyolefins, fillers and specific additives can be used. The resulting products can be applied at temperatures of 140 to 180 °C. The packaging and applicators are similar to those for PUR hot-melts. Table II-4.2 presents a starting formulation [5].

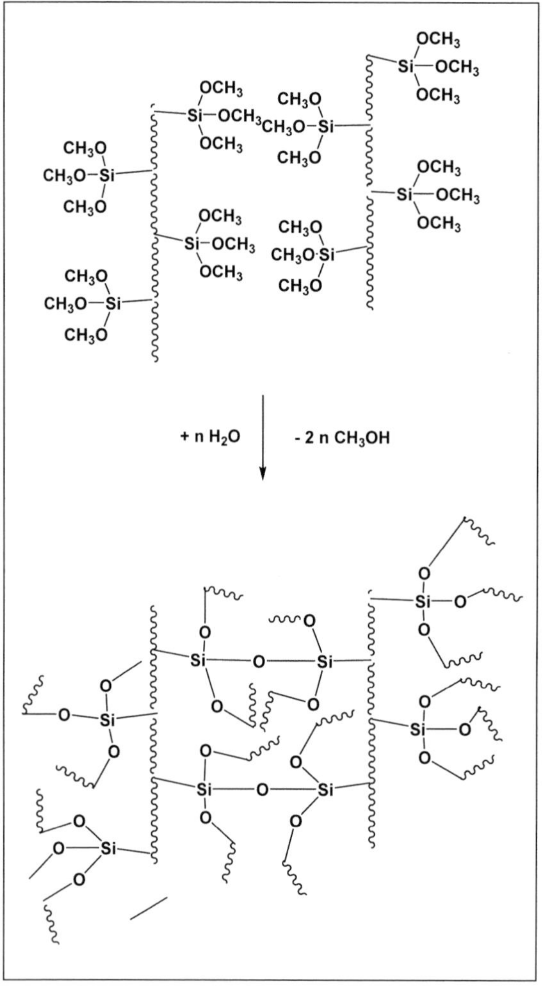

Figure II-2.144: Chemical structure and curing reaction of POR hot-melts

The product described here is not a typical hot-melt. It contains a large quantity of molecular sieve specifically intended for use as drying agent in insulating glass.

Evaluation

POR hot-melts complement the already widely accepted PUR hot-melts. The amorphous structure of the base polymers, combined with the low glass transition temperature, bestows long open times for processing and application. Due to the relatively high-molecular base polymers, they possess good hot-melt properties, set rapidly and develop high initial flexibility. In contrast to the PUR hot-melts, they do not suffer from bubbling during curing, as the small quantities of liberated methanol slowly are able to diffuse out of the polymer. POR hot-melts are not classified as dangerous goods and are highly stable under application conditions. Furthermore, the combination of silane functions and low-polarity polymer means that POR hot-melts adhere strongly to glass, ceramics and even polypropylene. However, their slow curing reaction is a disadvantage. It can take days or even weeks for the water needed for the curing reaction to diffuse through the low-polarity polymer.

Advantages

+ not classified as dangerous goods
+ good stability under application conditions
+ long open time
+ rapid setting
+ high green strength
+ good adhesion to glass, ceramics and PP

Disadvantages

- slow curing
- application temperatures higher than for PUR hot-melts.

Fields of application

In the manufacture **of utility vehicles,** POR hot-melts are used to bond sandwich elements because they set faster than 2-component PUR adhesives. The slower cure reaction is not critical to further processing of the elements, because the hot-melt property alone provides sufficient strength. In the **automotive industry,** they can often serve as an alternative to PUR hot-melts, which require a dangerous goods label [6]. Like PUR hot-melts, they are successfully employed in the **book-binding industry**. They are also used in the **insulating glass industry** on account of their low vapour-transmission rate [5].

A typical guide formulation for POR hot-melts is given in Table II-2.63.

Table II-2.63: POR hot-melt

Item	Raw material	Parts by weight
1	Silane functionalized poly-α-olefine	20 to 70
2	Butyl rubber	5 to 30
3	Poly-α-olefine	5 to 30
4	Carbon black	5 to 30
5	Molecular sieve, zeolithe, 3 Å	20 to 30
6	Catalyst	0.1 to 2
7	Organosilane	0.1 to 2

2.4.2.4 References

[1] Schindler, M:, Coating, 2003, 1, 22–24
[2] Stark, F., Adhäsion 9, 2002, 16–20
[3] company brochure JOWAT Klebstoffe, JOWAT News, D035, 05, 2003
[4] company brochure Degussa AG, Vestoplast,
[5] DOS 196 24 236
[6] Hürther, H. U., under automotivetechnology.net/editorials/Dupont.htm
[7] products.evonik.com/assets/37/69/VESTOPLAST_Product_Range_EN_EN_Asset_1203769.pdf

2.4.3 Silicones and silane-modified systems

Silicones are organosilicon polymers whose backbone consists only of Si-O linkages. They are properly called polyorganosiloxanes. Curing (also known as vulcanisation) occurs at room temperature in the presence of humidity: RTV-1 (room temperature vulcanising, 1-component) [9].

The highly simplified scheme in Figure II-2.145 shows that every terminal hydroxy group of the polydimethylsiloxane reacts with a trifunctional curing molecule to yield a tetrafunctional polymer, which is stable in storage and cures in the presence of humidity.

The methyl groups can be wholly or partially replaced by aryl or other groups.

Silicones have the largest share of the global sealants market (Figure II.2.146) [1].

Curing of RTV-1 silicones

The different types of curing mechanisms are acidic (HX = acetic acid) and neutral (HX = oxime, alcohol) (Figure II.2.146); alkaline types (HX = amine) now belong to the past.

Acetoxy types cannot be used on alkaline substrates (e.g. concrete) or on metals susceptible to corrosion (e.g. iron, zinc, copper, brass). Facade construction is therefore usually done with neutral systems.

The curing rate depends on the crosslinking system. Acetoxy and methoxy types crosslink very quickly, while others react slowly.

Skinning occurs from a few minutes up to half an hour. The curing rate depends on the joint width, humidity and temperature, and ranges from 1 to 2 mm per day.

Further ingredients of silicone sealants:
- plasticisers (non-reactive silicone oils)
- reinforcing (nanoparticle) fillers (fumed silica, carbon black)
- non-reinforcing fillers (chalk),
- coloured pigments

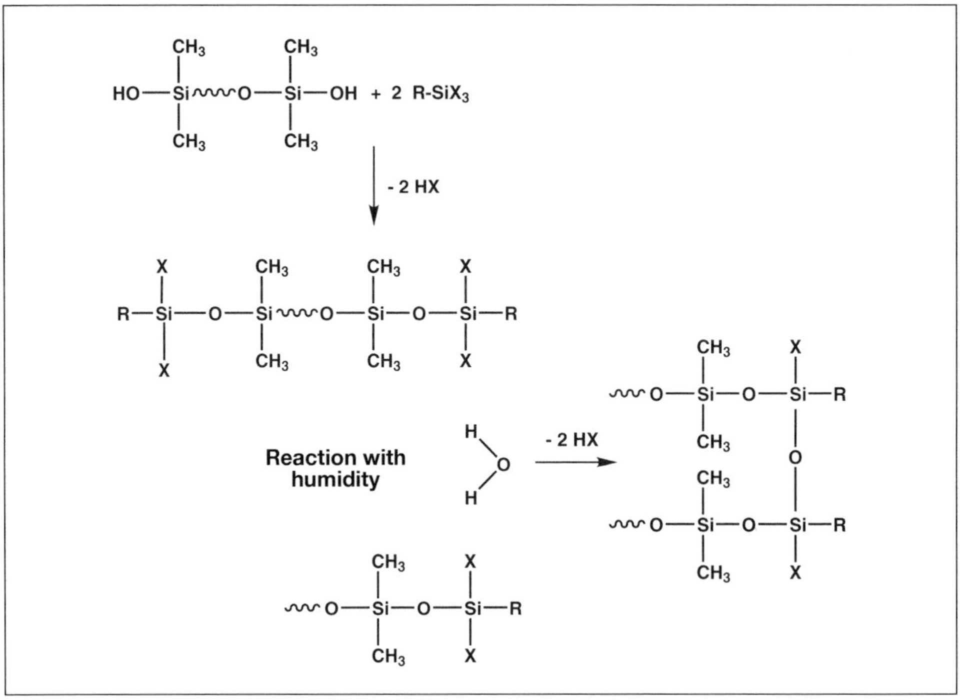

Figure II-2.145: Basic reactions of 1-component silicone sealants

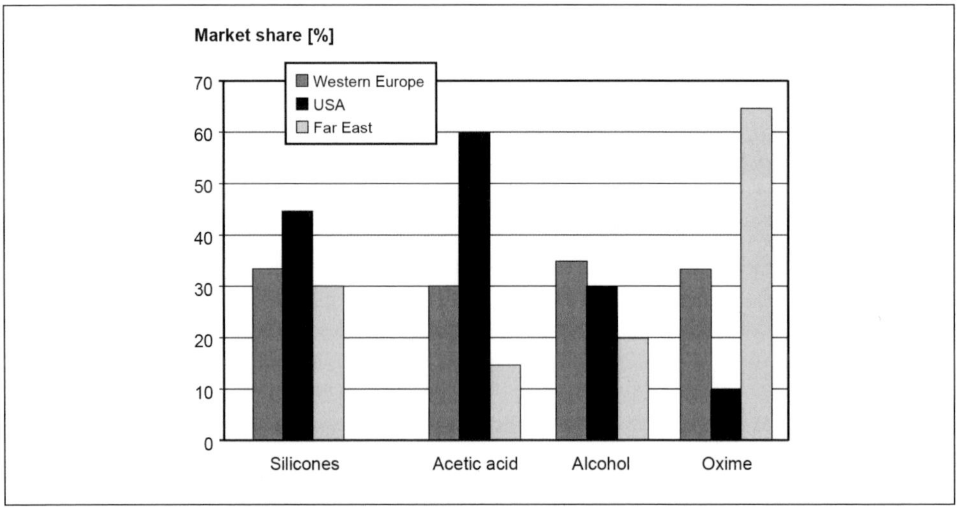

Figure II.2.146: Market share (%) of silicone sealants and relative share (%) of the various types of curing systems

- other additives (e.g. organotin compounds as catalyst and fungicide, silane adhesion promoters, desiccants and others)

Contaminated silicone sealant in warm, humid environments (e.g. sanitary areas) is attacked by black mould. Silicone for sanitary use therefore has a relatively high level of fungicide.

The maximum permissible permanent deformation is 25 %. Silicone sealants cannot be painted. They cannot be used in paint shops either, due to the risk of cratering.

2-component, condensation-curing systems (RTV-2) also exist, but these are less important and are used where the humidity is too low. Addition-curing RTV-2 silicones exist as well, but will not be discussed here. RTV-2 silicones usually cure rapidly.

Table II-2.64: Guide formulation for a moisture-curing, 1-component construction sealant based on silicone (RTV-1) [16]

Item	Raw Material	Parts by weight
1	Polydimethylsiloxane, hydroxyl-terminated	50 to 65
2	Polydimethylsiloxane, trimethylsilyl-terminated	20 to 30
3	Fumed silica	6 to 10
4	Methyl and/or ethyltriacetoxysilane	4 to 5
5	Dibutyltin diacetate*	0.01

* Item 5: dibutyltin diacetate is now classified as corrosive and harmful to the environment; if possible, an alternative catalyst should be used

Properties of silicones

By virtue of their being organosilicon polymers that have inorganic structural elements, silicones have some outstanding properties, relative to purely organic polymers.

Advantages

+ high resistance to heat (up to 150 °C, special types resistant up to 250 °C)
+ high flexibility at low temperatures (down to -40 °C)
+ good elastic recovery (resilience)
+ excellent weathering resistance
+ very good resistance to chemicals
+ low shrinkage during crosslinking
+ excellent adhesion to silicate materials (ceramics, enamel, glass, concrete and others) and to metals (especially aluminium)

Their drawbacks include the lack of coatability and – like all moisture-curing, 1-component systems – the dependence of the curing rate on the ambient conditions (temperature, humidity).

It should also be noted that despite the excellent resistance to condensed moisture, the barrier effect against water vapour is very low.

Applications of silicones [2]

- construction sealants
- sealing of
  - oil sumps of vehicle engines
  - steam irons
  - burners and boilers
- aerospace adhesives (for bonding silicon solar cells and ceramic parts to metal substrates)
- adhesive bonding of glass windows to stoving ovens and sealing cookers with silicone rubber.

Guide formulation

A guide formulation for a moisture-curing, 1-component construction sealant based on silicone (RTV-1) is presented in Table II-2.64 [3]. It is an acidic, acetoxy type.

The frequently high-molecular, hydroxyl-terminated polydimethylsiloxane crosslinks with the alkyltriacetoxy silane [R-Si(OOC-CH_3)$_3$, R = methyl and/or ethyl] in the presence of humidity. Dibutyltin diacetate catalyses this reaction. The non-reactive trimethylsilyl-terminated polydimethylsiloxane is mostly a liquid and acts as plasticiser. The fumed silica serves as rheological additive and reinforcing filler. Such sealants often have a high filler content, e.g. chalk (P/B ≈ 1), to reduce costs.

2.4.3.1 Silane-modified sealants and adhesives

Silane-modified sealants based on MS polymers (also called hybrid polymers) are shown in Figure II-2.147.

They are 1-component systems and cure by reacting with humidity. Skinning times vary with the system, but range from 10 to 20 minutes. The curing rate can be up to 5 mm/day; again, it is system-specific. Owing to their chemical structure MS sealants have good weathering and ageing resistance. Furthermore, they adhere to nearly everything, without the need for a primer. They are often the answer for critical substrates. Freshly applied MS sealants can be coated with water-based and solvent-borne paints.

In Japan, construction sealants based on MS polymers have a market share of 45 % [1].

Guide formulation

A guide formulation for a construction sealant based on MS polymers is presented in Table II-2.65 [1, 4].

The mechanical properties of the cured sealant can be adjusted by mixing two MS polymers having different functional groups (Table II-2.66). Note that the two silanes in the formulation (Figure II-2.148) have different roles: DAMO is an adhesion promoter whereas the VTMO serves as desiccant because it is more reactive to humidity. Moreover, it is best to dry pigment and filler before use [1], as humidity affects the storage stability of the sealant. An anti-ageing agent could also be added (e.g. HALS product).

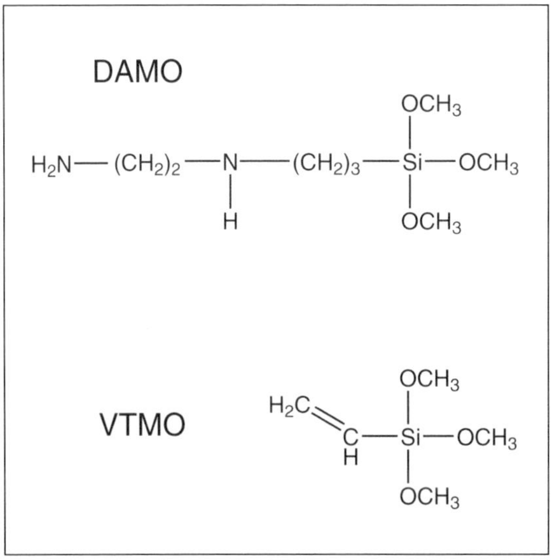

Figure II-2.148: N-Aminoethyl-3-aminopropyl trimethoxysilane (DAMO) and vinyl trimethoxysilane (VTMO)

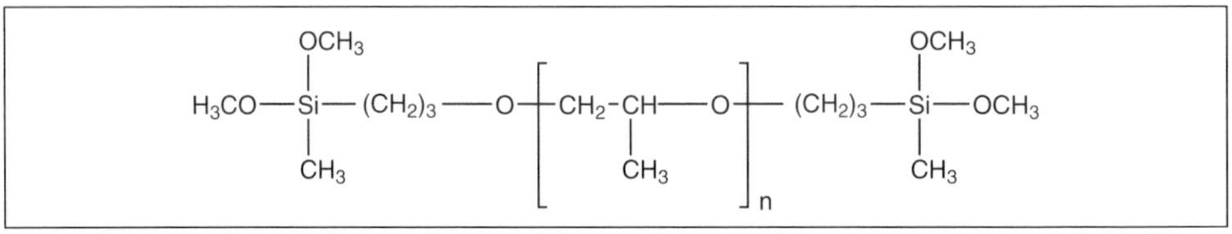

Figure II-2.147: MS polymer

Table II-2.65: Elastic adhesive based on MS Polymer 50 Shore A

Item	Raw material	Parts by weight
1	MS polymer	100
2	Plasticiser	55
3	Titanium dioxide, rutile	20
4	Rheological additive	5
5	Filler	180
6	Stabilizer	1
7	Stabilizer	1
8	Desiccant	2
9	Adhesion promoter	5
10	Catalyst	2
Total		371

Item 1: MS Polymer "S303H, Kaneka
Item 2: "Jayflex DIUP", ExxonMobil
Item 3: "Sachtleben R-FK-2", Venator Materials Corporation
Item 4: "Crayvallac SLX", micronised polyamide rheology modifier, Arkema
Item 5: "SILLITIN Z 86", natural combinations of silica and kaolinite, HOFFMANN MINERAL
Item 6: "Tinuvin 770", solid basic hindered amine light stabilizer, BASF SE
Item 7: "Tinuvin 327", light stabilizer, BASF SE
Item 8: "Dynasylan VTMO", vinyl trimethoxysilane (VTMO, Figure II-2.148), Evonik
Item 9: "Dynasylan AMEO" N-aminoethyl-3-aminopropyl-trimethoxysilane (DAMO, Figure II-2.148), Evonik
Item 10: organo-tin compound as crosslinking catalyst, e.g. dibutyltin bisketonate; e.g. "Metatin Catalyst 740", Actylis

Table II-2.66: Guide formulation for a construction sealant based on MS polymers

Item	Raw material	Parts by weight
1	MS polymers	25
2	Plasticiser	17
3	Titanium dioxide, rutile	3.0
4	Filler	50
5	Rheological additive	3.5
6	Adhesion promoter	0.5
7	Desiccant	0.7
8	Catalyst	0.3
Total		100

Item 1: mixture of two MS polymers; linear polyether with methyldimethoxysilane-end groups (Fig. III-3.14) and a higher-functional MS-polymer in the ratio 60:40; e.g. "S203H" and "S303H", Kaneka, Japan
Item 2: di-iso-undecyl phthalate, e.g. "Jayflex DIUP", ExxonMobil
Item 4: ground calcium carbonate, coated with stearic acid, average particle size 2 µm, specific surface area 5 m^2/g
Item 5: polyamide wax; e.g. "Crayvallac SLX" or "-SL", Cray Valley [13]
Item 6: N-aminoethyl-3-aminopropyl-trimethoxysilane (DAMO, Figure II-2.148)
Item 7: vinyl trimethoxysilane (VTMO, Figure II-2.148)
Item 8: organo-tin compound as crosslinking catalyst, e.g. dibutyltin bisketonate; e.g. "Metatin Catalyst 740", Actylis

In recent developments, a polyacrylate backbone is used instead of a polyether backbone (e.g. MS polymer), which can also crosslink with silane functions through condensation. However, other functional groups can also be used with this new type of raw material, which also allow radical crosslinking (Figure II-2.149) [10].

2.4.3.2 Silane-terminated polyurethanes

Adhesives/sealants based on silane-terminated polyurethanes (Figure II-2.150) exist, too. Unlike traditional moisture-curing, 1-component polyurethane sealants, these silane-terminated polyurethanes do not release carbon dioxide on curing with moisture.

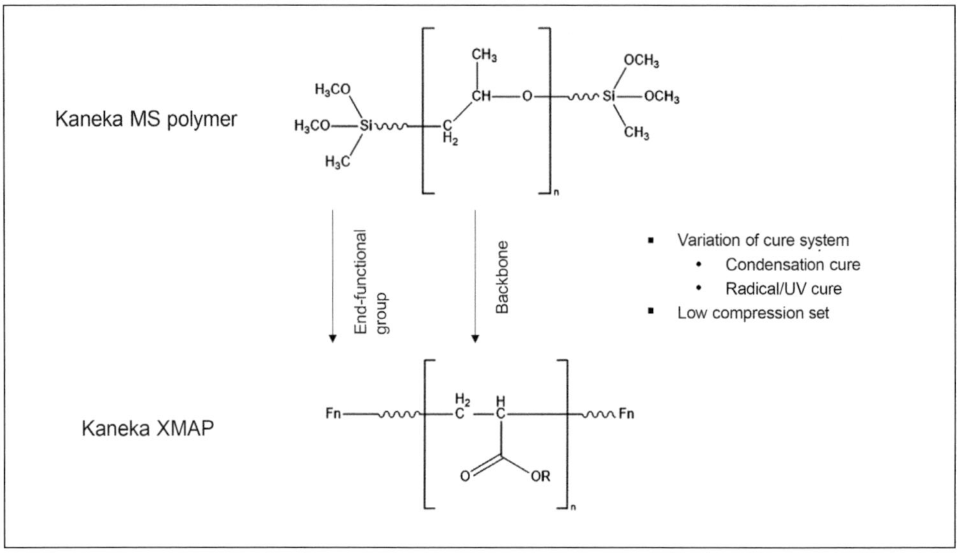

Figure II-2.149: Kaneka XMAP compared to Kaneka MS Polymer [10]

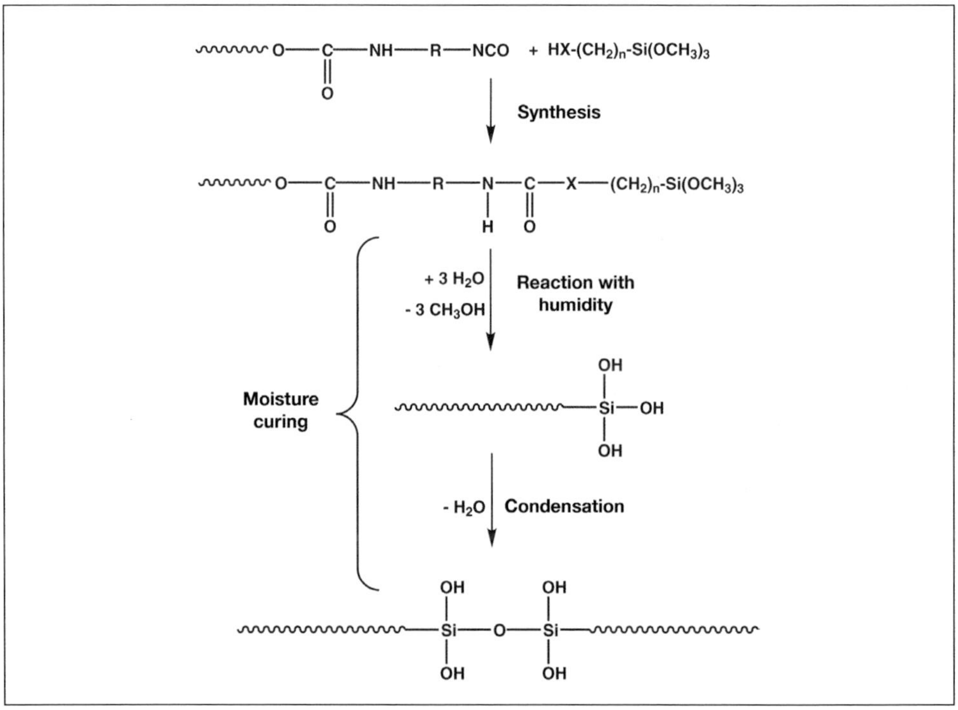

Figure II-2.150: Silane-terminated polyurethanes (-XH is -SH, -OH or -NHR)

Guide formulation

Instead of N-phenyl-aminopropyl trimethoxysilane, it is possible to use N-(n-butyl)-aminopropyl trimethoxysilane (Figure II-2.151), but reduced by 0.8 wt% (Table II-2.67) [5].

The NCO/OH ratio of the prepolymer in Table II-2.67 will now calculated for verification purposes. The molecular weights and functionalities of the two polyether polyols (Table II-2.67) enable the hydroxyl content (% OH) to be estimated as shown in Chapter II-2.2.1.5. That calculation returns OH equivalent masses of 2000 and hydroxyl contents of 0.85 % OH for the diol and the triol.

The NCO/OH ratio can now be calculated from the following equation:

NCO/OH ratio = wt % polyisocyanate · 17 % NCO
wt % polyol · 42 %OH

$$= \frac{4.8 \cdot 17 \cdot 33.6}{(28.4 \cdot 42 \cdot 0.85) + (4.8 \cdot 42 \cdot 0.85)} = 2.3$$

Thus, the large excess of isocyanate groups yields a relatively low-molecular, isocyanate-terminated prepolymer.

The data presented in the formulation in Table II-2.67 can be used to determine if addition of silane blocking agent (aminosilane) is stoichiometric.

The total amount of prepolymer is 28.4 + 4.8 + 4.8 = 38 pbw (parts by weight)

Isocyanate base value = 42 · 100 · % OH (for 100 pbw polyol)
17 · % NCO

$$\text{Isocyanate base value} = \frac{(42 \cdot 28.4 \cdot 0.85) + (42 \cdot 4.8 \cdot 0.85)}{17 \cdot 33.6} = 2.1 \text{ (pbw MDI)}$$

With the help of the isocyanate base value, the stoichiometric reaction of the two polyols with MDI computes formally to 2.1 pbw MDI. Hence, 4.8 - 2.1 = 2.7 pbw MDI remains formally. In reality, the large excess of MDI reacts with the polyols via just one isocyanate group (see Figure II-2.152).

2.7 pbw MDI is equivalent to 0.011 mol or 0.022 val (two NCO groups). The NCO equivalent mass is therefore calculated as follows:

NCO equivalent mass = mass prepolymer : isocyanate content (val) = 38 : 0.022 = 1727

Hence, the isocyanate content of the prepolymer = $\frac{42 \cdot 100\,\%}{1727}$ = 2.4 %

For complete reaction of these isocyanate groups, the following amount of N-phenyl-aminopropyl trimethoxysilane is required (molecular weight = H-active equivalent mass = 255.2 g/mol):

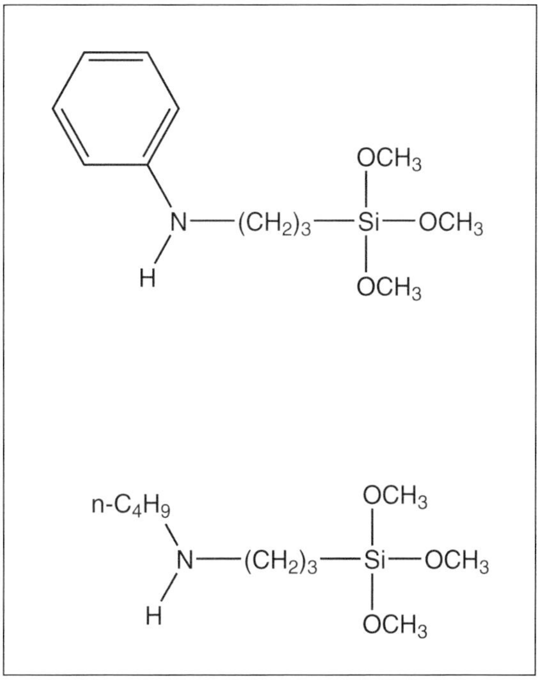

Figure II-2.151: N-phenyl-aminopropyl trimethoxysilane and N-(n-butyl)-aminopropyl trimethoxysilane

Table II-2.67: Guide formulation for a sealant based on a silane-terminated polyurethane [15]

Item	Raw material	Parts by weight
1	Polyether diol (molecular weight 4000)	28.4
2	Polyether triol (molecular weight 6000)	4.8
3	Diisocyanate	4.8
4	Aminosilane	4.2
5	Plasticiser	19.0
6	Filler	38.1
7	Rheological additive	0.7
Total		100

Item 1: "Acclaim Polyol 4200", OH equivalent weight approx. 2,000 g, Covestro
Item 2: "Acclaim Polyol 6300", OH equivalent weight approx. 2,000 g, Covestro
Item 3: diphenylmethane-4,4'-diisocyanate (MDI), 33.6 % NCO
Item 4: N-phenyl-aminopropyl trimethoxysilane (Figure II-2.151)
Item 5: diisodecyl phthalate
Item 6: calcium carbonate, coated with stearic acid, average particle size 0.38 µm
Item 7: fumed silica
Prepolymer synthesis: Heat item 1 to 3 in vacuum to 60 °C for 2 hours. Then add item 4 and heat for additional 2 hours at 60 °C under nitrogen

quantity of aminosilane (expressed in terms of 100 g prepolymer) = (100 · H-active equivalent mass) : isocyanate equivalent mass quantity of aminosilane (expressed in terms of 38 g prepolymer) $= \frac{38 \cdot 255.2\,\%}{1727} = 5.6\,\%$ (pbw). In other words, there is a deficit of aminosilane in the formulation (Table II-2.67).

To illustrate this, let us calculate the molar ratios of the reactants (Table II-2.67) (rounded to integers):

polyether diol : polyether triol : MDI : aminosilane ≈ 9 : 1 : 24 : 21 (see Figure II-2.152)

The small quantity of triol (Table II-2.67) yields a low crosslink density during curing and therefore confers high flexibility. However, the reaction of the aminosilane (Figure II-2.150) boosts the functionality of the prepolymer. This leads to a higher crosslink density and to a higher tensile strength (undesirable in this case) relative to the original polyurethane standard [5].

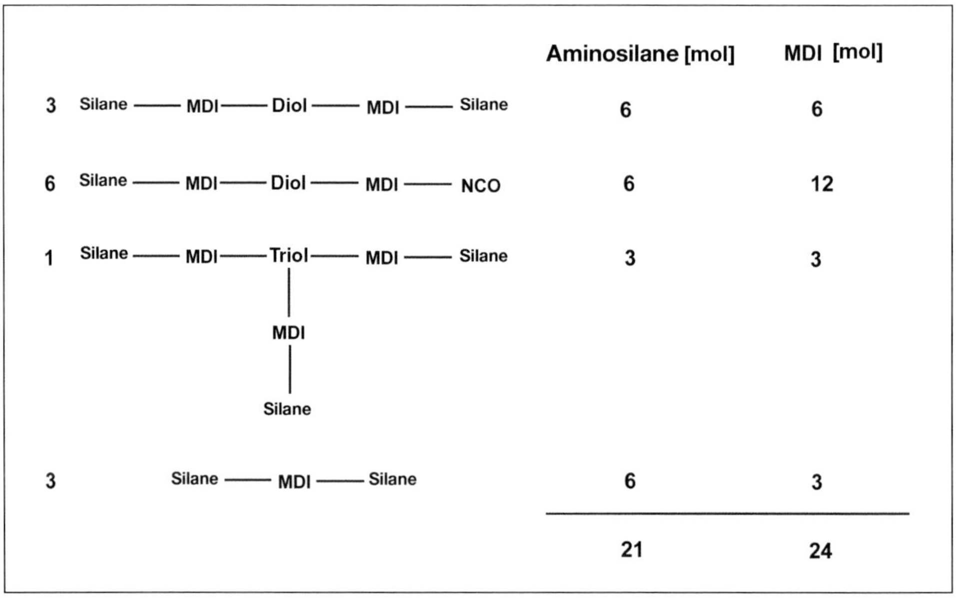

Figure II-2.152: Simplified diagram of the binder for a sealant based on silane-terminated polyurethane (Table II-2.67)

Table II-2.68: Guide formulation for a 1-component silane-terminated polyurethane adhesive with good adhesion properties on various substrates and improved thermal stability based on a commercially available raw material [8]

Item	Raw material	Parts by weight
1	SPUR+ 1015 prepolymer	66.70
2	SPUR+ 3040 prepolymer	33.30
3	Chalk	111.11
4	Thixotropic agent (SiO_2)	4.17
5	Colourant	1.39
6	Moisture scavenger	1.94
7	Adhesion promoter	1.94
8	Plasticizer	55.56
9	UV stabilizers	0.83
10	UV stabilizers	0.83
11	Catalyst	0.28

Item 1: silane-terminated prepolymer, Momentive
Item 2: silane-terminated prepolymer, Momentive
Item 3: calcium carbonate, coated with stearic acid, average particle size 0.38 µm
Item 4: fumed silica, "CAB-O-SIL TS-720", Cabot
Item 5: TiO_2, "Ti-Pure R-960", Chemours Titanium Technologies
Item 6: silane, "Silquest A-171", Momentive
Item 7: silane, "Silquest A-1120", Momentive
Item 8: DIDP, "Jayflex DIDP", ExxonMobil
Item 9: "Tinuvin 213", BASF SE
Item 10: "Tinuvin 662 LD", BASF SE
Item 11: DBTDL, dibutyltin dilaurate, "Fomrez SUL-4", Galata Chemicals

Table II-2.69: Guide formulation for a 1-component silane-terminated polyurethane sealant based on a commercial raw material [8]

Item	Raw material	Parts by weight
1	SPUR+ 1015 prepolymer	23.1
2	Plasticizer	18.5
3	Moisture scavenger	0.4
4	Chalk	55.5
5	UV stabilizers	0.5
6	Thixotropic agent (SiO_2)	0.2
7	Colourant	1.2
8	Adhesion promoter	0.58
9	Catalyst	0,05

Item 1: silane-terminated prepolymer, Momentive
Item 2: DIDP, "Jayflex DIDP", ExxonMobil
Item 3: "Silquest A-171", Momentive
Item 4: calcium carbonate, coated with stearic acid, average particle size 0.38 µm
Item 5: UV-Absorber P, 2-(2H-Benzotriazol-2-yl)-p-Kresol, Avokal GmbH
Item 6: fumed silica, "CAB-O-SIL TS-720", Cabot
Item 7: TiO_2, "Ti-Pure R-960", Chemours Titanium Technologies
Item 8: "Silquest A-1110" silane, Momentive
Item 9: DBTDL, dibutyltin dilaurate, Trigon

An alternative for the synthesis of silane-terminated polyurethanes, as summarised in Table II-2.67, is the use of finished raw materials, which allow the user easier access to this class of adhesives. Examples of such formulations are given in Table II-2.68, Table II-2.69 and Table II-2.70.

An overview of recent developments in **sealants** made with silane-crosslinking polymers is presented in [6]. The silanes used so far have been γ-silanes; this means that there are three methylene groups between the silane function and the amino group (Figure II-2.151). A more recent and more reactive class of silanes is that of the α-silanes, which have only one methylene group acting as spacer. Polyethers terminated with α-silanes allow faster-curing systems to be formulated [7].

Table II-2.70: Guide formulation for a 1-component silane-terminated polyurethane adhesive (75 Shore A) based on commercially available raw materials [11]

Item	Raw material	Parts by weight
Step 1		
1	Desmoseal S XP 2458	36.43
2	Mesamoll	4.88
3	Irganox 1135	0.49
4	Fibadur 04007284	0.10
5	CAB-O-SIL TS-720	1.97
6	SILLITIN Z 86	51.63
7	Dynasylan VTMO	2.20
8	DBU (Diazabicycloundecene)	0.10
Step 2		
9	Dynasylan 1146	1.10
10	Dynasylan 1505	1.10

Item 1: silane-terminated prepolymer, Covestro
Item 2: plasticizer, Lanxess
Item 3: antioxidant, BASF SE
Item 4: pigment, Finke Colors
Item 5: thixotropic agent, Cabot
Item 6: extender, Omya
Item 7: adhesion promotor, Covestro
Item 8: DIDP, Jayflex DIDP, ExxonMobil
Item 9: adhesion promotor, Covestro
Item 10: 3-aminopropylmethyldiethoxysilane, Covestro
Mixing: For the production of this adhesive a vacuum dissolver with side scraper is needed (vacuum dissolver PC-LBDV 7, PC Laborsystem).

Step 1:
The components are homogenized successively under stirring, if necessary, the filler may be added in several portions. The drying agent VTMO and the catalyst are added shortly before dispersion.
The dispersion has to be made under cooling (inside temperature: < 70 °C) under static vacuum (< 200 mbar) for:
- approx. 10 min at 2500 rpm (dissolver) + 250 rpm (butterfly) and
- approx. 10 min at 1000 rpm (dissolver) + 100 rpm (butterfly)

The water content after dispersion should be < 200 ppm.

Step 2:
The components are homogenized with the mixture from step1.
The dispersion has to be made under cooling (inside temperature: < 70 °C):
- approx. 10 min at 1000 rpm (dissolver) + 100 rpm (butterfly) under static vacuum (< 200 mbar))
- approx. 5 min at 1000 rpm [11]

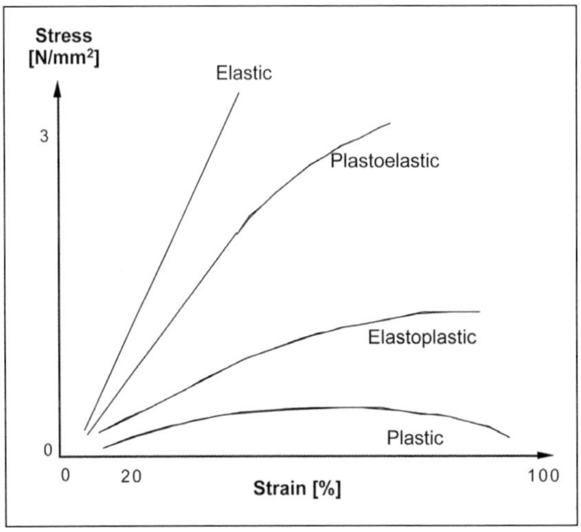

Figure II-2.153: Highly simplified stress-strain diagram of different sealants

2.4.3.3 Classification of construction sealants

Stress-strain behaviour

The stress-strain behaviour (Figure II-2.153) affords a rough qualitative way of classifying sealants as elastic, plastoelastic, elastoplastic and plastic.

Elastic recovery (resilience), which characterises the elastic portion of a sealant, allows a more precise distinction. Elastic recovery is at least 70 % for elastic sealants, 40 to 70 % for plastoelastic types, 20 to 40 % for elastoplastic types and less than 20 % for plastic types [6].

Application form

Construction sealants come in the following application forms:

- compressible, preformed sealants (elastic gaskets)
- plastic sealing compounds, e.g. butyls
  - kneadable
  - profiles
- elastic sealing tapes (bonded with an adhesive-sealant of the same type over the joint)
- self-levelling sealants (flat floor seals)
- extrudable, non-sag sealants (the focus of this chapter)

For use in building construction, 1-component sealants are always advantageous because they rule out the possibility of mixing errors by workers.

2.4.3.4 References

[1] Mack, H., Choosing the Right Silane Adhesion Promoters for SMP Sealants (2002) in adhesivesmag.com and contribution to the 2nd Organosilicon Days, Munich (2003)
[2] Langer, M., Adhäsion, 40, No. 1-2 (1996) p. 26 ff
[3] Wacker Silicones Info-Service (2004)
[4] One-Component Moisture Curing Methoxysilane Sealants (2001) in crayvalley.com
[5] Brochure from Degussa (now Evonik), "Sivento Silanes for Adhesives and Sealants", (9/2001)
[6] Pröbster, M., Adhäsion (2004) No. 1-2, p. 10–14
[7] Schindler, W., FARBE UND LACK, 112 (2006) No. 10, p. 44–49
[8] momentive.com/en-us/brands/spur
[9] De Buyl, F., "Silicone sealants and structural adhesives." International Journal of Adhesion and Adhesives 21.5 (2001): 411-422.
[10] kaneka.be/technology-products/liquid-polymers/kaneka-xmaptm
[11] www.hoffmann-mineral.com/guideformulation/file/1342

3 Pressure-sensitive adhesives (PSA)

Even though the term pressure-sensitive adhesive is defined as "adhesive which in a dry state is permanently tacky at room temperature and adheres readily to surfaces under light and brief pressure" in EN 923 [15], this section covers both the actual pressure-sensitive adhesive mass, and pressure-sensitive products such as adhesive tapes, labels, decals or self-adhesive die-cut parts.

3.1 Basic principles

Pressure-sensitive adhesives (PSAs) are permanently tacky products which adhere to substrates under the application of slight pressure. This chapter covers the following process steps involved in the manufacture of pressure-sensitive adhesives and pressure-sensitive adhesive products (see Figure II-3.1):

- the polymerization of the pressure-sensitive adhesive polymer
- its formulation to pressure-sensitive adhesive compounds e.g. by adding resins, fillers and stabilizers
- the coating to a carrier to produce roll goods
- the conversion to single- or double-sided adhesive tapes, labels, decals, die-cut parts, etc. resulting in pressure-sensitive adhesive products

In contrast to the adhesives discussed so far in this book, which solidify after their application, i.e. after application to an adherend and joining, whereby they must be in liquid form either through a physical process or a chemical reaction, the pressure-sensitive adhesives can be described as non-curing or non-solidifying adhesives. Nevertheless, they must be in liquid form

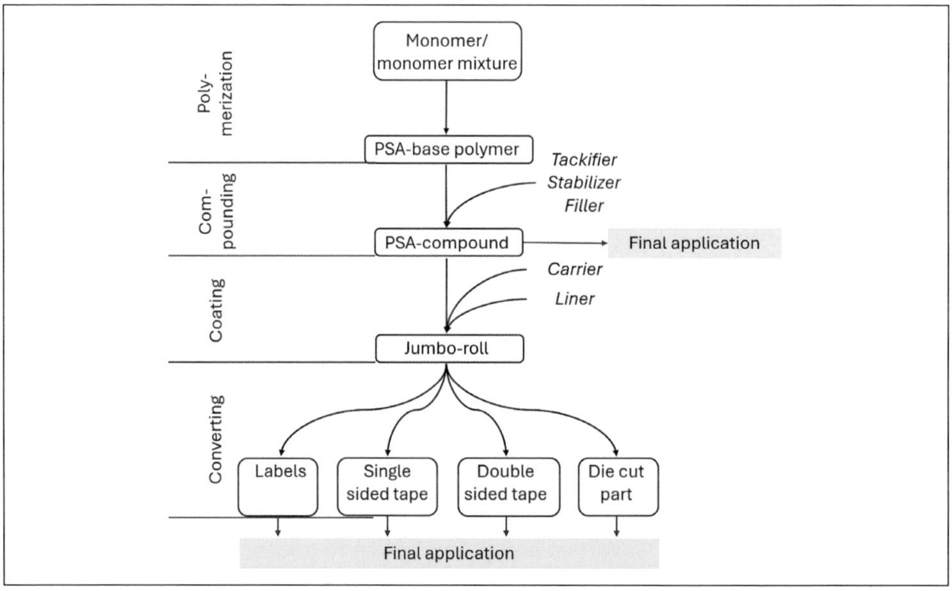

Figure II-3.1: Production steps for the manufacture of pressure-sensitive adhesive products

(solution in organic solvents, water-based dispersion or emulsion, hot-melt) to allow coating to a carrier in the manufacture of an adhesive tape, a label, etc.
They exhibit a combination of the properties of liquids and solids, called viscoelasticity allowing them to wet adherends upon contact, without the help of organic solvents or water. The plastic flow of a pressure-sensitive adhesive during wetting and development of adhesion to the substrate is promoted by a low glass transition temperature (T_g) and relatively low molecular weight. Pressure-sensitive adhesives usually have glass transition temperatures which are 40 – 70 K below application temperature (mostly ambient temperature) [4]. Advantageous, but not mandatory for good wetting and therefore for the development of adhesion, are a long contact time, high contact pressure and elevated temperature.

Pressure-sensitive adhesives have the ablility to transfer a certain amount of load. In simple terms, pressure-sensitive adhesives are designed in such a way that the ratio between flowability and flow resistance is adapted to the respective application. However, the performance of pressure-sensitive adhesives depends not only on the viscoelastic properties of the adhesive but also on its ability to build up adhesion to the adherend (Chapter I-2.2 and I-2.1). The adhesion of pressure-sensitive adhesives is achieved solely through physical interaction with the surface of the adherend; chemical bonds do not play any role. The increase in bond strength observed in many pressure-sensitive adhesives in the first approx. 24 hours (Figure II-3.2) is a result of the chain mobility still present in the viscoelastic state. Since the polymer chains have a certain mobility, the polar or polarisable groups of the polymer responsible for adhesion can orient themselves in such a way that additional interactions are formed with the surface of the adherent. (Figure II-3.3).

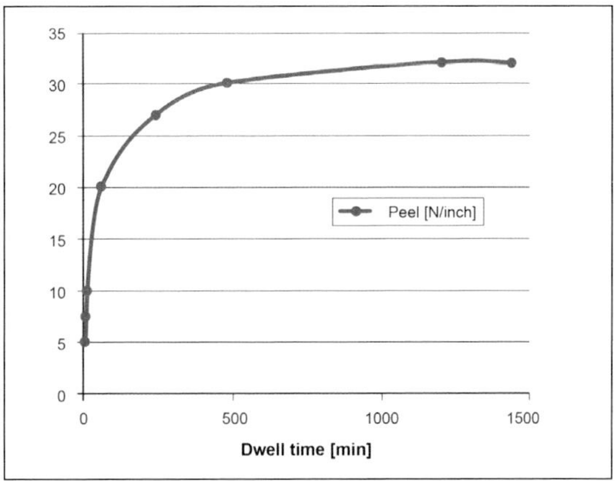

Figure II-3.2: Time-dependent strength build-up of pressure-sensitive adhesives

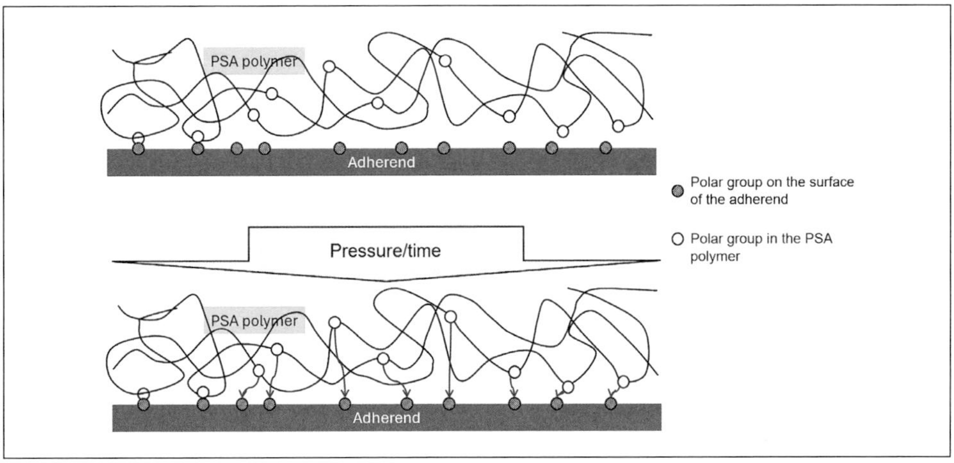

Figure II-3.3: Orientation of the polar or polarizable groups in the PSA polymer

Table II-3.1: Classification of pressure-sensitive adhesives by peel resistance (180°) [3]

Pressure-sensitive adhesive	**Peel resistance [N/25 mm]**
Extremely permanent	> 14
Permanent	10 to 14
Partly removable	6 to 8
Removable and repositionable	2 to 4
Easily removable	< 1

The adhesive strength of a pressure-sensitive adhesive bond is mostly characterized by its peel adhesion under a specified peel rate. The force required to peel a strip of adhesive tape from a specified substrate at a specified angle and speed (normally expressed in N/25 mm), together with the shear strength or creep resistance, are important properties characterizing pressure-sensitive adhesives. Depending on the application, permanent adhesion or, as in the case of masking tapes, protective films and sticky notes a residue-free removability may be desired. Table II-3.1 provides an overview of the corresponding peel adhesion values. In addition, there is the group of repositionable pressure-sensitive adhesives, defined as adhesives that can be removed and repositioned after bonding, e.g. for position correction [2].

In contrast to solidifying adhesives, where the shear strength is typically specified as the force at which a bond fails [MPa] or [N/mm^2], for pressure-sensitive adhesives the time [h] that the bond withstands a constant load [kg/mm^2] is usually specified as the so-called shear adhesion or holding power. For bonds that are exposed to elevated temperatures, the shear adhesion is determined either at a defined elevated temperatures or by using the so called SAFT-test at increasing temperatures, see Chapter IV-4.4 for details.

At this point, it should be noted that a further difference to the solidifying adhesives is the level of load that can be transferred per unit area. Except for more recent developments in which chemical cross-linking takes place after the adherends have been joined, triggered e.g. through temperature or moisture, the strength that can be achieved with pressure-sensitive adhesives is usually less than 0.5 MPa. Despite this low strength level compared to the adhesives discussed in Chapter II-2, pressure-sensitive adhesives have proven themselves in multiple applications in a wide variety of areas.

Typically, pressure-sensitive adhesives are used in the form of pre-manufactured single or double-sided adhesive tapes, labels, stickers, sticky note pads, die-cut parts or self-adhesive parts (referred to in this book as pressure-sensitive adhesive products). Single-sided adhesive tapes are often used, for example, to seal cartons, as office tape, as temporary masking tape or protective film that can be removed without leaving any residue, for marking or fixation purposes or as bandages in wound care. Double-sided adhesive tapes are used to join parts, whereby one of the adherends is often already provided with the double-sided adhesive tape during its manufacture. For the actual bonding process, just the liner protecting the adhesive layer must be removed before the part is then joined to the previously cleaned second adherend under light pressure. Labels are used for labelling and, where necessary, writing on parts. Just like masking tape and protective film, sticky notes are characterized by the fact that they can be easily removed without leaving any residue. The direct application of pressure-sensitive adhesive from drums or pails to the adherend immediately before joining is also possible, but an exception.

As pressure-sensitive adhesives are generally used in the form of single-sided but mostly double-sided adhesive tapes to produce adhesively bonded joints or in the form of single-sided pressure-sensitive labels, the coating process and the influences of the various materials on the properties of an adhesive tape should first be described before the various technologies for pressure-sensitive adhesive compounds are described.

3.2 Pressure-sensitive tapes

3.2.1 Construction of pressure-sensitive adhesive tapes

A distinction is made between single-sided, double-sided and transfer adhesive tapes. Figure II-3.4 shows their construction. These are usually placed on the market in the form of rolls, but occasionally also in the form of sheets or specially cut to the dimensions of the part to be bonded, e.g. by laser or die-cutting.

3.2.1.1 Carrier/backing

This is a flexible material to which the pressure-sensitive adhesive is applied in the production of adhesive tapes. With double-sided adhesive tapes, this is referred to as the carrier, with single-sided tapes as the backing.

The function of the carrier/backing is to improve the dimensional stability of the viscoelastic, pressure-sensitive film and thus facilitate the application of the adhesive tape. In addition, it can serve as a separating layer between the two layers of pressure-sensitive adhesive in double-sided tapes and thus to produce adhesive tapes whose two sides offer different adhesive properties adapted to the respective adherend to be bonded (often referred to as asymmetric double sides adhesive tape). An overview of materials frequently used as carrier can be found in Table II-3.2. In the case of single-sided adhesive tapes, the backing is often coated with a release coating on the side not carrying the pressure-sensitive adhesive, so that the adhesive film does not need to be covered by a release liner when the tape is produced as a roll.

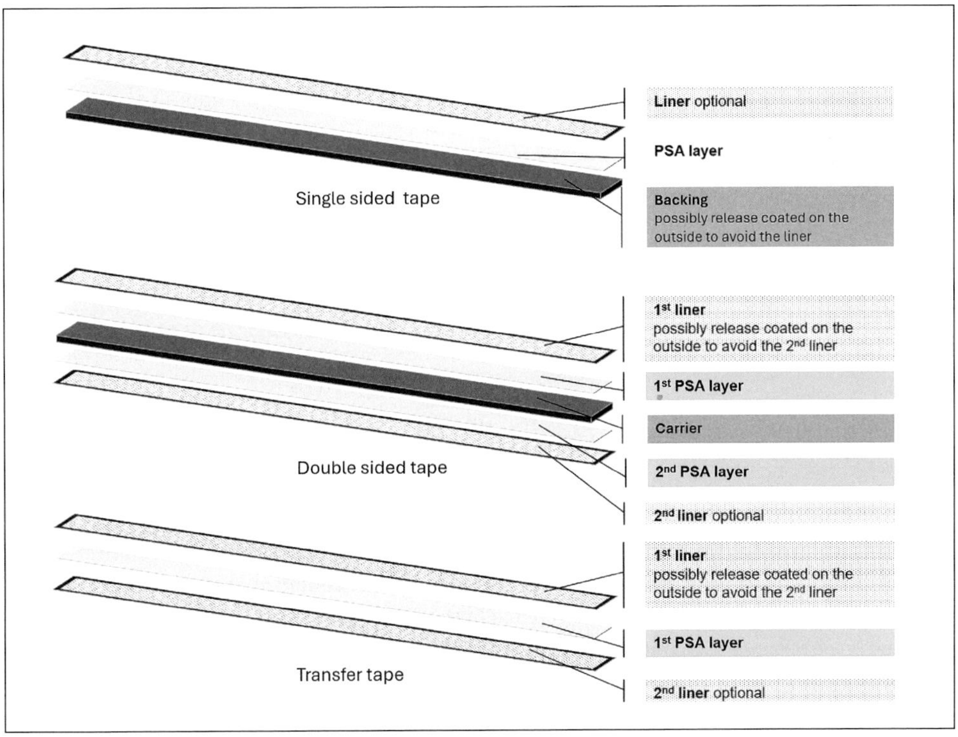

Figure II-3.4: Schematic construction of adhesive tapes

Table II-3.2: Materials frequently used as carrier/backing

Material	**Properties**	**Applications**
Polymeric film – Polyester – Polyethylene – PVC (might contain plasticizer!) – …	high tear and dimensional stability, even with low thickness flexible, thin media resistant, migration barrier against plasticizers and other chemicals PVC: contains plasticiser/shrinkage no tolerance compensation	profile bonding mirror bonding hook bonding bonding of emblems insulating tapes mounting of membrane switches
Paper/non-woven – Coated paper – Tea filter paper – Crepe	thin reasonable dimensional stability easy to tear no tolerance compensation good value	self-adhesive signs foam lamination masking tapes
Scrim Fibres connected to each other at intersections with a large distance between them 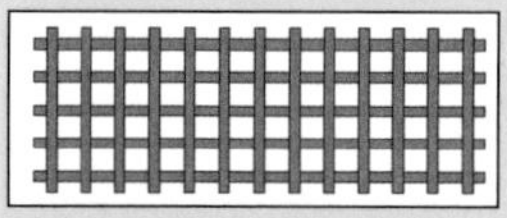	improved dimensional stability adhesive layer is available for both bonding sites as it stays capable of flow due to the wide meshes limited tolerance compensation	foam lamination
Textile-/glas fabric Woven fibre material 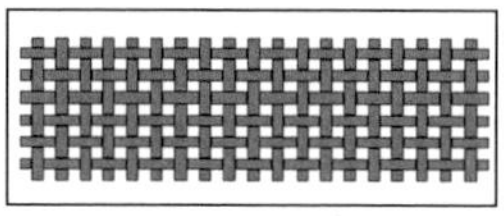	good dimensional stability tear resistant good repositionability safe, permanent anchorage of the adhesive with the carrier thick adhesive layers limited tolerance compensation	carpet mounting self-adhesive preparation of flexible materials gaffer tape
Foam – Polyethylene – Polyurethane – PVC – …	good conformability to uneven surfaces compensation of tensions compressibility compensation of different expansions of the objects to be bonded	mounting of emblems and decorative trims in the automotive sector mirror bonding hook bonding
Metal film – Aluminium – …	good dimensional stability tear resistant limited tolerance compensation	heating and ventilation

3.2.1.2 Release liner (liner)

The function of the release liner, often referred to as liner, is to protect the adhesive layer from soiling and environmental influences and to prevent the layers of an adhesive tape from sticking together on the roll or to prevent the adhesive layer of self-adhesive parts from unintentionally sticking to other parts, e.g. the packaging. Often used are release-coated paper, polymer films, e.g. made of PET, PE or PP films. PP films often do not require a release coating due to their low surface energy as well as creped materials. Paper is coated with silicone or polyethylene, while polymer films are usually coated with silicone. In order to avoid the use of a liner for single sided adhesive tapes a backing coated

Table II-3.3: Materials frequently used as liner

Release liner	**Construction**	**Properties**
Silicone-coated paper	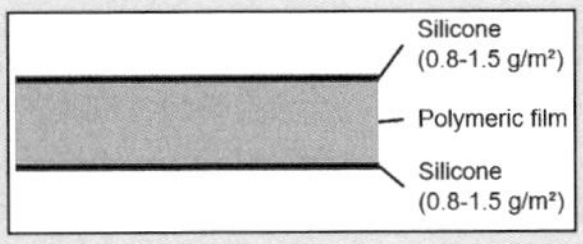	relatively inexpensive good tear resistance low moisture resistance differentiated release force possible
PE-coated paper The paper is coated with a thin PE layer to which a silicone layer is applied	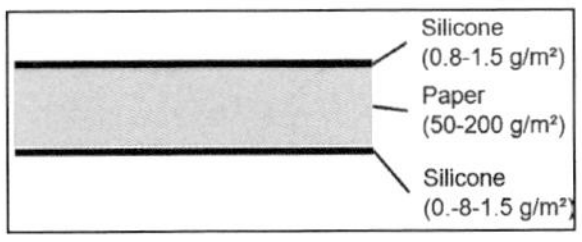	improved dimensional stability and moisture resistance differentiated release force possible
Silicone-coated polymeric film	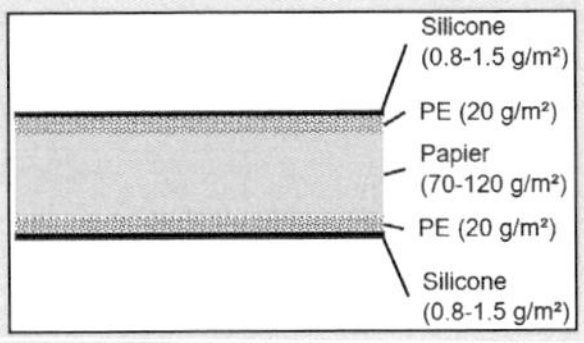	flexible and smooth tear-resistant anchoring of the silicone coating required to avoid silicone transfer
Creped polymeric film	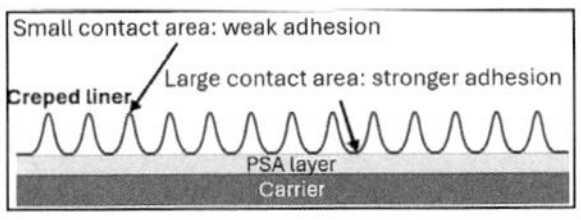	no silicone coating required differentiated release force via the size of the contact area

on one side by a release coating is used. In the case of transfer films in roll form, liners coated on both sides are used, whereby the coatings are selected so that the two sides of the liner differ in their release characteristics. This ensures that the adhesive layer remains on a defined side when the adhesive tape is unwinded. An overview of materials frequently used as carrier can be found in Table II-3.3.

3.2.1.3 Pressure-sensitive adhesive layer (PSA compound)

Pressure-sensitive adhesives can be produced based on different natural as well as synthetic polymers in different processes, which in turn differentiate between the form in which the pressure-sensitive adhesive is coated to the carrier. Figure II-3.5 provides an overview of the most important technologies while Figure II-3.6 gives an overview of the market share by technology and by chemistry. [46]

3.2.2 Comparison of general properties

It is not surprising that the general properties of pressure-sensitive adhesives are strongly determined by the respective base polymer, rubber, acrylate, polyurethane or silicone. Furthermore, the form, either solvent-based, aqueous as a dispersion or as a 100 % product, not only determines the coating process but also the

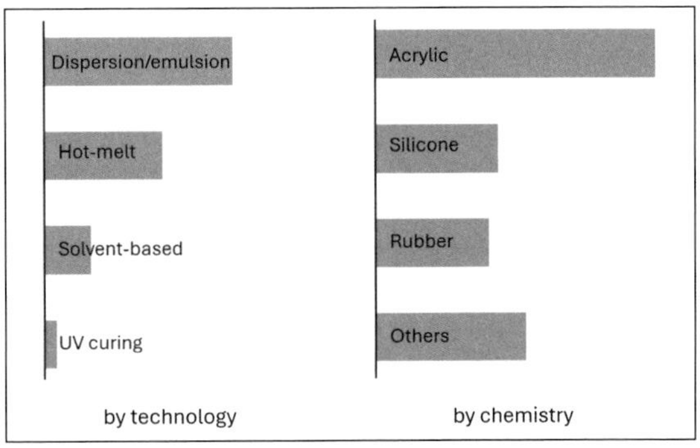

Figure II-3.6: Relative market share of pressure-sensitive adhesives by technology and chemistry [46]

Table II-3.4: Pressure-sensitive adhesives – typical technology-based properties

Polymer type	Advantages	Disadvantages
Acrylic	good UV resistance colourless transparent to clear films good ageing resistance cheaper than silicones	low initial tack limited adhesion to low energy surfaces (can be improved by resin modification) more expensive than rubber
Rubber Natural or synthetic	high initial tack good adhesion to low energy surfaces cost effective compared to acrylics and silicones	limited ageing resistance sensitive to UV low resistance to low temperatures significantly more expensive than acrylics and silicones
Silicone	very good adhesion to low-energy surfaces and silicone very good temperature and chemical resistance	low initial tack low peel adhesion

resulting properties of the bond. Table II-3.4 provides an overview of the typical characteristics of the various technologies. However, it should be noted that solvent-based systems as well as 100 % solid systems like hot-melts and UV curing systems generally offer advantages over water-based dispersion adhesives in terms of resistance to moisture and direct contact with water.

The properties that can be achieved with the respective technologies, together with their areas of application, are discussed in more detail in Chapter II-3.3.

3.2.3 Manufacturing of tapes

3.2.3.1 Coating

In the manufacture of adhesive tapes (Figure II-3.7), typically the release liner is coated with the pressure-sensitive adhesive as a solution, dispersion or hot-melt in the desired constant layer thickness, whereby defects such as air bubbles, pinholes, dirt stains, streaks caused by trapped dirt particles or gel particles must be avoided. Typical dry coating weights are 20 – 25 g/m^2 but depending on the required final properties both lower and higher coatings weights are also common. Coating is typically performed by using suitable application devices in a roll-to-roll process

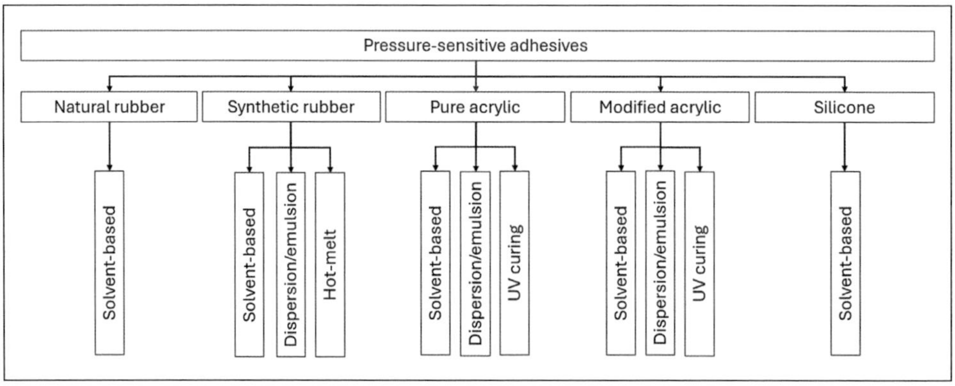

Figure II-3.5: Most common pressure-sensitive adhesive technologies

with a coating width of typically 500 to up to 1600 mm. Common technologies for coating with solvent-based pressure-sensitive adhesives are reverse roll, knife-over-roll and roll-over-roll. Aqueous dispersions or emulsions are usually applied using the wire-rod process while hot-melt pressure-sensitive adhesives are applied as a low viscous polymer melt using the slot-die process. A detailed description of these and other coating technologies can be found in [16].

After coating, the pressure-sensitive adhesive film must be dried to produce a uniform dry coating from the wet adhesive by evaporating the solvent mixture from solvent-based or the water from dispersion or emulsion pressure-sensitive adhesives. Forced air ovens are usually used to accelerate the drying process. In order to avoid the formation of a film that would hinder complete drying, or causing bubbles drying is usually carried out in stages, i.e. the oven has sections with different temperatures.

Coating speed is an important factor, especially in terms of productivity, with drying, i.e. the length of the drying channel, being the limiting factor in most cases. Solvent-based systems offer advantages over aqueous systems here, but for reasons of occupational safety and environmental protection, in addition to an appropriate explosion protection, at least thermal post-combustion and, better still, solvent recovery are required.

After drying the pressure-sensitive adhesive layer is laminated with the carrier. Depending on the carrier material and the pressure-sensitive adhesive corona pre-treatment of the carrier to ensure good adhesion of the pressure-sensitive adhesive is required.

A high-temperature coating with the pressure-sensitive adhesive as a low-viscosity, solvent-free melt is an option, too. As these hot-melt adhesives cool on the liner, their cohesion builds up.

Alternatively, reactive systems can also be used that contain photoinitiator and can be cross-linked extremely quickly by ultraviolet light (UV). The acrylic polymer is polymerized until a viscosity allowing coating is reached and coated to a liner with common coating technologies and finally polymerized by passing through a UV light tunnel. Compared to the adhesives described in Chapter II-2.3, a lower degree of cross-linking is aimed for to maintain the tacky surface [3]. The process is similar to the one shown in Figure II-3.7, instead of a drying oven a much shorter UV tunnel is used.

3.2.3.2 Converting

During converting, the large master rolls also names jumbo rolls (width up to 1.6 m, diameter 1.2 m, weight approx. 2 tonnes) coming from the coater are brought into the shape required for further use. After conditioning the master rolls under defined climatic conditions, edge trimming is carried out using a rewinder allowing not also cutting to a specific length but also cutting out of

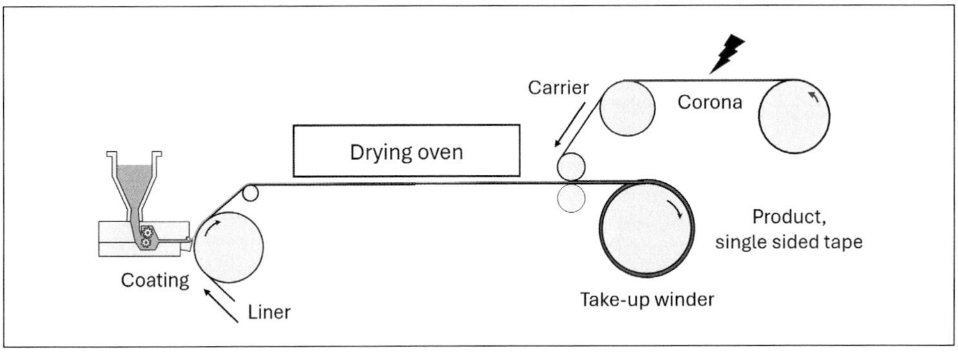

Figure II-3.7: Coating process on the example of producing a single sided pressure-sensitive tape using a solvent- or water-based adhesive

defects noticed during the coating process. Using a so-called rewind splitter, cutting to width can be carried out simultaneously, but this is only worthwhile for large quantities, as the process is fast but requires a long set-up time to adjust the up to 100 blades and the coil cores. The quality of the slitting depends on the knives used and the pressure-sensitive adhesive layer. It is obvious that the adhesive must have a consistency that does not lead to soiling of the knives. Rolls pre-cut to length, so called logs can be slitted without unwinding in the so-called lathe slitting process. Here, the log of an adhesive tape is rotated on a stationary mandrel at high speed while a circular blade cuts (slices) the material to the specified width. This process is used when small quantities of adhesive tape need to be cut to several different widths. For more details see: [17, 19].

When large quantities of identical shapes need to be produced quickly from a single material, such as in the production of labels or specially shaped double-sided adhesive parts, die-cutting is the process of choice. Specially manufactured punching tools, which consist of sharp blades mounted on a plate, are pressed onto the sheet lying on a base and the desired shape is cut out.

By optimizing the arrangement of the identical shapes on the sheet, cutting waste can be minimized and the number of punched parts obtained per surface area can be optimized. The process enables the rapid manufacture of standardized products that can be used in mass production. Almost any shape and size can be die-cut from a material. The possibilities range from individual pieces cut from a sheet or roll to "kiss-cut", a process in which the die-cut piece remains attached to a roll of release paper or other material.

If only small quantities or initially only sample parts are required, laser cutting is a cost-effective alternative without the need for expensive punching tools. The devices essentially consist of a computer-controlled plotter that carries a laser for cutting the materials instead of a pen. The safety measures required when using a laser are complied with by an appropriate design in a protective housing. The plotter can be fully set up in just a few minutes by importing digitalized drawings. Long set-up and waiting times for the tool are eliminated and potential changes to the design of the part can be realized in minutes avoiding the costs associated with manufacturing a new tool.

3.3 Formulation of the PSA-Compound

It is obvious that the adhesive properties of a pressure-sensitive adhesive composition are essentially determined by the structure of the polymer formed during polymerization, i.e. by its chemical basis, the monomer composition, the molecular weight distribution and its resulting glass transition temperature. Furthermore, the properties can also be specifically modified after polymerization by compounding, either by mixing different polymers or by adding other components like

- tackifier,
- plasticizer,
- extender,
- thickeners and thixotropic agents,
- filler,
- crosslinker,
- solvents,
- stabilizers, antioxidants,
- antifungal agents,
- surfactants, emulsifier, wetting agents, film formation agents.

3.3.1 Solvent-based pressure-sensitive adhesives

Even though the market share of solvent-based pressure-sensitive adhesives is significantly lower than that of water-based dispersion or emulsion pressure-sensitive adhesives and hot-melts due to the well-known occupational safety and environmental protection requirements associated with the use of solvents, they are still often used for demanding applications due to the properties that can be achieved. The solvent is only released during the coating process at the adhesive tape manufacturer, where solvent emissions into the environment can be minimized by using appropriate plant technology (post-combustion or solvent recovery); no solvents are released during the actual bonding process.

3.3.1.1 Solvent-based natural rubber PSA

Rubber latex is an aqueous dispersion of natural rubber (NR) from rubber trees. The bark of the trees is scratched, and the latex is collected in small containers (Figure II-3.8). The latex is coagulated in bowls, for example with formic acid, and the water is squeezed out. During the squeezing process, the last roll stamps a characteristic pattern onto the rubber sheets. The sheets are then dried (Figure II-3.9) and are sometimes smoked to speed-up the drying process [18, 20].

Before being dissolved, natural rubber (polymeric cis-1,4-isoprene) is mastificated using a two-roll rubber mill or internal mixer to achieve the required viscoelastic properties and to increase solubility. The degree of required milling depends on both the required properties of the resulting adhesive and the state of the rubber. As a natural raw material, the state can vary from batch to batch. By adding pentachlorothiophenol the process of mastification can be accelerated. The dissolution is a slow process therefore the rubber is often granulated to create a larger surface and thus to accelerate dissolving. Typical solvents are white spirit, hexane and heptane. Toluene was also widely used in the past, but due to health concerns it is practically no longer used today.

Figure II-3.8: Tapping latex from a rubber tree

Figure II-3.9: Drying rubber sheets

Table II-3.5: Starting point formulation for the solid part of a natural-rubber based pressure-sensitive adhesive [12]

Item	Raw material	Parts by weight (solids)
1	Natural rubber	55.0
2	Resin based on poly-β-pinene (softening point 70 °C)	41.2
3	Plasticising oil	2.7
4	Stabilising agent	1.1
Total		100

Item 4: polymeric trimethyldihydrochinoline

Table II-3.6: Natural rubber-based pressure-sensitive adhesive with increased heat resistance [47]

Item	Raw material	Parts by weight (solids)
1	Natural rubber	54
2	Polyterpene tackifier	35
3	Brominated phenol-formaldehyde resin	5
4	Zinc resinate accelerator	5
5	Antioxiodant	0.5
6	Antioxidant LC	0.5
7	Zinc dibutyldithiocarbamate	0.5
total		

Item 2: "Piccolyte S 115", Pinova, Inc.
Item 3: "SP-1056", SI Group, Inc/SI Group-Switzerland GmbH
Item 5: 2,5-di-tert-amylhydroquinone
Item 6: "Ionol LC", Raschig GmbH
Item 7: "Butyl Zimate SG", Vanderbilt Chemicals, LLC

As pure natural rubber is not very tacky, tackifier resins, e.g. rosin, terpene or hydrocarbon resin are added (up to 50 – 120 phr). Thus, tack and peel adhesion are increased up to a certain level of addition whereas creep resistance mostly decreases. Antioxidant additives (0.5 – 2 phr amine, phenol or dithiocarbamate) are added to improve aging resistance. If necessary, fillers (35 – 65 phr) are admixed to reduce costs and pigments to impart a colour. After the substrate has been coated, the double bonds of the polymer can be crosslinked e.g. by added zinc oxide/sulphur cross-linking system or by halogenated phenol-formaldehyde resin accelerated by zinc resinate in a controlled manner to increase heat distortion and solvent resistance. [21]

Guide formulations

A starting formulation for the solids content of a pressure-sensitive adhesive is presented in Table II-3.5 [12].

Table II-3.6 shows an example for a rubber-based pressure-sensitive adhesive with increased heat resistance. [47]

Evaluation

Solvent-based pressure-sensitive adhesives based on natural rubber are fairly inexpensive. They adhere strongly to non-polar substrates. However, the double bonds in the polymer lead to poor ageing resistance and yellowing.

Major applications are low-cost packaging tapes, masking tapes, low-cost double-sided tapes and inexpensive labels.

3.3.1.2 Solvent-based acrylic PSA

Even though solvent-free acrylate dispersions (see Chapter II-3.3.2) and UV cross-linking acrylates (Chapter II-3.3.4) have become increasingly important in the manufacture of pressure-sensitive adhesive products, solvent-based systems cannot always be dispensed with. Solvent-based acrylic pressure-sensitive adhesives are made by radical polymerization of the monomer mixture under atmospheric pressure in organic solvents at an elevated, well-controlled temperature. The reaction process corresponds in principle to that shown in Figure II-2.99 in Chapter II-2.3.1.1.

For production (see Figure II-3.10) the solvent or solvent blend is usually charged in the polymerization reactor equipped with a reflux condenser and brought to boiling point. The starter solution and the monomer mixture are added from separate storage tanks at a defined rate. The reaction process can be easily monitored by measuring the viscosity. For safety reasons, in the event of an uncontrolled exothermic polymerization reaction, the plant is equipped with a so-called catch tank in which the reaction mass is collected and thus prevented from being released into the environment.

By adding additional initiator, a post-reaction is then carried out to reduce the content of residual monomers. After cooling, the reaction mass is stabilized with an alcohol and can finally be formulated with a cross-linking agent, the addition of a wetting agent or to improve adhesion especially to low-energy plastics, by adding a tackifier resin solution.

To avoid the formation of undesirable gel particles which very likely will cause streaks during coating, polymerization is started with a high monomer concentration of up to 70 % and a low proportion of initiator under reflux conditions. The remaining monomer is added during the course of the reaction, whereby the addition of further initiator keeps the reaction going. The viscosity can be regulated by adding solvent.

The low glass transition temperature of acrylic polymers required to achieve pressure-sensitive properties is mainly achieved by using monomers based on esters of acrylic and methacrylic acid and alcohols with 4 to 12 carbon atoms. Longer carbon chains reduce adhesive strength because of a tendency to crystallize. With regard to the influence of chain branching, it has been shown that short-branched chains (C_1 to C_6) make the polymer harder, while this effect is reversed with longer chains.

Homopolymers are not suitable as pressure-sensitive adhesives. To achieve the desired divergent properties (tackiness, peel adhesion, shear strength), so-called base monomers with tackify-

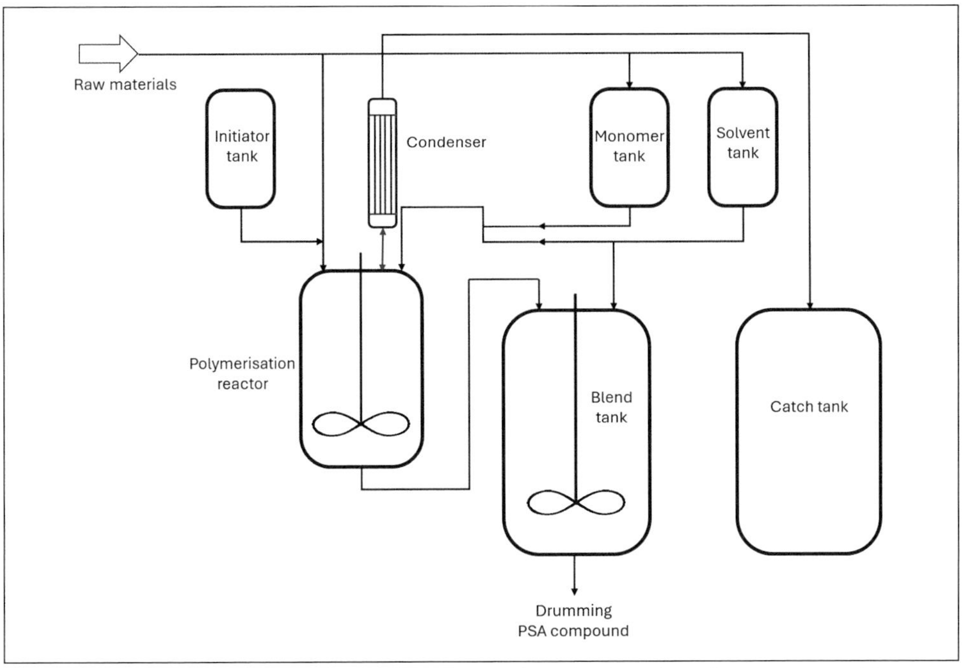

Figure II-3.10: Schematic design of a unit for solvent polymerization of acrylic pressure-sensitive adhesives

ing properties such as n-butyl acrylate (T_g -54 °C) and 2-ethylhexyl acrylate (T_g-70 °C) are combined with so called modifying monomers that have a higher glass transition temperature and therefore stiffening properties, such as vinyl acetate (T_g 28 °C), methyl acrylate (T_g 8 °C), methyl methacrylate (T_g 108 °C) or styrene (T_g 100 °C). Furthermore, smaller fractions of functional monomers like 2-hydroxyethylacrylate (T_g -15 °C), acrylic acid (T_g 106 °C) or methacrylic acid (T_g 228 °C) may be integrated for the purposes of cross-linking to improve cohesive strength or to improve adhesion. The glass transition temperature of the resulting polymer can be calculated from the monomer composition according to the following formula.

Equation II-3.1

$$T_{g\,(polymer)} = \left[\frac{1}{\left[\frac{c_{(1)}[\%]}{100 \cdot T_{g(1)}[K]} + \frac{c_{(2)}[\%]}{100 \cdot T_{g(2)}[K]} + \text{-----} + \frac{c_{(n)}[\%]}{100 \cdot T_{g(n)}[K]} \right]} \right] - 273.16\,K$$

where

$T_{g\,(polymer)}$: glass transition temperature of the polymer
C_n [%]: concentration by weight of the monomer
$T_{g(1)}$: glass transition temperature of the monomers

Modification with resins, plasticisers or fillers is possible, but not essential. In general pressure-sensitive acrylic polymers consist of about

- 50 – 90 % base monomers
- 10 – 40 % modifying polymers
- 2 – 20 % functional monomers

The solvent used for polymerization should have a good dissolving power for the monomer mixture as well as for the resulting polymer and the initiator used. It also has the function of preventing an excessive increase in temperature due to the highly exothermic reaction and should have a low chain transfer rate in order to achieve polymers with the desired high molecular weight and to avoid gel formation. Other important criteria for the selection of the solvents are, in addition to the lowest possible health and environmental hazard potential and the lowest possible costs, the possibility of solvent recovery after coating, as well as good drying properties (avoidance of blistering). Ethyl acetate, n-heptane, acetone and white spirit are frequently used solvents, while mixtures of these solvents are also often used. Toluene, which was frequently used for a long time, is hardly used today for health and safety reasons.

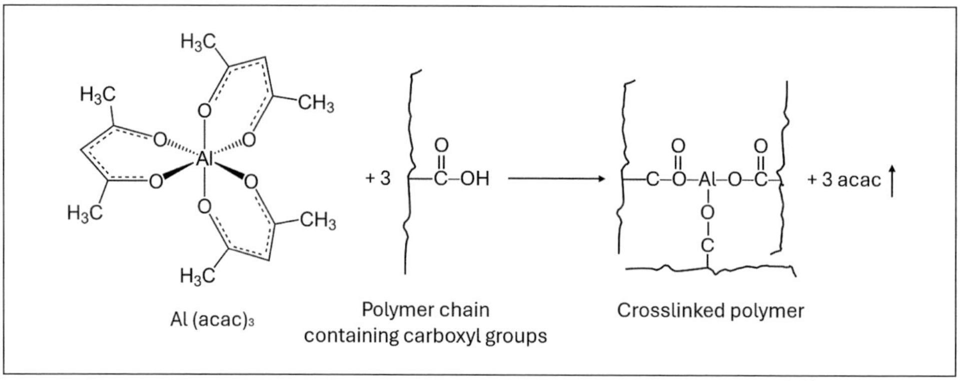

Figure II-3.11: Self-cross-linking by means of aluminium acetylacetonate ($Al(acac)_3$) [48]

Oil-soluble peroxides such as the dibenzoyl, di-tert-butyl or dilauryl peroxide already mentioned in Chapter II-2.3.1.1 are widely used to initiate the polymerization reaction, although they can occasionally lead to undesirable cross-linking and their rate of decomposition depends on the type of solvent used. Azo compounds such as 2,2'-azo-bisisobutyronitrile do not show these disadvantages. The initiator concentration is usually 0.05 – 2 %, based on the monomer content. A lower proportion leads to a higher molecular weight. The polymerization rate is directly proportional to the monomer concentration and is related to the square root of the initiator concentration.

Table II-3.7: Formulation example for a removable solvent-based acrylic PSA adhesive

Item	Raw material	Parts per weight
1	Butyl acrylate	36.4
2	Hydroxyethyl acrylate	1
3	Acrylic acid	1
4	Benzoylperoxide	0.5
5	Heptane	15
6	Ethylacetate	20
7	Ethanol	22
8	Isopropanol	1
9	Crosslinker	3
10	Cross-linking catalyst	0.1
Total		100

Item 4: "Perkadox L-W75", Nouryon Functional Chemicals B.V
Item : butylated melamine resin, "Cymel 27-804", Peninsula Polymers
Item 10: magnesiumbromide hexahydrate

Pressure-sensitive adhesives based on acrylic acid esters only cross-linked via hydrogen bonds or interpenetrating networks show significant property deficits, in particular an insufficient thermal-mechanical load-bearing capacity. By incorporating small amounts (0.1 – 0.3 % by weight) of difunctional monomers such as 1,6-hexanediol diacrylate, cross-linking is already achieved during copolymerization, which leads to an improvement in shear strength. However, the effect is limited, as an excessive increase in cross-linking density increases the risk of gel formation.

Alternatively, cross-linking can be achieved by incorporating small amounts of functional monomers, e.g. containing hydroxyl, amine, thiol or isocyanate groups. In a 2-component process, cross-linking can be achieved during the coating process by mixing in a reactive cross-linking agent that is adapted to the functional groups contained in the polymer (e.g. diisocyanate). However, these systems have the disadvantage that, depending on the reactivity of the system, there is a more or less pronounced increase in viscosity, which in turn affects the consistency of the resulting coating weight. It is also possible to select a crosslinker that practically does not react with the reactive groups of the incorporated functional monomer at room temperature but does react at the elevated temperature during drying.

Another process often used for the production of so-called self-cross-linking acrylate adhesives is the process via metal chelates such as aluminium acetylacetonate ($Al(acac)_3$). As $Al(acac)_3$) is a solid, it is usually added as a 3 – 5 % solution in ethyl acetate. The aluminium chelate complex is in equilibrium with the extremely reactive aluminium incompletely chelated with acetylacetone. The addition of acetylacetone shifts the equilibrium towards the saturated aluminium. During drying after coating, the acetylacetone evaporates (boiling point 144 °C), so that the equilibrium shifts towards the reactive aluminium that is incompletely chelated and cross-linking occurs as a result of the substitution of the acetylacetonate groups of the $Al(acac)_3$ by the carboxyl groups of the polymerised acrylic acid (Figure II-3.11). [48]

The adhesion of solvent acrylic adhesives, particularly to low-energy plastics, can be improved by adding tackifier resins. In order to reduce the temperature resistance as little as possible, the

Table II-3.8: Formulation example for a high shear and high temperature resistant solvent-based acrylic PSA adhesive

Item	Raw material	Parts per weight
1	2-ethylhexyl acrylate	20
2	Ethyl acrylate	10
3	Methyl acrylate	10
4	Acrylic acid	2
5	Benzoylperoxide	2
6	Toluene	15
7	Heptane	5
8	Ethylacetate	30
9	Isopropanol	5
10	Aluminium acetylacetonate	0.2
11	2,4-pentanedione	0.8
Total		100

Item 5: „Perkadox L-W75", Nouryon Functional Chemicals B.V

use of solid resin solutions is recommended. For example, hydrocarbon resins that are as colourless as possible, phenol-modified indene-coumarone resins and ester resins based on partially hydrogenated rosin are suitable.

Guide formulations

Table II-3.7 provides an example for a coater ready removable pressure-sensitive adhesive based on butylacrylate containing a melamine cross-linking system that reacts with hydroxyl and carboxyl groups of the polymer chains under cross-linking when exposed to heat in the drying zone.

Table II-3.8 shows an example for a coater ready permanent pressure-sensitive adhesive offering a high shear and temperature resistance. Cross-linking is achieved by aluminium acetylacetonate.

Evaluation

In addition to the general advantages already mentioned in Chapter II-3.2.2, solvent-based acrylic pressure-sensitive adhesives offer a significantly improved water resistance compared to water-based dispersion pressure-sensitive adhesives, which makes them ideal for outdoor applications in particular, but also for other applications with high water or moisture contact like clear bottle labelling where even whitening of the adhesive layer is not allowed.

3.3.2 Water-based PSA

Due to increasing environmental concerns and regulations, pressure-sensitive adhesives have increasingly been used in the form of aqueous polymer dispersions predominantly based on acrylate since the mid-1980s. [1]. Continuous improvements in product properties have enabled solvent systems to be increasingly replaced by aqueous systems in many, but not all, applications, making them the most important class of pressure-sensitive adhesives today (see Figure II-3.6). [46]

3.3.2.1 Dispersion PSA based on acrylic polymers

Like solvent-based acrylics, the properties of acrylic dispersion pressure-sensitive adhesives can be regulated via the composition of the monomers. Monomers which lead to low glass transition temperatures increase tack, e.g. 2-ethylhexyl acrylate and n-butyl acrylate. Monomers which lead to high glass transition temperatures increase cohesion, e.g. methyl methacrylate.

Polymerization process and formulation

Although acrylic dispersion pressure-sensitive adhesives are usually produced by emulsion polymerization, they can also be produced by solution polymerization and subsequent dispersion in water.

Emulsion polymerization process

For emulsion polymerization, the emulsifier is first dissolved in water, whereby micelles are formed from a certain concentration, in which the polymer chains later grow. After the monomers and the thermally decomposing initiator are added the mixture is heated above the decomposition temperature of the initiator (in most cases 50 – 90 °C). Initially, small polymer radicals are formed in the water phase which after addition of some of the poorly water-soluble monomers develop an affinity for the hydrophobic micelle interior and diffuse there, so that the reaction proceeds inside the micelle. It is also possible that a generated polymer particle first attaches emulsifier and thus forms a micelle around itself.

The micelle wall of emulsifier molecules acts in principle as a membrane that allows the penetration of further monomers into the micelle, whereby the polymerization in the micelle continues. As more emulsifier molecules are added, the micelle wall grows, so to speak. Provided that an excess of initiator is present, polymerization only comes to a stop when the monomer has been completely consumed. In order to achieve a low content of residual monomers, additional initiator is usually added towards the end, i.e. after a conversion of the monomers of at least 95 %. In this way, acrylic dispersions with a solids content of up to approx. 60 % can be produced with a viscosity that is still suitable for coating. If a solids content above 60 % is required, it is recommended, for example, to add an excess of emulsifier or so-called nano-emulsions, which are characterised by nanodroplets with a uniform particle size distribution (20 – 500 nm) and can be obtained by ultrasonication.

In contrast to solvent polymerization, where the molecular weight of the polymers is limited by the viscosity of the resulting polymer solution, the molecular weight and also the degree of cross-linking of the polymers generated by emulsion polymerization have practically no influence on the viscosity of the dispersion, so that the advantages of a higher molecular weight and a certain degree of cross-linking can be exploited. [22]

Monomers

As with solvent-based acrylates, the adhesive properties are essentially determined by the monomer composition and the ratio of low to high molecular weight polymer fractions. Essentially the same monomers are used as in solution polymerization. Monomers with carboxyilc groups as well as acrylic, methacrylic and itaconic acids and their salts stabilise the dispersion. Therefore, many acrylic dispersions have good pressure-sensitive adhesive properties, without added resin. Very tacky products can be obtained with the help of regulators. Many commercially available acrylic pressure-sensitive adhesive dispersions are coater-ready which means they can be used without any modification for coating substrates.

Emulsifiers (surfactants)

Typically, a mixture of anionic and non-ionic emulsifiers at a ratio of 2 – 5 % based on the monomer mixture is used. Of course, if mixtures of emulsifiers are used, the individual components must be compatible with each other. Common emulsifiers are:

- ethoxylated fatty alcohols (EO degree: 3 to 50, alkyl radical: C_8- to C_{36})
- ethoxylated mono-, di- and tri-alkylphenols (EO degree: 3 to 50, alkyl radical: C_4- to C_9)

- alkali metal salts of dialkyl esters of sulphosuccinic acid
- alkali and ammonium salts of
  - alkyl sulphates (alkyl radical: C_8 to C_{12})
  - ethoxylated alkanols (EO degree: 4 to 30, alkyl radical: C_{12} to C_{18})
  - ethoxylated alkylphenols (EO degree: 3 to 50, alkyl radical: C_4 to C_9)
  - alkylsulphonic acids (alkyl radical: C_{12} to C_{18}) and
  - alkylarylsulphonic acids (alkyl radical: C_9 to C_{18})

In the past alkyl phenol especially nonyl phenol and octyl phenol ethoxylates have been widely used as non-ionic emulsifiers. However, the REACH Regulation stipulates that these may only be used in a very few specific exceptional cases since 2021. As alternatives alkylphenol ethoxylate-free, biodegradable secondary alcohol ethoxylate emulsifiers such as "Tergitol 15-S-9 A", or "Ecosurf EH-9" from Dow can be used.

Examples of commercially available emulsifiers suitable for acrylic emulsion polymerizations are "Dowfax 2 A1" (alkyldiphenyloxide disulfonate from DOW), "Dextrol OC 50" (phosphate ester surfactant from Ashland), "Texapon NSO" ($C_{12}C_{14}$-Fatty alcohol ethersulfate, sodium salt from BASF), "Disponil FES 77" (fatty alcohol ether sulphate, sodium salt from BASF), "Lutensol AT 18" (ethoxylate of linear oleochemical alcohol from BASF) and "Emulphor NPS 25" (high-molecular weight ether sulphate from BASF). [22]

The emulsifier concentration influences the resulting molecular weight of the polymer. Increased emulsifier concentration especially in the start phase of the polymerization reaction increases the micelle concentration and therefore the number of particles formed. The molecular weight is inversely proportional to the number of particles formed. Emulsifiers do also influence pressure-sensitive properties. Depending on the type of emulsifier, these can act as plasticisers or adversely affect the adhesive properties or water resistance of the bond due to their tendency to migrate to interfaces. It is therefore preferable to use emulsifiers that are compatible with the polymers as they are less likely to migrate to the surface. Using emulsifiers which are fugitive or reactive and copolymerise is an effective method for preparing dispersion acrylic pressure-sensitive adhesives without migratory emulsifiers.

Initiators

In particular, water-soluble ammonium and alkali metal salts of peroxysulphuric acid, e.g. sodium peroxodisulphate, hydrogen peroxide or organic peroxides, e.g. tert-butyl hydroperoxide are used as initiators at approx. 0,1 – 10 % relative to the monomer mixture. In order to achieve the highest possible molecular weight, the initiator level should be kept as low as possible at the start of polymerization in order to keep the micelle concentration low, as already mentioned above. In addition, the common organic peroxides such as benzoyl peroxide are also used, sometimes as secondary catalysts. [22, 23]

Regulators

By adding regulators, e.g. in quantities of 0 – 0.8 parts by weight, based on 100 parts by weight of the monomers to be polymerised, the molecular weight can be reduced in order to obtain acrylic pressure-sensitive adhesive dispersions with a high tack for subsequent compounding. Suitable are, for example, compounds with a thiol group such as tert-butylmercaptan, thioglycolic acid ethyl acrylate, mercaptoethynol, mercaptopropyltrimethoxysilane or tert-dodecylmercaptan. [22]

Compounding

Modification of water-based acrylic pressure-sensitive adhesive dispersions for the purpose of optimising the properties is simple. However, due to the limited shear stability of dispersions, care must be taken during compounding to ensure that the individual raw materials are mixed homogeneously under low shear, e.g. using a propeller stirrer. Dissolvers and other high-speed mixers are generally unsuitable as they can cause the dispersion to break. Care must also be taken to avoid entrapped air and the formation of difficult to destroy foam.

Acrylic polymer dispersion

Pressure-sensitive adhesives with both high cohesion and good tack can be obtained, for example, by mixing different acrylate dispersions. Low-molecular weight acrylic dispersions are added specifically to improve tack. A large range of dispersions of varying tack and adhesive strength are available from reputable producers (e.g. from Alberdingk Boley, trade name "Alberdingk AC" [24], BASF SE, trade name "Acronal" [25], HB Fuller, trade name "Acryltac" [26] and from Vinavil S.p.A., trade names "Crilat" [27]).

Tackifier

Tack and adhesion can also be improved by adding tackifier resin dispersions that are compatible with the base polymers, offering cost advantage, too.

Important tackifiers are natural or chemically modified rosin resins. Rosin resins consist mainly of abietic acid (Figure II 3.12) or abietic acid derivatives. Rosin consisting about 90 % abietic acid and isomers and is obtained from certain tree resins. Various derivatives can be made from rosin resins by chemical reactions like esterification with methanol, ethanediol, diethylene glycol, triethylene glycol, 1,2,3-propanethiol or pentaerythritol, hydrogenation or implantation of acid groups with maleic anhydride, followed by hydrolysis of the anhydride ring (Figure II-3.13).

The tackifiers are added to the acrylate dispersions in the form of aqueous dispersions, unless they are water-soluble resins. The proportion of tackifier is usually 10 to 50 parts by weight based on 100 parts by weight of polymer. (solid/solid). [22]

Ready to use aqueous tackifier dispersions are available, too (e.g. from Teckrez, LLC, trade name "Teckros" [28] and from Kraton, trade name "Aquatac" [29]). Besides rosin-based tackifier resins hydrocarbon resin based aqueous

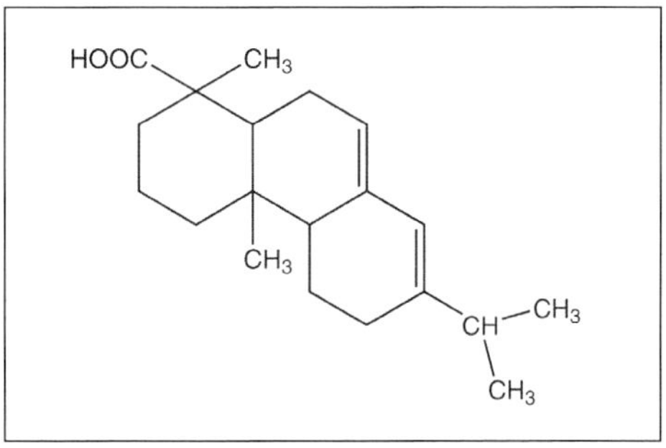

Figure II 3.12: Abietic acid

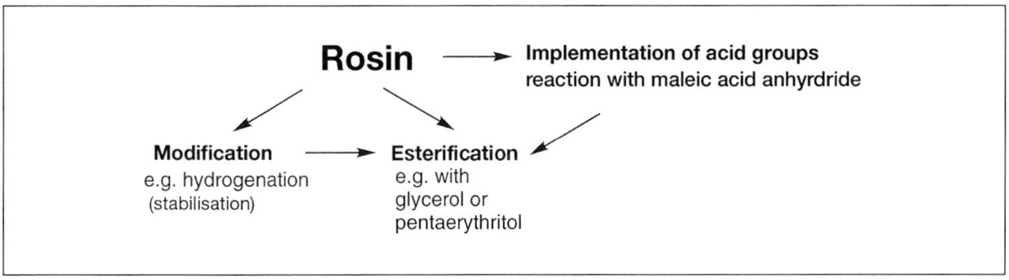

Figure II-3.13: Rosin based tackifier resins

Table II-3.9: Dispersion PSA for labels

Item	Raw material	Parts per weight
1	Aqueous dispersion of an acrylate copolymer containing carboxyl groups.	75
2	Tackifier resin dispersion	25
3	Antifoaming agent	0.1

Item 1: "Acronal V 215", BASF SE
Item 2: "Snowtack SE780GE, Lawter, Inc.
Item 3: "Foamaster WO 2323" BASF SE

Table II-3.10: General purpose acrylic pressure-sensitive adhesive [32]

Item	Raw material	Parts per weight
1	Acrylic dispersion, 60 % solids	100
2	Wetting agent, solution of sodium sulphosuccinate and isotridecanol ethoxylate in water	0.4 – 1.3
3	Neutralising agent, ammonia 10 % for pH 7.5 – 8.5	0 – 2
4	Defoamer	0.1 – 0.5
5	Tackifier dispersion, anionic hydrocarbon, 55 % solids	30
6	Rheology modifier, associative, alkali swellable acrylic dispersion	as required
7	Water	as required

Item 1: "Acronal A 220", BASF SE
Item 2: "Lumiten I-3200", Ashland
Item 4: "Drewplus L-108", Ashland
Item 5: „Tacolyn 1070 Resin Dispersion", Synthomer
Item 6: "Latekoll DS 6269", BASF SE

Table II-3.11: Starting formulation for general purpose acrylic dispersion pressure-sensitive adhesives [32]

Item	Raw material	Parts per weight
1	Acrylic dispersion, 60 % solids	60
2	Aqueous dispersion of polybutylacrylate	20
3	Polyvinyl isobutyl ether	20

Item 1: "Acronal 81D", BASF SE
Item 2: "Acronal 4D", BASF SE
Item 4: "Lutonal I 650" BASF SE

dispersions (e.g. from Synthomer PLC, trade name "Tacolyn" [30]) are suitable for the use with acrylic pressure-sensitive adhesive dispersion.

Other additives

In addition to the additives mentioned above, further additives such as thickeners/rheology modifiers preferably associative thickeners, defoamers, wetting agents, preservatives, plasticisers, pigments or fillers may be required in the formulation of acrylate dispersion pressure-sensitive adhesives, depending on the requirements profile. [22]

The function of preservatives is to protect the dispersion in the delivery container or in storage tanks against microbial damage. Due to their potential risk to human and animal health and to the environment, their use is regulated throughout Europe by the Biocidal Products Regulation [31] Accordingly, only products whose active substances are included in the positive list of approved biocidal active substances may be used as preservatives.

Guide formulations

Table II-3.9 shows an example of a formulation of a pressure-sensitive adhesive based on a commercially available acrylic dispersion and a resin dispersion.

Table II-3.10 shows a typical guide formulation for a general purpose acrylic pressure-sensitive adhesive, e.g. for the production of labels from polymer films used in packaging applications. [32]

Table II-3.11 shows a typical starting formulation for general purpose acrylic dispersion pressure-sensitive adhesives with aggressive tack, good initial adhesive strength and good film flexibility for bonding tapes. [32]

Further advice on the formulation, application and especially the testing of water-based pressure-sensitive adhesives is presented in [6, 22].

Evaluation

Compared to their solvent-based counterparts water-based acrylic dispersion pressure-sensitive adhesives are a more sustainable and safer alternative to solvent-based products for many applications, but have some disadvantages, particularly in terms of their resistance to water and moisture. Thus, three main phenomena, water whitening, loss of adhesion and cohesion and swelling, potentially occur on contact with water or high humidity. Especially for outdoor applications with a long service life expectation, solvent-based or 100 % systems offer advantages.

The main advantages of water-based dispersion pressure-sensitive adhesives are that they contain no volatile organic compounds (VOCs), are non-flammable, generally not harmful to health and have a low odour, which significantly reduces the requirements for occupational health and safety and environmental protection during their manufacture and processing into self-adhesive products. Further advantages are the high solids content of up to 60 – 70 % compared to around 20 % for solvent-based pressure-sensitive adhesives, the easy cleaning of machines and tools (as long as the adhesive has not yet dried to an irreversible film), the good adhesion to polar materials, the good temperature and ageing resistance (as long as the moisture load remains low) and the availability of many ready-to-coat products.

In addition to the limited water and moisture resistance already mentioned, the slow, energy-intensive drying process at elevated temperatures and the need for compounding with tackifier resins to improve adhesion to non-polar surfaces with the risk of a reduction in heat resistance are disadvantages that must be considered.

3.3.3 Hot-melt PSA

The basic principles of hot-melt adhesives in general have been discussed in Chapter II-1.3.1 of this book.

Various polymers can be used for the production of pressure-sensitive hot-melt adhesives. The most widely used technology is probably based on styrene block copolymers (SBC). In addition, pressure-sensitive hot-melt adhesives based on butyl rubber, but also ethylene vinyl acetate and atactic polypropylene are available. In addition to the base polymer, which mainly influences the cohesive properties but also other properties such as ageing and weathering resistance, they contain tackifier resins, plasticizer, possibly also fillers and stabilizer tailored to the respective base polymer.

3.3.3.1 Styrene isoprene butadiene block copolymer based hot-melt PSA

In contrast to random copolymers, block copolymers consist of segments of homopolymers connected to segments of other homopolymers. These are tri-block copolymers with polystyrene endblocks (partial molecular weight: 10,000 to 50,000 g/mol) and polyisoprene or polybutadiene centreblocks (partial molecular weight: 30,000 – 150,000 g/mol).

For pressure-sensitive adhesives purposes it is differentiated between linear triblock styrene butadiene styrene (SBS), styrene isoprene styrene (SIS) and styrene isoprene butdiene styrene (SIBS) block copolymers with different styrene content. Due to the high cost and difficulty in sourcing isoprene, SIBS is used to replace some of the isoprene content in standard SIS grades.

The triblock copolymers combine the positive properties of both polystyrene and polyisoprene respectively polybutadiene. Polystyrene possesses high cohesive strength whereas polyisoprene and polybutadiene exhibit good mechanical properties at low temperatures. The components of the copolymer are incompatible, a fact which has an impact on the morphology. Phase separation occurs at a microscopic level. To an extent depending on the ratio of the constituents, the copolymer consists of a continuous polybutadiene phase and a discontinuous, disperse polystyrene phase (diameter of the disperse regions: about 10 nm). The multiphase structure corresponds to that of a physically crosslinked system called a thermoplastic elastomer (Figure II-3.14).

The styrene fraction of commercial products varies from 15 – 40 wt%. On account of the incompatibility of the two polymer constituents, such block copolymers have two glass transition temperatures (polystyrene: about 100 °C, polybutadiene: about -80 °C). The upper glass transition temperature limits the heat-distortion resistance (if no additional cross-linking occurs), while the lower glass transition temperature characterises the low-temperature properties. [12, 13]

Relatively new are hydrogenated copolymers like SEBS (styrene-ethylene/butylene-styrene), and SEPS (styrene-ethylene/propylene-styrene). SEBS is produced through a selective hydrogenation of butadiene double bonds in SBS to obtain a copolymer with enhanced weathering and chemical stability, high heat stability (> 200 °C) and good low temperature properties. SEBS is used besides others to formulate pressure-sensitive adhesives for medical tapes meeting the requirements for medical tapes like no skin irritation during wear, good initial tack, low trauma on removal and compatibility with the common sterilization processes (ethylene oxide, gamma radiation and steam). [45]

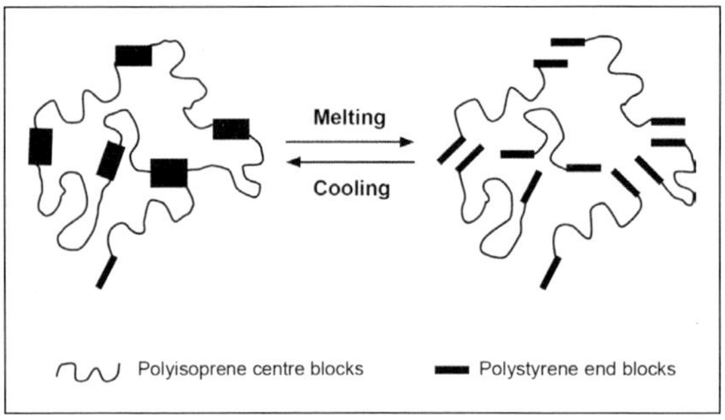

Figure II-3.14: Simplified diagram of a SIS tri-block polymer

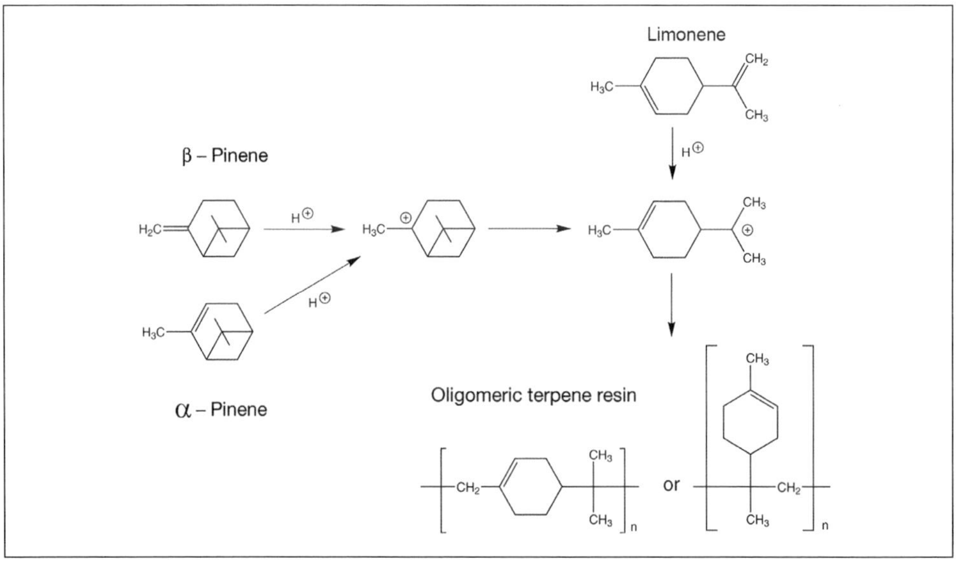

Figure II-3.15: Oligomerisation of terpens by Lewis acid induced Friedel-Crafts reaction

Formulations

For modification to yield hot-melt pressure-sensitive adhesives, tackifier resins (15 – 50 phr), plasticizers (up to 40 phr), liquid hydrocarbon oil (up to 15 %), fillers and different additives (predominantly for stabilisation; 1 – 5 phr) are added.

The choice of resins is heavily influenced by their compatibility with the respective phases. Aromatic resins, such as coumarone-indene resins and poly-α-methylstyrene resins, are very compatible with the polystyrene phase. They can raise or lower the glass transition temperature of the polystyrene phase in accordance with the composition. As they have comparatively low molecular weights, they reduce the application viscosity of the polymer melt. An increase in the glass transition temperature boosts cohesion and resistance to heat distortion. Aliphatic resins, such as terpene resins obtained by Lewis acid induced Friedel-Crafts reaction (Figure II-3.15) [7] and rosin resins (Figure II-3.13), are more compatible with the polybutadiene or polyisoprene phase. They increase tack and peel adhesion. Relatively new aromatically modified C_5 hydrocarbon resin shows excellent performance in formulations that have traditionally required styrenated terpenes resins.

Oils like naphthenic oils predominantly serve as plasticizers because of their compatibility with the centre block phase. Oils reduce hardness and modulus and increase tack.

Fillers (chalk, silicates, carbon black, etc.) reduce costs and modify rheology. Antioxidants/stabilizer employed include sterically hindered phenols and zinc dibutyl dithiocarbamate, while further additives can be UV absorbers, such as benzotriazoles.

Table II-3.12: Guide formulation for a SIS based pressure-sensitive adhesive

Item	Raw material	Parts by weight
1	Styrene-isoprene-styrene block copolymer, 19 % polystyrene	41.3
2	Low molecular weight tackifier, hydrocarbon resin	49.6
3	Resinous plasticiser (liquid hydrocarbon resin)	8.2
4	Sterically hindered primary phenolic antioxidant stabilizer	0.9
Total		100

Item 1: "Kraton D1114 PT", Kraton Corporation
Item 2: "Piccotac 1094 E", Synthomer
Item 3: "Picco A 10", Synthomer
Item 4: "Irganox 1010", BASF SE

Table II-3.13: SBC-based pressure-sensitive adhesive with good bond strength [34]

Item	Raw material	Parts by weight
1	Styrene-isoprene-butadiene-styrene copolymer, 19 % polystyrene	20
2	Styrene-butadiene-styrene copolymer, 33 % polystyrene	20
3	Aromatically modified C-5 hydrocarbon resin tackifier	49.8
4	Hydrotreated and highly refined, high viscosity naphthenic oil	10
5	Sterically hindered primary phenolic antioxidant stabilizer	0.2
Total		100

Item 1: "Kraton D1171", Kraton Corporation
Item 2: "Kraton D1118", Kraton Corporation
Item 3: "Wingtack STS", Cray Valley
Item 4: "Nyflex 222", Nynas
Item 5: "Irganox 1010", BASF SE

Properties	
Rolling ball tack [cm]:	1.4
Loop tack, stainless steel [N/25 mm]:	36
180° peel, stainless steel [steel [N/25 mm]:	24
Shear adhesion failure temperature, 500 g [°C]	84

Table II-3.14: SEBS-based hot-melt PSA for medical tapes [45]

Item	Raw material	Parts by weight
1	Linear SEBS, styrene content 15 %	18
2	Hydrogenated hydrocarbon resin (Tg 36 °C)	45
3	Hydrogenated hydrocarbon resin (Tg -23 °C)	8
4	Paraffinic oil	28
5	Antioxidant	1
Total		100

Item 1: "Calprene H6180X", Dynasol Group

Properties

Melt viscosity (Brookfield Thermosel) 120 °C [cp]	2290
Rolling ball tack [cm]:	18
Loop tack, stainless steel [N/25 mm]:	5.3
180° peel, stainless steel [steel [N/25 mm]	6.2

Hot-melt pressure-sensitive adhesives are produced in internal mixers (Banbury mixers), double-Z-kneaders or planetary mixers at temperatures of 120 – 170 °C. Higher temperatures may cause thermal damage of the hot-melt. Ideally, access of oxygen should be restricted.

Guide formulations

Table II-3.12 shows a guide formulation for a pressure-sensitive adhesive based on polystyrene-polyisoprene tri-block copolymer (SIS) as used, for example, in the production of packaging tapes. [14].

Table II-3.13 shows the composition of an SBC-based pressure-sensitive adhesive with good bond strength [34]

Table II-3.14 shows a guide formulation for a SEBS-based pressure-sensitive adhesive for medical tapes [45]

3.3.3.2 Butyl based hot-melt PSA

Pressure-sensitive adhesive based on butyl are predominately used for sealing operations both in the form of tapes, both coated to a liner but also as transfer tape and as bulk material for dispensing in a molten state. Important applications include primary seals in the manufacture of insulating glass and various sealing operations in the construction industry and in car body construction.

Butyl rubber (IIR) or in short just "butyl" is a copolymer composed of about 98 % isobutene and about 2 % isoprene, or pure polyisobutylene (PIB) (see Figure II 3.16). It should be noted that butyl rubber is produced by cationic copolymerization and the remaining C=C double bond from isoprene is in the *trans*-configuration. It is this double bond which enables butyl rubber to be crosslinked e.g. with sulphur or sulphur donor compounds increasing cohesive strength but reducing tack. If only a small part of the polymer is to be crosslinked, the addition of halogenated

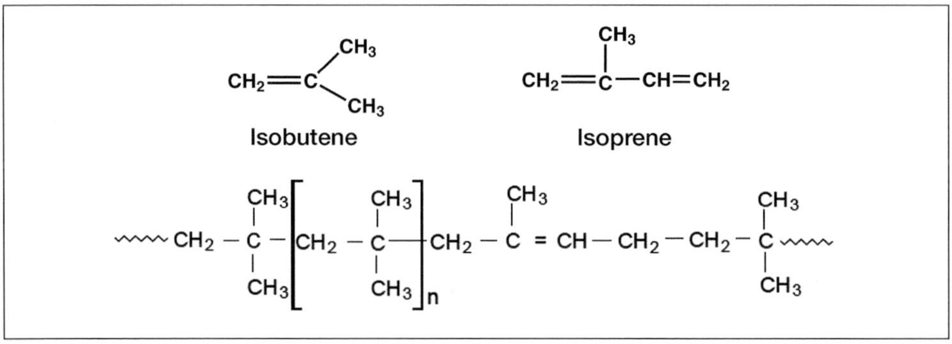

Figure II 3.16: Butyl rubber (IIR) as copolymer of isobutene and isoprene

(chlorinated or brominated) butyl rubber is a good option. The cross-linking is then performed by zinc oxide without the non-halogenated elastomer portion being included in the cross-linking and a butyl with increased cohesion is obtained. [35]

The molecular weight of IIR ranges from 40,000 – 5 million (and is most about 250,000) g/mol. PIB is manufactured in three different ranges of molecular weights:

- 300 – 4,000 g/mol: oily to honey-like liquids
- 40,000 – 120,000 g/mol: high-viscosity, highly tacky compounds
- > 300,000 g/mol: rubber-like elastic compounds (elastic only on exposure to short-term loads; flowable on exposure to long-term loads)

Table II-3.15: Starting point formulation for butyl-PSA backed carpet tiles [35]

Item	Raw material	Parts by weight
1	Butyl rubber, low unsaturated/ low viscosity grade	24.9
2	Aliphatic hydrocarbon tackifier resin	24.9
3	Petrolatum, m.p. 57 °C	12.5
4	Petroleum-derived micro-crystalline wax	37.4
5	Sterically hindered primary phenolic antioxidant stabilizer	0,3
Total		100

Item 1: "Exxon Butyl 065", ExxonMobil
Item 2: "Excorez 1304 resin", ExxonMobil
Item 3: "Indopol H100", Ineos (Biesterfeld)
Item 4: "Be Square 175", NuCera Solutions
Item 5: "Irganox 1010", BASF SE

Owing to the aliphatic and non-polar character of IIR and PIB (no tertiary C-H bonds), these show good resistance to ageing (heat, light, oxidation); furthermore, the densely packed molecular structure gives rise to low gas permeability and low water absorption. These properties are very important for adhesives and especially insulating glass sealants.

Tack can be adjusted by selectively adding PIBs of low, medium and high molecular weight. Furthermore, (compatible) esters of resin acids or terpene resins can be added. The addition of fillers such as chalk, talc, etc. reduces the price of raw materials. Carbon black is often added as a filler, which further improves resistance to light and oxygen.

The disadvantage of PIB and IIR (non-crosslinked) is cold flow; they are therefore less suited for adhesive bonds or seals under constant load. Their good adhesion to metals should be emphasized. PIB sealants even absorb small quantities of oils and so there is no need for substrate surfaces to be absolutely oil-free.

Production usually takes place in heated vacuum kneaders or Banbury mixers, whereby care must be taken to ensure that the material has a sufficiently high viscosity at all times, otherwise there is a risk that a homogeneous mixture will not be obtained. Accordingly, the addition of liquid raw materials must be carried out in several steps.

The melt viscosity can be easily customised by adding thermoplastic ingredients, including resins and waxes. Petrolatum and waxes are the most effective viscosity reducers. Resins, waxes and thermoplastics such as low-density polyethylene, ethylene vinyl acetate and crystalline as well as amorphous polypropylene increase penetration value and cohesive strength. LDPE and crystalline polypropylene in particular increase the softening point of the mixture. [37]

Guide formulations

Table II-3.15 shows a starting formulation for a butyl-based pressure-sensitive adhesive suitable for the manufacture of self-adhesive carpet tiles.

Table II-3.16: Starting formulation for an automotive butyl tape sealant [37]

Item	Raw material	Parts by weight
1	Partially cross-linked butyl rubber	36.4
2	Carbon black	32.7
3	Polybutene, MW app. 900	20.0
4	Paraffinic oil, 55 % of Paraffin hydrocarbon element	10.9
Total		100

Item 1: "Kalar 5215", HB Fuller
Item 2: "N-347", Penta Carbon
Item 3: "Indopol H100", Ineos (Biesterfeld)
Item 4: "Sunpar 2280", Sunoco

Table II-3.17: Generalized formulation of a typical EVA based hot-melt PSA [36]

Item	Raw material	Parts by weight
1	EVA-copolymer	30 – 50
2	Tackifier resin	30 – 50
3	Plasticizer	0 – 20
4	Filler	0 – 5
5	Antioxidant	0.5

Table II-3.18: EVA-based hot-melt formulation for a pressure-sensitive label adhesive [14]

Item	Raw material	Parts by weight
1	EVA-copolymer, 40 % VA-content	30
2	Tackifier resin, glycerol ester of gum rosin	40
3	Liquid aliphatic low molecular weight hydrocarbon resin	29
4	Sterically hindered primary phenolic antioxidant stabilizer	1
Total		100

Item 1: "Evanatne 40-55", SK Functional Polymer
Item 2: "Permalyn 5095", Synthomer
Item 3: "Piccotac 1020-E", Ineos (Biesterfeld)
Item 4: "Irganox 1010", BASF SE

Table II-3.16 shows a guide formulation for a butyl tape for use in automotive construction showing excellent adhesion to many common substrates, excellent weathering and low permeability to moisture and gases.

3.3.3.3 Ethylene vinyl acetate based hot-melt PSA

EVA copolymers are predominately used for solvent (Chapter II-1.2.3.4) or water-based (dispersion) (Chapter II-1.2.3.4) as well as hot-melt adhesives (see Chapter II-1.3.2.1) in permanent applications where pressure-sensitive properties are not required. However, EVA amorphous copolymers containing a high proportion of vinyl acetate have been used for compounding early hot-melt pressure-sensitive adhesives. Their properties depend largely on molecular weight and vinyl acetate content of the polymer. The main deficiency is the low shear resistance especially at elevated temperature.

Guide formulations

Formulations according to the generalized formulation shown in Table II-3.17 offer initially good tack and peel adhesion but these might drastically decrease on aging. The reason is the incompatibility of tackifier resins with the EVA copolymer. Following the introduction of block-copolymers, EVA-based hot-melt adhesives were replaced by corresponding adhesives.

Table II-3.18 shows an EVA-based hot-melt formulation for a pressure-sensitive label adhesive.

3.3.3.4 Atactic polypropylene based hot-melt PSA

Table II-3.19: Generalized formulation of a hot-melt PSA based on atactic polypropylene [36]

Item	Raw material	Parts by weight
1	Atactic polypropylene (16,000 – 20,000 molecular weight)	78
2	EVA copolymer	5
3	Polyterpene tackifier resin	17
Total		100

Atactic polypropylene is a comparatively inexpensive raw material that is also only suitable for the formulation of pressure-sensitive adhesives to a limited extent. The pressure-sensitive adhesive properties are achieved by formulation with tackifier resins and relatively small amounts of EVA copolymer to improve shear resistance. A generalized formulation for a hot-melt pressure-sensitive adhesive is provided with Table II-3.19. Reducing the proportion of the polyterpene tackifier resin improves removability. [8, 36].

3.3.3.5 General evaluation of hot-melt pressure-sensitive adhesives

One advantage of hot-melt pressure-sensitive adhesives is that they must first be melted during the coating process before they are applied directly to the carrier or liner, a sufficient temperature resistance provided. In contrast to solvent or dispersion pressure-sensitive adhesives, however, no drying is required for the application medium (solvent or water) to evaporate. The fact that no drying section is required and that solvent systems require post-combustion or solvent recovery means that not only investment costs and space but also energy costs are saved.

The adhesive properties of hot-melt adhesives are limited due to the restriction to only a few base polymers (essentially only SBC and butyl are of importance), see Table II-3.4. A further disadvantage is the significant reduction in viscosity when heated, which can only be influenced to a limited extent by the formulation, and the resulting limited service temperature of the bond produced with hot-melt pressure-sensitive adhesives.

3.3.4 UV curing pressure-sensitive adhesives

In addition to the three most important PSA technologies described before – water-based, solvent-based and hot-melt adhesives – which together cover by far the majority of global demand, the relatively new UV curing pressure-sensitive adhesives have gained importance thanks to the advancements in properties and sustainability. A distinction is made here between the two technologies.

- relatively low-viscosity room temperature coatable systems that crosslink after application at room temperature by means of UV radiation. These systems are sometimes called UV curing "syrup" formulation

and

- hot-melt adhesive systems that require heating to facilitate the coating process. UV sensitive chemistry is built into the hot-melt, allowing subsequent cross-linking to improved shear and elevated-temperature properties

From a polymer point of view both acrylic polymers with photoreactive groups that are polymerized into the polymer backbone and styrene-block-copolymers containing free vinyl groups requiring an additional photoinitiator system are possible, but it seems that the use of acrylates and modified acrylates predominates. [5, 38, 39]

The basic principles of the UV-induced curing of acrylic substances by radical polymerization has been described in Chapter II-2.3.4.

3.3.4.1 UV curing room temperature coatable pressure-sensitive adhesives ("syrup")

The main components of the systems that can be coated at room temperature ('syrup') are the elastomer, the adhesive additives, the monomer or monomer mixture and the photoinitiator or photoinitiator mixture. Combinations of mono- and difunctional aliphatic urethane acrylate oligomers, for example, are used as elastomer components. The molecular weight and the glass transition temperature, which is well below room temperature, provide the pressure-sensitive adhesive with the required plasticity under slight pressure when applying the pressure-sensitive adhesive film and the elasticity required for good peel adhesion.

Saturated polyester co-resins are used as adhesion promoter. Compared to the oligomer, they have a lower molecular density and a higher glass transition temperature, which allows the elastomer good mobility within the system and thus further improves the plasticity required during application and the elasticity needed for good peel adhesion.

The addition of reactive monomers primarily serves to adjust the viscosity and the levelling properties during the coating process and to control surface wetting.

When selecting the photoinitiator, care must be taken to ensure that its absorption spectrum matches the emission spectrum of the UV lamp. On the other hand, it should be considered that the wavelength of the UV light used influences the penetration depth of the UV light. The longer the wavelength, the deeper the UV light penetrates the material. Short-wave UV radiation promotes surface hardening, which increases the shear strength but reduces the adhesion property in the form of the tack. Conversely, long-wave UV light preserves the tack, which may lead to reduced shear strength due to the low degree of cross-linking in the surface area. The objective must therefore be to achieve the optimum balance for the respective application by selecting the appropriate photoinitiator system in conjunction with the choice of the UV lamp.

Other additives used include stabilizers to prevent premature cure due to low-level light exposure, antioxidants, plasticisers, defoamers, wetting agents and small amounts of fillers or pigments. For the latter, care must be taken to ensure that they do not excessively reduce the UV transparency of the adhesive.

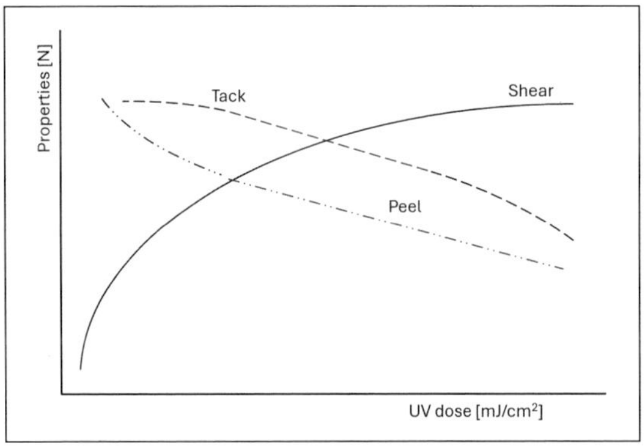

Figure II-3.17: Tack, peel and shear strength as a function of UV radiation dosage [41]

Guide formulations

The guide formulations shown in the following Table II-3.20 to Table II-3.23 show the basic procedure when it comes to adapting the properties of a pressure-sensitive adhesive to the special requirements of an application. It should be noted at this point that the important properties of UV curing pressure-sensitive adhesives such as tack, peel adhesion and shear strength can also be influenced by changing the radiation intensity, i.e. by the power of the

Table II-3.20: High tack UV curing pressure-sensitive adhesive [40]

Item	Raw material	Parts by weight
1	Monofunctional, aliphatic urethane acrylate, diluted in 2-ethyl hexyl acrylate (T_g = -16 °C)	17
2	Inert co-resin, modified polyester resin diluted in urethane acrylate (T_g = -18 °C)	55
3	2-(2-ethoxyethoxy)ethyl acrylat (T_g = -53 °C)	22
4	Liquid photoinitiator mixture (absorption peaks at 253 and 368 nm)	4
5	Liquid photoinitiator 2-Methyl-1-(4-methylthiophenyl)-2-morpholinepropane-1-one (absorbance peak at 307 nm)	2
Total		100

Item 1: "Genomer 4188/EHA", Rahn AG
Item 2: "Genomer 6043/M22", Rahn AG
Item 3: "Miramer M170", Rahn AG
Item 4: "Genocure LTM", Rahn AG
Item 4: "Genocure PMP", Rahn AG

Resulting properties

Viscosity	1,500 mPas
Loop Tack, 25 x 25 mm	15 N/625 mm^2
Peel adhesion	11 N/25 mm
Static shear (25x 25 mm, 1 kg, room temperature)	1.25 h
Static shear (25x25 mm, 80 °C)	0.0 h

Table II-3.21: High peel and good heat resistant UV curing pressure-sensitive adhesive [40]

Item	Raw material	Parts by weight
1	Monofunctional, aliphatic urethane acrylate, diluted in 2-ethyl hexyl acrylate (T_g = -16 °C)	53
2	Inert co-resin, modified polyester resin diluted in monofunctional urethane acrylate (T_g = +12 °C)	25
3	Isodecyl acrylat (T_g = -60 °C)	14
4	Difunctional oligoamine (T_g = -45 °C)	4
5	Liquid photoinitiator mixture with absorption peaks at 253 and 368 nm (Genocure LTM)	4
Total		100

Item 1: "Genomer 4188/EHA", Rahn AG
Item 2: "Co-Resin 02-819/M22", Rahn AG
Item 3: "Miramer M130", Rahn AG
Item 4: "Genocure 5275", Rahn AG
Item 4: "Genocure LTM", Rahn AG

Resulting properties

Viscosity	10,000 mPas
Loop Tack, 25 x 25 mm	7 N/625 mm^2
Peel adhesion	30 N/25 mm
Static shear (25x 25 mm, 1 kg, room temperature)	no failure
Static shear (25x25 mm, 80 °C)	24 h

radiation source, the distance of the UV source to the pressure-sensitive adhesive mass or by the coating speed (Figure II-3.17).

Table II-3.20 shows a guide formulation for an UV-curable liquid pressure-sensitive adhesive offering a high tack but poor shear performance. The high tack value of the formulation is based on the low functionality and low T_g of the ingredients contained. Item 1, a monofunctional aliphatic urethane acrylate offers ideal PSA elasticity properties. Item 2 and item 3 provide the low glass transition temperature which, together with the low cross-linking of this PSA (only monofunction-

Table II-3.22: Good peel, high shear and high heat resistant UV curing pressure-sensitive adhesive [40]

Item	Raw material	Parts by weight
1	Monofunctional, aliphatic urethane acrylate, diluted in 2-ethyl hexyl acrylate (T_g = -16 °C)	52
2	Difunctional, aliphatic urethane acrylic diluted in 2-ethyl hexyl acrylate (T_g = -15 °C)	22
3	Polyester acrylate (T_g = +28 °C)	4
4	Isodecyl acrylate (T_g = -60 °C)	14
5	Difunctional oligoamine (T_g = -45 °C)	4
6	Liquid photoinitiator mixture with absorption peaks at 253 and 368 nm	4
Total		100

Item 1: "Genomer 4188/EHA", Rahn AG
Item 2: "Genomer 4269/M22", Rahn AG
Item 3: „Polyester acrylate 03-849", Rahn AG
Item 4: "Miramer M130", Rahn AG
Item 5: "Genomer 5275", Rahn AG
Item 6: "Genocure LTM", Rahn AG

Resulting properties

Viscosity	97,000 mPas
Loop Tack, 25 x 25 mm	5 N/625 mm²
Peel adhesion	18 N/25 mm
Static shear (25x 25 mm, 1 kg, room temperature)	no failure
Static shear (25x25 mm, 80 °C)	no failure

al monomers are used), results in a very high initial adhesion characterised by a high loop tack value. However, the formulation shows only marginal cohesion, which is reflected in the low shear strength value and low heat resistance.

To improve the shear strength, the proportion of monofunctional aliphatic urethane acrylate can be increased, as shown in the formulation in Table II-3.21. The co-resin used is an inert modified polyester resin diluted in monofunctional urethane acrylate, which has a higher T_g of +12 °C. A difunctional oligoamine is added to increase the degree of cross-linking. The improvement in peel adhesion results from the combination of increased cross-linking ability, T_g and elasticity of the adhesive.

Table II-3.22 shows a formulation in which both the degree of cross-linking and the T_g of the adhesive were increased to achieve a further improvement in shear and temperature resistance. For this purpose, the inert adhesive resin used in the formulation in Table II-3.21 was replaced with a di-functional aliphatic urethane acrylate and 4 % of a trifunctional polyester acrylate was added.

Although the increased cross-linking and increased glass transition temperature have a negative effect on the adhesion properties, they improve the desired target properties to a considerable extent. The addition of isodecyl acrylate monomer also has a positive effect on heat resistance.

The formulation shown in Table II-3.23 offers a pressure-sensitive adhesive with well-balanced properties including excellent adhesion and shear strength. This was achieved by re-incorporating the co-resin already used with the formulation from Table II-3.21, which improves the elasticity properties of the oligomeric component while reducing cross-linking density resulting in improved peel adhesion properties without compromising the excellent shear strength and heat resistance.

3.3.4.2 UV curing hot-melt pressure-sensitive adhesives

The main components of these products are acrylate copolymers with chemically bound photoreactive groups. The advantage of this approach is that the photoinitiator cannot migrate. The monomer composition of UV curable acrylic hot-melts is determined by the adhesive properties required for the intended end use application. In general, the same monomers as mentioned in

Table II-3.23: UV curing pressure-sensitive adhesive offering well-balanced properties [2, 40]

Item	Raw material	Parts by weight
1	Monofunctional, aliphatic urethane acrylate, diluted in 2-ethyl hexyl acrylate (T_g = -16 °C)	40
2	Difunctional, aliphatic urethane acrylic diluted in 2-ethyl hexyl acrylate (T_g = -15 °C)	10
3	Inert co-resin, modified polyester resin diluted in monofunctional urethane acrylate (T_g = +12 °C)	25
4	Difunctional oligoamine (T_g = -45 °C)	3
5	Isodecyl acrylate (T_g = -60 °C)	18
6	Liquid photoinitiator mixture with absorption peaks at 253 and 368 nm	4
Total		100

Item 1: "Genomer 4188/EHA", Rahn AG
Item 2: "Genomer 4269/M22", Rahn AG
Item 3: "Co-Resin 02-819/M22", Rahn AG
Item 4: "Genomer 5275", Rahn AG
Item 5: "Miramer M130 ", Rahn AG
Item 6: "Genocure LTM", Rahn AG

Resulting properties

Viscosity	6,000 mPas
Loop Tack, 25 x 25 mm	7,5 N/625 mm2
Peel adhesion	28,5 N/25 mm
Static shear (25x 25 mm, 1 kg, room temperature)	no failure
Static shear (25x25 mm, 80 °C)	no failure

Chapter II-3.3.1.2 are used. Benzophenone is often used as photoreactive group, which is often introduces in the form of an acrylate monomer (often 2-ethylhexyl acrylate) with a benzophenone side group during polymerization. The production can be carried out using the known processes of free-radical solvent, emulsion or bulk polymerization. The copolymers can then be processed into hot-melt adhesives by removal of the solvent, coagulation of the latex or melt processing of the pure polymers. [42]

The molecular weights are typically in the range of 200,000 g/mol. Together with the typically strictly linear structure processing with conventional hot-melt coaters at moderate temperature

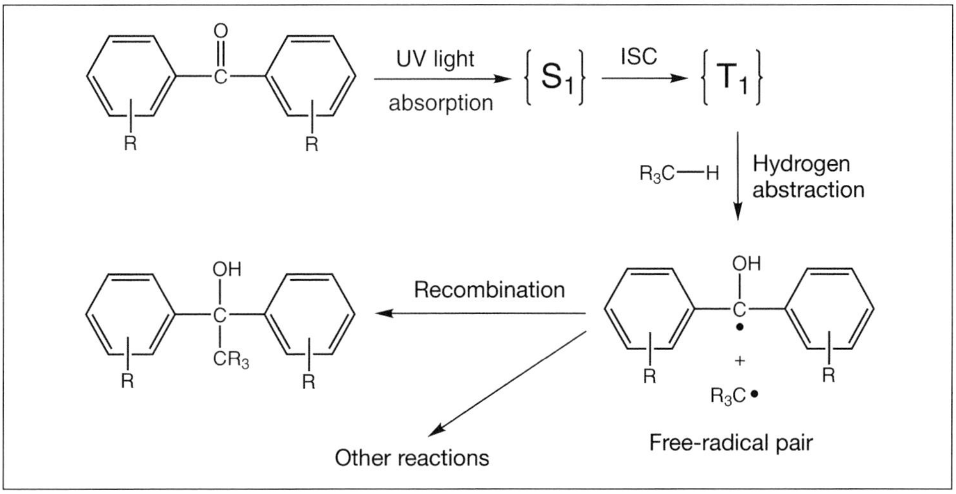

Figure II-3.18: Mode of action of benzophenone as a photoinitiator [43]

Table II-3.24: Guide formulation for a UV curing acrylic hot-melt pressure-sensitive adhesive for permanent labels [10].

Item	Raw material	Parts by weight
1	UV cross-linkable, acrylic hot-melt polymer	70
2	Tackifier resin, hydrogenated rosin ester	30
Total		100

Item 1: "acResin A203UV", BASF SE
Item 2: "Foral 85", Synthomer

of about 100 – 150 °C is possible. After coating, cross-linking takes place using UV light. Similar to the before described liquid UV curing acrylic pressure-sensitive adhesives, which can be applied at room temperature, the properties can be customised by selecting the curing parameters (Figure II-3.17).

Figure II-3.18 shows the mechanism of action for benzophenone as photoinitiator. Through absorption of UV-C light, benzophenone is excited to the singlet-1 state (S_1 in Figure II-3.18). The lifetime of the S_1 state (10^{-10} – 10^{-7} s) is sufficient to allow intersystem crossing (ISC) to the triplet-1 state (T_1). The electron spin reverses, which is why the T_1 state appears to be a diradical. Bimolecular (intermolecular) reactions are possible in the T_1 state on account of its long life time of 10^{-5} – 10 s. A potential reaction in the T_1 state is abstraction of a hydrogen atom to yield a free-radical pair (Figure II-3.18). This free-radical pair can recombine or react chemically in other ways. Figure II-3.19 shows that UV curing acrylics undergo 90 % recombination, which leads to cross-linking and boosts cohesion [9, 10, 43]

Guide formulations

Such copolymers are commercially available under the trade name "acResin" from BASF SE, for example. The properties required for a particular application can be achieved not only as mentioned above by adjusting the cure conditions but also by combining different copolymers, adding tackifier resins, crosslinker additives and additional photoinitiator.

With regard to tackifier resins there is the limitation that only resins which do not interfere with free-radical cross-linking of the UV curing acrylics are suitable. This means they shall not contain C=C double bonds and furthermore should not absorb UV radiation, especially in the range used for curing the adhesive.

Table II-3.24 shows a simple guide formulation for a UV curing acrylic pressure-sensitive adhesive for firmly adhering labels. The adhesive is applied at about 130 °C to label stock (coat weight:

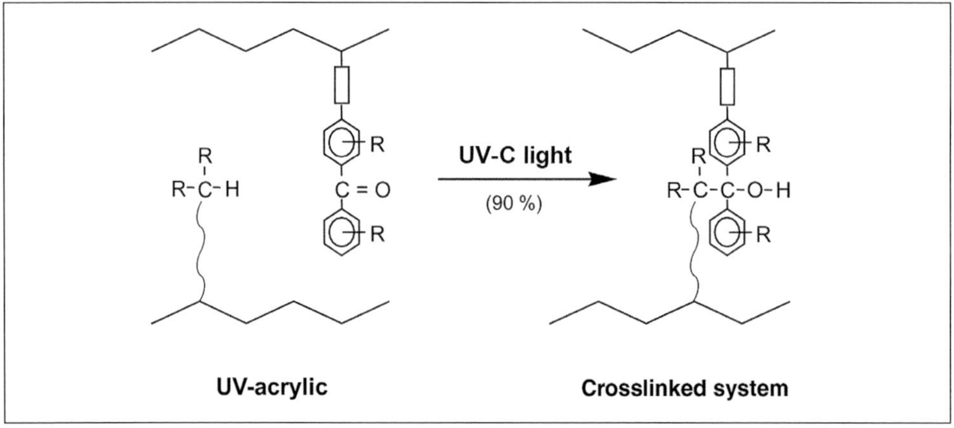

Figure II-3.19: Cross-linking of UV curing hot-melt pressure-sensitive adhesives [43]

20 g/m^2). UV curing is performed with one lamp (60 W/cm) and a line speed of 20 m/min; this is equivalent to a UV radiation dose of 280 mJ/cm^2. After 20 min, the crosslinked pressure-sensitive adhesive has a peel adhesion of 22 N/25 mm (paper tears) [10].

Table II-3.25 shows an optimized guide formulation with regard to shear strength and heat resistance of the cross-linked hot-melt pressure-sensitive adhesive. The addition of an unsaturated oligomer like di-, tri- or polyfunctional acrylates or urethane acrylates facilitates cross-linking. These unsaturated oligomers have a relatively low molecular weight of usually less than 3,000 g/mol. The addition of a cationic sulphonium salt photoinitiator improves both the UV thick-film curing properties as well as curing speed while the addition of a so-called photosensitiser, due to its ability to absorb and transfer light energy that cannot directly excite the photoinitiator to the initiator, accelerates polymerization and ultimately improves the properties of the cured products.

Table II-3.25: Optimized formulation showing improved shear strength and heat resistance of the UV-cured hot-melt pressure-sensitive adhesive

Item	Raw material	Parts by weight
1	UV cross-linkable, acrylic hot-melt polymer, T_g -39 °C	93.75
2	Unsaturated oligomer, di-tri-methylolpropane tetraacrylate	5.0
3	Cationic photoinitiator	0.75
4	Photoinitiator, mixture of 2-isopropylthioxanthone and 4-isopropylthioxanthone	0.5
Total		100

Item 1: "acResin A260UV", BASF SE
Item 2: "SR 355", Sartomer
Item 3: „Irgacure 250", BASF SE
Item 4: "SpeedCure ITX ", Lambson (Sartomer)

Resulting properties	
Recommended coating temperature	120 – 140 °C
Viscosity (Thermosel #27 @ 2.5 rpm, 130 °C)	app. 80,000 mPas
Peel adhesion	18 N/25 mm
Static shear (25x25mm, 1 kg, 70 °C)	>168 h
Sear Adhesion Failure Temperature (SAFT)	> 200 °C
(25x25 mm, 1 kg, 40 → 200 °C @ 0,5 °C/min)	120 – 140 °C

3.3.4.3 Evaluation of UV curing pressure-sensitive adhesives

One significant advantage of UV-curable pressure-sensitive adhesive is that they don't contain any solvent or water. Therefore, this product eliminates the need for highly regulated solvent extraction systems or for time- and energy-consuming oven-drying steps. The liquid products described in Chapter 3.3.4.1 allow coating by means of existing lines just retrofitted with a UV lamp, the hot-melts described in Chapter 3.3.4.2 can be processed with standard equipment for standard hot-melt pressure-sensitive adhesives again retrofitted with a UV lamp. As a result, UV curing pressure-sensitive adhesive in general offer in comparison to conventional products a higher productivity and lower manufacturing costs. In addition, the rapid cure improves dimension accuracy and allows products showing good thermal, water and chemical resistance. Another non-negligible advantage is that the properties of the pressure-sensitive adhesive film can be modified to a certain extent by varying the UV curing conditions and thus adapted to the properties required for the respective application.

By changing the manufacturing process and using renewable feedstock instead of fossil raw materials such as naphtha and natural gas, UV curing acrylate copolymers with built-in photoinitiator groups are now available. These products are characterised by a cradle-to-gate carbon footprint (PCF) of zero kg CO_2 e/kg without having to accept compromises in performance and quality compared to their conventional counterparts and therefore help the adhesives industry to cut carbon emissions while retaining the performance. [44]

Table II-3.26: Examples for pressure-sensitive adhesive applications

	Repositionable products	Removable products	Permanent products
Basic properties	Low bond strength The adhesive system is not damaged during removal allowing	medium to high bond strength can be removed without leaving any residue mostly single use	relatively high bond strength for semi-structural applications good to high temperature resistance
Typical areas of application	Sticky notes Resealable closure stickers (e.g. for paper tissue packaging) etc.	masking tapes hook and poster fastening systems Wound bandages and transdermal therapeutic systems (e.g. nicotine patches) stickers, decals labels self-adhesive films, (surface protection)	exterior car mirrors car protection and trim mouldings damping and stiffening elements in vehicle construction mirror bonding applications in window and facade construction mobile phones tamper evident tapes or labels fleet marking packaging tapes, etc.

3.4 Evaluation of pressure-sensitive adhesives (general)

The manifold and sometimes quite different properties that can in principle be achieved with pressure-sensitive adhesives have led to a large number of applications in various markets despite some disadvantages. Compared to the adhesives discussed in Chapter II-1 and Chapter II-2, which solidify through a physical process or a chemical reaction, pressure-sensitive adhesives offer the great advantage of a certain immediate strength, so that fixation until a certain handling strength is achieved is not required. On the other hand, the achievable shear strength is significantly lower with a maximum of only approx. 0.5 MPa and must often be compensated for by a correspondingly enlarged bond area.

Some pressure-sensitive adhesives are particularly suitable for applications where easy removability without leaving any adhesive residue on the adherends is an important criterion. Accordingly, a distinction is made between removable, re-positonable adhesives and permanent products. Table II-3.26 lists a few of the multiple applications pressure-sensitive adhesives are used in.

Today, the three main PSA technologies are water-based (WB-PSA), solvent-based (SB-PSA) and hot-melts (HM-PSA). The global demand for these types of PSAs totals ≈ 93 %, with the rest being UV-curable hot-melts (UVHM-PSA) and silicone-based PSAs. Broadly speaking, SB-PSAs and UVHM-PSAs tend to offer "high" performance, HM-PSAs offer medium performance and WB-PSAs offer lower performance at the lowest cost. Each of these technologies has its advantages, but also distinct disadvantages: low solvent/plasticizer resistance for HM-PSAs, high concentration of volatile organic compounds (VOC) for SB-PSAs and large equipment footprints required for WB-PSAs[38] .In contrast to the adhesives discussed in Chapter II-1 and Chapter II-2, which solidify through a physical process or a chemical reaction, pressure-sensitive adhesives are usually applied in the form of adhesive tapes or other dimensioned parts already provided with a pressure-sensitive adhesive layer. This has the advantage that no complex application technology is required and component soiling due to unintentional transfer of the adhesive can be practically ruled out. On the other hand, tolerances in the adhesive joint thickness can only be compensated for to a very limited extent. The realisation of curved adhesive is also difficult with adhesive tapes and requires the use of corresponding die-cut parts, whereby considerable scrap can sometimes be produced during their manufacture depending on the geometry of the bond line to be realised.

The use of asymmetrical double-sided adhesive tapes, i.e. adhesive tapes in which the two sides of the liner are coated with different adhesives, each with properties adapted to the part to be bonded, offers the possibility of securely joining adherends that differ greatly in their adhesive properties.

3.5 New developments

As pressure-sensitive adhesives have already proven their suitability as a fixation in the use of solidifying adhesives, it is logical to combine the immediate strength properties of pressure-sensitive adhesives with the high strength offered by adhesives cured by a chemical reaction in one product. New developed reactive pressure sensitive adhesives utilise the mechanisms of chemically cross-linking adhesives (Chapter II-2) to combine the low but immediate strength of conventional pressure-sensitive adhesives with the high strength that can be achieved after a certain cure time with chemically curing adhesives. Such products based on epoxy resins of corresponding molecular weights that crosslink when heated via thermally activatable curing agents (Chapter II-2.2.2) or products based on polyurethanes in which further cross-linking takes place via blocked isocyanates when heated (Chapter II-2.2.1) are already commercially available. However, both require temperature-controlled storage and transport conditions to prevent premature curing. While the epoxy systems offer reasonable surface tack, this is (so far) lacking in the polyurethane systems so that a certain contact pressure is required in addition to temperature during cure. Moisture-curing products based on cyanoacrylate technology (Chapter II-2.3.5) are also available. These require protection from moisture during transport and storage, but show good adhesion even at room temperature, so that additional fixation is usually not necessary. The initial adhesion is achieved as for pressure-sensitive adhesives by applying slight pressure and cure to achieve higher strength values is triggered as with cyanoacrylate adhesives by a reaction with humidity.

3.6 References

[1] Aydin, O., Adhäsion, 40 (1996) No. 11, p. 26–29
[2] Milker, R., Czech, Z., Adhäsion, 47 (2004) No. 9, p. 18–23
[3] Müller, B.,Poth, U., Coatings Formulation, Vincentz Network (2006)
[4] Gierenz, G., Karmann, W., Adhesives and Adhesive Tapes, Wiley-VCH (2001)
[4] Malik, R., Advances in Radiation-Curable PSAs (2001) in www.adhesivesmag.com
[6] Distler, D., Wässrige Polymerdispersionen – Synthese, Eigenschaften, Anwendungen (aqueous polymer latexes - synthesis, properties, applications), Wiley-VCH (1999) p. 127 - 150
[7] Mildenberg, R., Zander, M., Collin, G., Hydrocarbon Resins, Wiley-VCH (1997)
[8] Benedek, I., Pressure-Sensitive Formulation, VSP (2000) p. 430
[9] Meyer-Roscher, B., Barwich, J., Düsterwald, U., Zosel, A., Radiation Curable Acrylic Systems in Handbook of Pressure-Sensitive Adhesive Technology (ed. D. Satas), 3rd ed. (1999); see also www.basf.de
[10] „acResin“ – Die Innovation der Klebstoffe ("acResin“ – the innovation of adhesives), brochure of BASF SE
[11] Czech, Z., Milker, R., Farbe & Lack, 110 (2004) No. 2, p. 18–20
[12] Satas, D., Handbook of Pressure Sensitive Adhesive Technology, 3rd edition, Satas & Associates (1999)

[13] Skeist, I., Handbook of Adhesives, 3rd edition, Chapman & Hall (1990)
[14] Brochure Eastman (former Hercules) Kayser & Mackay (1998)
[15] EN 923:2015, Adhesives – terms and Definitions
[16] Cohen, E., Gutoff, E., Modern coating and drying technology, VCH Publishers Inc., 1992
[17] Satas, D., Rienau J.H., in D. Satas editor, Handbook of Pressure sensitive adhesive Technology, 2nd Edition Van Nostrand Reinhold 1989
[18] https://de.wikipedia.org/wiki/Naturkautschuk, accessed 2024-06-25
[19] https://www.can-dotape.com/slitting/
[20] https://en.wikipedia.org/wiki/Natural_rubber, accessed 2024-06-25
[21] Butler, G. L., in Satas, D., editor, Handbook of Pressure sensitive adhesive Technology, 2nd Edition Van Nostrand Reinhold 1989
[22] EP 1 479 744 B1, https://patentimages.storage.googleapis.com/8c/22/4b/ced540229c1d41/EP1479744B1.pdf
[23] Satas, D., Acrylic Adhesives, in D. Satas editor, Handbook of Pressure sensitive adhesive Technology, p. 396 ff, 2nd Edition Van Nostrand Reinhold, New Yorj 1989
[24] https://www.alberdingk-boley.de/en/Products/Water-Based-Dispersions/Acrylics
[25] BASF brochure Innovative solutions for adhesive raw materials https://insights.basf.com/files/pdf/Adhesives%20Brochure.pdf
[26] https://www.hbfullerproducts.com/products/acryltac-sp-409
[27] Vinavil brochure Polymers for industry https://adhesives.specialchem.com/tech-library/brochure/
[28] https://teckrez.com/dispersions/
[29] https://kraton.com/products/aquatac/
[30] https://www.synthomer.com/product-results
[31] Directive 98/8/EC of the European Parliament and of the Council https://eur-lex.europa.eu/legal-content/EN/TXT/?uri=celex%3A31998L0008
[32] https://adhesives.specialchem.com/formulation/acrylic-psa-for-permanent-paper-tackified-labels
[33] https://adhesives.specialchem.com/formulation/simplified-waterborne-acrylic-psa-for-tapes-with-soft-and-hard-backing-a-by-basf
[34] https://adhesives.specialchem.com/formulation/sbc-hot-melt-psa-for-oriented-polypropylene-a
[35] Higgins, J. J., Jagisch, F.C., Stucker, N.E. in D. Satas editor, Handbook of Pressure sensitive adhesive Technology, 2nd Edition Van Nostrand Reinhold 1989
[36] Satas, D., Miscellaneous Polymers, in D. Satas editor, Handbook of Pressure sensitive adhesive Technology, p. 518 ff, 2nd Edition Van Nostrand Reinhold, New York 1989
[37] https://adhesives.specialchem.com/formulation/butyl-automotive-tape-sealant
[38] Orilall, C., Weinert, Z., Scholte, J., Dong, C., Klang, J., Recent progress in UV-/EB-Curable PSA, https://uvebtech.com/articles/2019/recent-progress-in-uv-eb-curable-pressure-sensitive-adhesives/ accessed 2224-08-02
[39] Petrie, E., Advancements in UV Cure Pressure Sensitive Adhesives: Entering the Circular Economy, https://adhesives.specialchem.com/tech-library/article/advancements-in-uv-cure-pressure-sensitive-adhesives accessed 2024-08-02
[40] Roches, S. des, Murphy, W., adhäson kleben + dichten 10/2006 p. 20, Vieweg Verlag, Wiesbaden
[41] Chech, Z., Butwin, A., Milker, R., Photoinitiatoren und ihre Schrumpfwirkung, adhesion kleben+dichten 04/2009, p. 34, Vieweg Verlag, Wiesbaden
[42] EP 3 091 043 B1
[43] UV-Curable Acrylic Hot Melts for PSAs, ASI Industry, May 2022, https://www.adhesivesmag.com/articles/82987-uv-curable-acrylic-hot-melts-for-psas accessed 2024-08-30
[44] BASF press release, July 4, 2023 https://myindustryworld.basf.com/emea/en/discover/press_releases/basf-now-offering-its-high-performance-uv-acrylic-hotmelt-acresi accessed 2024-12-08
[45] ASI Adhesives & Sealants eMagazine December 2024, accessed 2024-12-08
[46] https://www.marketsandmarkets.com/Market-Reports/pressure-sensitive-adhesives-market-1135.html
[47] Butler, G. L., Natural Rubber Adhesives, in D. Satas editor, Handbook of Pressure sensitive adhesive Technology, p. 260 f, 2nd Edition Van Nostrand Reinhold, New York, 1989
[48] Czech, Z., Vernetzung von Haftklebstoffen auf Polyacrylatbasis, Szczecin, 1999

Part III Future trends and challenges

1 Introduction

The revision of Part II of this book, which, in addition to describing the various types and families of adhesives and sealants, also provides information on how to formulate these adhesives, required an update of the guide formulations given there. Not only did these often need to be adapted to reflect technical progress, but it was also found that some of the raw materials used were no longer available, partly due to the tightening of chemical legislation (e.g. REACH).

There is no doubt that, in the context of limited resources, a shift from a linear economy to a circular economy [1, 2] is essential. Together with the protection of health and the environment and the achievement of the 'UN Sustainable Development Goals' (SDGs) [3], these are important challenges that require a fundamental change in thinking and, of course, also affect the chemical industry and, by extension, the adhesives and sealants industry.

Even though all member states of the United Nations have agreed to the SDGs, country-, or region-specific differences are to be expected in measures taken for the implementation of the goals. This applies not only to the more general SDGs, but also to the measures for transformation to a circular economy and the adjustments to environmental and health protection legislation.

In addition, it is to be expected that the technical requirements for bonded joints and thus also for adhesives will continue to change as a result of increased requirements for existing products, but also due to new types of products, and will certainly become more stringent in many cases.

2 Impact on the adhesives industry

As in the past, various legal, technical, ecological or economic factors will continue to have a significant impact on the adhesives and sealants industry and the requirements placed on the adhesives and sealants used.

The European Chemical Industry Council (CEFIC), for example, has stated in its position paper [4] that it supports the objectives of the European Green Deal [16], which means that the way in which and what is produced will have to change. In doing so, global competitiveness must be maintained. As Europe is leading the way in this respect, the following potential changes, and the associated risks and opportunities are based on what is currently known about the European sustainability agenda.

2.1 Health and environmental protection

To further improve the protection of human health and the environment from hazardous chemicals the European Commission aims to revise the rules as currently set out in the REACH Regulation (EC No 1907/2006) [17] and to encourage innovation to develop safe and sustainable alternatives. The European Commission has started to prepare the legislative revision proposal(s) as a priority for 2023. The REACH revision is currently expected to enter into force between 2025 and 2027. It is expected that the implementation of the REACH revision will be subject to certain transition periods. [5]

2.1.1 Registration of polymers

Polymers are currently exempted from registration under REACH. However, Article 138(2) of the REACH Regulation [17] provides for the possibility of extending the registration requirements to certain polymers if the risk posed by these polymers can be demonstrated and practical and cost-effective ways of selecting 'polymers requiring registration' (PRR) can be developed.

The European Commission is currently working on identifying polymers for registration, proposing grouping criteria, defining the registration process and assessing the costs and benefits of polymer registration. [13, 14]

2.1.2 Increased registration requirements

For endocrine disrupters (EDs), (other) substances with critical hazard properties, including effects on the nervous and the immune systems and low tonnage substances, the information requirements for registration should be increased.

2.1.3 New SVHC categories

New SVHC categories are expected for very persistent and very mobile (vPvM) substances. [5]

2.1.4 Generic risk management approach (GRA)

The proposed 'generic risk management approach' for fast-track restrictions, aims to subject certain chemicals to regulatory measures based solely on their intrinsic hazard characteristics, regardless of where and how the chemicals are used and whether a risk actually exists. In other words, no risk assessment is carried out, so that the existing conditions of use and risk management measures to ensure a safe use are not taken into account. Neither the committees of the European Chemicals Agency (ECHA) nor the Member States are involved in this process. [5, 6]

2.1.5 Mixture assessment factor/mixture allocation factor (MAF)

Furthermore, in line with the Chemicals Strategy for Sustainability (CSS) [7, 8], the European Commission plans to introduce regulatory requirements to address the problem of 'combined exposure to chemicals', i.e. exposure to unintentional mixtures of different chemicals in the environment. A so-called 'mixture assessment factor'/'mixture allocation factor (MAF)' would be applied to all risk assessments of chemicals to ensure that these unintentional mixtures remain safe for humans and the environment.

According to a CEFIC study [9], around 90 % of all unintentional mixtures monitored do not give rise to concern and a further approximately 5 % of the cases giving rise to concern are already covered by the existing legislative framework and are either already subject to authorisation or are on the priority list of the Waste Framework Directive. Only about 5 % of the identified mixtures of concern are not covered by the current legislative framework.

CEFIC is currently working on a "decision tree" to be implemented into the current regulatory framework to support the decision-making process whether a MAF needs to be applied for a specific substance when running an environmental risk assessment. [9]

2.1.6 Essential use concept

Although the European Commission's "essential use concept" is not currently implemented in the REACH legislation, it should be mentioned here.

Hazardous chemicals can have serious effects and cause significant damage to human health and the environment. To avoid this harm, and also the resulting costs to society, to promote innovation for non-toxic alternatives and to achieve a clean circular economy, the European Commission has announced in the 'Chemicals Strategy for Sustainability' (CSS) [15] that the most harmful substances will be phased out in non-essential uses, in particular in consumer products, and minimised and substituted as far as possible in all uses. Essential uses are defined as certain uses that are essential to society. These include, for example, climate change mitigation, the digital transformation, health protection, security and defence.

According to the concept of 'essential uses', if the use of a harmful substance is necessary for health and/or safety and/or is essential for the functioning of society, and if there are no acceptable alternatives, it may continue to be used for that purpose for a certain period of time. [10, 11, 12]

2.1.7 Climate change

In line with the 13th goal of the UN Sustainable Development Goals: 'Take urgent action to combat climate change and its impacts'[3] and due to the expected changes in the product requirements resulting from the observed climate change, existing relevant European standards should be reviewed as part of the regular revision process to determine whether changes to the standard are necessary due to the impacts of climate change. This applies in particular to products in the area of critical infrastructure with a long expected lifetime. [20]

It is to be expected that this will lead to adjustments in the requirements for bonded or sealed joints and therefore for the adhesives and sealants used, which in turn will require an adjustment of the adhesive and sealant formulations.

2.1.8 Impact on the adhesives and sealants industry

Although many questions remain unanswered regarding the various initiatives, it is to be expected that more substances will disappear from the market, either because their use in blends leads to more stringent labelling that is not accepted by the adhesive users, because the registration effort and costs are too high in relation to sales and are therefore discontinued, or because their use is restricted to specific essential uses and their production is therefore no longer profitable.

Polymer registration: Polymers are the substances that enable adhesives and sealants to function and are therefore essential raw materials for the adhesives and sealants industry. In some cases, they are produced during the manufacture of adhesives and sealants, and in other cases, they are formed only during the chemical hardening of adhesives and sealants. According to the Association of the European Adhesives & Sealants Industry (FEICA), the information provided during a potential future polymer registration should be limited to the

- identification of manufacturer/importer
- CAS number (if available)
- chemical name
- tonnage band

with the information being treated as confidential business information by ECHA. [13, 14]

Generic risk management approach (GRA): This could lead to a situation where professional users (e.g. employees in the construction or repair industry) would no longer be allowed to work with adhesives and sealants containing substances with hazardous properties.

Mixture assessment factor (MAF): A general adoption of the requirements for risk assessments resulting from the MAF would unduly increase the registration burden and would consequently jeopardise the diversity of substances essential for the formulation of adhesives and sealants.

Essential uses concept: In principle, it is to be welcomed that the use of certain harmful substances is not to be banned outright, but that their use may continue under specified conditions for certain essential applications until equivalent but less harmful alternatives are available. However, the question of the criteria and evaluation procedures remains unanswered. So far, the criteria are rather vague and open to interpretation, which raises several questions and concerns. For example, it is unclear how it will be ensured that

- decisions on essential uses are taken by a politically accountable body, closely linked to a scientific assessment and only if an unacceptable risk is identified for the use;
- the body has the necessary competence and authority to take decisions and is responsible for them;
- the decision-making process is transparent and involves representatives of all stakeholder community, including industry and civil society, to ensure the legitimacy of the process.

2.2 Circular economy

The overall goal of a circular economy is to conserve resources by keeping materials in the economic cycle for as long as possible, thereby minimising the amount of waste generated. Reducing resource consumption and maintaining the value of raw materials and products are the top priorities. This is reflected in the so-called R-framework with its hierarchically structured, mutually complementary and co-existing 9 R-strategies (Figure III-2.1). Even if the circularity of a product is currently often assessed only on its recyclability and perhaps reparability, the other R-strategies should not be neglected, especially as they are higher up in the hierarchy, at least as far as R9 recycling is concerned.

The repairability and recyclability of products made from different materials is closely linked to the ability to separate these products. In the case of recycling, the separation into as pure material streams as possible is a pre-

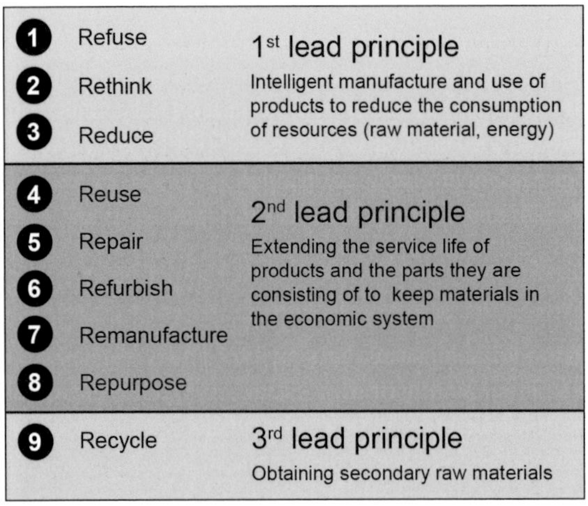

Figure III-2.1: The 9 R-strategies underlying the circular economy

requisite for the recovery of high quality recyclates. Repair often requires the disassembly of the product and therefore separation of the joints in order to replace defective components. Adhesively bonds are often mistakenly thought to be inseparable or non-demountable, and are therefore mistakenly thought to make recycling or repair more difficult.

DIN/TS 54405 [19] describes the various methods that are usually available for separating and recovering adherends from bonded joints and shows that, provided that this requirement is taken into account in the design of the bonded component and the selection of the adhesive, it is entirely possible to separate bonded joints. DIN/TS 54405 [19] is currently being transferred to an international standard at ISO, which is expected to be published as a draft (ISO/DIS 21037) at the end of 2025. It is planned to supplement the standard with examples that have been tried and tested in practice for a long time (such as the repair or replacement of bonded windscreens in vehicles and the removal of labels from returnable bottles), but also with new developments (such as the use of a special primer that expands under defined heat exposure and thus causes the bond to separate).

The demand for adhesives that can be deliberately separated under defined conditions is expected to increase. Adhesive manufacturers can support this development by creating innovative solutions and thus secure or increase their market share.

With its ability to permanently join components made of virtually any conceivable material combination, bonding is seen by experts as an enabler of the circular economy. Modern lightweight construction is inconceivable without bonding technology. Lightweight construction not only means that thinner sheet metal can be used in the production of vehicles, for example, while maintaining the same stiffness and strength properties, thus saving raw materials. It also increases energy efficiency in use by reducing the mass to be moved.

Adhesive bonding also extends the life of products, allowing them to be used more intensively or for longer, which ultimately reduces the use of resources.

The so-called energy transition, as a not insignificant part of the European Green Deal, is virtually impossible to implement without adhesive bonding technology. On the one hand, adhesives are essential for the development of alternative energies (solar and wind energy), but also for e-mobility. [18]

Just as with the development of e-mobility has created a need for adhesives with new, special requirements for the production of battery boxes, it is to be expected that the formulators of adhesives and sealants will be faced with other new, challenging tasks in the future as a result of the circular economy.

In Part II of this book, reference has already been made on a number of examples where raw materials for adhesives and sealants are manufactured on the basis of renewable raw materials, such as rosin esters as tackifier resins in hot-melt or pressure-sensitive adhesives, castor oil as a polyol in polyurethane adhesives or the derivatives of cashew nut shell oil in epoxy adhesives. It is to be expected that with the transformation to a circular economy, further new ways of producing adhesive raw materials, such as from recyclates or by binding CO_2, will be used. As these raw materials do not necessarily have the same properties as their conventional petroleum-based counterparts and therefore cannot simply be substituted on a one-to-one basis, the formulators of adhesives and sealants will also be challenged to find solutions. [18]

When evaluating the use of adhesive bonding technology in the manufacture of a product with regard to the influence on the goals of a circular economy, it is possible to run into conflicting goals, for example between recyclability and the resources used in manufacturing and/or use. In this case, the assessment requires a holistic approach, taking into account the entire life cycle of the product and the graduated importance of the R-strategies. The use of adhesive technology must always be assessed holistically in order to determine its impact on the life cycle assessment!

2.3 References

[1] European Parliament, Circular economy: definitions and benefits, https://www.europarl.europa.eu/topics/en/article/20151201STO05603/circular-economy-definition-importance-and-benefits, accessed 2025-01-03

[2] DIN, Circular thinking in standards, https://www.din.de/resource/blob/1141718/ec85d8193accc5792372f7f1cb2b246a/circular-thinking-in-standards-standards-for-the-implementation-of-r-strategies-data.pdf accessed 2024-01-03

[3] United Nations, https://www.un.org/sustainabledevelopment, accessed 2025-01-03

[4] CEFIC The EU Chemical Industry Transition Pathway, https://transition-pathway.cefic.org/#our-industry, accessed 2025-01-03

[5] REACHLaw, REACH Revision, https://www.reachlaw.fi/reach-revision/, accessed 2025-01-03

[6] FEICA, Generic approach to risk management (GRA), https://www.feica.eu/our-projects/reach/GRA, accessed 2025-01-03

[7] ECHA, Chemicals Strategy for Sustainability, https://echa.europa.eu/hot-topics/chemicals-strategy-for-sustainability, accessed 2025-01-03

[8] VCI, VCI position on Chemical Strategy for Sustainability, https://www.vci.de/ergaenzende-downloads/2020-11-09-vci-position-en-eu-chemicals-strategy.pdf, accessed 2025-01-03

[9] CEFIC Mixture allocation factor, https://cefic.org/policy-matters/mixture-allocation-factor/, accessed 2305-01-03

[10] European Commission, Limiting most harmful chemicals to essential uses, https://ec.europa.eu/commission/presscorner/detail/en/ip_24_2151, accessed 2025-01-03

[11] European Commission, Guiding criteria and principles for the essential uses concept in EU legislation dealing with chemicals, https://op.europa.eu/en/publication-detail/-/publication/90926c62-0365-11ef-a251-01aa75ed71a1/language-en, accessed 2023-01-03

[12] CEFI, How to introduce the 'Essential Uses' concept under REACH, https://cefic.org/app/uploads/2021/05/2021-05-How-to-introduce-the-'Essential-Uses-Concept-under-REACH-Concept-paper.pdf, accessed 2025-01-03

[13] FEICA, Polymers Requiring Registration (PRR), https://www.feica.eu/our-projects/reach/polymers-requiring-registration, accessed 2025-0103

[14] Defining a reasonable, safe and efficient system for the notification of polymers, https://www.feica.eu/information-center/, accessed 2025-01-03

[15] European Commission, Chemicals strategy, https://environment.ec.europa.eu/strategy/chemicals-strategy_en, accessed 2025-01-03

[16] European Commission, The European Green Deal, https://commission.europa.eu/strategy-and-policy/priorities-2019-2024/european-green-deal_en, accessed 2023-01-03

[17] REACH Regulation, https://eur-lex.europa.eu/legal-content/EN/TXT/?uri=CELEX%3A02006R1907-20221217, accessed 2025-01-03

[18] B. Mayer, A. Groß, Circular economy and adhesive bonding technology, Fraunhofer Verlag, 2020,

[19} DIN/TS 54405, Construction adhesives – Guideline for separation and recycling of adhesives and substrates from bonded joints, Beuth Verlag, Berlin

[20] European Commission Implementing Decision on deciding to make a standardisation request to the European standardisation organisations pursuant to Article 10 (1) of Regulation (EU) No 1025/2012 of the European Parliament and of the Council in support of implementation of the EU Strategy on Adaptation to Climate Change [COM(2013) 216 final], https://www.etsi.org/images/files/ECMandates/M526.pdf, accessed 2025-01-03

Part IV Adhesives and sealants formulation approach

1 Basic principles

The development of a new adhesive or sealant usually starts with the creation of a requirement profile for the adhesive or sealant to be developed. Given the similarities between adhesives and sealants, the term "adhesive" is used interchangeably for both in the following text. This is done for the sake of simplicity and to improve readability.

The quality of a bonded or sealed joint, i.e. in particular its mechanical properties and durability, is determined by the adhesive, the intended application method, the material and the surface quality of the adherends used, and the forces acting on the bond, which are also highly dependent on the joint design. When designing or selecting an adhesive for a specific application, the whole system must be considered. (Figure IV-1.1)

It is advisable to involve all relevant stakeholders in the definition of the requirements for the adhesively bonded part and, in particular, for the adhesive bond in order to finally define the (preferably measurable) properties to be fulfilled by the adhesive. It is a trivial fact that the adhesive must show sufficient adhesion to the adherends. This involves determining the requirements

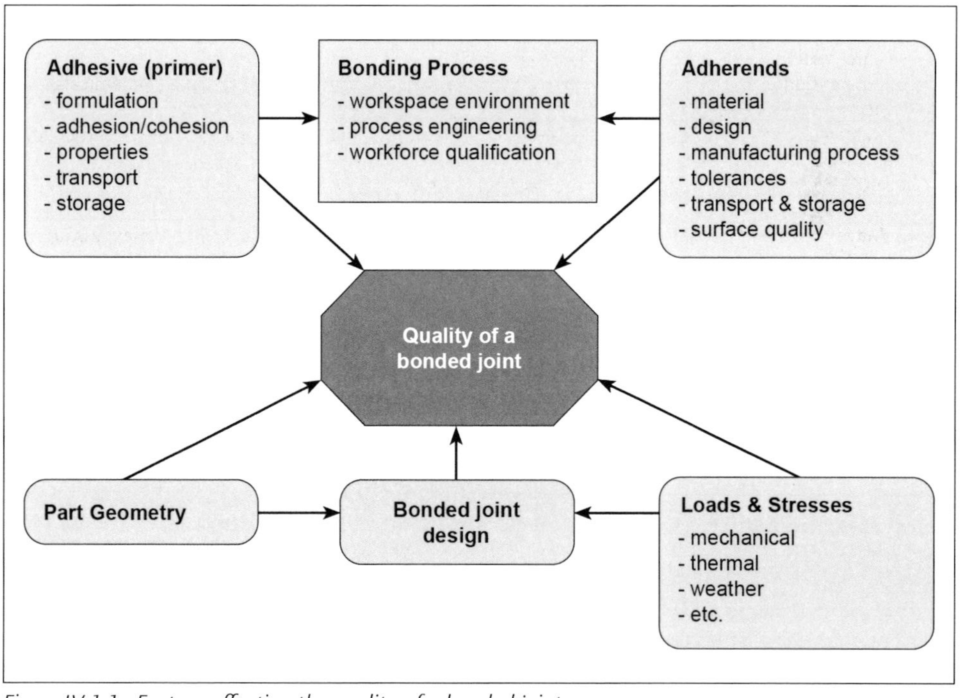

Figure IV-1.1: Factors affecting the quality of a bonded joint

resulting from the intended application of the bonded part and the design of the bonded joint, considering the expected service life. The focus here is on the requirements in relation to the stresses that can realistically be expected during the service life of the part, i.e. the required adhesive strength, taking into account the type, level, duration and frequency of occurrence (see Figure IV-1.2). The climatic conditions, i.e. the temperature application range (minimum and maximum temperature as well as the time course of temperature changes), the respective relative humidity, possible radiation exposure (e.g. UV light) and possible contact with water, operating materials, chemicals, etc. must not be disregarded. Additional required properties such as fire safety, food safety, electrical or thermal insulation or conductivity, and acoustic decoupling may be necessary and must be considered.

In the future, requirements for the circularity of adhesively bonded products will become increasingly important. Although the removability of adhesive bonds [1] for the purpose of repair, reuse or recycling of the adhesive is desirable, other requirements that contribute significantly to the circularity of a product, such as extended product life, improved material or energy efficiency, must not be neglected. In summary, it will be increasingly important in the future to combine the requirements for safety, technological performance, resource efficiency and long-term durability of adhesively bonded products and structures, including their repairability and recyclability, with the social requirements of a sustainable circular economy in the best possible way. [2]

Further requirements include the number of units to be produced, the desired cycle times and any preferences regarding the application technology and curing conditions intended for the bonding process. The system technology required for adhesive production must also be considered.

Last but not least, regulatory requirements such as the various national or company-specific lists of banned chemicals must also be taken into account.

1.1 Adhesion properties

The basics of adhesion have already been covered in Chapter I-2 of this book. It was explained therein that the adhesion properties of adhesives are essentially determined by the chemical composition of the respective adhesive polymer, i.e. the monomer composition.

When defining the requirements profile for an adhesive to be developed, the choice of materials for the adherends should already be made with bondability in mind. However, for various reasons it may be necessary to use a particular material. In this case, the evaluation of the various methods described in Chapter I-2.3 for a specific surface modification to improve bondability should be considered, as these methods offer the possibility of significantly facilitating the development of the adhesive and helping to keep the cost of the adhesive within an acceptable range.

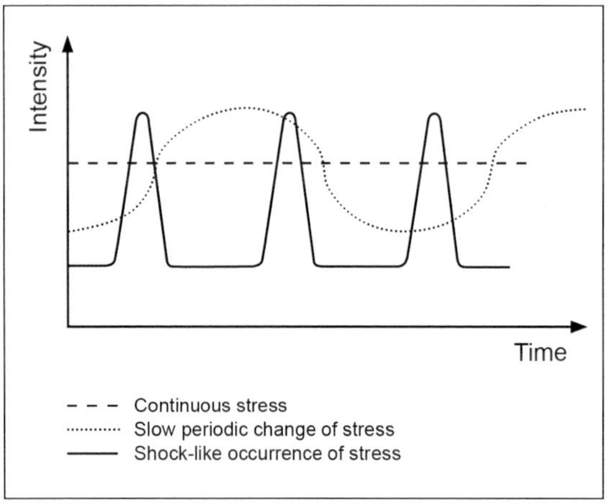

Figure IV-1.2: Different temporal progressions of stresses

1.2 Cohesion of adhesives and viscoelasticity

All materials, including adhesives, deform under the influence of loads. Ideally, when tensile deformation Δl, proportional to a force F, occurs, as shown in Figure IV-1.3, the deformation can be described by Hooke's law (Equation 1) (where k is the spring constant for tensile deformation.)

$F = k \cdot \Delta l$ (Equation 1)

As the force and the resulting deformation vary with the geometry of the substrate, Hooke's law is normalised to read:

$\sigma = E \cdot \varepsilon$ (Equation 1a)

where $\sigma = F/A$ is the tensile stress in [MPa] and $\varepsilon = \Delta l/l$ is the elongation in [%].

E is the modulus of elasticity (Young's modulus), an important characteristic property of the material.

Where shear deformation occurs at an angle γ, as shown in Figure IV-1.4, the following equations apply by analogy:

$F = k' \cdot \Delta l$ (Equation 2)

and

$\tau = G \cdot \gamma$
(Equation 2a)

where $\tau = F/A$ is the shear stress in [MPa], k' is the spring constant for shear deformation and $\gamma = \Delta/h$ is the shear strain in [%]. G is the shear modulus and, again, is a characteristic property of the material.

Tensile deformation and shear deformation are related as follows:

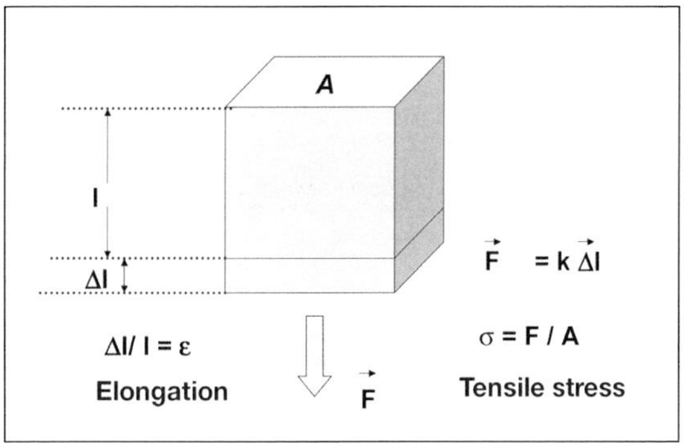

Figure IV-1.3: Tensile deformation

→ $E \approx 3G$
(Equation 3)

Ideal Hookean materials would exhibit a strictly linear stress-strain behaviour with the Young's modulus E or shear modulus G as the slope. In reality, polymers only follow Hooke's law for low deformations. At increasing deformation, the elongation is no longer proportional to the applied force. Further deformation causes the materials to break. Figure IV-1.5 shows the corre-

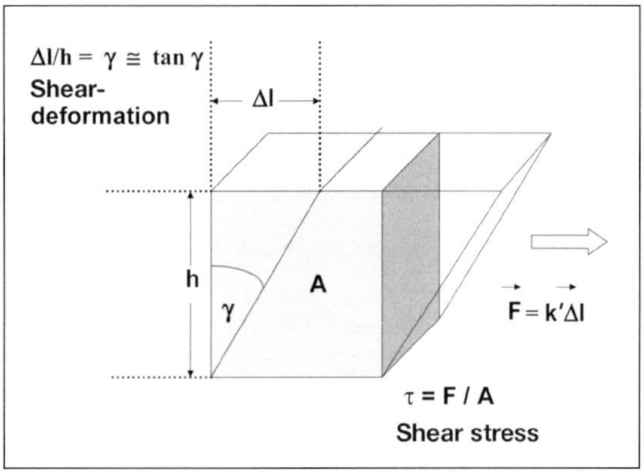

Figure IV-1.4: Shear deformation

sponding stress-strain curves for adhesives with different mechanical properties. The ideal case, namely a linear slope, is obtained only for very small deformations. The slope of the linear part corresponds to the Young's modulus E. High strength adhesives are characterized by a high elastic modulus (up to 2000 MPa) at very low elongation at break, typically of just a few percent. Elastic adhesives or sealants are characterized by a low Young's modulus (typically up to 1 MPa) and high elongation (up to approx. 1000 %). Products often named adhesive sealants or used for so call "elastic bonding" application show with up to 5 MPa a somewhat higher modulus but a reduced elongation in the range of about 150 – 300. Tough adhesives show well balanced mechanical properties with a Young's modulus of a few 100 MPa still offering a decent elongation of up to 100 %. As a rule of thumb, it can be said that for a given adhesive chemistry the higher the modulus the lower the elongation and vice versa.

The plastic creep of polymeric materials under continuous load corresponds to a flow process characteristic of fluids. Real polymers behave more elastically or plastically, depending on their structure and the nature of the load. This is known as **viscoelasticity**. When a force is applied quickly for a short time, polymers behave elastically; when the force is applied for a long time, they behave more plastically. Highly crosslinked polymers tend to creep less under continuous load than non-crosslinked thermoplastics do. The elastic component of the modulus in a shear deformation is described by the storage modulus G'. The elastic component is a measure of the elastic recovery of a material when the load is removed. The plastic part of the modulus, which is responsible for flow, is described by the loss modulus G''. It is a measure of irreversible deformation processes under load. The temperature and frequency dependent (rate of deformation) loss factor $\tan \delta = G''/G'$ is therefore a characteristic for describing the viscoelastic behaviour of polymeric materials. A detailed description of these properties is beyond the scope of this book, so the interested reader is referred to [18] and [19].

The cohesive properties can be improved in a targeted way by formulating the polymers with appropriate fillers. For more detailed information, see the Chapter IV-1.4

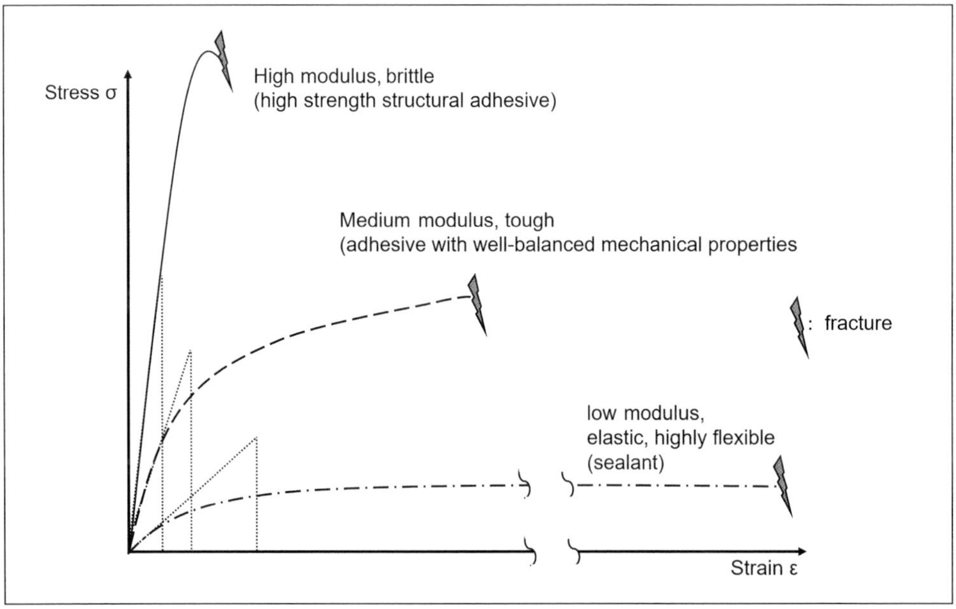

Figure IV-1.5: Stress-strain curves for adhesives/sealants with different mechanical properties, E is the slope of the linear part of the curve

1.3 Thermomechanical properties

The mechanical properties of polymers, and thus also those of the adhesives discussed in this book, change to a greater or lesser extent with temperature. At low (material-specific) temperatures, the polymer is frozen in a glass-like, brittle state. When heated, there is initially a moderate decrease in strength until the glass transition temperature (T_g) is reached, at which the properties change in some cases significantly. Depending on their structure, polymeric materials exhibit different behaviour when heated. This means, the glass transition can occur in a narrow temperature range or over a wide temperature range. The extent of the change in properties also depends on the structure of the polymer. A highly crosslinked thermoset material will lose some of its strength and consistency until it reaches the glass transition temperature, with a slight decrease in the Young's modulus. At the glass transition temperature, there will then be a significant decrease in strength and in the Young's modulus until it finally decomposes when heated further. An amorphous, wide-meshed polymer shows a transition to a rubber-like elastic state when the glass transition temperature is reached, its modulus drops by about five orders of magnitude and its elasticity increases. When heated further, the modulus increases slightly (entropy elasticity) and finally drops rapidly at the start of decomposition (Figure IV-1.6). Crystalline polymers largely retain their strength up to the crystalline melting point. From the melting point onwards, they soften quickly. Crosslinked polymers then exhibit a rubber-like elastic behaviour, while non-crosslinked polymers melt to form viscous fluids (Figure IV-1.7). Knowing the glass transition temperature of polymeric materials, or the temperature range in which the glass transition occurs and the extent of loss of strength that occurs, is therefore of great importance in adhesive bonding technology, especially if it occurs in the temperature range in which the adhesively bonded product is intended to be used.

The glass transition temperature is typically measured by means of DSC (differential scanning calorimetry) or dynamic mechanical analysis (DMA). For DSC the test specimen and a reference specimen are heated up and the heat flow needed to maintain identical temperatures is quantified in a diagram. For DMA, a sample is subjected to a periodic load in one of several types of deformation (bending, tension, shearing and compression) by measuring the modulus as a function of temperature information on phase transitions are obtained. For details on DSC and DMA see section Chapter IV-4.2.1

The phase transition temperatures of polymeric materials can vary within a wide range, in accordance with their structure and formulation. Cross-linking (e.g. phenolic resins) and bulky side

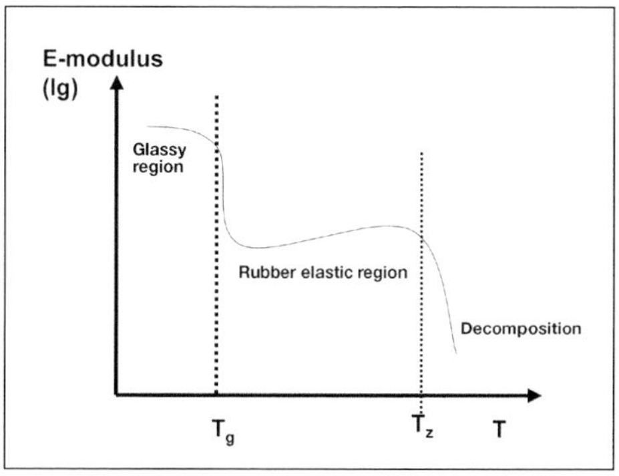

Figure IV-1.6: Thermomechanical curve for an amorphous crosslinked polymer (schematically)

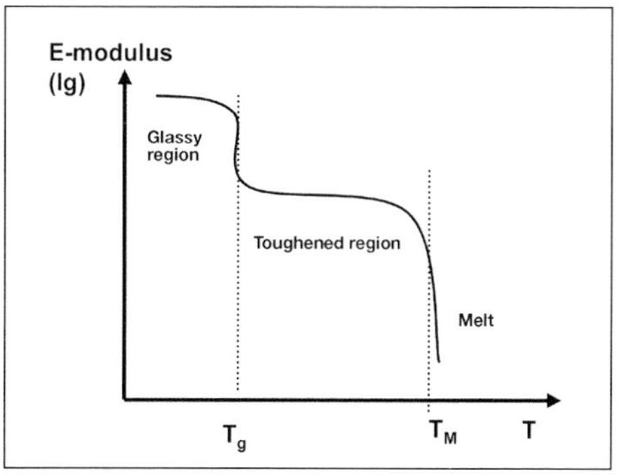

Figure IV-1.7: Thermomechanical curve for a crystalline non- crosslinked polymer, schematically

Table IV-1.1: Glass transition and melting temperatures of selected polymers

Polymer	**Glass transition temperature T_g [°C]**	**Melting temperature T_m [°C]**
Silicone	-127	
Polyethylene (HD)	-125	144
Polypropylene oxide	-75	
Polyisobutene	-73	
Natural rubber	-73	
Polychloroprene	-40	50
Polypropylene (isotactic)	-8	170
Polyvinyl acetate	32	
Polyamide-6	55	220
PVC	81	
Polystyrene	100	
Polymethyl methacrylate	105	

groups (e.g. polystyrene) raise the glass transition temperature. A regular structure (e.g. isotactic PP) or intermolecular hydrogen bonds (e.g. polyamides or polyurethanes) favour crystallinity. Long sidechains can act as internal plasticizer and lower the glass transition temperature (e.g. polyacrylates of long-chain alcohols). Similarly, addition of compatible plasticisers reduces the glass transition temperature. To an extent depending on their structure, compatible resins can increase or lower the glass transition temperature on hot-melt pressure-sensitive adhesives (see Chapter II-3.3.3). Table IV-1.1 shows the glass transition temperatures of important polymers for adhesives and sealants [2].

1.4 Properties affected by fillers

The flow properties of an adhesive during application are one of the most important requirements for an adhesive. For example, in applications where the adhesive is applied in bead form to inclined surfaces or possibly even overhead, both a sufficient sag resistance and good pumpability are essential. On the other hand, for other applications self-levelling properties may be required, for example in potting applications. In addition, other properties such as thermal or electrical conductivity, insulation or simply a specific colour may be required.

In many cases, the addition of appropriate fillers, can contribute to achieving the desired properties. As most fillers are more economical than the materials used to formulate adhesive polymers, the use of fillers is often motivated by a desire to reduce costs.

Fillers are usually relatively inert solid, particulate, mostly mineral materials that are incorporated into adhesives or sealants to achieve certain properties. The predominantly used mineral fillers are also referred to as mineral flours due to their usually small particle size. The properties achievable with the various available fillers are determined by the

- chemical composition
- geometry of the particles, especially
  - the particle size distribution
  - the aspect ratio (ratio of the length of the filler particle to its cross-sectional diameter)
- specific surface area
- oil absorption value (the amount of oil, in grams, needed to just make a flowing paste out of 100 g of the powdery filler). For non-porous particles, the oil absorption value is directly related to the specific surface area of the material)

- surface chemistry
- specific gravity

During the production of mineral fillers, unsaturated terminal silicon and oxygen atoms are formed when molecular bonds break. These react with water molecules from the air to form hydroxyl groups, to which further water molecules can attach. Since the water molecules cannot be completely removed during the subsequent drying process, even at low pressure, elevated temperatures and long drying times, the bond between the filler and the polymer matrix is weakened.

Surface treatment of the mineral filler with silanes or silane-based compounds as shown in Figure IV-1.8 optimises when selecting a silane that is adapted to the polymer system the compatibility between polymer matrix and filler particles, so that the system-improving properties of the inorganic filler can be utilised to the full. [3]

1.4.1 Processing properties

Flow properties – sag resistance

For many applications, a good pumpability combined with a good sag resistance of the dispensed adhesive or sealant is desired. Put simply pumpability is for a given diameter and length of the hose or pipe mainly determined, on the one hand, by the viscosity of the adhesive and on the other hand by the shear forces applied by the pump.

The degree of sag resistance depends on how quickly the thixotropy is restored after the shear forces generated by the application process cease, i.e. the non-sag property is restored. Fillers increase the viscosity and depending on the type of filler used also the non-sag properties. However, the ratio between the two depends heavily on the type of filler.

Equipment wear

Adhesives and sealants are often processed from larger containers of up to 200 litres or more via special application systems, often in automated processes. Here, it is important to select fillers with low hardness, i.e. low abrasiveness, in order to minimise excessive wear of the equipment, especially of feed pumps, valves and nozzles.

Control of bond line thickness

Glass spheres or polymeric micro balloons, incorporated into the adhesive formulation are used as an internal shim, assuring a minimum bond line. Here the diameter of the micro balloons is the positive stop that will prevent too a thin a bond-line.

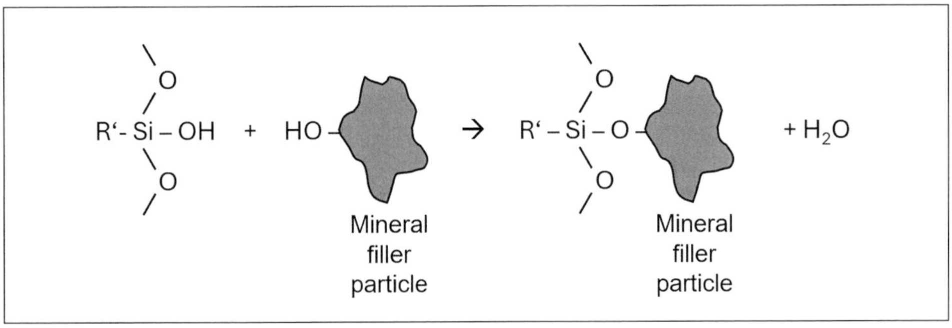

Figure IV-1.8: Silane reaction at the surface of mineral particles

1.4.2 Functional properties

Cohesion

Fillers are not only suitable to improve processing but can depending on their structure, also influence the cohesive properties of the cured adhesive. Fibrous or flake-like fillers show a certain reinforcing effect, while fillers with a lamellar structure such as kaolin, mica or graphite can be used to guide the failure mode towards cohesive fracture.

Shrinkage and durability

The degree of shrinkage during solidification of an adhesive, which in the case of water- or solvent-based systems is caused by the evaporation of the water or solvent or, in the case of chemically curing systems, by the curing reaction, can affect the durability of a bond or seal. Depending on the extent of the shrinkage, stress is created in the adhesive bond, both at adhesive-adherend-interfaces and in the adhesive itself, which can weaken the bond. Since fillers do not shrink by themselves, the addition of fillers reduces the volume fraction of shrinking components and thus the overall degree of shrinkage of the adhesive or sealant.

The shrinkage that occurs during the thermal cure of adhesives can be compensated for by adding fillers that expand upon heating (see sub-chapter "Specific gravity").

Reduction of exothermicity

The heat released during the cure of reactive, highly exothermic systems can lead to irreversible thermal damage of the polymer, especially in the case of thick layers of adhesive. In case of two-component systems, a comparatively small increase in temperature caused by the exothermic curing reaction may already shorten the pot life.

Analogous to the reduction of shrinkage, fillers reduce depending on the amount added exothermicity through the above-described dilution effect. In addition especially when fillers showing a good thermal conductivity (see sub-chapter "Thermal and/or electrical conductivity"), their addition will improve heat dissipation.

Specific gravity

Most mineral fillers have a higher specific gravity than the adhesive polymers, so their use in adhesives results in an increase in specific gravity which is proportional to the specific gravity of the filler and the filler loading.

Some fillers, such as hollow glass spheres and plastic micro balloons, have a lower specific gravity than the base resin and can therefore be used to reduce the specific gravity of the formulated adhesive. Special plastic micro balloons consist of a thermoplastic polymer shell filled with an organic liquid, typically a solvent. If these are added to an adhesive that cures at elevated temperatures, the polymer shell softens as the adhesive is heated and the organic liquid evaporates at the same time, causing the micro balloons to expand in volume, which leads to a defined and fairly uniform expansion of the adhesive. Expanding fillers can be used to compensate for the shrinkage that occurs when adhesives are cured by heat

Provided that the adhesive polymer exhibits robust adhesion to the polymer shell and both the polymer shell and the contained organic liquid are selected in accordance with the cure conditions, the loss of strength in comparison to the unexpanded adhesive is only marginal.

Since adhesive and sealant prices are usually quoted on a weight basis, but the requirement for a particular application is characterised by volume, it is essential to know the specific gravity in order to compare the respective adhesive costs.

Coefficient of linear thermal expansion (CLTE)

Differences in the thermal expansion characteristics between the adherends and the adhesive lead to significant mechanical forces acting on the assembly, which can have a detrimental effect on its durability. The utilisation of an elastic adhesive that is capable to compensate the forces through its deformability represents a potential solution. However, it is important to note that such highly deformable adhesives often exhibit a low cohesive strength, which may prove inadequate for a given application. An alternative approach is to adjust the thermal expansion characteristics of the adhesive by adding special fillers, thus achieving a better match with that of the adherends. In cases where two adherends made from different materials are involved, it is advisable to adjust the coefficient of thermal expansion of the adhesive to a value that lies between the coefficients of the two adherends.

Thermal and/or electrical conductivity

Polymers and thus the adhesives formulated from them usually have an electrically insulating effect and do not exhibit any particular thermal or electrical conductivity. The opposite properties required for certain applications, in particular in the electrical and electronics industry, can be achieved by formulating the polymers with appropriate fillers. The major differences between the various requirements in terms of the extent of the electrical or thermal conductivity should be considered.

If for example soldering processes in microelectronics are to be replaced by bonding, because the former involves the risk of a thermal damage of electronic components due to the temperature load, then a high level of electrical conductivity, on the scale of that offered by metals, will of course be needed. Typically, silver flakes or powders are used at a high degree of filling (in the range of about 90 % by weight) so that a direct contact between the individual particles is achieved. Silver offers compared to copper an advantage in conductivity stability. In other applications, a significantly lower electrical conductivity is sufficient to achieve for example coatability of an adhesive in the cathodic dip coating process (see Chapter II-1.4), as it can be achieved, for example, by a combination of the significantly cheaper fillers like carbon black and graphite.

In addition to the thinning effect already described before in sub-chapter "Reduction of exotherm", the use of thermally conductive fillers can promote the dissipation of heat generated by the exothermic chemical curing of adhesives or sealants, thus preventing the development of stress or thermal damage to the adhesive.

As the power density of electronic components continues to increase, the thermal interface between these heat-generating electronic components and the heat sink becomes a crucial factor. To ensure the optimal functioning and longevity of electronic modules, it is essential to enhance the heat flow to avoid thermal damage of the components. Due to their intrinsic composition, metal-filled adhesives and sealants exhibit not only a good electrical but also a good thermal conductivity. However, this property can not only be a significant financial burden, especially when silver is used, but can also be unacceptable when electrical insulation is required. Consequently, silver may not be the material of choice for adhesives with the sole objective of thermal conductivity.

If the objective is heat dissipation and electrical conductivity is undesirable, or even electrical insulation properties are required, a variety of fillers are available. The selection of the best filler is mainly driven by the degree of required thermal conductivity. Standard fillers, like calcium carbonate or talc, permit thermal conductivities of approximately 0.2 – 0.3 W/K m, which presents benefits over a purely mechanical contact not totally avoiding air pockets. The use of fillers with increased thermal conductivity, including magnesium oxide, aluminium oxide, boron nitride, alu-

Table IV-1.2: Properties of commonly use thermal conductive materials [7]

Mineral		Thermal conductivity [W/K m]	Electrical resistance [Ω m]
Electrical insolating	Magnesium oxide	30	1020
	Aluminium oxide	30	10^{12}
	Boron nitride	30 ⊥, 400 ‖	10^{13}
	Aluminium nitride	180	10^{15}
	Aluminium silicate	14	10^{13}
	Zinc oxide	30	10^{11}
	Zinc sulphide	27	10^{11}
Other fillers		**Thermal conductivity [W/K m]**	**Electrical resistance [Ω m]**
Electrical conductive	Copper	400	1.7×10^{-8}
	Aluminium	235	2.2×10^{-8}
	Iron	80	1.0×10^{-7}
	Graphite	150	2.5×10^{-6}
	Conductive carbon black	15	5×10^{-5}
	Silver	430	16×10^{-3}
	Carbon nanotubes	6000	10^{3}
	Silicon	150	10^{3}

minium nitride, aluminium silicate, zinc oxide, zinc sulphide and those with a somewhat reduced electrical resistance like carbon-nanotubes, silicon or graphite, makes it possible to achieve higher thermal conductivities of up to approx. 5 W/K m.

Table IV-1.2 provides an overview about the properties of commonly used thermal conductive filler materials.

The achievable thermal conductivity is influenced as shown in Figure IV-1.9 by different parameters. However, in addition to the thermal conductivity, the fillers also have a strong influence on other properties. With an increasing filler content, not only does the thermal conductivity increase proportionally, but at the same time, processability is impaired by a significant increase in viscosity and abrasiveness, which should not be neglected. Furthermore, due to their relatively high price, these fillers also influence the price of the product. When formulating a thermally conductive adhesive or sealant, a compromise must therefore always be made between thermal conductivity, processability, mechanical properties and material costs. [20]

Fire resistance

In general, cured adhesives and sealants are combustible materials due to their polymer composition, with their flammability properties being determined by their composition. Fillers already show a certain reduction in flammability simply due to their dilution effect. However, these may not always be sufficient to meet application-specific fire protection requirements, so that special flame retardants must be added.

The effect of most flame retardants is based on one or several chemical or physical processes that occur in the condensed or gaseous phase. Examples of flame retarders are:

- inorganic flame retardants such as aluminium hydroxide or magnesium hydroxide
- halogen-containing (mainly bromine or chlorine) flame retardants such as hexabromocyclododecane
- organophosphorus compounds such as trialkyl phosphates are used, for example
- synergist which are enhancing the effectiveness of flame retardants like antimony trioxide

Colour

Colouring to achieve for example a colour match with the respective substrate is much more common for sealants than for adhesives. In the case of adhesives, colouring is often limited to masking batch to batch changes in colour caused by colour fluctuations in the raw materials. To enable a visual inspection of the mixing quality of two-component adhesives, albeit a rather rough one, the two components are often dyed differently. In general, inorganic pigments are more effective than organic ones, so that the desired effect is usually achieved with relatively small quantities of approx. 1 – 3 %. Titanium dioxide is often used together with colourants to reduce colour intensity or to achieve the required hiding power. For ease of use, it is advisable to use masterbatches of pigment blends offered from different manufacturers at a wide variety of colours.

1.4.3 General properties of used fillers

The most frequently used types of filler are listed in the following and described in terms of their general characteristics.

Calcium carbonate [3, 6]

Calcium carbonate ($CaCO_3$) is a mineral that can be exploited through mining operations on a global scale. It occurs in a variety of forms, including limestone, marble, calcite, chalk and dolomite. To remove contamination, for example by abrasive quartz, grades resulting from a precipitation process are preferred. Different suppliers offer products derived from various geological

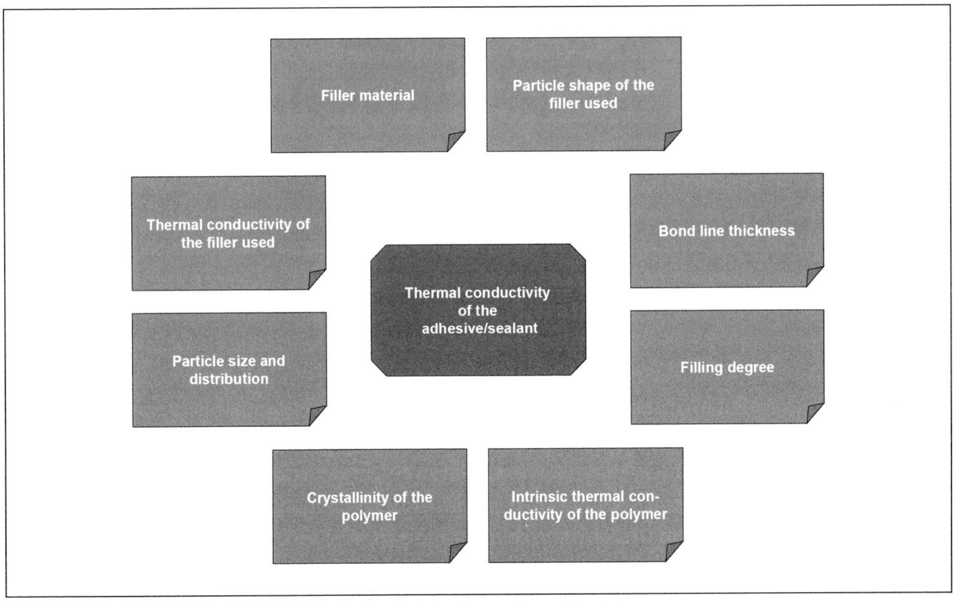

Figure IV-1.9: Important parameter influencing thermal conductivity

sources. In addition to the origin, the types mainly differ in their particle size and particle size distribution. To facilitate dispersibility in hydrophobic systems, grades coated with calcium stearate or stearic acid are used.

In addition to its relatively low price, the thixotropic effect of calcium carbonate, combined with its low hardness of around 3 – 4 (Mohs), are the main reasons for its use at concentrations of up to 50% by weight. Ultrafine precipitated grades with a particle size below 100 nm exhibit the most pronounced thixotropic effect. It is important to note that the thixotropic effect is by far not as pronounced as that achieved by special thixotropic agents, such as fumed silica, which are marketed at significantly higher prices.

Talc [3, 6, 12]

Chemically, talc is a hydrated magnesium silicate ($Mg_3Si_4O_{10}(OH)_2$), with the actual composition varying depending on the mining site. It mainly consists of thin platelets with an aspect ratio of 20 : 1 to 5 : 1 that are mainly white, however, it can also be grey, light to dark green or yellowish in colour due to impurities. Talc is available in different particle sizes, usually between 10 and 75 µm, but also in a correspondingly more expensive fine quality with 1 – 10 µm. Depending on the grade used talc offers a certain thickening effect together with some improvement of sag resistance. The low hardness (1 Mohs) and its inertness against most chemical reagents and acids are further advantages. Due to its low price talc is frequently used in the formulation of adhesives and sealants to reduce the cost of the formulation with just a minimal effect on physical properties. Due to its lamellar structure and high aspect ratio, talc offers barrier properties and fair, water and chemical resistance together with a certain reinforcing effect. Compared to particulate fillers, lamellar talc is triggering an increase in stiffness, tensile strength and creep resistance both at ambient and elevated temperatures.

It is suspected that the use of talc contributes to certain diseases, mainly ovarian and lung cancer. Therefore, appropriate safety measures must be taken when handling talc in powder form. This concerns the use in production of adhesives, but also during potential grinding operations of cured adhesive.

Kaolin [3, 6, 11]

Kaolin clay, also known as China clay, is a fine, natural, white, fine-grained and well-crystallised clay mineral with a layered structure. The octahedral aluminate (AlO_6) and tetrahedral silicate (SiO_4) layers are connected by oxygen atoms. Kaolinite, the main constituent of kaolin, is a hydrated aluminium silicate with a hexagonal plate-like crystal structure. From a chemical perspective, kaolinite is therefore an aluminium salt of silicic acid

Kaolin is obtained from naturally occurring clays and refined into an industrial raw material by means of elaborate processing techniques. Accompanying minerals are separated by wet-mechanical classification into different grain fractions using washing drums, cyclone cascades and centrifuges, and then dewatered by sedimentation and filtration as well as drying. Some kaolin products are additionally refined by bleaching and magnetic separation.

Kaolin clay crystals are hydrated, having 12 % crystal-bound water which is the reason for their plasticity. Calcining, by heating kaolin to temperatures above 1000 °C removes that water and destroys the plasticity. Furthermore chemical, morphological and mineralogical changes occur resulting besides others in an increase in hardness.

In addition to its main application in the ceramics industry, kaolin is used as a filler, e.g. in adhesives and sealants to reduce costs but also to control the degree of shrinkage upon hardening, to increase viscosity, for reasons of dimensional stability and as a pigment.

Mica [6, 15]

The most prominent feature of mica is its structure of tetrahedral and octahedral layers, with each octahedral layer sandwiched between two tetrahedral layers. These 'TOT sandwiches' are only weakly bonded to each other via large, low-charged interlayer cations. Mica has a low Mohs hardness of 2 (parallel to the layers) to 4 (all other directions). The colour varies from white to brownish; green or pink are rarer. There are different types of mica on which the light muscovite (K Al_2 [$AlSi_3O_{10}(OH)_2$]) and the dark phlogopite (K $Mg^{2+}{}_3$ [$AlSi_3O_{10}(OH)_2$]) are used industrially.

In adhesive and sealants mica-type fillers are used to improve sag resistance and dimensional stability. Thanks to their high aspect ratio they provide impermeability resulting good moisture resistance, low coefficient of expansion, and good electrical resistance.

Bentonite clay [6, 17]

Bentonite is a colloidal, both hydrophilic and organophilic clay mineral that is formed by weathering of volcanic ash. Montmorillonite, a sodium aluminium silicate with a variable amount of crystal water is the main component, making up for 60 – 80 % of the clay, Quartz, calcite, pyrtite mica and in particular biotite and feldspar, each of which can account for up to 5 %, may also be present depending on the mining site. Depending on the mineral admixtures and the purity, the colour varies between white, yellow, greenish, grey or brown.

Although the main component montmorillonite is with a Mohs hardness of 1 to 2 one of the least hard minerals the significantly higher hardness of the other minor components gives bentonite a certain abrasiveness.

Due to its very high specific surface area of 400 - 600 m^2/g, bentonite is used as a rheology modifier in adhesives and sealants as well as in emulsions, for example to increase thixotropy and achieve high sag resistance. However, its use is limited by its abrasiveness.

Due to the high alkalinity of bentonite, it can accelerate or even catalyse the epoxy curing reaction. Therefore, its reactivity must be considered when calculating the stoichiometric mixing ratio of resin and curing agent.

Calcium silicat (Wollastonit) [12, 13, 14, 40]

Wollastonite consists of the elements calcium, silicon and oxygen and has the chemical formula $Ca_3(SiO_3)_9$. The crystal structure consists of $(SiO_3)^{2-}$-chains that are linked to each other by the calcium cations. While pure wollastonite is bright white, impurities such as iron, magnesium, manganese, aluminium, potassium, sodium or strontium, which replace calcium in the mineral structure, can lead to grey, cream, brown, pale green or even red shades, depending on the type and quantity.

To be useful for many different applications, wollastonite must be processed; the processing should result in concentrates that consist of 97 – 99 percent pure wollastonite. After extraction from the mine, the ore is processed by crushing, grinding and milling. During these processes, wollastonite, due to its unique cleavage properties, breaks down into plate-like or needle-like particles, with aspect ratios of 3:1 to 20:1, and grades with different particle sizes and aspect ratios are obtained.

Wollastonite is largely inert and has a moderate hardness of 4.5 – 5.5 on the Mohs scale. The crystal structure and physical properties of wollastonite are stable up to about 1,120 °C, which makes wollastonite a substitute for asbestos, e.g. in thermal insulation applications and in friction linings. It is also used in plastics, adhesives and sealants to increase stiffness, to prevent stress cracks caused by shrinkage and for pigmentation.

Fused silica [3, 6]

Fused silica (SiO_2) is produced synthetically by melting and resolidifying of quartz in an electric arc at approximately 2,000 °C. Most notably is the extremely low coefficient of thermal expansion ($0.5 \cdot 10^{-6}$/K). Despite its high Mohs hardness of 6.5, fused silica is used as a functional filler in adhesives, especially epoxy-based ones. Due to the low coefficient of thermal expansion, fused silica-based fillers are used in epoxy-based potting compounds for applications in the electrical and electronics industry to minimise the different temperature-related dimensional changes between the potting compound (typically $65 \cdot 10^{-6}$/K) and the metal parts of the electronic components (typically $12 – 24 \cdot 10^{-6}$/K), thus preventing cracking of the potting compound or damage to the components. The use of silanised fused silica fillers significantly improves the integration of the filler particles into the polymer matrix, further reducing the risk of cracking (see Figure IV-1.8).

Quartz-containing fine dusts with grain sizes <5 µm (respirable crystalline silica) must be viewed critically from an occupational safety perspective. If these fine dusts are regularly inhaled in large quantities for many years, there is a risk that the cumulative effect will trigger lung disease (silicosis). Therefore, appropriate occupational safety measures must be taken, such as working with closed equipment if possible, providing good ventilation in the working area and, if the release of dust nevertheless cannot be ruled out, suitable respiratory protection. [21, 22]

Fumed silica [4, 5]

Fumed silica (SiO_2) also known as pyrogenic silica consists entirely of amorphous silicon dioxide (SiO_2) particles, which are aggregated into larger units. Due to its high efficiency, although it is relatively high in price, fumed silica is in regard of the number of applications certainly the most widely used rheology modifier in adhesive and sealant formulation.

To produce pyrogenic silicon dioxide, sand is first reduced with carbon. The resulting silicon is then converted with chlorine into silicon tetrachloride. In the final step, a homogeneous mixture of silicon tetrachloride, oxyhydrogen gas ($H_2 + O_2$) and an inert gas is subjected to high-temperature pyrolysis in a cooled combustion chamber. Releasing hydrochloric acid initially droplet-like silicon dioxide particles are formed in the flame, which then accumulate in the colder areas of the flame in a chain-like manner. These so-called secondary particles ultimately combine to form larger hydrophilic aggregates (tertiary particles) with a very high specific surface area. If required, hydrophobicity can be achieved by a subsequent surface treatment with silanes or siloxanes.

Fumed silica is widely used to achieve certain rheological properties, especially a fast recovery of viscosity and therefore a high sag resistance after shear thinning processes like dispensing of an adhesives. The exceptionally high surface area to mass ratio enables the formation of many particle interactions in liquid formulations, thereby contributing to the development of a three-dimensional physical network. This network is the basis for the rheological control offered by the use of fumed silica.

Optimal rheological properties are achieved by using grades with an optimal specific surface area and the right hydrophilic or hydrophobic properties. Since certain rheological properties can often be achieved already with very small amounts of fumed silica, which only marginally affect transparency, it is possible to formulate transparent adhesives with adequate sag resistance.

In general, grades with a larger specific surface area result in more pronounced rheological effects, while grades with a smaller specific surface area are easier to disperse. Hydrophobic grades are typically used for polar adhesive formulations e.g. in epoxy- or polyurethane formulations and hydrophilic grades are used for non-polar formulations, e.g. in rubber- or silicone-based sealants as well as in solvent and water-based adhesives even at higher pH-values.

In addition to its rheology-controlling effect, fumed silica offers further advantages in certain adhesive and sealant systems. For example, it has been shown that in 1-component polyurethane adhesives, fumed silica can increase elongation at break and improve adhesion to wood-based materials, and thus lap shear strength.

The typical amount of fumed silica required depends largely on the effects to be achieved, the grade and the adhesive/sealant matrix. Generally, the amount added is in the range of 1 – 5 wt. %. The incorporation of fumed silica into the adhesive or sealant usually takes place in the last manufacturing step under vacuum and at low shear rates, in order not to complicate the mixing of the other components by a high product viscosity and in addition to avoid irreversibly partly destroying the structure of the fumed silica particles. Particularly in the case of high loading levels, a stepwise addition is recommended due to the high volume. This prevents the formation of highly viscous adhesive domains within the still low-viscosity mixture. The high-viscosity domains can make it difficult to achieve a homogeneous mixture.

To avoid air bubbles in the adhesive or sealant, it has proven effective to first de-aerate the fumed silica powder added under vacuum, i.e. with the agitator stationary and only switch on the agitator after a sufficient vacuum has been reached. This prevents the air trapped between the fumed silica particles from being mixed into the adhesive mass and valuable fumed silica particles from being sucked into the vacuum pump due to their very low specific gravity.

Although the air can in principle also be removed subsequently by agitating in vacuum, this is not only complicated by the high product viscosity that has already been reached, but also by the inevitable foaming of the mass.

Carbon black [8, 9]

It should be clarified that the term 'soot' refers to combustion products formed under undefined conditions from a wide variety of carbonaceous materials, that have nothing in common with carbon black manufactured under defined conditions to achieve specified properties.

Approximately 98 % of carbon black produced worldwide is manufactured using the furnace process, whereby hydrocarbon raw materials are converted into carbon black at temperatures of around 1,600 °C in a furnace. The resulting grades, as well as those manufactured using other processes, are products whose properties are adapted to the intended application.

In adhesives carbon black is used, for example, for pigmenting, sometimes combined with the provision of UV protection, for giving the polymer matrix conductivity or antistatic properties, for reinforcing curing rubber adhesives or due to their relatively high specific surface area as a rheological additive to increase viscosity and improve sag resistance. In this way, the proportion of significantly more expensive fumed silica required to reach certain rheological properties may be reduced in pigmented adhesives or sealants with carbon black.

Conclusion

Due to the large number of different substance classes that can be used as fillers and the even larger number of commercially available products, the selection of the right filler or filler system to meet a specific set of requirements or for a specific application is anything but trivial and requires experience and a good knowledge of the various filler products available on the market. It is therefore crucial to consider the possible effects of fillers in the respective adhesive formulation with regard to the desired properties. Fillers can certainly improve certain properties, but they can also lead to a deterioration of other characteristics. It is therefore important to weigh the expected advantages against the possible deterioration of other properties.

2 Stress analysis of adhesive joints

2.1 Preliminary note

While the main function of a bond is the transmission of mechanical loads, the main function of a sealant is to provide a seal against a specific medium. In both cases, this function should be maintained over the entire service life. A detailed analysis of potential influences and their effects on the joint is therefore crucial. Since not only adhesives but also sealants have to transfer mechanical forces, albeit usually to a lesser extent, the same applies to both: the design of the bonded or sealed joint is of particular importance.

Even if the design cannot be tailored exclusively to the requirements of the adhesive or sealant, it is still advisable to take adhesive-related requirements into account as early as possible in the product design process.

2.2 Mechanical loads

2.2.1 Static loads

The level of the mechanical force acting on the bond or seal is, without question, of crucial importance. In addition, it must be differentiated between static loads which are constant over time, and dynamic loads, which vary over time. For loads that are constant over time, a distinction is made between those that are applied for a limited period of time and those that are applied permanently. In the case of dynamic loads, the frequency of occurrence and the temporal load curve are of importance. In particular, impact-type loads (e.g. crash) place different demands on the adhesive or sealant than slowly changing ones.

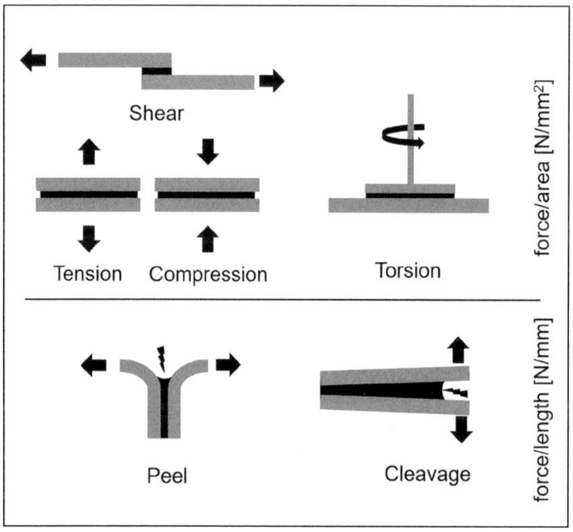

Figure IV-2.1: Basic load modes

Complex loads acting on real construction components can be reduced to a combination of **four basic loads** (see Figure IV-2.1). These are compression, in which substrates are pressed together; tension, where substrates are pulled apart perpendicular to the bond; shear; and peel. While in the first three cases the force acts over the entire adhesive surface (unit: [N/mm^2] or [MPa]), peel forces act only along a line that is more or less narrow depending on the mechanical properties of the adhesive (unit: [N/mm] or [N/cm]). Therefore, from the point of view of load bearing properties, peel forces are unfavourable and should be constructively avoided.

2.2.1.1 Analysis of shear deformation

Influence of the joint area

As expected, the load bearing capacity of an adhesive joint increases with increasing bond area (assuming all other parameters remain the same). However, for overlapping bonds subjected to shear, the load bearing capacity is not proportional to the length of the overlap.

Figure IV-2.2 and Figure IV-2.3 schematically show the influence of the joint area on load-transfer ability [23].

The reason for this is that the adherends stretch to certain extent under load depending on their material and thickness, which leads to an elongation of the adhesive layer at the overlap ends and thus to a higher mechanical load on the adhesive, while the load in the middle area is lower. Figure IV-2.4 shows the stress distribution for real substrates deformed under shear load.

Presuming there are no other defects the failure of an adhesive joint will always start where the highest stress peaks are. Consequently, increasing the joint area will usually improve load bearing capacity; however, such an increase might not always be reasonable and has virtually no effect once the joint area reaches a certain size. Where overlap is long, the maximum transferable load is determined only by the stress peaks at the ends of the joint.

However, increasing the overlap width leads to a proportional increase in bond strength. This can be of great importance for designing adhesively bonded structures [23].

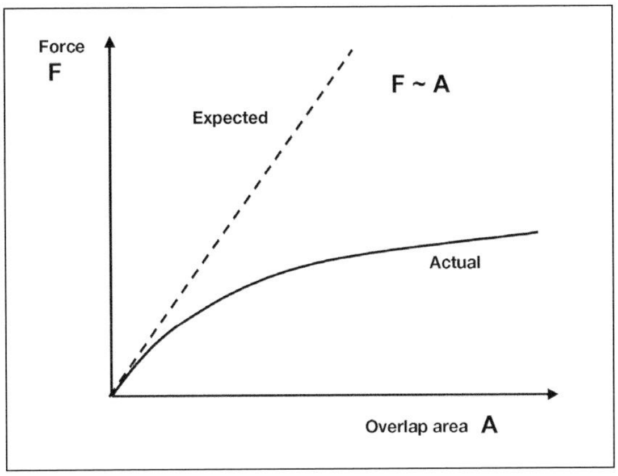

Figure IV-2.2: Load transfer of function of bond area when overlap area is changed

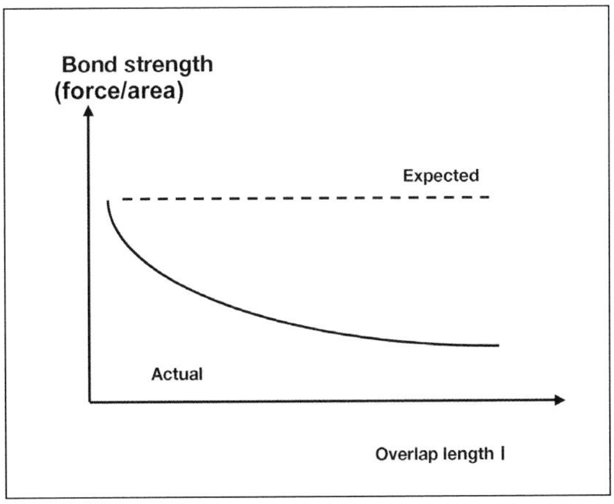

Figure IV-2.3: Bond strength as function of overlap length

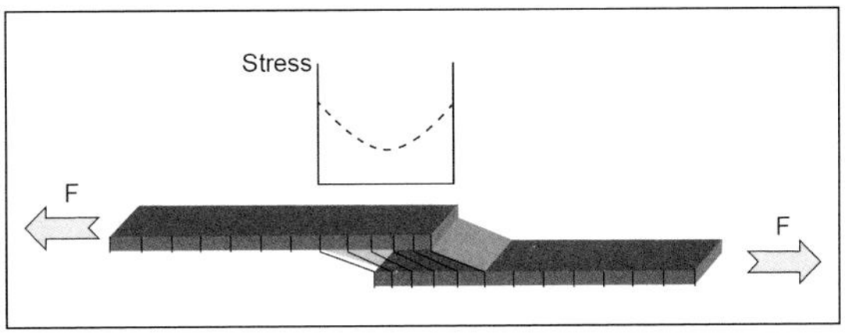

Figure IV-2.4: Stress peaks under shear load caused by stretching of the adherends under loads

Influence of adherends

With increasing thickness, assuming a constant force, the strain of the adherends decreases, so that an increase in the adherend thickness reduces the resulting stress peaks. The same effect can be achieved by changing to stiffer adherends. If the shear forces act eccentrically, as in the case of the tensile lap shear test specimen, the result is a distortion of the adherends leading consequently to unfavourable peel forces at the end of the joint. (Figure IV-2.5). Under load, thick adherends exhibit less tendency to distort and therefore have a positive effect on bond strength. Figure IV-2.6 shows the influence of adherend thickness on bond strength. The modulus of the substrates, too, exerts a significant influence on the bond strength. High modulus adherends can transfer loads without substantial deformation over the whole joint area. This produces a more uniform stress distribution and reduces stress peaks at the ends of the overlap. With low-modulus substrates, higher stress peaks will occur at the ends of the joint area due to substrate deformation and cause premature bond failure (Figure IV-2.7). In some cases, a failure in the adherend right at the end of the overlap occurs. Generally, the greatest bond strength is provided by a combination of high-modulus adhesives and high modulus adherends. [23–25]. On the other hand, the thermal expansion of the adherends, especially when different materials are bonded, may require the use of low modulus adhesives which offer a significantly higher elongation at break compared to high-strength, high-modulus adhesives [26].

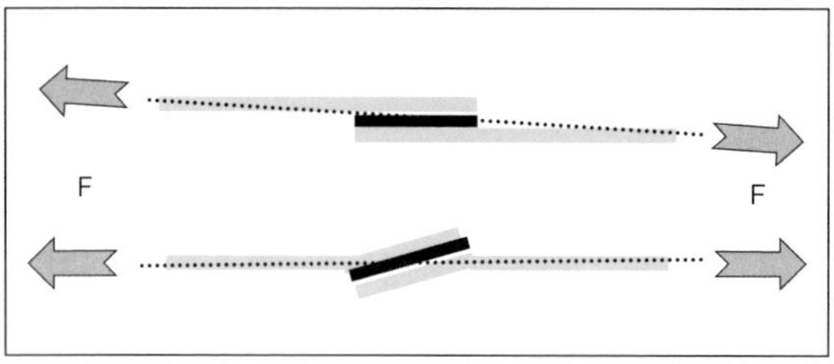

Figure IV-2.5: Distortion of adherends under shear load

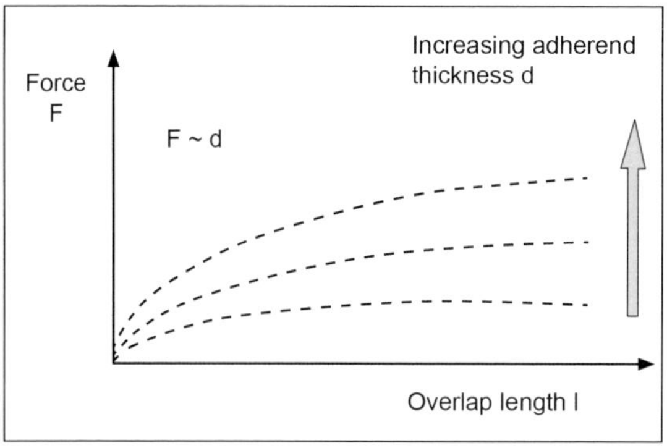

Figure IV-2.6: Influence of adherend thickness on bond strength

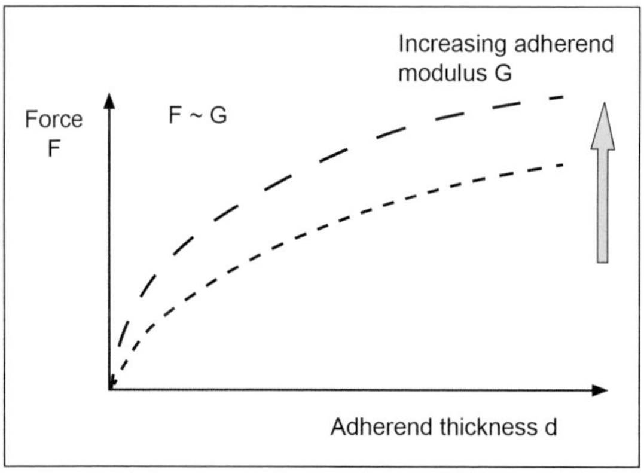

Figure IV-2.7: Influence of adherend modulus on bond strength

Influence of bond line thickness

The influence of the adhesive layer thickness on transferable loads is quite complex and illustrated in Figure IV-2.8. Where the gaps are very thin (less than approx. 0.05 mm), surface roughness might lead to undesirable contacts between the adherends and also wetting defects might occur. Such thin adhesive layers can only be achieved with very low-viscosity adhesives. The highest bond strength is obtained with adhesive layers ranging from 0.05 to 0.2 mm. This is of great importance for adhesively bonded constructions. Such thin adhesive layers are quite flexible under bending stress because the inner and outer radii of curvature become very similar. The situation is comparable to that of a glass fibre which, unlike a thick glass plate, does not break easily when bent. The decrease in the bond strength of thicker layers is attributed to reduced hindrance of transverse contraction and increasing internal stress due to shrinkage of the adhesive layer [26].

Nevertheless, there may be good reasons for choosing a significantly thicker adhesive joint. Such reasons are, for example, the necessity to compensate for dimensional tolerances of the parts to be joined and thus to prevent direct contact between the parts to be joined and thus, for example, contact corrosion or optical surface defects. Other reasons are the compensation of relative movements of the adherends during use, e.g. as a result of the effects of different thermal expansion or component distortion due to mechanical loads. In addition, when peel forces are applied, the forces are transmitted over a larger area when elastic low modulus adhesives are used than when high modulus adhesives are used.

Put simply, these effects can be attributed to the fact that, when subjected to a certain load, the elastic low modulus adhesives are subject to a corresponding deformation, but this completely disappears when the load is removed. [26]

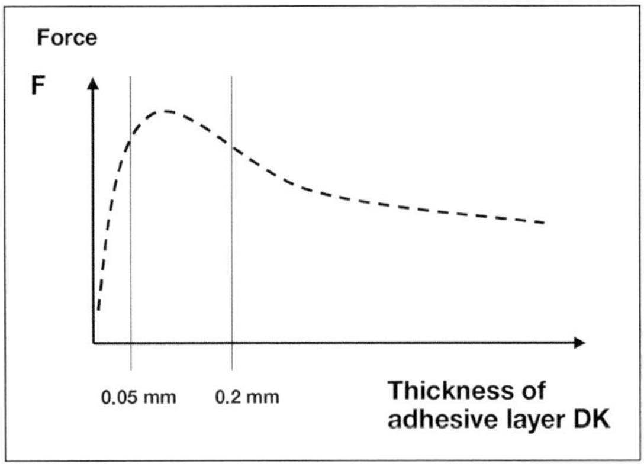

Figure IV-2.8: Influence of bondline thickness on bond strength

2.2.1.2 Analysis of peel deformation

Influence of acting force direction

As a general rule, the design of the bond should ensure that the force is applied perpendicular or parallel to the joint surface, i.e. as symmetrical as possible. When a tensile force acts asymmetrical on an adhesive joint, unfavourable peel or cleavage stresses will result (see Figure IV-2.9). The more asymmetrical the force acts, the less favourable the bond design.

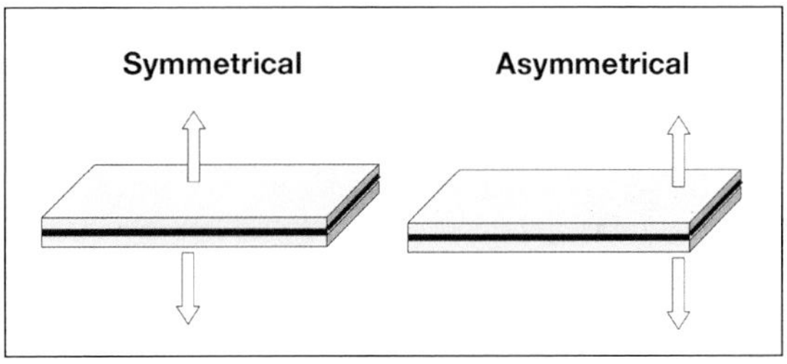

Figure IV-2.9: Symmetric and asymmetric stresses

In such cases e.g. under peel the applied load is no longer distributed over the whole bond area but is concentrated along a narrow line at the edge of the joint. Due to the bending of the peeled-off adherend, a compressive force acts on the adhesive joint near the fracture line, in addition to the tensile force (Figure IV-2.10). Theoretically, the forces during peeling only act along a line, but in practice, depending on the mechanical properties of the adhesive used, a linear area is affected. Since the forces are only transmitted in

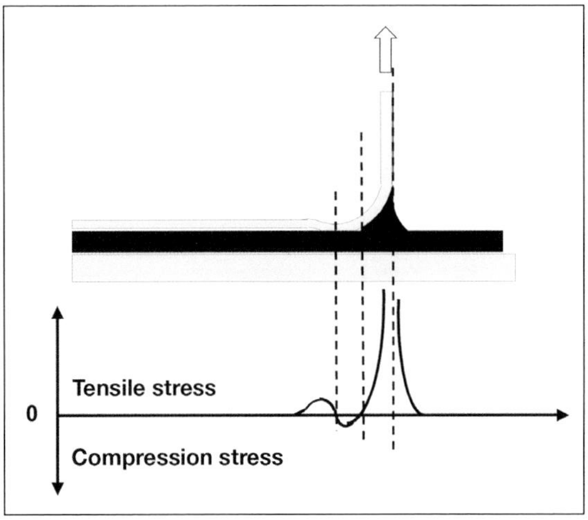

Figure IV-2.10: Stress distribution under peel load

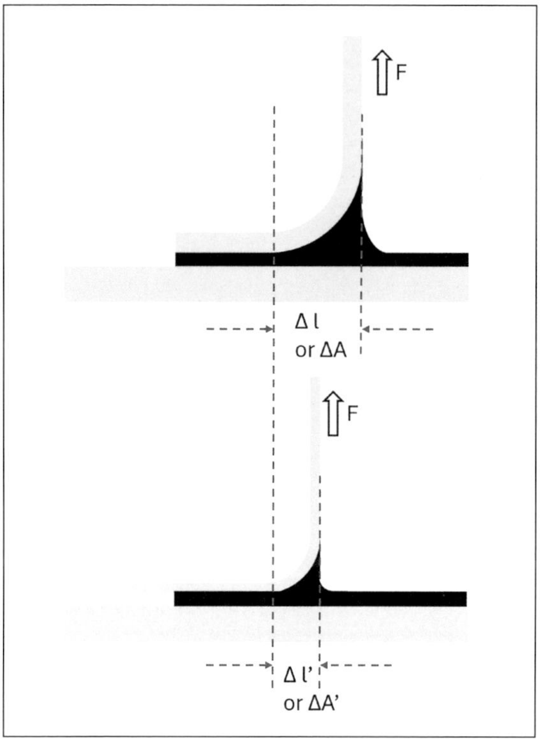

Figure IV-2.11: Stress distribution under peel load

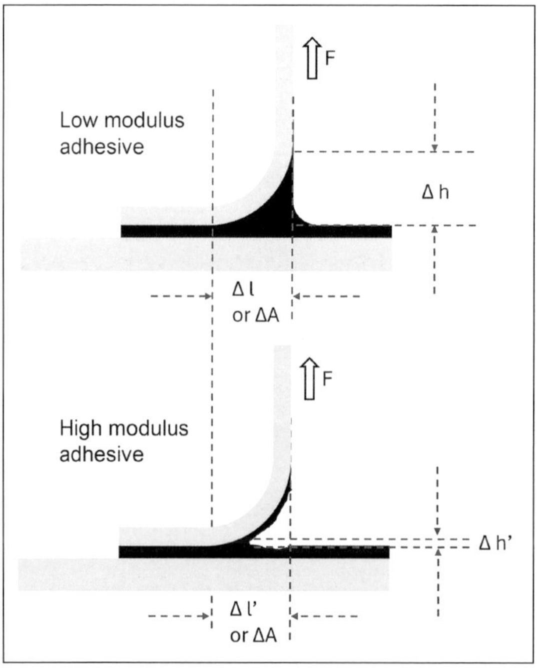

Figure IV-2.12: Influence of the adhesive modulus (elasticity) on peel resistance

this limited area of the adhesive layer, the size of the total bond area is not relevant for the load bearing capacity and adhesive defects or inhomogeneities in the adhesive layer are particularly critical. [23]

Influence of adherends

Under load, thin and flexible adherends exhibit worse load transfer capability than thicker ones. Loads are transferred across a small area ΔA because, during peeling, the peeled adherend is not just bent but rather is deformed in a certain radius which depends on its thickness. As shown in Figure IV-2.11, thicker substrates generate a lager radius than thin ones, and so a larger effective joint area is available for transferring loads. An increase in adherend modulus leads to the same result. Peeling increasingly turns into cleavage and a larger area becomes available for load transfer. [23]

Influence of adhesives

The composition of the adhesive greatly influences with the resulting mechanical properties the peel resistance. High molecular weight polymers and crosslinked polymers are typically characterized by a high cohesion, leading to high modulus and high bond strength values in the event of cohesive failure. Fillers, especially fibres, can increase peel resistance because their geometry allows to distribute the peel forces over a larger area. However, the peel resistance may be limited by the occurrence of adhesive failure.

Elastic, low modulus adhesives show, compared to non-elastic, high modulus adhesive more favourable properties under peel. As illustrated in Figure IV-2.12 for a given radius of separation, they can utilise a larger effective area for load transfer. During peel, the adhesive layer can "yield a certain distance" before it tears.

This is a fundamental problem in bonding. On the one hand, the ability to transfer high loads is desirable, which requires adhesives with a high modulus, especially when the available bonding area is limited. On the other hand, these adhesives are sensitive to the occurrence of peel or cleavage stresses, which cannot always be ruled out. Thus, a compromise often has to be made between load-bearing capacity and elasticity.

Influence of adhesive layer thickness

Regarding peel resistance, the influence of adhesive layer thickness is of minor importance. The peel resistance increases in proportion to the layer thickness until it reaches a limiting value. Thick layers exhibit higher absolute deformation, and so load transfer occurs within a larger area. Overly thick adhesive layers can change cohesive failure to adhesive failure [23].

Influence of peel angle

Typically, peel resistance decreases with increase in peel angle (0 to 140°), after which it often increases slightly again up to an angle of 180°. Figure IV-2.13 shows the typical characteristics of this angle dependence. Small peel angles correspond more to shear deformation, while peel angles of 90° correspond to normal stress, and are thus critically important to an adhesive joint [27, 28].

2.2.2 Dynamic loads

Dynamic loads on an adhesive bond cannot be significantly influenced by the design. This makes it all the more important to match the material properties of the adhesive to the requirements of the application.

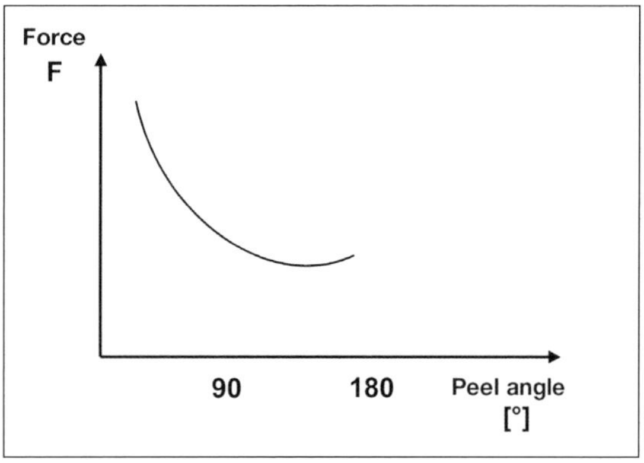

Figure IV-2.13: Influence of peel angle on peel resistance

2.2.2.1 Influence of the rate of deformation

The mechanical strength of adhesives and sealants – this holds for both shear and peel deformation – depends heavily on the rate of deformation. Over a wide range, rapid deformation yields higher strength values, but if the rate of deformation becomes too high, the bond strength will decrease again due to embrittlement. Figure IV-2.14 shows the relationship between peel rate and peel resistance [27]. Many polymers are sensitive to impact. Impact is defined as a very rapid, brief movement to which viscoelastic materials (typically

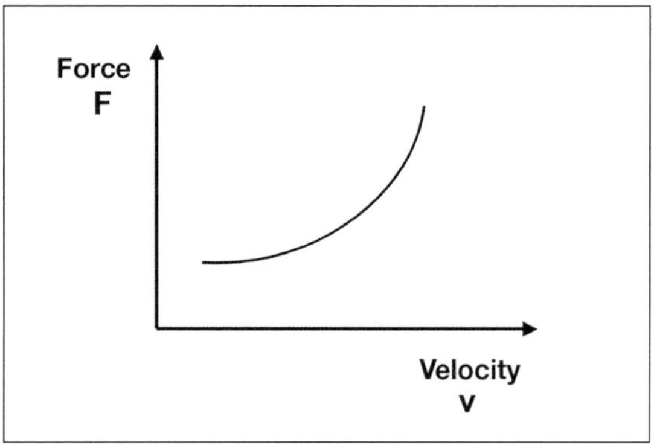

Figure IV-2.14: Influence of peel rate on peel resistance

polymers) react as if they have a higher glass transition temperature. This temperature/time superposition principle describes the correlation between the rate of deformation and the mechanical properties of a polymer-based material. At low rates of deformation, viscous and rubber-elastic behaviour dominates whereas, at high rates, energy elasticity dominates. In that case, the materials become brittle and can fail. Under impact, energy is rapidly transferred to the joint area (unit of impact resistance: energy/area in $[N \cdot m/m^2]$ = force/length in [N/m]). Clearly, peel resistance and impact resistance are physically related. Products having good peel resistance usually also have good impact resistance.

Good impact resistance can be obtained either by using polymers of low glass transition temperature, in which case only moderate bond strength values can be achieved (e.g. sealants or pressure-sensitive adhesives), or by using a multi-phase structure, in which the matrix confers mechanical strength through its high glass transition temperature and the dispersed polymer phase of low glass transition temperature boosts the impact resistance (see also Chapter II-2.2.1, Polyurethanes; Chapter II-2.2.2, Epoxies; Chapter II-2.3.1, Acrylates; Chapter II-2.4.1, Phenolic resins). Polymers formulated in this manner are said to be "toughened".

2.2.2.2 Periodic loads

Long lasting periodic loads can affect the durability of an adhesive joint because the viscoelastic properties of the polymers can lead to irreversible processes (material fatigue). The important parameters of periodic load are the frequency, the amplitude and the duration or number of repetitions. The long-term behaviour of the joint can be predicted from a knowledge of the loss factor ($\tan \delta = G''/G'$) of both the adhesive and the substrates. Periodic loads on an adhesive of high loss modulus generate a great deal of heat energy, which can damage the polymer. The effect of the amplitude of periodic deformation can be estimated if the elastic region of deformation is known. Elongation above the elastic region should at any rate be avoided.

2.3 Analysis of ageing influences

The properties of materials, including adhesives and their associated adherends, can be irreversibly or reversibly changed by external influences such as temperature, contact with gases and fluids, including water, radiation and combinations thereof. They are subject to aging. If these effects adversely affect the function, i.e. the load-bearing capacity of a bond, it is important to be aware of them and, in particular, to identify critical conditions in order to take them into account when designing the bond. This naturally applies to the selection of the materials of adherends and of adhesive.

2.3.1 Thermal effects

2.3.1.1 Influence of temperature on shear and peel resistance

Adhesives and sealants are viscoelastic materials that have specific phase transition temperatures. For polymers and therefore for adhesives the probably most important one is the glass transition temperature (T_g). Below the transition temperature polymers, have a high mechanical strength but are often brittle. Close to the glass transition temperature, the mechanical characteristics (Young's-, G-modulus) significantly decreases resulting in a transition from the glassy to a rubbery state (Figure IV- 2.15). This is associated with a change in the mechanical properties (E-, G-modulus) often by several orders of magnitude. A temperature increase above T_g is causing energy-elastic materials like adhesives to soften. A decrease in temperature below the T_g-range increases the

Table IV-2.1: Coefficients of linear thermal expansion of selected materials [29, 30]

Material	CLTE [$10^{-6}K^{-1}$]	Material	CLTE [$10^{-6}K^{-1}$]
Aluminium	23.1	polyethylene (HD-PE)	150 – 200
Iron	11.8	polypropylene (PP)	150
Carbon steel	10.8	polyvinylchlorid (rigid)	70 – 100
Stainless steel	10.1 – 17.3	polyamide (PA)	60 – 150
Silver	18	polycarbonate (PC)	60 – 70
Brass	19	polyoxymethylene (POM)	70 – 130
Copper	17	polymethylmethacrylate (PPMA)	75 – 80
Glass	8.5	polytetrafluorethylene (PTFE)	100 – 160
Borosilicate glass	3.3	wood (oak)	8
Enamel	8.0 – 9.5	Concrete	12

modulus, but it can also lead especially under impact conditions to a decrease in bond strength due to embrittlement. Some T_g data are provided for comparison purposes in Table IV-1.1.

To achieve consistent bond properties, it is recommended to use an adhesive with a T_g outside the temperature application range of the bonded part. In addition to the value of the glass transition temperature (usually defined as the inflection point of the graph showing the dependence of the modulus on temperature), both the temperature range over which the change occurs, and the extent of the change are of interest.

2.3.1.2 Coefficients of linear thermal expansion

Most materials expand as the temperature rises. As an adhesive joint usually consists of different materials, stress can occur within the adhesive joint due to different coefficients of linear thermal expansion (CLTE) of the materials involved. In the worst case, the bond may fail completely. Plastics (including adhesives and sealants) in general show a higher coefficient of thermal elongation than metals or stone when heated. Therefore, thick and elastic adhesive layers are better at compensating stress within the joint than are thin layers of a stiff adhesive.

Table IV-2.1 provides thermal expansion data for selected materials. The cubic or volume coefficient of thermal expansion is roughly 3α. The change in length resulting from heating, e.g. of an adherend, can be calculated using the following approximation equation:

$$\Delta l = \alpha \cdot l_o \cdot \Delta T$$

where

- Δl is the variation in length of the element
- α is the coefficient of linear thermal expansion
- l_o is the length of the element
- ΔT is the temperature difference

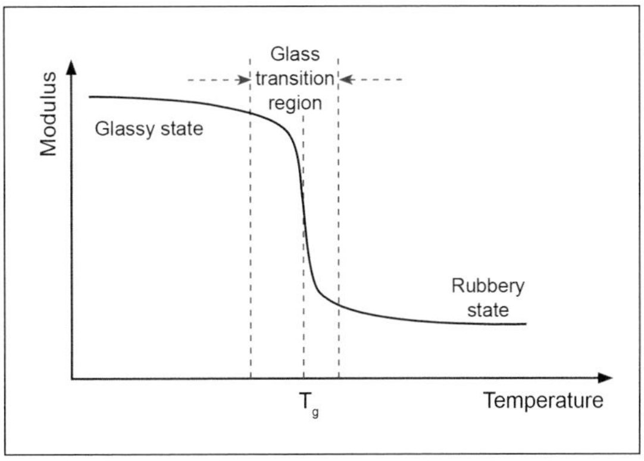

Figure IV- 2.15: Modulus of an adhesive as function of temperature

Table IV-2.2: Thermal resistance of adhesives (typical data)

Base polymer for the adhesive	Limit for permanent thermal exposure [°C]	Limit for short-term thermal exposure [°C]
Polyurethane	120	180
Epoxy	180	250
Acrylate	150	200
Cyanoacrylate	100	140
Phenolic resin	200	350

2.3.1.3 Degradation of organic polymers

All organic polymers decompose above a certain temperature. Therefore, there are always restrictions on the heat resistance of joints made with adhesives or sealants based on organic polymers. Nonetheless, a rough classification is still possible for structural adhesives. Table IV-2.2 indicates the limits on typical types of adhesives for permanent and brief thermal exposure. These figures are very rough and so conclusions must be drawn very carefully. They ignore effects arising from specific formulations and they do not provide any timeframes for exposure. They will also depend heavily on the presence of oxygen or moisture during thermal exposure and whether the bond is simultaneously exposed to mechanical loads. Table IV-2.2 reveals the fundamental disadvantage of adhesive joints compared with mechanical fastening systems or welding, which can withstand temperatures of several hundred degrees Celsius.

It should be noted at this point that thermoplastic polymers usually soften considerably well below their decomposition temperature, which leads to a corresponding loss of strength.

2.3.2 Analysis of exposure to different fluids

Many adhesives fail under permanent exposure to humid conditions. In practice, the adhesive joint may also be exposed to such fluids as fuel, gasoline, solvents and cleaning liquids. A certain degree of protection can be afforded by over-painting the adhesive joint, such as is done in the automotive industry. The ageing effects of exposure are attributable to different processes:

2.3.2.1 Swelling

In contrast to thermoset adhesive, thermoplastic adhesives show a certain solubility in certain solvents or are swollen by them. Elastomers show no solubility but are swollen by some solvents. Swelling is a reversible process, but it is always combined with a reduction in mechanical strength (modulus) and a change of dimensions. In this regard, then, liquids which cause swelling act like plasticisers. For adhesives applications, swelling must therefore be minimised. Swelling is promoted by liquids of polarity similar to the corresponding polymer. Thus, water will cause only highly polar polymers to swell extensively. The degree of swelling is also determined by the crosslink density of the polymer. As water is capable of coming into contact with almost any adhesive joint, highly polar polymers are less suitable and highly crosslinked polymers are preferred to elastomers or thermoplastics for resistance to liquids. Due to their polar structure, polyamides exhibit quite high water-uptake in the order of 1 – 4 % (at 23 °C, 50 % r.h.), the exact figure depending on the type of polyamide. As shown in Table IV-2.2, polymers can vary substantially in their water uptake [31].

Table IV-2.3: Water uptake of different polymers

Polymer	**Water uptake [%] at 23 °C, 50 % r.h.**	**Remarks on polymer structure**
Polyethylene	<<0.1	non-polar, semi-crystalline
Polystyrene	0.2	non-polar
Polyvinylchloride	0.2	low polarity
Polymethylmethacrylate	2	ester groups
Polyethylenterephthalate	0.35	ester groups
Polyamide 6	3	crystalline
Polyamide 12	1	less polar
Epoxies	0.7	crosslinked

2.3.2.2 Hydrolysis

Some polymers hydrolyse in the presence of water. Hydrolysis is usually catalysed by acids or bases. Elevated temperatures, too, can accelerate it. Polymers whose chain just consists of carbon (e.g. polyolefins, PVC, polystyrene) are resistant to hydrolytic degradation. Those which have hydrolysable groups in the side-chain like e.g. acrylates, polymethacrylate or polyvinyl acetate exhibit only slight hydrolytic degradability, because hydrolysis does only affect the side chain but not degrade the chain of the polymer backbone. Polymers which have hydrolysable groups in the polymer backbone like e.g. polyester, polyamides or polyurethanes have a certain sensitivity to hydrolytic degradation. A high degree of crystallinity can reduce the tendency to hydrolyse, examples are cellulose and polyoxymethylene, many polyamides and polyesters. The hydrolysis of polyesters is an autocatalytic reaction, because acids formed by hydrolysis boost further hydrolysis. This is a major consideration for adhesive joints because decomposition products usually cannot be removed.

2.3.2.3 Sub-surface migration

In bonded joints, the boundary layer between the adhesive and the surface of the adherend is often susceptible to water penetration. Water which migrates into the boundary layer can irreversibly destroy adhesion and cause corrosion in case metallic adherends are used. Such diffusion processes sometimes occur very slowly and are particularly pernicious as they often lead to total, irreversible loss of adhesion after some time. However, very good adhesion will minimise this. Hence, specific surface treatment methods can minimise sub-surface migration. Furthermore, the water may be prevented from diffusing into the adhesive layer by overpainting the joint.

2.4 Anisotropic effects

Many materials, such as metals, are isotropic, i.e. their properties are not directional. As soon as fibrous structures become involved, though, anisotropy can arise, i.e. the properties can become directional. This applies to many polymer-based materials, such as fibre-reinforced plastics serving as adherends, and to fibre-filled adhesives. In plastics processing, such as injection moulding, fibres are already given a preferred orientation by the injection nozzle. One of the most important materials exhibiting anisotropy is wood with the mechanical strength along the fibre is roughly ten times as high as in the perpendicular direction. Due consideration is thus needed in adhesively joined structures.

3 Bond design rules and recommendations

Table IV-3.1 summarises the rules for a reliable adhesively bonded structure discussed in the preceding chapters.

Table IV-3.1: Design rules for adhesively bonded joints

Rules for designing adhesive joints
Clean or pre-treat substrates adequately
Prefer thick substrates
Prefer high-modulus substrates
Use the maximum bonding area
Use the maximum width of the bonding area
Avoid peel loads
Design loads to be tensile or shear
Reinforce critical points of stress
Use toughened products if impact is to be expected
Prefer adhesive layers with a uniform thickness
Consider the hydrolytic and swelling properties of the adhesive if contact with water is expected
Protect adhesive joint from access of water
Protect adhesive joint from exposure to light
Adjust temperature range of application to phase transitions of the adhesive
Consider coefficients of thermal expansion
Consider substrate anisotropy

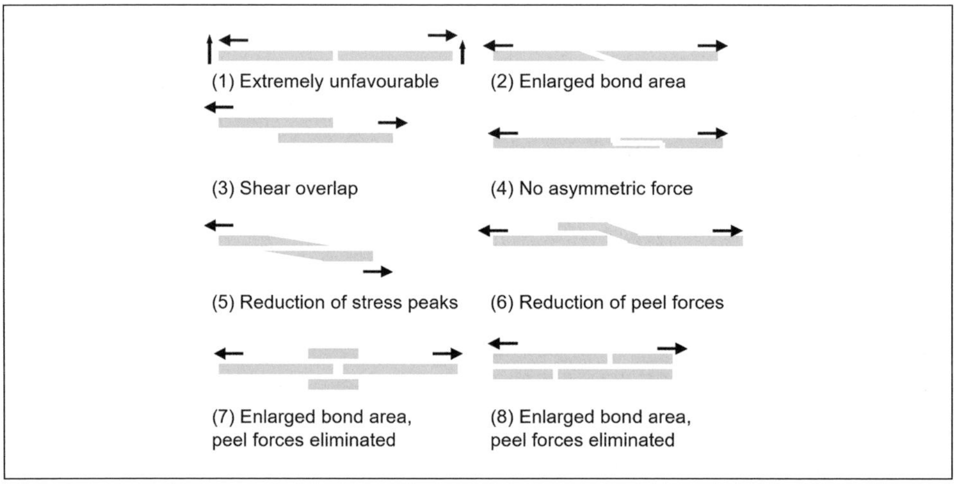

Figure IV-3.1: Butt and lap joints

Examples

Adherends can be joined favourably or less favourably. The butt joint (1) in Figure IV-3.1 is quite unfavourable due to the small bond area and the high risk of peel loads in the event of lateral forces. The bevelling of the adherends in the joint area shown in (2) leads to a more favourable design. A shear overlap joint (3) increases the bond area and leads to a better result. However, the asymmetrical direction of force is disadvantageous. The improvements in joint design shown in (4) and (6) minimise peel forces by introducing the forces symmetrically into the adhesive joint. Bevelling the adherends (5) minimises stress peaks, as the thinner parts of the substrates can be stretched more easily. The enlargements of the bond area (7) and (8) lead to further improved results.

Corner joints Figure IV-3.2 are critical for adhesives applications if lateral forces can lead to peel stress (1) and (2). Depending on the direction of the acting forces, peel stress can be reduced or avoided by a number of different measures. In many cases, enlarging the small available bond area shows only a limited effect. Figure IV-3.2 shows some examples for optimizing the bond design to improve bond performance.

Alternatives to unfavourable bond designs showing peel forces (1) and (3) are provided in Figure IV-3.3.

In mechanical engineering, round profiles, such as shaft-hub joints and pipes, often have to be prepared. Figure IV-3.4 indicates suitable measures for increasing the bond area and minimising the influence of peel forces.

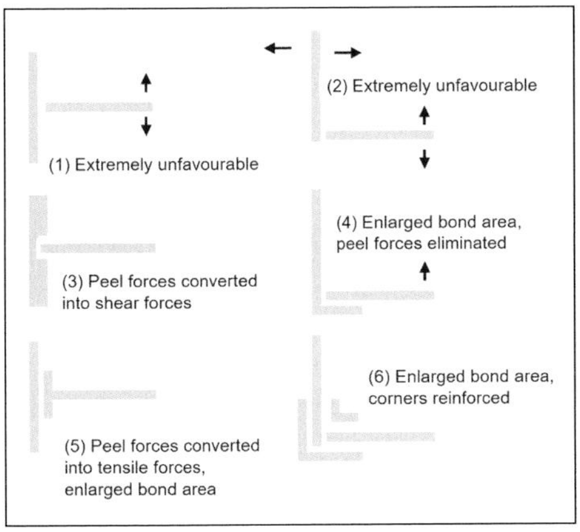

Figure IV-3.2: Ways to improvement corner joints

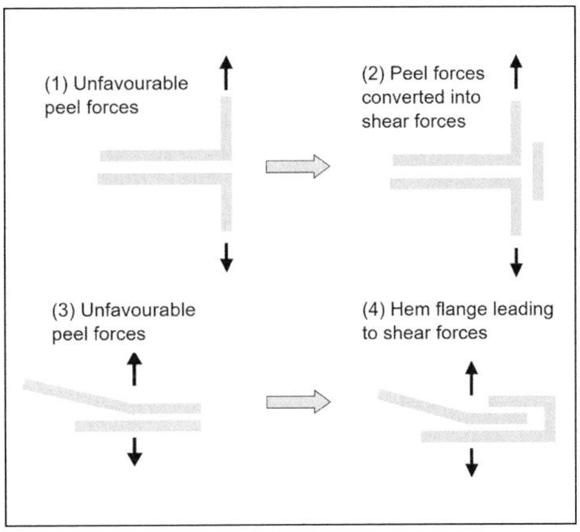

Figure IV-3.3: Avoidance of peel forces

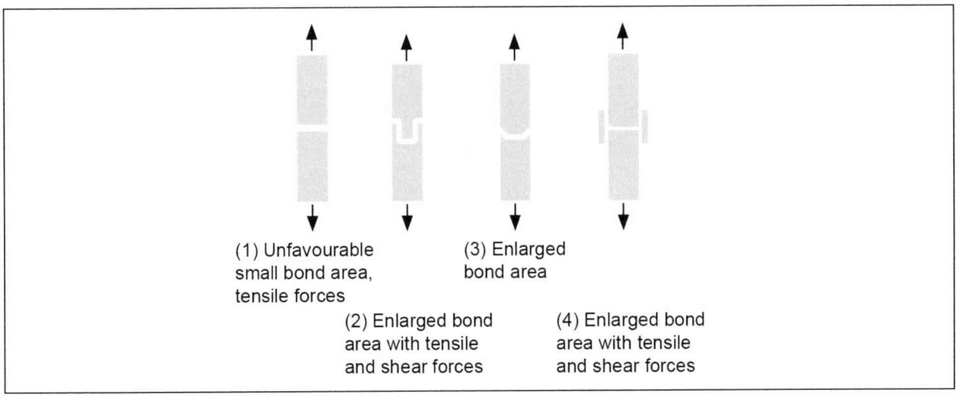

Figure IV-3.4: Joints design for coaxial parts

4 Testing of adhesive and bonded joints

When discussing the subject of testing adhesives, the following different objectives must be distinguished:

1. providing basic information on adhesive properties to support adhesive selection and quality assurance in the production of adhesives by the manufacturer or as part of an incoming goods inspection by the adhesive user
   - characterization of adhesives as delivered state
   - characterization of adhesives in the solidified state
   - characterization of adhesives and bonded joints regarding bond strength, adhesion properties and resistance against stress, media, environmental conditions, etc.
2. the verification that the load-bearing capacity over the entire projected lifespan of a bond is greater than the actual loads that occur (demonstration of usage safety)
3. quality assurance of the bonding process through accompanying tests, such as the preparation of work samples

Given the target audience of this book, the formulators of adhesives and sealants, the focus is on test methods falling into category 1. This is not to say that the test methods of categories 2 and 3 are of less importance. On the contrary, these tests are of great importance for ensuring the required quality of bonded products [32]. Even though this verification process is primarily the responsibility of the adhesive user and cannot always be carried out solely by means of tests according to standardised procedures but requires component-specific testing that takes into account the entire respective bonding process, the standardised procedures in particular can support the formulation of adhesives during the application-specific development of adhesives or sealants.

Before describing the most important and commonly used test methods in the following chapters, it should be noted that when reporting the values obtained, all the important parameters under which the test was carried out must be specified. This is the only way to ensure that the results from different sources can be compared. Usually, this information is provided in the section of the respective standard that deals with the test report.

Unfortunately, the technical data sheets for adhesives often refer to the adhesive manufacturer's own standard and no longer, as in the past, to a national (e.g. BS, ASTM, etc.), European (EN) or international (ISO) standard. Even if the in-house standards are often based on EN or ISO standards, in order to compare the technical datasheets of adhesives or sealants from different manufacturers, the details of the test methods used must be questioned.

Due to the specifics of pressure-sensitive adhesives test methods for testing this group of adhesives are covered in Chapter IV-4.3.

Hartwig Lohse, Matthias Popp: Formulating Adhesives and Sealants, 2nd revised Edition

4.1 Characterization of adhesives as delivered state

The data generated by such tests are either used for quality assurance but may also provide information regarding the suitability of the respective adhesive for as certain bonding process.

EN 542 – Adhesives – Determination of density

The standard specifies two methods that can be used to determine the density of any adhesive. One of the methods listed is used for liquid adhesives, while the other is used for solid, highly viscous or pasty adhesives. The density (mass per unit volume) of a sample is measured at 23 °C as the ratio of the mass of a given volume of the adhesive to the mass of the same volume of a reference liquid of known density, multiplied by the density of the reference liquid.

ISO 1675 – Plastics – Liquid resins – Determination of density by the pycnometer method

The test method specifies a procedure for determining the density of liquid resins using a pycnometer.

ASTM D 1875 – Standard test method for density of adhesives in fluid form

This test method covers the measurement of density (weight per gallon) of adhesives, and components thereof, when in liquid form. It is particularly applicable where the fluid has too high a viscosity or where a component is too volatile for a specific gravity balance determination.

EN 827 – Adhesives – Determination of conventional solids content and constant mass solids content

The standard describes methods for determining the proportion of non-volatile components. For this purpose, a defined mass of adhesive is placed in a container with a defined geometry and stored in a heating cabinet at a defined temperature for a defined period of time. The test parameters are determined depending on the type of adhesive. Alternatively, drying can be carried out until a constant weight is reached.

EN 1245 – Adhesives – Determination of pH

This standard specifies an electrometric method for determining the pH of adhesives, their base components and related products using a pH meter equipped with a combined glass and silver reference electrode.

ISO 976 – Rubber and plastics – Polymer dispersions and rubber latices – Determination of pH

The standard describes a method for determining the pH of polymer dispersions and rubber latices (natural and synthetic) using a pH meter with a combined glass and silver reference electrode. The method is also suitable for adhesives based on the above-mentioned substances.

EN 12092 – Adhesives – Determination of viscosity

The standard describes six methods for determining the viscosity of adhesives
- rotational viscometer method
- cone and plate viscometer method

- falling ball viscometer method
- glass capillary viscometer method
- cup-pouring method
- extrusion method

and provides information on which systems are suitable for which types of adhesives.

EN ISO 3219-1 – Rheology – Part 1: Vocabulary and symbols for rotational and oscillatory rheometry

EN ISO 3219-2 – Rheology – Part 2: General principles of rotational and oscillatory rheometry

Part 1 of the series of standards defines general terms used in the field of rotational and oscillatory rheometry.

Part 2 of the series of standards defines the general principles of rotational and oscillatory rheometry.

ISO 2555 – Plastics – Resins in the liquid state or as emulsions or dispersions – Determination of apparent viscosity using a single cylinder type rotational viscometer method

This standard describes a method for determining the apparent viscosity of resins in a liquid state using a single-cylinder rotational viscometer. The method can be used for viscosity measurements in the range 0.02 Pa s – 60,000 Pa s. The method is applicable to both newtonian and non-newtonian fluids, although the measured apparent viscosity depends on the velocity gradient to which the fluids are subjected during the measurement.

ASTM D 1084 – Standard test methods for viscosity of adhesives

The standard describes three methods for determining the viscosity of self-levelling adhesives.

ASTM D 4402 – Standard test method for viscosity determination of asphalt at elevated temperatures using a rotational viscometer

This is originally used to determine the apparent viscosity of asphalt at handling or application temperatures but is also widely used to characterise hot-melt adhesives. It is also known as "Brookfield Thermosel".

EN ISO 14678 – Adhesives – Determination of resistance to flow (sagging)

The standard describes various methods for measuring the tendency of adhesives to flow from a joint after application and before hardening. The extent of the flow is naturally influenced by the properties of the adhesive, but also by the applied tension, the duration of the bonding or setting process, the temperature and the surface of the adherend (roughness, surface energy).
The methods described are based on three types of tests:

- measuring the flow of the adhesive from a vertical surface or a surface inclined at 60° to the horizontal. This type simulates the conditions before joining
- measuring the flow of the adhesive from a circular opening or a gap. This type simulates the conditions after joining in the adhesive joint until it solidifies
- simulating the conditions in the adhesive joint during the cross-linking of a structural adhesive

EN ISO 1133-1 – Plastics – Determination of the melt mass-flow rate (MFR) and melt volume-flow rate (MVR) of thermoplastics – Part 1: Standard method

This standard describes two procedures for determining the melt mass-flow rate (MFR) and the melt volume-flow rate (MVR) of thermoplastic materials. Procedure A is used to determine the mass. Procedure B is a measuring method for the length of the extrudate.

For plastic material that exhibit a high rheological sensitivity to a time- or temperature-dependent history that the sample has undergone during the test and/or that are sensitive to moisture part 2 of the standard is recommended.

EN 1238 – Adhesives – Determination of the softening point of thermoplastic adhesives (ring and ball)

The standard describes a method for determining the softening point of hot-melt adhesives. Normally, hot-melt adhesives do not change from a solid to a liquid state at a specific temperature, but their viscosity decreases as the temperature rises. For the test, a metal ring of defined dimensions is filled with hot-melt adhesive. A ball is placed on the ring and sample holders are used to place the ring in an inert liquid bath, which is then heated at a predetermined heating rate. The temperature at which the adhesive, deformed by the ball weight, first reaches a certain position below the ring (light barrier) is given as the softening point, see Figure II-1.40.

EN 1240 – Adhesives – Determination of hydroxyl value and/or hydroxyl content

The standard describes a titrimetric method for determining the hydroxyl group content of organic materials, such as resins, polyols used as raw materials in the formulation of adhesives and sealants as well as of adhesives e.g. for defining the stochiometric mix ration of 2-component polyurethane adhesives.

EN 1241 – Adhesives – Determination of acid value

The standard describes a titrimetric method for determining the content of acidic components in organic materials, such as resins, polyols and adhesives.

EN 1242 – Adhesives – Determination of isocyanate content

Titrimetric wet-chemical determination of the content of isocyanate groups in organic materials, e.g. resins, adhesives.

EN 924 – Adhesives – Solvent-borne and solvent-free adhesives – Determination of flash point

The standard describes a method for determining the flash point of adhesives. The method can be used for solvent-based and solvent-free adhesives, but not for aqueous adhesives containing less than 10 % solvent.

EN ISO 13736 – Determination of flash point – Abel closed-cup method

This standard specifies a procedure for the determination of the flash point of combustible liquids in a closed cup, with flash points between -30.0 °C and at least 75.0 °C.

EN ISO 2719 – Determination of flash point – Pensky-Martens closed cup method

This standard describes methods for determining the flash point of flammable liquids, liquids containing suspended solids, liquids that tend to form a surface film under the test conditions, biodiesel and other liquids in the temperature range from 40 °C to 370 °C using a closed cup apparatus according to Pensky-Martens.

ISO 10364 – Structural adhesives – Determination of the pot life (working life) of multi-component adhesives

The standard describes various methods for determining the pot life of multi-component adhesives. Some methods are particularly suitable for quality control in the context of incoming or outgoing goods inspection. Others allow statements to be made as to whether the pot life of an adhesive is suitable for a specific bonding process. The different methods described for measuring this property do not necessarily provide identical results.

EN 14022 – Structural adhesives – Determination of the pot life (working life) of multicomponent adhesives

This standard describes five methods for determining the pot life of adhesives, each of which is based on particular circumstances; the flow behaviour of the adhesive in question and its reaction speed are particularly important.

EN ISO 10363 – Hot-melt adhesives – Determination of thermal stability

The standard describes a procedure for determining the thermal resistance, in particular when processing non-reactive hot-melt adhesives from a melting tank (see Chapter II-1.3.1). A sample of the hot-melt adhesive is heated under specified conditions, with samples taken at regular intervals and analysed for viscosity (Brookfield according to ISO 2555) and softening point (ring-and-ball method, DIN 1238).

EN ISO 11357 – Plastics – Differential scanning calorimetry (DSC)

This series of standards describes various differential scanning calorimetry (DSC) methods for the thermal analysis of polymers and polymer blends. The standard is used to observe and quantitatively determine various phenomena or properties such as:

- physical transformations (glass transition, phase transformations such as melting or crystallisation, polymorphic transformations, etc.)
- chemical reactions (polymerization, cross-linking and vulcanisation of elastomers and thermosets, etc.)
- oxidation stability
- heat capacity

EN ISO – 11358 – Plastics – Thermogravimetry (TG) of polymers

The standard describes the general conditions for and the performance of thermogravimetric analyses (TGA) of polymers in liquid and solid form. This measurement provides information especially about physical phenomena, such as

- phase transitions
- thermal decomposition

4.2 Testing of adhesive and sealants in the solidified state

The testing of adhesives or sealants in the solidified state can be carried out either on test specimens made solely from the respective adhesive or sealant, or on samples made from the adhesive or sealant combined with one or more adherends. While in the first case, information is obtained only on the properties of the solidified material (in particular the cohesive properties), tests on bonded joints include the specific interactions leading to adhesion between the adhesive or sealant and the respective adherend.

4.2.1 Testing of pure adhesives and solids in the solidified state

The testing of adhesives or sealants in a solidified state on test specimens made solely from the respective adhesive or sealant is used to characterise the properties of the solidified adhesive and does not reflect the specific interactions with the adherends.

EN ISO 527 – Plastics – Determination of tensile properties – Part 1: General principles

Part 1 of this series of standards specifies the general principles for determining the tensile properties of plastics and plastic composites under specified conditions. Several different types of test specimens are defined according to the different types of materials, which are described in more detail in the following parts of ISO 527.

The methods are used to examine the tensile deformation behaviour of test specimens and to determine the tensile strength, tensile/Young's modulus, as well as elongation at break under specified conditions.

ISO 48-4 – Rubber, vulcanized or thermoplastic – Determination of hardness – Part 4: Indentation hardness by durometer method (shore hardness)

ASTM D 2240 – Standard test method for rubber property durometer hardness (shore hardness)

The method described in these ISO and ASTM standards, like many other hardness tests, is based on the determination of the depth of an indentation in the material (adhesive/sealant) created by a given force on a standardized indenter (durometer). The depth depends on the hardness of the material, its viscoelastic properties, the shape of the indenter and the duration of the test. To a certain extend the method may be used for a rough comparison of the mechanical properties of adhesives.

EN ISO 11357 – Plastics – Differential scanning calorimetry (DSC)

This series of standards describes various differential scanning calorimetry (DSC) methods for the thermal analysis of polymers and polymer blends. The standard is used to observe and quantitatively determine various phenomena or properties such as:

- physical transformations (glass transition, phase transformations such as melting or crystallization, polymorphic transformations, etc.)
- chemical reactions (polymerization, cross-linking and vulcanisation of elastomers and thermosets, etc.)

- oxidation stability
- heat capacity

EN ISO 11358 – Plastics – Thermogravimetry (TG) of polymers

The standard describes the general conditions for and the performance of thermogravimetric analyses (TGA) of polymers in liquid and solid form. This measurement provides information especially about physical phenomena, such as
- phase transitions
- thermal decomposition

ISO 11359 – Plastics – Thermomechanical analysis (TMA)

Thermomechanical analysis (TMA) consists of determining the deformation of a test specimen under constant load as a function of temperature and/or time.

ISO 6721-11 – Plastics – Determination of dynamic mechanical properties (DTMA) – Part 11. Glass transition temperature

Dynamic thermomechanical analysis (DTMA) is a technique for measuring viscoelastic properties of materials by applying mechanical and thermal forces. This method helps in characterising materials under various environmental conditions. Part 11 of the ISO 6721 row of standards describes the determination of the glass transition temperature.

EN ISO 22007 – Plastics – Determination of thermal conductivity and thermal diffusivity

In addition to the general principles, the multipart standard describes various methods for determining the thermal conductivity of plastics. These can usually also be applied to adhesives and sealants.

Light and weathering resistance

The influence of light, especially UV radiation, is generally of minor importance for the characterisation of adhesives, since the adhesives are usually covered by the two adherends. The influence of light must be considered only for very thick adhesive joints, as used for elastic bonding, or for transparent adherends such as glass, PMMA and PC. On the other hand, the influence of light on sealants should not be underestimated as they are often exposed to direct sunlight.

For adhesives photodegradation is especially critical, because it could damage the adhesive layer directly at the phase boundary and lead to irreversible loss of adhesion. In one of the most important adhesive applications in the automotive industry, windscreen bonding, the adhesive layer is typically protected from harmful radiation by the ceramic coating on the glass and additionally by a carbon-black-filled "glass primer" (see also Chapter I-2.3.1).

For sealants as well as for elastic bonding applications the situation is different, because sealant gaps typically have dimensions in the order of centimetres and therefore the sealant surface is directly exposed to light. Light, especially its high-energy UV portion, can initiate free-radical reactions and bring about irreversible damage to polymeric materials. On exposure to light, many elastomer-based sealants have a tendency to discolour, embrittle and crack. Other polymers experience degradation and the attendant consequences. Figure IV-4.1 shows cracks in the surface of a polysulphide expansion joint exposed to weathering for about 40 years. The embrittlement is limited to the surface layer – the interior is still elasto-plastic. Figure IV-4.2 shows two scanning electron micrographs of the sealant's surface. The plan view shows filler particles in the surface laid open by photooxidative polymer degradation. The cross section shows that the damage caused by weathering is only on the surface and that the sealant in the interior is still intact.

Test methods for light and weather resistance basically distinguish between natural weathering and artificial weathering in special test chambers. The ageing processes which occur during artificial and natural weathering cannot be expected to correlate with each other because of the large number of factors which influence these processes. Definite relationships can only be expected if the important parameters (distribution of the irradiance over the photochemically relevant part of the spectrum, temperature of the specimen, type of sprinkling and sprinkling cycle, and relative humidity) are the same in each case or if their effect on the sealant is known. While the conditions can be kept constant in artificial weathering test, this is generally not the case with natural weathering. The parameters acting on the test specimens are not only subject to natural seasonal fluctuations, which can certainly be seen as an advantage, but can also vary from year to year.

ISO 2810 – Paints and varnishes – Natural weathering of coatings – Exposure and assessment

This international standard, which was originally developed to determine the weather resistance of coatings but can also be used for adhesives and sealants, describes the parameters to be considered when selecting the type of natural weathering and the method of natural weathering to determine weather resistance (direct weathering or weathering behind window glass). In addition to the geographical location of the test location which determines the climatic conditions to be

Figure IV-4.1: Surface of a polysulphide sealant after about 40 years exposure to weathering

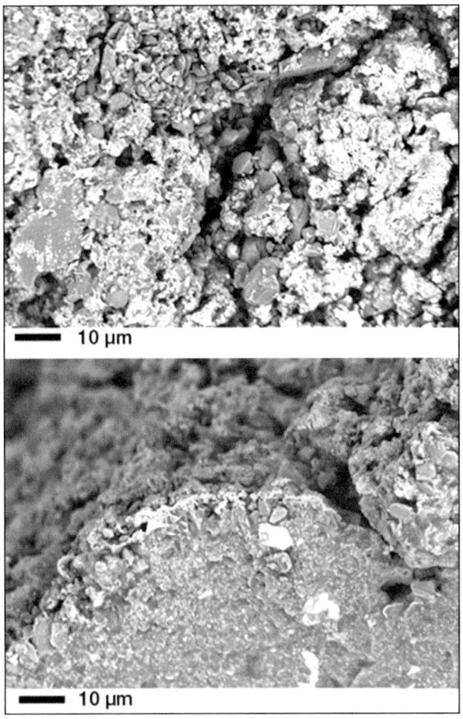

Figure IV-4.2: Scanning electron micrographs of Figure IV-2.1.
Top: surface
Botton: cross section

expected, beside others, the duration of the exposure (typically a year or longer), the orientation of the test specimens with regard to the cardinal directions and the inclination angle of the test are relevant test parameters.

ISO 11341 – Paints and varnishes – Artificial weathering and exposure to artificial radiation – Exposure to filtered xenon-arc radiation

and

ISO 4892-2 – Plastics – Methods of exposure to laboratory light sources – Part 2: Xenon-arc lamps

Although ISO 11341 was originally developed for determining the weather resistance of coatings and ISO 4892-2 for plastics, both are also used for adhesives and sealants. In the standards for testing weather resistance in special test chambers, the light generated by xenon arc lamps is adjusted using suitable filters to the spectral distribution of global solar radiation in the ultraviolet and visible range or, with another filter, to the spectral distribution of global solar radiation in the ultraviolet and visible range behind a 3 mm thick glass windowpane. The test chamber is further equipped with a device for wetting the test specimens allowing to simulate specific conditions.

Testing for volatile components

Fogging is the emission of volatile organic compounds (VOC) from materials. These substances can condense on surfaces and thereby reduce the amount of light that is transmitted or reflected. Examples include the diffuser and reflector lenses in headlights and also the windscreens of vehicles. Furthermore, VOCs can be harmful to health or can be perceived as unpleasant just by virtue of their intrinsic odour. With adhesives, fogging can be caused by plasticisers, residual monomers and specific additives.

EN ISO 17071 – Leather – Physical and mechanical tests – Determination of fogging characteristics

In this method, material samples are exposed to an elevated temperature in a closed chamber with a cooled cover plate. In one method, quantification of the amount of condensed VOC is carried out by measuring the reflection or the light-scattering properties of the cooled plate with the condensed volatile components. In the other method, quantification is carried out gravimetrically. Both methods are also used to characterise the fogging behaviour of adhesives.

The German Association of the Automotive Industry (VDA) has developed a series of methods for determining the fogging behaviour of materials.

VDA 270 – Determination of the odour characteristics of trim materials in motor vehicles

The standard describes a method for testing the odour of materials and components of the vehicle interior and parts that come into contact with the air flow supplied to the vehicle interior. The test is used to assess odour behaviour under the influence of temperature and climate. For the test, samples are conditioned in a glass vessel for 24 hours and then evaluated by three testers according to a six-point scale. In addition to the point scale evaluation, there is usually also a description of the odour sensation.

VDA 276-3 – Determination of the total concentration of hydrocarbon compounds in test chambers and vehicle interiors with a flame ionization detector (FID)

This standard describes a method for determining the total concentration of hydrocarbon compounds released from a material in a test chamber, where detection is carried out using a flame ionisation detector.

VDA 278 – Thermal desorption analysis of organic emissions for the characterization of non-metallic materials for automobiles

The standard describes a method for quantifying the fogging behaviour of adhesives and bonded joints using thermodesorption. In this process, the substances released from a heated component are separated using gas chromatography and identified by their mass spectra (GC/MS). The method not only allows the total emission to be determined, but also the individual components to be evaluated and quantified, which ultimately facilitates targeted improvement of a product with regard to its emission behaviour. The method is recommended by the "Industrieverband Klebstoffe e.V." for testing the fogging of adhesive tapes.

4.2.2 Destructive testing of adhesively bonded joints

Since adhesive joints cannot be tested completely non-destructively, destructive tests are of great importance not only in product development and quality assurance of adhesives and sealants, but also in the selection of adhesives or sealants for a specific application. During development of adhesives or sealants, the characterisation of adhesives by their key properties, as well for screening different adhesives or sealants for a specific application is typically performed.

The purpose of destructive testing of bonded joints is to provide typical data about the cohesion, the mechanical properties, the load bearing capacity and the adhesion properties of a certain adhesive. Some of the test methods can be used as part of the verification of bonded joints to demonstrate that the load-bearing capacity of the bonded joint is always greater than the actual load that occurs under all real conditions during the entire expected lifetime.

Adhesive bonds are also tested to predict the suitability of an adhesive for the intended application. The objective is to ensure the durability of a bond over the intended service life (which can be very long, such as at least 25 years for wind turbines or more than 30 years for rail vehicles), taking into account the actual loading scenarios that occur. According to EN ISO 21368 Adhesives – Guidelines for the fabrication of adhesively bonded structures and reporting procedures suitable for the risk evaluation of such structures [32], this can be accomplished by

1. demonstrating that the occurring stresses are lower than the stress resistance
2. assembly or component testing
3. documented experience or
4. a combination of methods 1 – 3

When applying method 1, strengths determined according to standardised procedures are often compared with the forces expected either on the basis of calculations or estimates.

The results are more valuable the more closely the bonding conditions correspond to those of the intended application. This includes not only the type of forces that actually occur, but also, in particular using adherends which are as close to reality as possible, including material, surface treatment and the curing conditions corresponding to the intended process. It is obvious that testing according to method 1 requires a detailed knowledge of all the influences acting on adhesive joints during their lifetime. In daily practice, such test programmes are developed on the basis of experience, to the best of one's knowledge and often in close cooperation between adhesive manufacturer and user.

Verification based on tests according to standardised procedures alone will only be sufficient in a few simple cases, so that their use is typically the screening of different adhesive formulations and have to be supplemented by tests according to methods 2 or 3 when it comes to validate a certain adhesive. Since the bonding process, similar to welding, for example, is a so-called "special process", i.e. bonded joints cannot be tested completely non-destructively, a certain risk will always remain.

Besides the bond strength the kind of failure is an important criterion. According to EN ISO 10365 "Adhesives – Designation of main failure patterns" [33] a distinction is made between different modes of failure.

- **Cohesive failure (CF)** (Figure IV-4.3)
  The adhesive fails more or less uniformly within the adhesive layer, leaving residues of the adhesive on both substrates.
  → adhesion > cohesion
  Cohesive failure is usually desirable because, on the one hand the mechanical properties (E-modulus, G-modulus, elongation, etc.) of the adhesive can be used to a certain extent in the design of a bonded structure and, on the other hand, metallic substrates, for example, are protected from corrosion in the event of bond failure. The chemist should check if a further increase in cohesion might lead to a further strengthening of the joint. This can be achieved by increasing the crosslink density of the polymer.
- **Surface close cohesive failure (SCF)** (Figure IV-4.4)
  In surface close cohesive failure or also known as slight cohesive failure or thin layer cohesive failure, the adhesive does not break symmetrically in the middle of the joint but breaks very close to the phase interface of one of the adherends.
  → adhesion > cohesion
  Surface close cohesive failure is sometimes difficult to recognise, especially if adherend and adhesive are similar in colour because the thin remaining layer can become nearly invisible and so it resembles adhesive failure.
  Surface close cohesive failure can indicate surface effects on the substrate, such as dirt, weak oxide layers, thin polymer layers (weak boundary layers) of low-molecular polymer fractions, or other components bleeding from the adhesive during solidification. In these cases, proper pretreatment of the substrates is recommended.

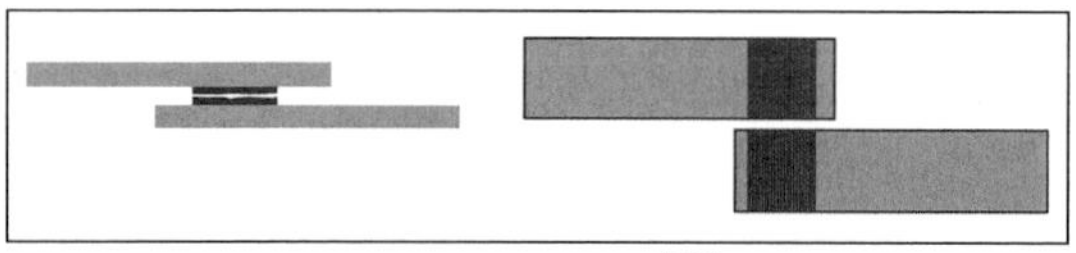

Figure IV-4.3: Cohesive failure (CF) [33]

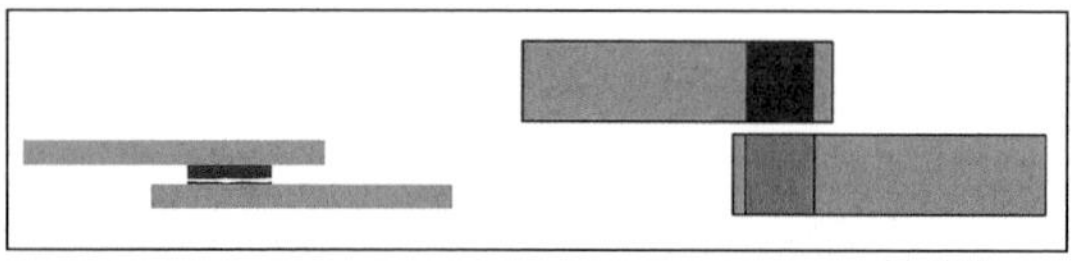

Figure IV-4.4: Surface close cohesive failure (SCF) [33]

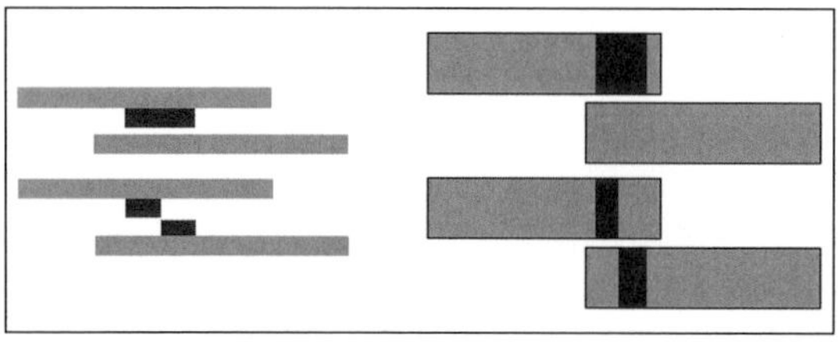

Figure IV-4.5: Adhesive failure (AF) [33]

- **Adhesive failure (AF)** (Figure IV-4.5)
  The bond has failed in the boundary layer between adhesive and adherend. With the naked eye no or just marginal residues of the adhesive are visible on the surface of the adherend.
  → cohesion > adhesion

Adhesive failure is usually undesirable because it is an indication for an insufficient wetting of the adherend by the adhesive resulting in poor adhesion. Experience shows that within a series of test specimens prepared and tested identical the standard deviation for adhesive failure is significantly greater than for cohesive failure. In addition, adhesive failure is often an early indication of less-than-ideal long-term durability, especially when exposed to moisture. In case of an adhesive failure on metallic adherends the surface is no longer protected from corrosion by the adhesive layer. It might be worthwhile for the formulating chemist to check whether a reduction of bond strength is acceptable and to reduce the level of cohesion by adding certain fillers with a layered structure like graphite to the extent that the adhesive fails cohesively.
On the other hand, adhesive failure may be tolerable if the bond strength does not decrease, even after exposure to severe ageing conditions. Phenolic adhesives often used successfully for bonding aluminium in the aerospace industry are such an example.

- **Substrate failure (SF)** (Figure IV-4.6)
  When the test specimen is tested one of the adherends breaks. The break usually occurs either close to the end of the overlap zone or close to the clamps used to fixture the specimen in the tensile testing machine.
  Test specimens that fail due to fracture of the adherend only allow the conclusion that both adhesion and the cohesion are greater than the breaking force determined. It should be noted that a fracture of the adherend can be provoked by choosing a large bonding surface or thin adherends.
- **Cohesive substrate (adherend) failure (CSF)** (Figure IV-4.7)
  The bond fails due to material tearing out of the surface of the adherend.
  As with test specimens that fail due to the substrate (adherend) failure, it can only be stated that both the adhesion and the cohesion of the used adhesive are greater than the determined breaking force. It should be noted that the selection of a large bonding area can provoke material tear. This failure mode occurs, for example, on wood and fibre-reinforced plastics, particularly in SMC and BMC.
- **Other potential failure patterns**
  - **Sub-surface migration (SSM)** (Figure IV-4.8)
    Sub-surface migration is characterised by cohesive fracture in the centre of the bond with an encircling strip of adhesive failure.
    Sub-surface migration occurs when media such as moisture or various chemicals attack the adhesion zone, i.e. the interface between the adhesive and the surface of the adherend. The process is starting from the edges of the bond and significantly reducing adhesion.
  - **Corrosion (COR)**
    If moisture is causing sub-surface migration on metallic adher-

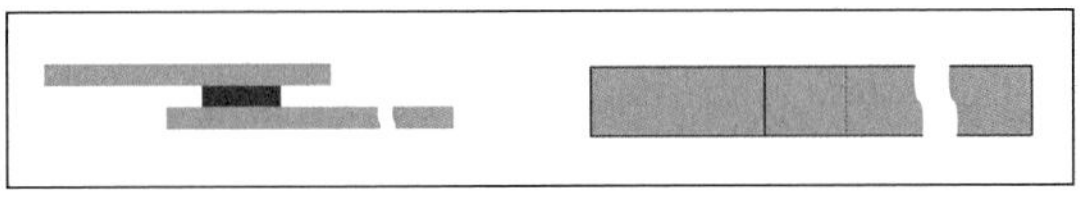

Figure IV-4.6: Substrate (adherend) failure (SF) [33]

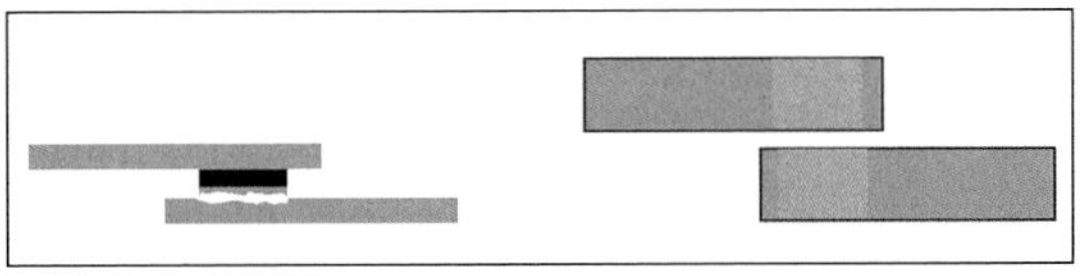

Figure IV-4.7: Cohesive substrate (adherend) failure (CSF) [33]

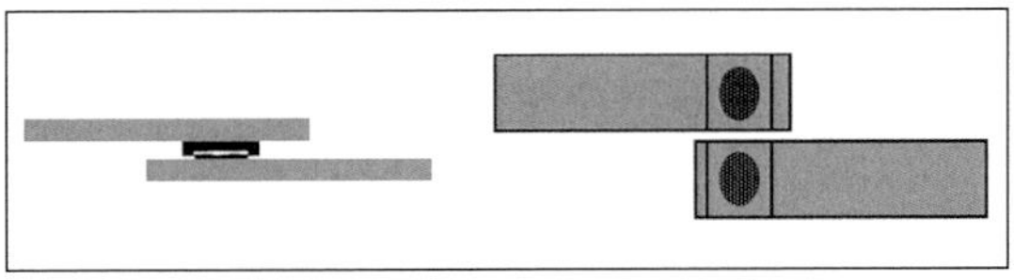

Figure IV-4.8: Sub-surface migration (SSM)

ends like steel, zing-coated steel or aluminium corrosion may occur, e.g. red rust on steel and white rust on galvanised surfaces.

- **Failure modes on coated surfaces**
  In the case of coated adherends, it should be noted that the fracture can occur in the coating itself, in the boundary layer between coating and base material of the adherend, or at the boundary layer between adhesive and coating. For the evaluation of the result and the conclusions to be drawn from it, a precise indication of the failure location is required.
- **Mixed failure (MF)**
  It is not unusual for different types of failure to occur simultaneously in one test specimen. This is then referred to as a mixed failure. In this case, it is important to determine the respective proportions of the various failure types. EN ISO 10365 [33] describes a rather extensive procedure for this. In many cases, it is sufficient and, with a certain amount of practice, also quite reproducible to estimate the respective proportions in 5-percent steps.

Ideally, the fracture surface should be evaluated on the original test specimens shortly after the destructive test has been conducted. If the destructive tests are carried out, for example, as part of a customer-specific adhesive or sealant development project, it is advisable to agree on the criteria for assessing the fracture pattern, in particular the distinction between cohesive failure (CF) and surface close cohesive failure (SCF) on the one hand and adhesive failure (AF) and surface close cohesive failure (SCF) on the other. Coloured photographs are a valuable addition to the results documented in the test report.

In Chapter IV-2 it has been shown that the dimensions (length of overlap, substrate thickness, etc) significantly influence the bond strength. The same is true for the rate at which an adhesive joint is destroyed, because polymers are viscoelastic materials, and their properties are time dependent. To ensure that the data generated are comparable, it is important to follow standardised and accepted test methods. The most important test methods are briefly described below.

4.2.2.1 Quasi static tests

In principle, a distinction is made here between tests under surface loads such as shear, tension or torsion and more linear loads such as those that occur under peeling or cleavage conditions.

Surface loads

When designing adhesively bonded joints, the basic objective is to transfer the resulting loads across a sufficiently large surface area, taking into account the loads and the adhesive strength. It is therefore not surprising that such tests are of particular importance in the development of adhesives and their qualification for a specific application. The upper part of Figure IV-2.1 shows the different types of surface-area load distribution

EN 1465 – Adhesives – Determination of tensile lap-shear strength of bonded assemblies

Figure IV-4.9 shows the preferred dimensions of the tensile shear test specimen according to EN 1465. The adherends are joined in an overlap geometry, with the tensile forces acting as parallel to the joint area as possible.

During the test the tensile force is increased until the joint fails. The ratio of maximum force to bonded area (F/A) is given as the tensile lap shear strength with the unit [N/mm^2] or [MPa]. In North America, the unit [psi] (pounds per square inch) is still commonly used. The following applies:

$1\ \text{N/mm}^2 = 1\ \text{MPa} \approx 145\ \text{psi}$

Even though the results of the tensile lap shear test are influenced not only by the adherends used but also by the effects shown in Figure IV-2.4 (formation of stress peaks by stretching of the adherends under loads) and Figure IV-2.5 (formation of peel forces at the overlap ends due to eccentric force introduction), and accordingly, when comparing tensile lap shear results, special attention must be paid to the respective test conditions, the tensile lap shear test is probably the most widely used method. It is performed for quality control purposes, during development of new adhesive systems and screening different adhesive for a particular application. In technical data sheets, tensile lap shear strength data are used to characterise adhesives.

In addition to EN 1465, the tensile shear test is also described in ISO 4587, ASTM D 1002 and ASTM D 3163.

ISO 11003-2 – Adhesives – Determination of shear behaviour of structural adhesives – Part 2: Tensile test method using thick adherends

In order to achieve the most uniform shear stress distribution possible and to minimise other types of stress, such as peel stresses as can occur in a tensile lap shear test according to EN 1465, the test is carried out on specimens consisting of thick, rigid steel adherends with a reduced overlap length (Figure IV-4.10).

The tensile force is recorded as a function of elongation in a stress-strain diagram, which is useful for determining the shear modulus from the slope of the graph at low elongation. These material data can serve as a basis for calculating adhesive joints. In the example shown in Figure IV-4.11 for a relative substrate shift of 0.005 mm and a thickness of the adhesive layer of 0.2 mm, a shear stress of 25 MPa is recorded. With tan $\gamma \approx \gamma$ at low elongation, the shear strain is $\gamma = 0.05$, and with $G = \tau / \gamma$, the shear modulus is $25/0.05 = 500$ MPa.

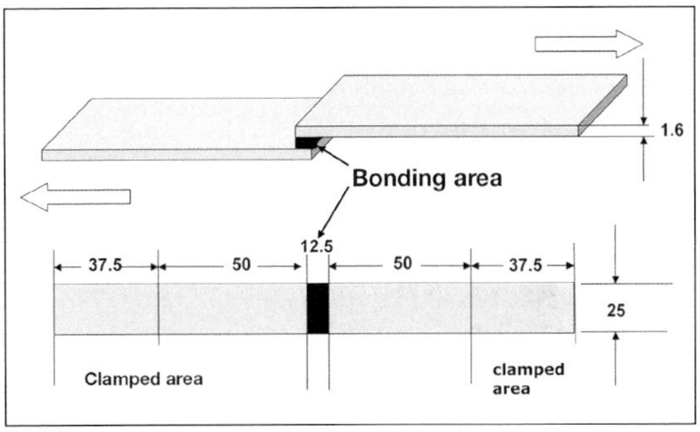

Figure IV-4.9: Dimensions of the tensile lap shear test specimen according to EN 1465

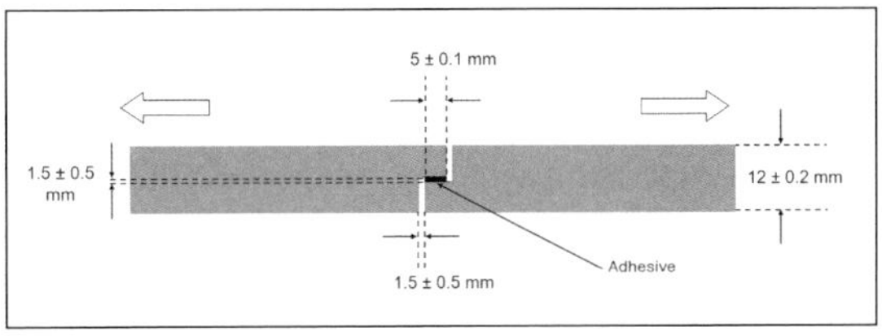

Figure IV-4.10: Dimensions of the thick adherend tensile lap shear test specimen according to ISO 11003-2

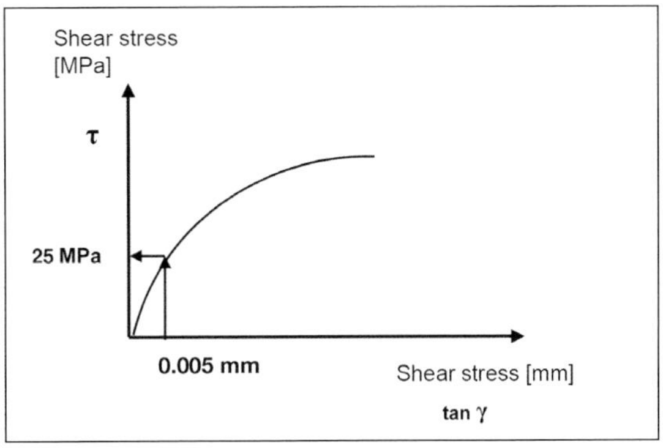

Figure IV-4.11: Example of a shear stress – shear strain diagram

EN 15870 – Adhesives – Determination of tensile strength of butt joints

In this test, a bond between two square or round parts, which are butt-joined at right angles to their longitudinal axis, is subjected to a tensile force acting along the specimen axis until it breaks. The recorded result is the determined breaking strength.

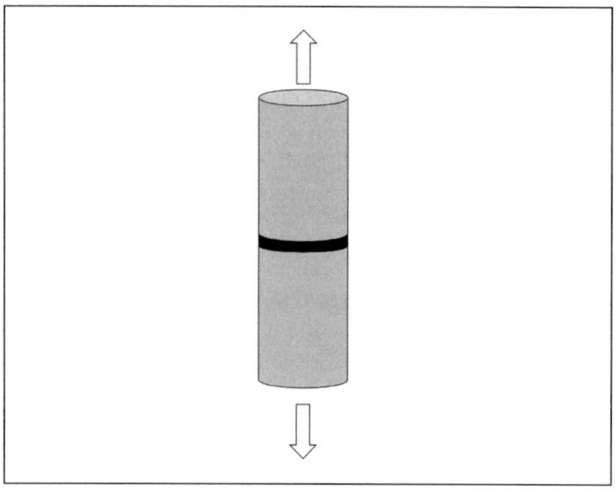

Figure IV-4.12: Principle of determining the tensile strength of butt bonds according to EN 15870

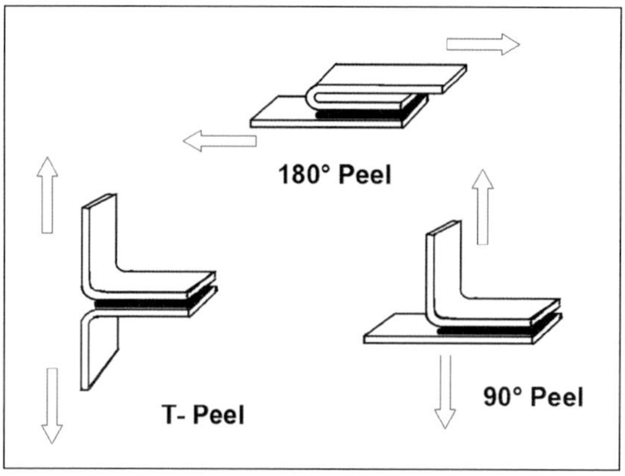

Figure IV-4.13: Peel test variants

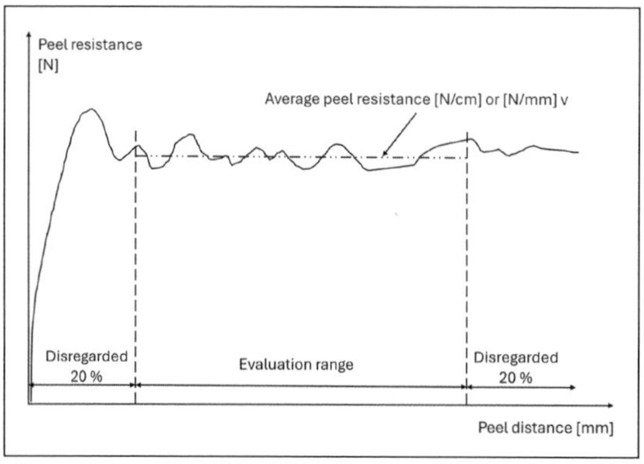

Figure IV-4.14: Peel diagram

Peel and cleavage loads

Even if adhesive bonds should be designed in such a way that peel or cleavage stresses (see lower part of Figure IV-2.1) are largely avoided, they often cannot be avoided completely, so that the knowledge about the behaviour under peel or cleavage stresses is quite important. Due to the linear force introduction, tests under peel or cleavage stress allow a better differentiation between different surface conditions of the adherends compared to those with a surface load introduction. This is particularly helpful when selecting surface treatment processes to improve adhesion.

The peel angle plays a crucial role in peel tests, so it must be kept as constant as possible during the test. A basic distinction is made between peel tests performed at an an angle of 90 or 180° and those performed in T-peel mode (Figure IV-4.13). The mechanical properties, rigid or flexible, of the adherends are also of influence on the test result.

During the test, the two adherends are separated from each other at a constant speed at the angle specified by the method, and the force required, normalised to the width of the test specimens, is recorded as a function of the peel distance in a peel diagram. The peel resistance is given as the force (F) per length (l) in [N/mm] or [N/cm], where (l) corresponds to the width of the peel strip In North America, the unit [pli] (pounds per length inch) is still frequently used, where 1 [N/cm] corresponds to 0.57 [pli]. When evaluating the peel diagram, the values of the first and last 20 % of the peel distance are usually discarded and the mean value of the peel forces determined in the evaluation area is taken as peel resistance.

For rigid substrates, the T-Peel test of DIN EN 28510 (DIN 53282) and DIN EN 14173 can be used while ASTM 1876-72 is better for more flexible substrates. Typically, the initial tear strength and the relative peel resistance of further peeling is documented.

EN ISO 11339 – Adhesives – T-peel test for flexible-to-flexible bonded assemblies

This standard describes a T-peel test that can be used to determine the peel resistance of adhesives when bonding two adherends. It is particularly suitable for metallic adherends but can also be used for other flexible adherends.

Before testing, the two non-bonded ends of the flexible adherends must be bent in opposite directions so that each is perpendicular to the bond, forming a T-shaped specimen. During the test, the adherends are separated from one open end of the bond at an almost constant speed, so that the separation progresses along the adhesive layer.

EN ISO 8510-2 – Adhesives – Peel test for a flexible-bonded-to-rigid test specimen assembly

The 180° peel test described in this standard is used to determine the peel strength of bonds between a flexible and a rigid adherend. A strip of the flexible adherend is bonded to the rigid adherend with the adhesive to be tested. After the adhesive has solidified, the adherends are separated at a constant test speed, starting from the open part of the bond, using a tensile testing machine, while recording the peel force, so that the separation progresses in the longitudinal direction of the bonded adherends.

ASTM D 6862 90° – Standard test method for 90 degree peel resistance of adhesives

This standard described a test method for the determination of the peel resistance of bonds between a rigid and a flexible adherend, where the peel angle is constant at 90°. The peel angle of 90° is achieved by the use of a test set-up which consists of a platform which is smoothly driven horizontally at the same speed as the crosshead of the tensile testing machine moves vertically assuring that a 90° peel angle is maintained between the flexible and the rigid adherend. The platform shall have means for clamping the fixed (rigid) adherend. Figure IV-4.15 shows the principle of the test fixture.

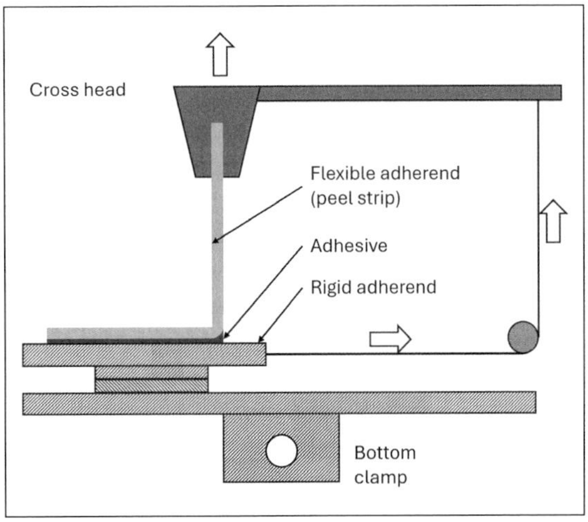

Figure IV-4.15: Test fixture according to ASTM D 6862

EN 1464/ISO 4578 – Adhesives – Determination of peel resistance of adhesive bonds – Floating roller method

This standard is used to determine the peel resistance of bonds between a rigid and a flexible adherend, where the peel angle is constant at 90°. Compared to ASTM D 6862 a different device is used to assure the consistent 90° peel angle by peeling off the flexible adherend via a deflection roller using a tensile testing machine as illustrated in Figure IV-4.16.

ISO 21194 – Elastic adhesives – Testing of adhesively bonded joints – Bead peel test

The test method described allows the evaluation of the adhesion of elastic adhesives and sealants (having a minimum elongation at break of 100 % and a modulus of elasticity of maximum 10 MPa) on various substrates. However, the test does not provide quantitative information on the peel resistance.

The principle of the test is illustrated with Figure IV-4.17. The adhesive or sealant to be tested is applied as a bead to the adherend, with the adhesion at one end of the bead being prevented, for example, by a strip of release film. After complete solidification, the end of the bead that is not attached to the adherend is lifted with pointed pliers and a sharp knife is used to cut at an angle of about 45° in the pulling direction down to the adherend surface. At the same time, the pliers are used to pull the bead off at an angle of 130° – 160°. The pull-off force is increased until a crack occurs. As failure progresses in the adhesive, the adhesive bead is continuously re-cut during the pull-off process. These incisions must be continued at the acute angle between the adhesive bead and the surface of the adherend up to the bonding surface, with an interval of about 3 sec between each incision, during which the material is further loaded. This procedure must be repeated until a piece of adhesive bead of at least 50 mm is pulled off.

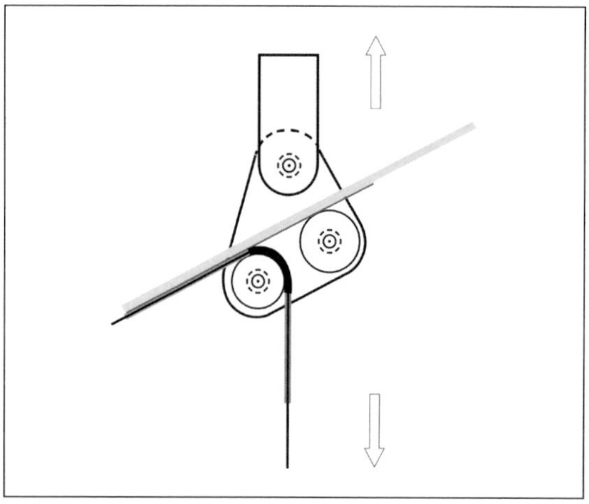

Figure IV_4.16: Floating roller peel test

The failure pattern shall be evaluated in accordance with ISO 10365 [33]. In the case of a multi-layer structure of the adherend (varnish, primer, etc.), it shall be taken into account that the failure may also occur in the coatings or their interfaces.

In this way, the effect of various coatings or surface pre-treatments of the adherend on the adhesion achieved by an adhesive can be compared. This test can also be used for process monitoring and quality assurance accompanying production.

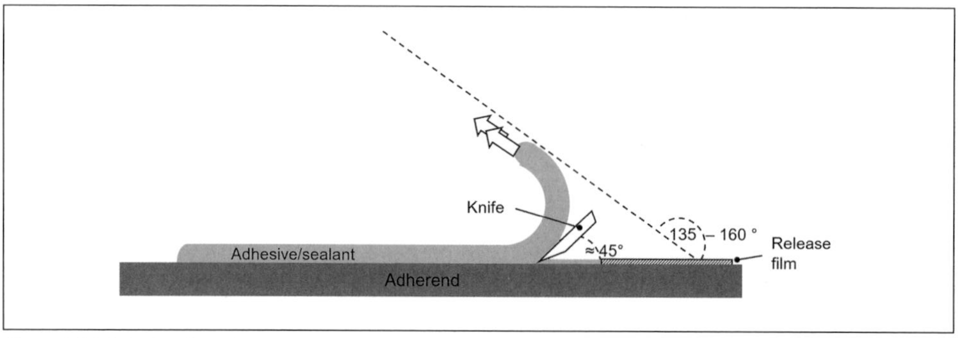

Figure IV-4.17: Bead peel test

Determination of cleavage strength of bonded joints

The principle of the cleavage peel test is to split a test specimen consisting of two bonded parts along the bond line by applying a tensile load with the force application points being arranged eccentrically with respect to the centre of the bonded area (Figure IV-4.18).

While **ISO 15107 – Adhesives – Determination of cleavage strength of bonded joints** and **ASTM D 1062 Standard test method for cleavage strength of metal-to-metal adhesive bonds** describe the performance of the test with metallic adherends, **ASTM D 3807 – Standard test method for strength properties of adhesives in cleavage peel by tension loading (engineering plastics-to-engineering plastics)** describes the test performed on plastic adherends.

Valuable information can be derived from the test results, i.e. the force required to separate the bond (cleavage strength [N/mm]) and the type of failure according to ISO 10365 [33]. The latter allows, for example, a good differentiation of different surface preparations of the adherends.

It has to be mentioned that the cleavage peel test can only be used for direct comparison of either different adhesives or different surface pre-treatment methods. In particular, when using plastic substrates, the test results depend on the rigidity of the adherends, i.e. their Young's modulus and thickness.

4.2.2.2 Tests of the resistance against static long-term loads

Testing for creep resistance

Adhesives are viscoelastic materials. Under permanent stress, irreversible creep processes, including total failure after prolonged periods, may occur.

EN 15336/ISO 15109 – Adhesives – Determination of the time to rupture of bonded joints under static load

The EN as well as the quite similar ISO standard describes a method for determining the time to failure of an adhesive bond statically loaded under shear under certain conditions. In this test, the overlapped bonded test specimen is loaded in its longitudinal axis by weights of different masses and the time to fracture of the bond and the mode of failure are determined. The results are presented logarithmically by plotting the respective stress (mass of the weight) against the average time to fracture. In regard to the results, a distinction is made between the short-term creep strength $\tau_{B,t}$, i.e. the force per unit area F that causes a detachment after a certain time t, and the long-term creep strength τ_{∞}, i.e. the maximum force per unit area that does not cause a detachment.

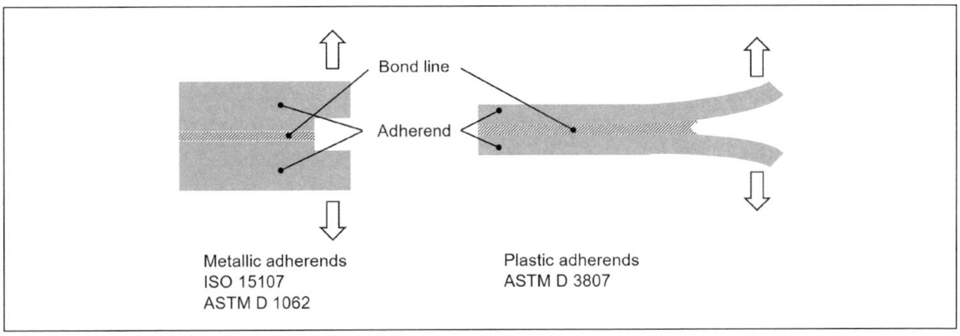

Figure IV-4.18: Cleavage peel test

The dead load test provides information about the durability of an adhesive joint under continuous, permanent load. The creep resistance of an adhesive is also a measure of the viscous part (plasticity) of the modulus in a composition. Therefore, the dead load test provides information about the true long-term resistance under load. It does not provide information about the behaviour of the adhesive under brief, fast-acting forces.

ASTM D 2293 – Standard test method for creep properties in shear by compression loading (metal-to-metal)

To determine the creep resistance, the overlap-bonded test specimen is clamped in a special spring-loaded testing device that allows by means of a tensile testing machine to apply and lock a certain compression load so that the compression load continues to affect the test specimen even after the testing device has been removed from the tensile testing machine. The deflection resulting from the compression load is recorded.

ASTM D 2294 – Standard test method for creep properties in shear by tension loading (metal-to-metal)

To determine the creep resistance, the overlap-bonded test specimen is clamped in a special spring-loaded testing device that allows by means of a tensile testing machine to apply and lock a certain tensile load so that the tensile load continues to affect the test specimen even after the testing device has been removed from the tensile testing machine. The deflection resulting from the tensile load is recorded.

Wedge test

ISO 10354 – Adhesives — Characterization of durability of structural-adhesive-bonded assemblies — Wedge rupture test

The wedge test was introduced by Boeing to investigate the ageing characteristics of adhesives in aircraft applications. As shown in Figure IV-4.19, bonded specimens are placed under stress by driving a wedge asymmetrically between the substrates. The resulting crack propagation over time is documented. The wedge test provides information about the behaviour of the adhesive joint under asymmetrically acting forces and under peel stress. Additionally, it enables the crack behaviour to be assessed. It yields information about behaviour under permanent load, but not about load-transfer ability.

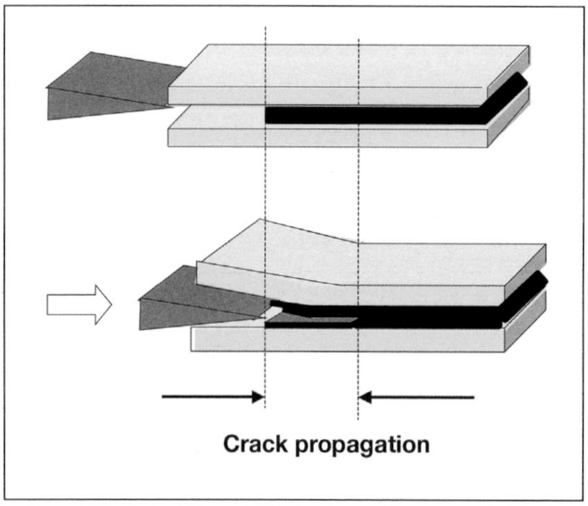

Figure IV-4.19: Wedge test

The wedge test is used for rigid and thick substrates, where peel tests are more difficult to perform. A modified, more demanding variant additionally exposes the specimens to specific environmental conditions (moisture and temperature including contact to certain chemicals).

4.2.2.3 Tests of the resistance against dynamic loads

In the case of dynamic loads, a distinction is made between very fast, i.e. sudden, loads and those resulting from a constantly changing load effect. Since adhesives react differently under rapidly and impactfully applied mechanical loads than under the more normal static loading conditions described so far, investigating the behaviour of bonded joints subjected to impact loads is of crucial importance. It is equally important to gain insights into the behaviour of adhesive bonds under oscillating loads (e.g. vibration resistance).

Impact strength tests

The impact test provides information about the behaviour of an adhesive joint under the influence of fast and short-acting (crash) forces. The effect of such forces on a bonded joint and therefore the crash resistance of an adhesive can be tested by pendulum impact testers or by means of a drop tower.

ASTM D 950 – Standard test method for impact strength of adhesive bonds

Figure IV-4.20 shows the principle of the test. The test specimen consists of a 25 mm^2 square, 10 mm thick adherend block which is bonded to a larger base placed in the grip of a pendulum impact tester. The pendulum is positioned so that the face of the upper adherend is struck by the pendulum hammer as closely as possible to the bond.

The impact strength is calculated as the energy absorbed in the generation of the test piece failure, divided by the bonded area of the test piece, and expressed as an area-related energy in joules per square metre [J/m^2]. This method is only partially suitable for differentiating between different types of adhesives. However, the effects on impact resistance of differences in adhesive formulation within a single type of adhesive can be recognised. There is no correlation between the measured force and the adhesive's fracture energy.

EN ISO 11343 – Adhesives – Determination of dynamic resistance to cleavage of high-strength adhesive bonds under impact wedge conditions – Wedge impact method

This standard describes a wedge peel impact test method, suitable to measure the resistance of an adhesives to cleavage fracture at a defined speed. For the test, a wedge of defined dimensions is driven by means of a pendulum at a defined speed into the bond of a test specimen (Figure IV-4.21). The test causes the connection to break and the adherends to separate. The test provides data suitable to calculate the dynamic resistance to cleavage necessary to bring an adhesive joint to failure.

energy/area = force/length
[$N{\cdot}m/m^2$], [J/m^2], [N/m]

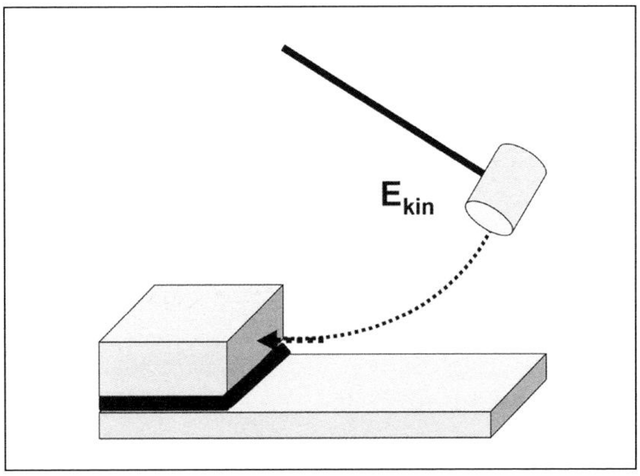

Figure IV-4.20: Impact block shear test

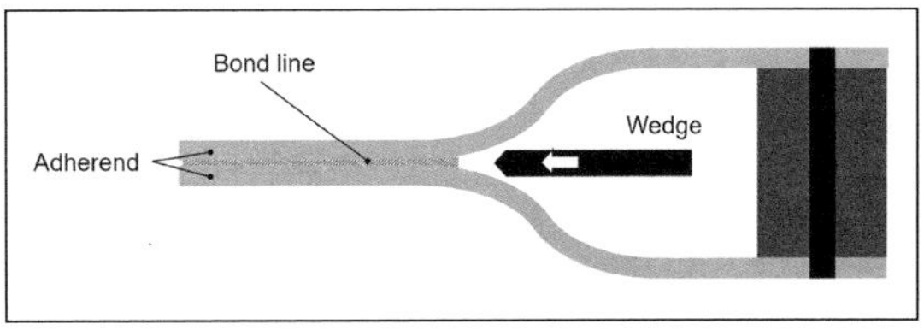

Figure IV-4.21: Test specimen according to ISO 11343

Drop tower testing

For this purpose, test specimens, usually box beams made of steel, are aligned on a platform and a weight is dropped onto the test specimen from a defined height. The test is recorded using high-speed cameras, which allow a video analysis and a precise observation of the material and component behaviour under crash-relevant conditions. On the one hand, the failure of the adhesive bond is important; if it bursts almost completely (Figure IV-4.22) or if the adhesive bond holds and the impact energy is absorbed by a fold-like deformation of the box beam. The extent of the crushing (delta L of the box beam) provides valuable information: the lower the crushing, the higher the energy absorption.

Fatigue resistance test

Determining the fatigue strength, i.e. the decrease in the strength of a bonded joint due to long-term periodic alternating loads, is of particular importance for the design of components that are subjected to such loads during their life cycle. For this purpose, tensile lap shear test specimens are subjected to a load that periodically alternates between a low and a high level. In addition to both force levels, the frequency and the number of load cycles have a significant influence on the test result.

EN 9664 – Adhesives – Test methods for fatigue properties of structural adhesives in tensile shear

The standard describes a procedure for determining the fatigue strength of adhesives under tensile shear loading. A tensile shear test specimen according to EN 1465 is subjected to the cyclic action of an alternating tensile stress, i.e. the load applied by an appropriate machine alternates at a constant frequency between a high and a low force level (see Figure IV-4.23). This means that a static stress (τ_m) is superimposed by a periodically alternating stress (τ_a). The number of cycles to failure is determined.

4.2.2.4 Non-destructive testing

Non-destructive test methods are still of secondary importance. One advantage of them is that they can even be performed on the final bonded construction without destroying it. They are often less costly (which is also related to the fact that the part is not destroyed) and some are quick and simple to perform.

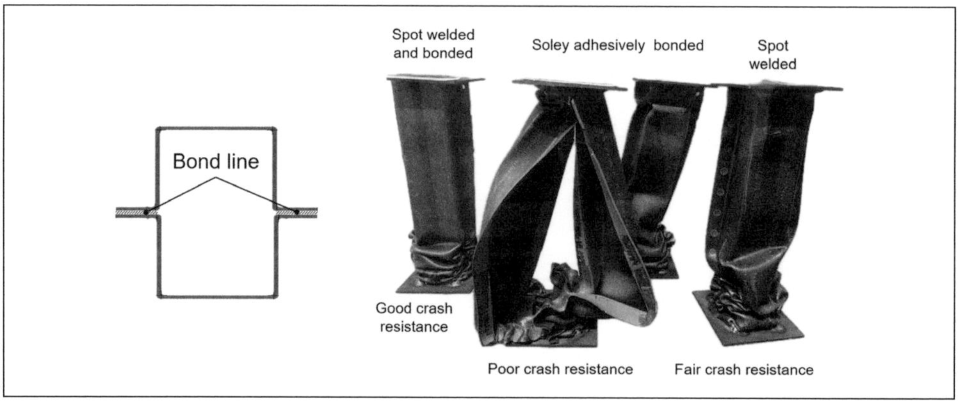

Figure IV-4.22: Box beam cross section (left). Bonded and spot-welded box beams after drop tower testing (right)

They are mainly used to answer questions related to quality control, as well as for determining local defects, such as inhomogeneities arising from mixing errors, wrong local temperatures during curing, and pinholes in the adhesive layer. Their greatest disadvantage is that there is currently no non-destructive testing method for adhesive bonds that can provide reliable information about the adhesion between the adhesive layer and the adherend. Furthermore, defects must be of a certain size in order to be detected. It is therefore not surprising that non-destructive test methods are only used in exceptional cases in the development of adhesives and sealants.

A distinction is drawn between acoustic, electric, thermal and spectroscopic methods [35].

Acoustic methods

In the simplest case, the bond can be examined by means of a tap hammer. Bonding defects are noticeable by a different tapping sound.

More sophisticated methods use variable-frequency ultrasound (e.g. 1 to 5 MHz) to examine the bonded joint. The recorded resonant frequency can reveal local inhomogeneities. An alternative is to use ultrasonic pulses, since local inhomogeneities produce a different type of echo.

Thermal methods

The surface of a bonded component is heated briefly in the area of the adhesive bond, e.g. using an IR emitter. After a short time, the surface is scanned, for example, using an IR camera, which in principle registers the surface temperature. Local inhomogeneities in the adhesive layer, such as air inclusions, lead to slight temperature differences due to the resulting difference in thermal conductivity. These temperature differences become clearly visible in the IR image when they exceed a certain size.

Electrical methods

Adhesives are dielectric materials. A metal joint therefore behaves like a capacitor, with the capacitance being influenced by the thickness of the adhesive layer and by local inhomogeneities (e.g. impedance spectroscopy).

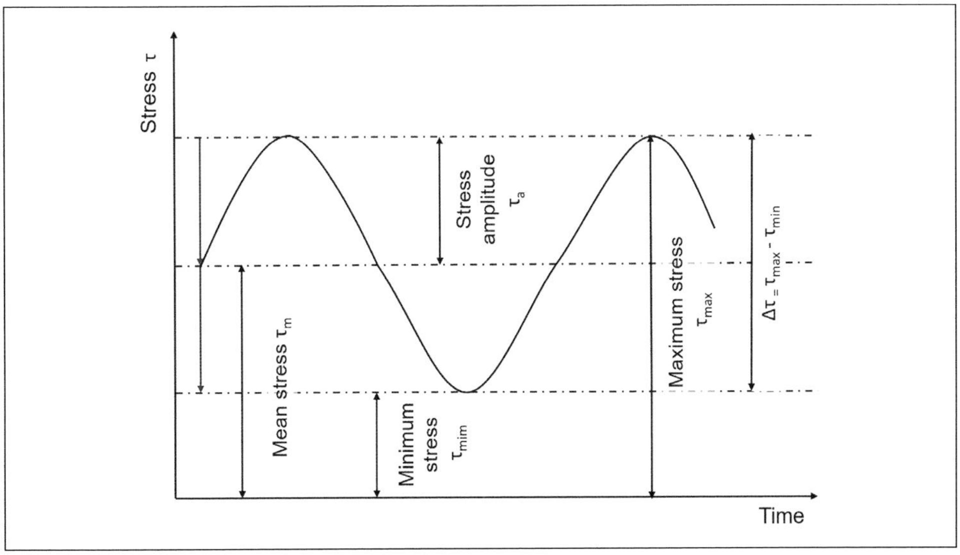

Figure IV-4.23: Fatigue stress cycle

Methods involving electromagnetic radiation

X-ray procedure

Adhesive joints can be investigated for inhomogeneities with X-ray radiation if they contain metal powders or other contrast media.

Computerised tomography (CT)

Similar to medical diagnostics, computerised tomography is also suitable to provide detailed internal images of a bonded part. Even though the equipment required is quite sophisticated and the time needed to obtain sufficiently high-resolution images increases with the size of the component to be examined, CT has proven to be useful, for example, in the examination of bonded electronic components. When the images are analysed on the computer, the entire component is displayed in three dimensions and can be rotated as required to view it from all directions. Furthermore, cross-section images can be taken and examined. Unlike microscopic examination of polished cross-section samples, there is no longer any need to place the cross-section at the exact position of a defect, which is unknown at the time of sample preparation.

4.2.2.5 Durability testing

Since adhesively bonded or sealed parts are often expected to have a long service life, sometimes under harsh conditions, methods for predicting this service life are indispensable tools both for formulating adhesives and sealants and for selecting suitable products for a specific application. Therefore, ageing effects of the adhesive or sealant used and the adherends, as well as any effects resulting from the combination of the two, must be taken into account. In order to keep development times as short as possible, methods are preferred that make it possible to draw conclusions about a product's performance over its entire expected life cycle after the shortest possible testing times.

Typically, standardised methods of destructive testing (e.g. tensile lap shear specimens according to DIN EN 1465) are used in the development of adhesives and sealants, and in some cases non-destructive methods are used as well. The corresponding test specimens are exposed under specified conditions and tested afterwards. The results are compared with those of a test without exposure and evaluated. In the case of destructive testing, the usual criteria are strength and failure mode (according to DIN EN ISO 10365 [33]).

Determining the temperature influence

Both high as well a slow temperature may have an effect on the adhesive's or sealant's properties. These effects can be reversible or irreversible.

Testing for irreversible effects

The specimens, typically in tensile lap shear, peel or cleavage geometry are stored at a defined temperature, in some cases also at defined relative humidity (ISO 291, Plastics — Standard atmospheres for conditioning and testing, provides a good overview about common test conditions) for a certain period of time, reconditioned to ambient conditions and tested. Compared to specimens stored under standard climate conditions, loss of plasticiser or degradation of the polymer, and even post-reaction may be observed. The storage conditions (temperature and time) are selected according to the designated application.

ISO 19212 – Adhesives – Determination of the temperature dependence of shear strength

Compared to samples tested at room temperature, embrittlement can be observed at low temperatures. Above about 100 °C, however, most adhesives show a decrease in strength along with an increase in adhesion failure.

According to ISO 19212, the test can be carried out on tensile lap shear test specimens but can also be performed on other test geometries.

Determining the influence of different thermal expansion properties of the materials in bonded joints

If the influence of the different coefficients of thermal expansion of the adherend materials or between the adhesive and the adherend on the bond is to be investigated, tests carried out on small, standardised test specimens do not adequately reflect the forces occurring on larger real components. The dimensions of the real components must therefore be taken into account.

Determining the resistance against moisture (water)

A high moisture load can have a negative effect on the properties of adhesives and sealants. Depending on their formulation, adhesives and sealants show a more or less pronounced water absorption. The absorbed water can, act as a plasticiser and change the cohesion properties. In most cases, this is reversible, i.e. after re-drying, the original values are restored. Irreversible damage due to moisture is caused by sub-surface migration of moisture into the boundary layer between the adhesive or sealant and the adherend, resulting in damage to the adhesion. This often results in a typical failure pattern also referred to referred to as undermining in ISO 10365; in the case of metallic parts, it may be accompanied by corrosion. As illustrated with Figure IV-4.24 the effect of sub-surface migration heavily depends on the geometry of the joint because, with small joint areas, border diffusion effects play a more important role.

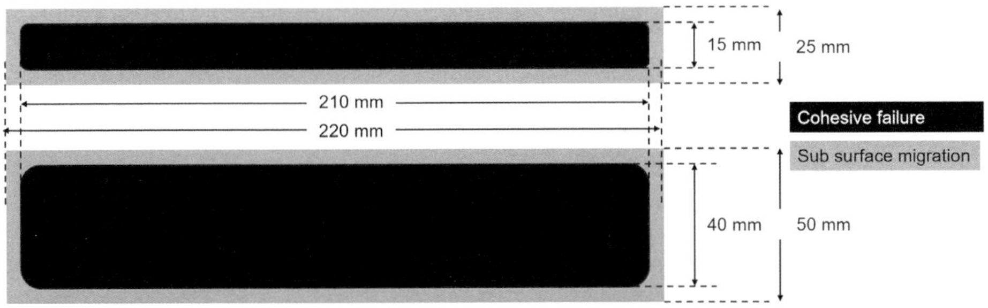

	Small specimen	Large specimen
Total bond area	A_t = 220 x 25 = 5,500 mm²	A_t = 220 x 50 = 11,000 mm²
Sub surface migration SSM	5 mm	5 mm
Bond area SSM	A_{UM} = 2,350 mm²	A_{UM} = 2,600 mm²
Bond area CF	A_{CF} = 210 x 15 = 3,150 mm²	A_{CF} = 210 x 40 = 8,400 mm²
Load bearing before test	27,500 N	55,000 N
Load bearing after test	15,750 N	42,000 N
Residual strength	57 %	76 %

Figure IV-4.24: Influence of bond dimension on the loss of bond strength due to sub-surface migration

DIN 53287 – Testing of adhesives for metals and adhesively bonded metal joints – Determination of the resistance to liquids

The standard describes the test for determining the resistance of bonded test specimen to fluids in general. The changes in properties (e.g. tensile lap shear strength and failure mode) caused by the effect of fluids at different temperatures are usually determined by comparing fluid-soaked samples with samples stored in a normal climate.

The result depends heavily on the geometry of the joint because, with small joint areas, border diffusion effects play a more important role. In addition, overpainting the bonded substrates can improve the result.

EN ISO 6270-2 – Paints and varnishes – Determination of resistance to humidity – Part 2: Condensation (in-cabinet exposure with heated water reservoir)

Test specimens, e.g. tensile lap shear specimens, are exposed to an atmosphere of 40 °C and approximately 100 % humidity in a test chamber for a specified time, resulting in condensation of water on the test specimens. The test chamber is equipped with a heated water-filled base tray. Heating the water causes water to evaporate, which then condenses on the chamber walls and the test specimens and controlling the temperature in the test chamber. Afterwards, the specimens are reconditioned to ambient conditions and tested.

EN ISO 9227 – Corrosion tests in artificial atmospheres – Salt spray tests

In the case of metallic adherends, the effect of sub-surface migration caused by moisture and the resulting weakening of the bond can be increased by substances, such as road salt. ISO 9227 describes a test in which the test specimens (usually tensile lap shear test specimens) are exposed to a salt spray generated by spraying a sodium chloride solution (50 ± 5 g/l, pH 6.5 – 7.2) in a test chamber at 35 °C. The spray solution should not be applied directly to the test specimens but should be atomised in the chamber.

Durability tests combining different environmental conditions – time-lapse tests

While in the past, the static tests described above, e.g. in accordance with DIN EN ISO 6720-2 were used, today so-called climate change or time-lapse tests, in which the test specimens are periodically exposed to changing conditions, are increasingly used. To keep the testing times as short as possible and thus also to enable short development times, the stresses acting on an adhesive bond or seal under real-life conditions over the expected service life are intensified and time compressed.

When defining the test conditions, use is made of the fact that the processes involved in the ageing of adhesives and sealants, like all chemical and biological reactions, are accelerated by an increase in temperature. This means that a test temperature moderately above the maximum temperature that occurs in practice is selected. A rule of thumb says that a temperature increase of 10 K accelerates a chemical reaction by a factor of 2 – 4. However, it must be noted that the useful extent of the acceleration is limited. In addition to the desired acceleration of the ageing reactions, a too high temperature might allow reactions to take place that would not occur under normal operating conditions of the component to be tested due to their activation energies.

When compressing the time, the periods during which the component is exposed to non-critical conditions, e.g. normal room conditions, are partly replaced by critical ones, e.g. increased temperature, media influence, mechanical load, etc. The resulting lack of 'recovery phases' undoubtedly has an influence and must be taken into account.

To avoid excessively stringent test conditions and thus requirements for the adhesive or sealant that are too high and do not reflect reality, thereby unnecessarily complicating adhesive development, it is recommended to continuously compare the results obtained in the laboratory with the findings from tests on parts used under real conditions. Conversely, this approach would also avoid test conditions that are too weak, i.e. pretending a good resistance to ageing that does not exist in reality. This means that the conditions should be as stringent as necessary, but not more stringent than necessary – anything can be tested to destruction.

When formulating adhesives and sealants, it is common to use time-lapse tests that are well established in the respective market segment and are largely based on the experience of the respective adhesive users.

In addition to the test methods described below as examples, there are a number of others, some having the status of a national, European or international standard, but also those that have been developed by adhesive users specifically for their area of application. Some of these methods only differ slightly in the test conditions. In the past, these methods were developed independently by the respective users, and the results obtained were continuously compared with the behaviour of parts that had aged under real life conditions. Over time, extensive data sets have been collected that provide a good basis for assessing the long-term durability of bonded joints in future components with a similar structure and field of application. A harmonisation of the various, albeit only slightly different, test conditions would question the value of these data sets and the comprehensive further use of the test expertise often developed over decades.

EN ISO 11997-1 – Cycle B (formerly known as VDA 321 – 415)

In this especially for metal-to-metal bonds developed test, the test specimens are subjected to a 24-hour salt spray test according to EN ISO 9227 followed by a 4-day condensation water climate change test and a 2-day recovery phase t ambient conditions This cycle is usually repeated 10 times.

BMW PR 308.1 [34]

This test is in particular used for plastic-to-plastic interior and exterior parts where salt spray exposure is generally not critical. Even though the corrosive effect of salt can have a negative impact on the adhesion to the metal when bonding plastic to metal, the test is also used for interior parts where aluminium is bonded to plastics.

The test specimens are subjected to the temperature-humidity profile shown in Figure IV-4.25 in an automatic climatic test chamber. The temperature-humidity profile is repeated several times where the number of repetitions depends on the respective application area in the vehicle (e.g. interior, exterior). After reconditioning to ambient conditions, the specimens are tested and evaluated against unaged specimens by comparing bond strength and failure mode.

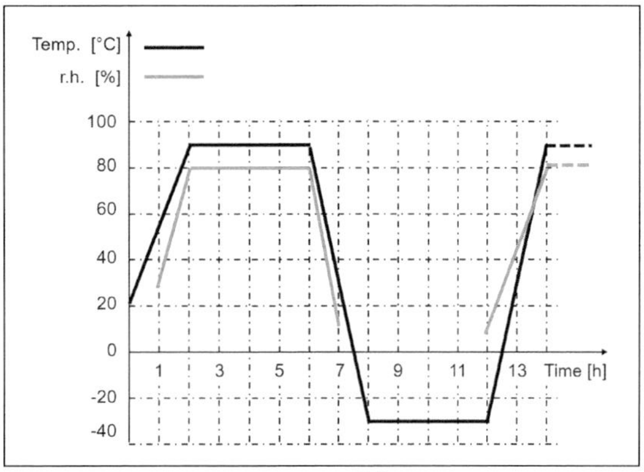

Figure IV-4.25: BMW PR 308.1 [34]

EN ISO 13523-27 – Coil coated metals – Test methods – Part 27: Resistance to humid poultice (Cataplasm test)

In contrast to the two aforementioned climate change tests, the Cataplasm test requires considerably less sophisticated equipment. No special climate test chamber is needed, just a normal heating oven and a freezer.

The test specimens (usually again tensile lap shear test specimens in accordance with DIN EN 1465) are wrapped in cotton wool, alternatively white paper wipes can be used, placed in a tightly closed plastic bag together with a certain amount of demineralised water. After sealing, the bags are stored at 70 °C for a specified time (usually 7, 14 or 21 days). After removal of the bag from the oven, the test specimens are removed from the bag, the cotton wool is cleared away and the specimens are transferred to a freezer maintained at -20 °C. After 2 hours, the samples are removed from the freezer and, after a further 2 hours of reconditioning at room temperature, the destructive test is carried out at room temperature. As is usual for ageing tests, the results (bond strength and failure mode) are evaluated in comparison to unaged test specimens.

The high temperature of 70 °C, together with the high humidity, and the fact that substances released from either the adhesive/sealant or the adherends, put a great deal of stress on the bond. When the samples are then stored at -20 °C, moisture that has penetrated either into the adhesive/sealant or into the interface between the adhesive/sealant and the adherend will freeze under volume expansion and cause additional stress to the bond. Last but not least, the reconditioning time of 2 hours is quite short, so it can be assumed that at the time of destructive testing, the redrying of the test specimens has not yet been completed.

In this respect, the cataplasm test is considered a very harsh test, perhaps maybe too harsh for some applications. Because of this and the relatively short time required, it is often used as a screening test. A negative result does not necessarily mean that the adhesive or sealant is unsuitable for an application, whereas a positive result indicates very good moisture resistance.

EN 60068-2 – Environmental testing – Part 2-14: Tests – Test N: Change of temperature

In contrast to the slow heating and cooling rates of the test according to BMW PR 308.1, so-called thermal shock tests involve significantly higher heating and cooling rates, in which the test specimens alternate between two test chambers which are set to different temperatures. The lower and upper temperature (e.g. -40/+120 °C) and the short transfer time from one chamber to the other (e.g. 2 – 3 min, 20 – 30 sec or <10 s) depend on the most extreme temperature change conditions to be expected in the respective application. The time spent at the upper and lower temperatures is selected depending on the heat capacity of the test pieces so that the test piece is at temperature equilibrium with the current chamber before changing to the respective other chamber. Such tests are used, for example, on electrical components of motor vehicles that are either encapsulated in a polymer or mounted in a bonded housing and are located in a position exposed to water sprays, e.g. in the engine compartment. The test allows to examine, among other things, the resistance of the adhesive bond against the mechanical stresses resulting from the different thermal expansion behaviour of the adherends.

4.3 Testing of pressure-sensitive adhesives (PSA)

In addition to the national, European (EN) and international (ISO) standardisation organisations, some trade organisations of adhesive tape and label manufacturers are also dealing with the creation of standards for the characterisation of pressure-sensitive adhesives and the self-adhesive articles such as adhesive tapes. The objective of the collaboration between ISO and CEN and the major trade organisations associated in the Global Tape Forum (GTS) (see Figure 4-IV.26) was to internationally harmonise the various standards created by the respective organisations for testing and characterising standards for adhesive tapes.

In addition to CEN and ISO, PSTC and AFERA offer manufacturers and users of pressure-sensitive adhesives and adhesive tapes a range of standards for determining the performance of adhesive tapes. [36, 37]. The most important of these in the context of pressure-sensitive adhesive development are:

- tack
- adhesion measured as peel strength
- shear strength measured as creep resistance

and are briefly summarised hereafter.

In general, the test results depend heavily on the climatic conditions (temperature and relative humidity) during storage of the adhesive tapes to be tested as well as during preparation and testing of the test specimens. This also requires that the adhesive tapes to be tested must be adequately acclimatised for a sufficient period of time before the test specimens are prepared.

4.3.1 Tack testing

Tack is the property of a material by which it forms a bond of measurable strength immediately (in less than a second) upon contacting another surface [38]. This involves flow and wetting pro-

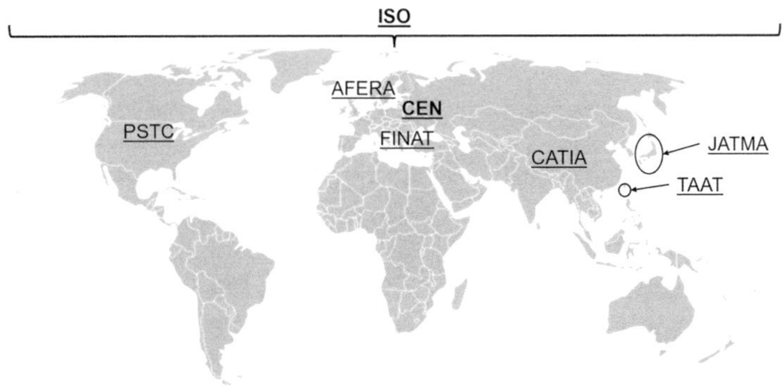

CEN	*Comité Européen de Normalisation*	ISO	*International Organization for Standardization*
PSTC	*Pressure Sensitive Tape Council*	CATIA	*China Adhesives and Tape Industry Association*
TAAT	*Taiwan Regional Association of Adhesive Tape Manufacturers*	JATMA	*Japan Automobile Tyre Manufacturers Association*
AFERA	*European adhesive tape association*		

Figure 4-IV.26: Standardization in the field of pressure-sensitive adhesives

cesses on the part of the adhesive polymer. Tack properties are influenced mainly by the glass transition temperature of the polymer, low-molecular fractions and potentially added tackifier resins. Such resins can increase the glass transition temperature of the polymer, so that the adhesive shows increased adhesion in the event of fast deformation. Typically, the tack of an adhesive formulation increases with increase in resin content up to a certain limit. When the resin concentration is too high, the tack decreases again. Tack can be measured by different methods, but the results are not always transferable [36, 37, 39].

4.3.1.1 Finger tack

Although the experienced PSA-specialist will get a first rough idea about the degree of the tackiness of a pressure-sensitive adhesive by touching the surface with a finger this test does not provide any values and can only serve as a rough highly subjective test. However, it can act as a preliminary test. Finger tack can also be measured as wet tack by applying the wet adhesive between thumb and finger and estimating the force needed for separation.

4.3.1.2 Loop tack

EN 1719 – Adhesives for paper and board, packaging and disposable sanitary products – Tack measurement for pressure-sensitive adhesives – Determination of loop tack

GTS 6007 – Loop tack of adhesive tapes

Afera 5014 – Loop tack of adhesive tapes

PSTC-16 – Loop tack

To determine the loop tack, a loop of a pressure-sensitive adhesive tape (width 24 mm) is brought by means of a tensile testing machine in contact with a substrate, usually a polished stainless steel plate, forming a 24 mm x 24 mm contact area, with the only applied force being the weight of the adhesive tape itself. Immediately after full-surface contact is achieved, the tape is removed again from the stainless steel plate with the required force being registered. It hast to be considered that the results may vary with the carrier material of the tape used because the tack depends on the contact pressure which is influenced by the stiffness of the tape.

Alternatively, as shown in Figure IV-4.27, the adhesive can be applied to the plate and the uncoated loop is brought into contact. The force needed to remove the loop is a measure of the tack. As in this case the tack between adhesive and loop material and not as with the first described between adhesive and stainless steel is measured the obtained results might differ. Again, the results vary with the tape material because the formation of adhesion depends on the surface properties of the to be wetted tape.

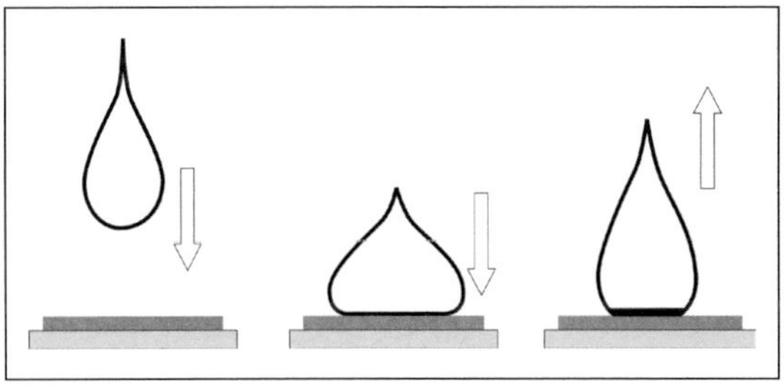

Figure IV-4.27: Loop tack test

4.3.1.3 Quick stick

EN 1945 – Self adhesive tapes – Measurement of quick stick

Afera 4015 – Quick stick

PSTC-5 – Quick stick of pressure-sensitive tapes

The adhesive (coated on a flexible tape) is placed on the mating substrate, typically stainless steel without application of any significant pressure. Light pressure is applied using a specified roll (Figure IV-4.28) and is then peeled off at an angle of 90°. The peel resistance is a measure of the tack. The results depend on the (rolling) pressure applied and the contact time. The quick stick method yields quite reproducible results and is therefore frequently used for quality control.

4.3.1.4 Polyken probe tack

ASTM D 2979 – Standard test method for pressure-sensitive tack of adhesives using an inverted probe machine

In this test method, the tip of a cleaned stainless steel rod (diameter 5 mm, surface roughness 250 – 500 nm) is brought into contact with the pressure-sensitive adhesive layer at a controlled speed, under a specified pressure, for a short time and at a specified temperature. Thereafter, the bond formed between the probe and the adhesive is broken at a controlled rate by pulling back the stainless steel rod. The tack is measured as the maximum force required to break the bond (Figure IV-4.29).

4.3.1.5 Rolling ball tack

EN 1721 – Adhesives for paper and board, packaging and disposable sanitary products – Tack measurement for pressure-sensitive adhesives – Determination of rolling ball tack

PSTC-6 – Tack rolling ball

ASTM 3132-17 – Standard test method for tack of pressure-sensitive adhesives by rolling ball

As shown in Figure IV-4.30, a steel ball of defined dimensions, weight and surface quality rolls down an inclined plane at a defined angle from a defined height onto an adhesive layer. When first hitting the pressure-sensitive adhesive layer the ball has a defined kinetic and rotational energy but is slowed down according to the tackiness of the adhesive. The distance the ball travels before it stops is a measure of the tack; the shorter the distance the

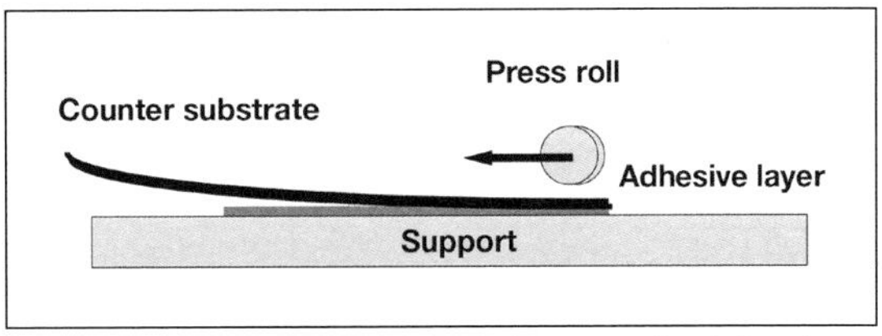

Figure IV-4.28: Quick stick – application of a defined light pressure to the tape

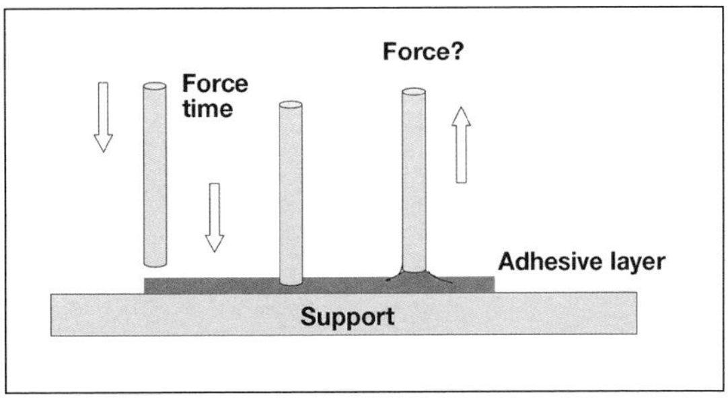

Figure IV-4.29: Polyken probe tack

higher the tack. The measurement is easy to perform, but results are sometimes difficult to reproduce, because small deviations in execution, especially in the procedure for cleaning the steel ball can lead to significantly different results. Round robin tests have shown that the method provides good results when, for example, different pressure-sensitive adhesives are tested in a direct comparison at the same time by the same operator. However, the comparability of tests carried out at different times and/or in different laboratories is poor.

4.3.2 Peel resistance

EN ISO 29862 – Self adhesive tapes – Determination of peel adhesion properties

GTS 6004 – Measurement peel adhesion of adhesive tape

Afera 5001 – Measurement of peel adhesion of adhesive tape

PSTC 101 – International standard for peel adhesion of pressure-sensitive tape

The test specimens used for testing peel adhesion are prepared by bonding the flexible pressure-sensitive tape (peel strip) to a rigid test panel typically made of stainless steel under controlled conditions. Since, as their name suggests, pressure-sensitive adhesives develop adhesion when pressure is applied the contact pressure used to prepare the test specimens has a significant influence on the test result. For this reason, the standards prescribe the use of a roller specified in terms of weight and material. During the rolling process is performed at a prescribed speed, only the dead weight of the roller is effective, so that the rolling process can be carried out reproducibly, especially when using a motor-driven roller.

After a specified dwell time (waiting time between preparation of the test piece and the start of the stripping process), the peel strip is pulled off at a specified speed, with the force applied being recorded. This is done either at an angle of 180° or, for less flexible peel strips, at an angle of 90°, using a tensile testing machine.

The mean peel force is determined from the peel diagram as described above with Figure IV-4.14 and is stated in the test report together with the so-called crack initiation value, i.e. the force required at the beginning of the peel process to separate the peel strip from the test panel for the first time, and the type of failure according to ISO 10365 [33].

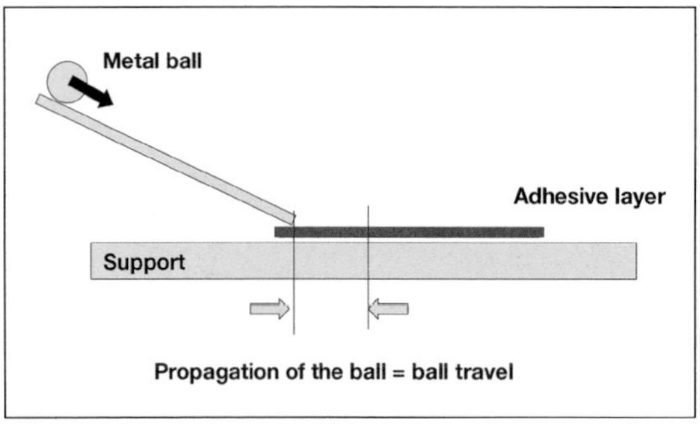

Figure IV-4.30: Rolling ball test

Peel resistance of pressure-sensitive adhesives serves as a measure of adhesion, provided adhesive failure occurs. Since the adhesion of adhesives to a given material is determined by its ability to wet the adherend, peel adhesion can be influenced by modifying the rheological properties of the pressure-sensitive adhesive polymer and its glass transition temperature during the formulation of pressure-sensitive adhesives.

4.3.3 Static shear/creep resistance

ISO 29863 – Self adhesive tapes – Measurement of static shear adhesion

GTS 6006 – Static shear adhesion test

Afera 5012 – Static shear adhesion of adhesive tapes

PSTC-107 – Static shear adhesion of adhesive tapes

The test specimens used for the shear adhesion test (Figure IV-4.31) are prepared by bonding the flexible pressure-sensitive adhesive tape to a rigid test plate, usually made of stainless steel. The provisions set out in Chapter IV-4.3.2 must be observed.

The test specimens are attached to a test fixture at an angle of 2° from the vertical and the PSA tape is loaded with the specified weight. Under controlled climatic conditions (typically 23 ± 1 °C and 50 ± 5 % relative humidity) either the time until the total failure of the bond or until it has slipped a certain distance is determined. Alternatively, the extent of the slippage after a certain time can also be recorded. As with the peel adhesion test, the type of failure is also recorded.

The shear or creep resistance of the adhesive can be adjusted via the glass transition temperature and the crosslink density of the polymer.

GFT 6001 – Shear adhesion failure temperature (SAFT) of adhesive tapes

Afera 5013 – Shear adhesion failure temperature (SAFT) of adhesive tapes

PSTC-17 – Shear adhesive failure temperature (SAFT) for pressure-sensitive adhesive tapes

To determine the SAFT, the static shear test according to GFT 6006 described above is carried out in a programmable heating furnace, whereby the furnace temperature rises at a constant, controlled rate and the time until the bond fails is recorded.

4.3.4 Accelerated aging of pressure-sensitive adhesives

Adhesive bonds made with adhesive tapes can be tested for their resistance to temperature, moisture and chemicals as described in Chapter IV-4.2.2.5. For example, exposure of peel or shear test specimens under condensation conditions according to EN ISO 6270-2 or in the BMW PR 308.1 alternating test has proven valuable for obtaining information on the resistance to moisture over time, even at elevated temperatures.

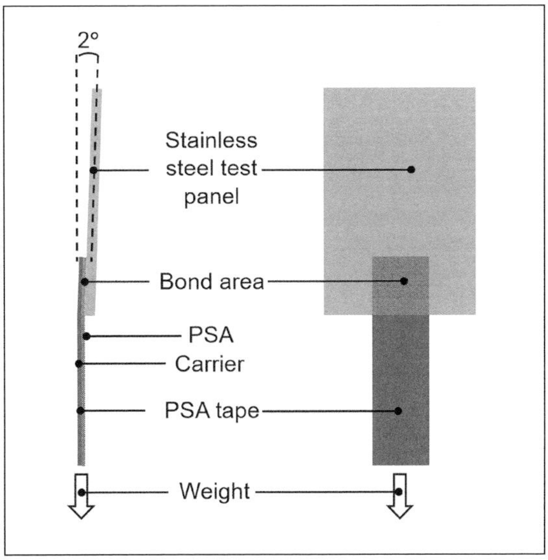

Figure IV-4.31: Static shear test

Afera 4026 – Accelerated ageing of adhesive tapes

This test method describes procedures for testing the shelf life of adhesive tapes by determining adhesive properties such as tack, peel adhesion and static shear strength after storage at elevated temperatures (40 or 65 °C) and comparing them with those obtained on unaged adhesive tape.

The measurement of the force required to unwind a roll of adhesive tape (EN 1944, Self-adhesive tapes – Measurement of unwind adhesion at low speed) carried out before and after heat ageing, also provides valuable information about the shelf life of the tape rolls.

4.4 References

[1] ISO 21037 Adhesives – Guideline for separating adhesively bonded joints enabling repair and improving recycling (under development)
[2] B. Mayer, A. Gross, Circular Economy and Adhesive Bonding Technology, Fraunhofer Verlag, 2020
[3] Quarzwerke GmbH, Company brochure “Product range” https://www.quarzwerke.com/fileadmin/quarzwerke/Broschueren/HPF_23_Productrange_DE_EN_Web.pdf accessed 2024/10
[4] Wacker Chemie AG, Company Brochure “Strength with HDK” https://www.wacker.com/cms/de-de/products/insights/hdk-pure-diversity.html accessed 2024/10
[5] Wikipedia, “Pyrogenes Siliciumdioxid” https://de.wikipedia.org/wiki/Pyrogenes_Siliciumdioxid accessed 2024/10
[6] Edward M. Petrie, Fillers and Extenders for Adhesives & Sealants: Selection and Formulation Tips, SpecialChem https://adhesives.specialchem.com/selection-guide/fillers-and-extenders-for-adhesive-sealant-formulation accessed 2024/10
[7] Quarzwerke GmbH, Company brochure “Silatherm”, https://www.quarzwerke.com/fileadmin/quarzwerke/Broschueren/QW-Brosch-Product_Silatherm-2019_Web.pdf accessed 2024/10
[8] KG Deutsche Gasrußwerke GmbH & Co, Webpage, https://gasruss.de/index.php/de/produkte accessed 2024/10
[9] Orion Engineered Carbons S.A., Company brochure “Specialty carbon blacks” https://orioncarbons.com/wp-content/uploads/2023/04/22_06_21_td_0112_farbrusstabelle_emea_web_2.pdf accessed 2024/10
[10] Wikipedia “Kaolinite” https://en.wikipedia.org/wiki/Kaolinite accessed 2024/10
[11] Wikipedia “Kaolin” https://de.wikipedia.org/wiki/Kaolin accessed 2024/10
[12] Wikipedia “Talc” https://en.wikipedia.org/wiki/Talc accessed 2024-10
[13] Imerys S.A. Wollastonite, https://www.imerys.com/minerals/wollastonite accessed 22024-10
[14] Wikipedia, Wollastonite, https://en.wikipedia.org/wiki/Wollastonite accessed 2024/10
[15] Wikipedia Glimmergruppe, https://de.wikipedia.org/wiki/Glimmergruppe accessed 2024/10
[16] Phantom Plastics, https://phantomplastics.com/fillers-for-cte-clte-modification-of-plastics/ accessed 2024/10
[17] Steine & Minerale, Bentonit https://www.steine-und-minerale.de/gesteine/b/bentonit.html accessed 2024/10
[18] H. G. Elias: Makromoleküle, Wiley-VCH; 6th edition (2000)
[19] S. R. Hartshorn: Structural Adhesives, Plenum Press (1986)
[20] Klebtechnik Dr. Hartwig Lohse e.K., Basic investigation for effectiveness of heat conducting materials, https://how-do-you-glue.de/en/case-studies/project-1/ accessed 2024-10
[21] Euroquarz, Crystalline Silica safely, https://www.euroquarz.com/quarzfeinstaub accessed 2024/10
[22] GESTIS Substance database, Quartz, https://gestis-database.dguv.de/data?name=004110 accessed 2024/10
[23] G. Habenicht: Kleben, Grundlagen, Technologien, Anwendungen, Springer Verlag, 4th edition 2002. Page 366ff
[24] S. R. Hartshorn, Structural adhesives, Plenum Press, 1986
[25] W. Brockmann et al., Klebtechnik, Wiley-VCH, 2001
[26] B. Burchardt et al., Elastic Bonding, verlag moderhne industrie
[27] D. Satas: Handbook of Pressure Sensitive Adhesives Technologies, 3rd edition, Satas & Associates, 1999
[28] R. Hüther, Ein Beitrag zur Klärung der Adhäsionsmechanismen von Haftklebstoffen, Verlag Shaler (1995)
[29] Wikipedia, Thermal expansion, https://en.wikipedia.org/wiki/Thermal_expansion#Coefficients accessed 2024-12-10

[30] Wikipedia, Ausdehnungskoeffizient, https://de.wikipedia.org/wiki/Ausdehnungskoeffizient accessed 2024-12-10

[31] Baur, Brinkmann et al., Saechtling Kunststoff Taschenbuch, 30th edition, Carl Hanser Verlag, 2007

[32] ISO 21368:2022-03 Adhesives – Guidelines for the fabrication of adhesively bonded structures and reporting procedures suitable for the risk evaluation of such structures

[33] EN ISO 10365 Adhesives – Designation of main failure patterns

[34] H. Gleich, H. Lohse in M. Doobe, Kunststoffe erfolgreich kleben, Springer Vieweg, 2018

[35] G. Habenicht: Kleben, Grundlagen, Technologien, Anwendungen, Springer Verlag, 4th edition 2002, p. 691ff

[36] https://pstc.org/resources/test-methods/ accessed 2025-01-01

[37] https://afera.com/technical-centre/afera-test-methods/ accessed 2025-01-01

[38] ASTM D 3195 Standard test methods for Loop Tack

[39] F. H. Hammond jr in Satas, Handbook of pressure sensitive adhesive technology, Van Nostrand Reinhold, New York, 1989

[40] USGS, Wollastonite – A Versatile Industria Mineral, https://pubs.usgs.gov/fs/fs-0002-01/fs-0002-01.pdf accessed 2024/10

General literature

To facilitate access to adhesive and sealants technology for beginners, some recommended standard books are listed below. This list is inevitably incomplete. Further literature is cited at the end of each section.

Adhesives and sealants

- H. Onusseit, Adhesive Technology – Basic Principles, Beuth Verlag GmbH, Berlin, 2012
- G. Habenicht, Kleben – Grundlagen, Technologie, Anwendung, 6th ed., Springer-Verlag 2009
- E. M. Petrie, Handbook of Adhesives and Sealants, McGraw-Hill, New York 2000
- E. W. Flick, Adhesive and Sealant Compound Formulations, 2nd ed., Noyes Publ. 1984
- W. Brockmann, P. L. Geiß, J. Klingen, B. Schröder, Klebtechnik, Wiley-VCH 2005
- I. Skeist, Handbook of Adhesives, 3rd ed., Chapman & Hall 1990
- P. R. Hartshorn, Structural Adhesives, Plenum Press, 1986
- W. C. Wake, Adhesion and the Formulation of Adhesives, Applied Science Publishers Limited 1976
- G. Zeppenfeld, D. Grunwald, Klebstoffe in der Holz- und Möbelindustrie, 2nd ed., DRW 2005
- R. Jordan, Schmelzklebstoffe, Bände 4a–4c, Hinterwaldner Verlag 1985–1987
- M. J. Satriana, Hot-melt Adhesives, Manufacture and Applications, Noyes Data Corporation 1974
- Henry R. Dunning, Pressure Sensitive Adhesives, Formulations and Technology, 2nd ed., Noyes Data Corporation 1977
- D. Satas, Handbook of Pressure Sensitive Adhesive Technology, 3rd ed., Satas & Associates 1999
- I. Benedek, Pressure-Sensitive Formulation, VSP 2000
- E. Baust, Praxishandbuch Dichtstoffe, Industrieverband Dichtstoffe 1988
- H. Kittel(†), Lehrbuch der Lacke und Beschichtungen, 2nd ed., Vol. 7, Produkte für das Bauwesen, Beschichtungen, Bauklebstoffe, Dichtstoffe (editor H. Reul) S. Hirzel Verlag 2005
- R. M. Evans, Polyurethane Sealants, Technomic Publishing AG 1993

Raw materials

- D. Distler, Wässrige Polymerdispersionen – Synthese, Eigenschaften, Anwendungen, Wiley-VCH 1999
- R. Mildenberg, M. Zander und G. Collin, Hydrocarbon Resins, Wiley-VCH 1997
- H. Lucke, Aliphatic Polysulphides, Hüthig & Wepf 1994
- M. Szycher, Szycher's Handbook of Polyurethanes, CRC Press 1999
- U. Meier-Westhues et al., Polyurethanes, Vincentz Network 2019
- C. A. May, Epoxy resins, Marcel Dekker, inc., New York and Basel 1988
- B. Müller, Understanding Additives, 2nd ed., Vincentz Network 2019

Encyclopedias

- Ullmann's Encyclopedia of Industrial Chemistry, 5th ed., VCH
  Vol. A1: Adhesives, p. 221–226 (1985)
  Vol. A23: Sealing Materials, p. 499–513 (1993)
- Kirk-Othmer, Encyclopedia of Chemical Technology, 4th ed., Wiley
  Vol. 1: Adhesives, p. 445 ff (1991)
  Vol. 21: Sealants, p. 650–666 (1997)

- Encyclopedia of Polymer Science and Engineering, 2nd ed., Wiley
  Vol. 1: Adhesion and Bonding, p. 476–546 (1985)
  Vol. 1: Adhesive Compositions, p. 547–577 (1985)
  Vol. 15: Sealants, p. 131–145 (1989)
  G. Oertel (editor), Kunststoffhandbuch, Vol. 7, Polyurethane, 3rd ed., Hanser Verlag 1993

Reference books

- Römpp Chemie Lexikon (several volumes), Thieme Verlag, latest edition, also online: http://www.roempp.com/de/formate/encyclopedias/roempp/format/inhalt/chemie-mit-lacke-und-druckfarben.html
- Industrieverband Klebstoffe e.V., Handbuch Klebtechnik, Vieweg Verlag, latest edition
- E. Karsten (editor O. Lückert), Lackrohstofftabellen, Vincentz Network, 10th ed. 2000 (some raw materials for adhesives are also listed in this book)
- O. Lückert, Pigment+Füllstoff Tabellen, Vincentz Network, 6th ed. 2002 (many sealant formulations contain fillers)

Journals

- Adhäsion, Vieweg Verlag, Wiesbaden
- Coating, Verlag Coating Thomas & Co., St. Gallen (CH)
- FARBE UND LACK, Vincentz Network, Hanover
- European Coatings Journal, Vincentz Network, Hanover

New media and internet

- Informationsserie 27 (CD-ROM) "Kleben/Klebstoffe", editor Fonds der Chemischen Industrie 2001
- www.ivd-ev.de (Industrieverband Dichtstoffe)
- www.klebstoffe.com (Industrieverband Klebstoffe)
- www.adhäsion.com (Journal Adhäsion)
- www.adhesivesmag.com (American journal)
- www.specialchem4adhesives.com
- www.feica.com (Association of European Adhesives and Sealant Manufacturers

Authors

Dr Hartwig Lohse was born in 1955. Following his graduation from the University of Hamburg with a doctorate in chemistry, he engaged in a three-year period of employment in the development and technical service of sewing thread lubricants. In this capacity, he dealt with the subject of bonding as an alternative to plying. From 1987 onwards, he was employed by Tivioli Werke AG (now part of Sika AG) in Hamburg, where his primary focus was on the development and technical service of polyurethane adhesives, particularly for use in the field of car body construction. In 1996, he took on the role of European technical manager for adhesives at Ashland Specialty Chemicals, a position that included providing support to the adhesive development team based in the United States. In 2009, as a result of the financial crisis, he founded the consulting company Klebtechnik Dr. Hartwig Lohse e.K., which supports companies along the entire value chain of adhesive bonding in technical matters, such as adhesive selection and development, process planning and optimisation, troubleshooting and workforce qualification.

In addition to this, he is a member of the German Adhesives Association (IVK e.V.), where he serves as the chair of the Technical Committee on Structural Bonding and Sealing and is a member of the Technical Board. Furthermore, he contributes to the standardisation of bonding technology by participating in various DIN, CEN and ISO committees.

Dr Matthias Popp was born in Hamburg, Germany, in 1966. After graduating from high school in 1986, he studied chemistry at the University of Bremen. After completing his dissertation in 1997, he worked for Siegling GmbH, a medium-sized company in the textile coating industry in Hanover. This led to his first contacts in the field of adhesive technology and his subsequent employment at the Fraunhofer Institute for Manufacturing Technology and Advanced Materials IFAM from 2000 to 2006, where he focused on the formulation and instrumental analysis of adhesives. From 2006 to 2010, Dr Matthias Popp worked as a product development specialist at 3M in the automotive and aerospace sectors, where he gained particular experience in the field of epoxy adhesives. In 2010 he moved back to Bremen, Germany, where he is again working in the field of adhesive formulations at Fraunhofer IFAM. In the early 2000s, he also worked as a lecturer at the Bremerhaven University of Applied Sciences and has been a speaker for the Vincentz Network since 2010.

Index

Symbols

A

Index

B

C

F

N

O

P